Methods in Enzymology

Volume 419
ADULT STEM CELLS

METHODS IN ENZYMOLOGY

EDITORS-IN-CHIEF

John N. Abelson   Melvin I. Simon

DIVISION OF BIOLOGY
CALIFORNIA INSTITUTE OF TECHNOLOGY
PASADENA, CALIFORNIA

FOUNDING EDITORS

Sidney P. Colowick and Nathan O. Kaplan

QP601
M49
V. 419

Methods in Enzymology

Volume 419

Adult Stem Cells

EDITED BY

Irina Klimanskaya

Robert Lanza

ADVANCED CELL TECHNOLOGY
WORCESTER, MASSACHUSETTS

NO LONGER THE PROPERTY
OF THE
UNIVERSITY OF R.I. LIBRARY

ELSEVIER

AMSTERDAM • BOSTON • HEIDELBERG • LONDON
NEW YORK • OXFORD • PARIS • SAN DIEGO
SAN FRANCISCO • SINGAPORE • SYDNEY • TOKYO
Academic Press is an imprint of Elsevier

Academic Press is an imprint of Elsevier
525 B Street, Suite 1900, San Diego, California 92101-4495, USA
84 Theobald's Road, London WC1X 8RR, UK

This book is printed on acid-free paper. ∞

Copyright © 2006, Elsevier Inc. All Rights Reserved.

No part of this publication may be reproduced or transmitted in any form or by any
means, electronic or mechanical, including photocopy, recording, or any information
storage and retrieval system, without permission in writing from the Publisher.

The appearance of the code at the bottom of the first page of a chapter in this book
indicates the Publisher's consent that copies of the chapter may be made for
personal or internal use of specific clients. This consent is given on the condition,
however, that the copier pay the stated per copy fee through the Copyright Clearance
Center, Inc. (www.copyright.com), for copying beyond that permitted by
Sections 107 or 108 of the U.S. Copyright Law. This consent does not extend to
other kinds of copying, such as copying for general distribution, for advertising
or promotional purposes, for creating new collective works, or for resale.
Copy fees for pre-2006 chapters are as shown on the title pages. If no fee code
appears on the title page, the copy fee is the same as for current chapters.
0076-6879/2006 $35.00

Permissions may be sought directly from Elsevier's Science & Technology Rights
Department in Oxford, UK: phone: (+44) 1865 843830, fax: (+44) 1865 853333,
E-mail: permissions@elsevier.com. You may also complete your request on-line
via the Elsevier homepage (http://elsevier.com), by selecting "Support & Contact"
then "Copyright and Permission" and then "Obtaining Permissions."

For information on all Elsevier Academic Press publications
visit our Web site at www.books.elsevier.com

ISBN-13: 978-0-12-373650-5
ISBN-10: 0-12-373650-1

PRINTED IN THE UNITED STATES OF AMERICA
06 07 08 09 9 8 7 6 5 4 3 2 1

Working together to grow
libraries in developing countries

www.elsevier.com | www.bookaid.org | www.sabre.org

ELSEVIER   BOOK AID
            International   Sabre Foundation

To Richard Latsis, the Teacher
–Irina

Table of Contents

Section I. Ectoderm

Section II. Mesoderm

Contributors to Volume 419

Article numbers are in parentheses following the names of contributors.
Affiliations listed are current.

ANTHONY ATALA (17), *Wake Forest University School of Medicine, Wake Forest Institute for Regenerative Medicine, Winston-Salem, North Carolina*

GEORG BARTSCH, JR. (17), *Wake Forest University School of Medicine, Wake Forest Institute for Regenerative Medicine, Winston-Salem, North Carolina*

PAOLO BIANCO (6), *Department of Experimental Medicine and Pathology, La Sapienza University, Rome, Italy*

MATTHEW BJERKNES (14), *Department of Medicine, University of Toronto, Toronto, Ontario, Canada*

RALPH L. BRINSTER (11), *University of Pennsylvania School of Veterinary Medicine, Laboratory of Reproductive Physiology, Philadelphia, Pennsylvania*

HAL E. BROXMEYER (18), *Department of Microbiology and Immunology, Medicine, and Pediatrics, Indiana University School of Medicine, Indianapolis, Indiana*

DEBORAH A. BUFFINGTON (9), *Innovative BioTherapies, Inc., Ann Arbor, Michigan*

ANTONIN BUKOVSKY (10), *Laboratory of Development, Differenciation and Cancer, The University of Tennessee Graduate School of Medicine, Knoxville, Tennessee*

HAZEL CHENG (14), *Department of Medicine, University of Toronto, Toronto, Ontario, Canada*

SCOTT COOPER (18), *Microbiology and Immunology, Medicine, and Pediatrics, Indiana University School of Medicine, Indianapolis, Indiana*

PAOLO DE COPPI (17), *Wake Forest University School of Medicine, Wake Forest Institute for Regenerative Medicine, Winston-Salem, North Carolina*

DAWN M. DELO (17), *Wake Forest University School of Medicine, Wake Forest Institute for Regenerative Medicine, Winston-Salem, North Carolina*

YUVAL DOR (13), *Hebrew University, Hadassah Medical School, Department of Cellular Biochemistry and Human Genetics, Jerusalem, Israel*

RYAN R. DRISKELL (12), *Center for Gene Therapy, University of Iowa, Iowa City, Iowa*

JOHN F. ENGELHARDT (12), *Center for Gene Therapy, University of Iowa, Iowa City, Iowa*

ANDY J. FISCHER (3), *Department of Neuroscience, Ohio State University, Columbus, Ohio*

JOHN D. GEARHART (16), *Institute for Cell Engineering, Department of Obstetrics and Gynecology, Johns Hopkins School of Medicine, Baltimore, Maryland*

URSULA M. GEHLING (8), *Department of Medicine, University Hospital Hamburg-Eppendorf, Hamburg, Germany*

STAN GRONTHOS (5), *Mesenchymal Stem Cell Group, Institute of Medical and Veterinary Science, South Australia, Australia*

ix

ROBERT G. HAWLEY (7), *Department of Anatomy and Cell Biology, The George Washington University Medical Center, Washington, DC*

TERESA S. HAWLEY (7), *Flow Cytometry Core Facility, The George Washington University Medical Center, Washington, DC*

H. DAVID HUMES (9), *Department of Internal Medicine, Division of Nephrology, University of Michigan School of Medicine, Ann Arbor, Michigan and Innovative BioTherapies, Inc., Ann Arbor, Michigan*

DAVID A. INGRAM (18), *Department of Microbiology and Immunology, Medicine, and Pediatrics, Indiana University School of Medicine, Indianapolis, Indiana*

CANDACE L. KERR (16), *Institute for Cell Engineering, Department of Obstetrics and Gynecology, Johns Hopkins School of Medicine, Baltimore, Maryland*

SERGEI A. KUNZETSOV (6), *Department of Craniofacial and Skeletal Diseases, National Institutes of Health, Bethesda, Maryland*

HE LIU (5), *Peking University School of Stomatology, Beijing, China*

XIAOMING LIU (12), *Center for Gene Therapy, University of Iowa, Iowa City, Iowa*

LAURA E. MEAD (18), *Department of Microbiology and Immunology, Medicine, and Pediatrics, Indiana University School of Medicine, Indianapolis, Indiana*

JON M. OATLEY (11), *University of Pennsylvania School of Veterinary Medicine, Laboratory of Reproductive Physiology, Philadelphia, Pennsylvania*

MAYUMI ODA (15), *Animal Resource Sciences and Veterinary Medical Sciences, Laboratory of Cellular Biochemistry,*

The University of Tokyo, Bunkyo-ku, Tokyo, Japan

CHRISTIE ORSCHELL (18), *Department of Microbiology and Immunology, Medicine, and Pediatrics, Indiana University School of Medicine, Indianapolis, Indiana*

P. ARTUR PLETT (18), *Department of Microbiology and Immunology, Medicine, and Pediatrics, Indiana University School of Medicine, Indianapolis, Indiana*

PRITHI RAJAN (2), *Center for Neuroscience and Aging, Burham Institute for Medical Research, La Jolla, California*

ALI RAMEZANI (7), *Department of Anatomy and Cell Biology, The George Washington University Medical Center, Washington, DC*

THOMAS A. REH (3), *Neurobiology and Behavior Program, Department of Biological Structure, University of Washington, School of Medicine, Seattle, Washington*

BRENT A. REYNOLDS (1), *Queensland Brain Institute, University of Queensland, Brisbane, Australia*

RODNEY L. RIETZE (1), *Queensland Brain Institute, University of Queensland, Brisbane, Australia*

MARA RIMINUCCI (6), *Experimental Medicine, University of L'Aquila, L'Aquila, Italy*

PAMELA GEHRON ROBEY (6), *Department of Craniofacial and Skeletal Diseases, National Institutes of Health, Bethesda, Maryland*

SETH J. SALPETER (13), *Hebrew University, Hadassah Medical School, Department of Cellular Biochemistry and Human Genetics, Jerusalem, Israel*

MICHAEL J. SHAMBLOTT (16), *Institute for Cell Engineering, Department of Obstetrics and Gynecology, Johns Hopkins*

School of Medicine, Baltimore, Maryland

SONGTAO SHI (5), *Center for Craniofacial Molecular Biology, USC School of Dentistry, Los Angeles, California*

KUNIO SHIOTA (15), *Animal Resource Sciences and Veterinary Medical Sciences, Laboratory of Cellular Biochemistry, The University of Tokyo, Bunkyo-ku, Tokyo, Japan*

PETER L. SMITH (9), *Innovative BioTherapies, Inc., Ann Arbor, Michigan*

EVAN SNYDER (2), *Center for Neuroscience and Aging, Burnham Institute for Medical Research, La Jolla, California*

EDWARD SROUR (18), *Department of Microbiology and Immunology, Medicine, and Pediatrics, Indiana University School of Medicine, Indianapolis, Indiana*

MARTA SVETLIKOVA (10), *Laboratory of Development, Differenciation and Cancer, The University of Tennessee*

Graduate School of Medicine, Knoxville, Tennessee

SATOSHI TANAKA (15), *Animal Resource Sciences and Veterinary Medical Sciences, Laboratory of Cellular Biochemistry, The University of Tokyo, Bunkyo-ku, Tokyo, Japan*

TUDORITA TUMBAR (4), *Department of Molecular Biology and Genetics, Cornell University, Ithaca, New York*

IRMA VIRANT-KLUN (10), *IVF Laboratory, The University Medical Center Ljubljana, Ljubljana, Slovenia, Slovenia*

ISABELLE WILLSON (10), *Laboratory of Development, Differenciation and Cancer, The University of Tennessee Graduate School of Medicine, Knoxville, Tennessee*

MERVIN C. YODER (18), *Department of Microbiology and Immunology, Medicine, and Pediatrics, Indiana University School of Medicine, Indianapolis, Indiana*

Preface

Stem cells are of great interest to scientists and clinicians due to their unique ability to differentiate into various tissues of the body. In addition to being a promising source of cells for transplantation and regenerative medicine, they also serve as an excellent model of vertebrate development. In the recent years, the interest in stem cell research has spread beyond the scientific community to the public at large as a result of heated political and ethical debate.

There are two broad categories of stem cells – "embryonic" and "adult." Embryonic stem cells – also known as "*pluripotent*" stem cells – are derived from preimplantation-stage embryos and retain the capacity to grow in culture indefinitely, as well as to differentiate into virtually all the tissues of the body. Adult stem cells are found in most tissues of the adult organism; scientists are beginning to learn how to isolate, culture, and differentiate them into a range of tissue-specific types (and are thus considered *multipotent*).

Growing stem cells in culture and differentiating them on demand requires specific skills and knowledge beyond basic cell culture techniques. We have tried to assemble the most robust and current techniques (including both conventional and novel methods) in the stem cell field and invited the world's leading scientists with hands-on expertise to write the chapters on methods they are experts in or even established themselves. Volume 418, "Embryonic Stem Cells," offers a variety of know-how from derivation to differentiation of embryonic stem cells, including such sought-after methods as human embryonic stem cell derivation and maintenance, morula- and single blastomere-derived ES cells, ES cells created via parthenogenesis and nuclear transfer, as well as techniques for derivation of ES cells from other species, including mouse, bovine, zebrafish, and avian. The second section of this volume covers the recent advances in differentiation and maintenance of ES cell derivatives from all three germ layers: cells of neural lineage, retinal pigment epithelium, cardiomyocytes, haematopoietic and vascular cells, oocytes and male germ cells, pulmonary and insulin-producing cells, among others.

Volume 419, "Adult Stem Cells," covers stem cells of all three germ layers and organ systems. The methods include isolation, maintenance, analysis, and differentiation of a wide range of adult stem cell types, including neural, retinal, epithelial cells, dental, skeletal, and haematopoietic cells, as well as ovarian, spermatogonial, lung, pancreatic, intestinal, throphoblast, germ, cord blood, amniotic fluid, and placental stem cells.

Volume 420, "Tools for Stem Cell Research and Tissue Engineering," has collected specific stem cells applications as well as a variety of techniques, including gene trapping, gene expression profiling, RNAi and gene delivery, embryo culture for human ES cell derivation, characterization and purification of stem cells, and cellular reprogramming. The second section of this volume addresses tissue engineering using derivatives of adult and embryonic stem cells, including important issues such as immunogenicity and clinical applications of stem cell derivatives.

Each chapter is written as a short review of the field followed by an easy-to-follow set of protocols that enables even the least experienced researchers to successfully establish the techniques in their laboratories.

We wish to thank the contributors to all three volumes for sharing their invaluable expertise in comprehensive and easy to follow step-by-step protocols. We also would like to acknowledge Cindy Minor at Elsevier for her invaluable assistance assembling this three-volume series.

IRINA KLIMANSKAYA
ROBERT LANZA

Foreword

As stem cell researchers, we are frequently asked by politicians, patients, reporters, and other non-scientists about the relative merits of studying embryonic stem cells *versus* adult stem cells, and when stem cells will provide novel therapies for human diseases. The persistence of these two questions and the passion with which they are asked reveals the extent to which stem cells have penetrated the vernacular, captured public attention, and become an icon for the scientific, social, and political circumstances of our times.

Focusing first on the biological context of stem cells, it is clear that the emergence of stem cells as a distinct research field is one of the most important scientific initiatives of the 'post-genomic' era. Stem cell research is the confluence between cell and developmental biology. It is shaped at every turn by the maturing knowledge base of genetics and biochemistry and is accelerated by the platform technologies of recombinant DNA, monoclonal antibodies, and other biotechnologies. Stem cells are interesting and useful because of their dual capacity to differentiate and to proliferate in an undifferentiated state. Thus, they are expected to yield insights not only into pluripotency and differentiation, but also into cell cycle regulation and other areas, thereby having an impact on fields ranging from cancer to aging.

This directs us to why it is necessary to study different types of stem cells, including those whose origins from early stages of development confers ethical complexity (embryonic stem cells) and those that are difficult to find, grow, or maintain as undifferentiated populations (most types of adult stem cells). The question itself veils a deeper purpose for studying the biology of stem cells, which is to gain a fundamental understanding of the nature of cell fate decisions during development. We still have a relatively shallow understanding of how stem cells maintain their undifferentiated state for prolonged periods and then 'choose' to specialize along the pathways they are competent to pursue. Achieving a precise understanding of such 'stemness' and of differentiation will require information from as wide a variety of sources as possible. This process of triangulation could be compared to how global positioning satellites enable us to locate ourselves: signal from a single satellite tells us relatively little, and precision is achieved only when we acquire signals from three or more. Similarly, it is necessary to study multiple types of stem cells and their progeny if we are to evaluate the outcome of cellular development *in vitro* in comparison with normal development.

An answer to the question of when stem cells will yield novel clinical outcomes requires us to define the likely therapeutic achievements. Of course, adult and neonatal blood stem cells have been used in transplantation for many years, and it is likely that new sources and applications for them will emerge from current studies. It is less likely, however, that transplantation will be the first application of research on other adult stem cell types or of research on the differentiated progeny of embryonic stem cells. This reflects in part the degree of characterization of such progeny that will be needed to ensure their long-term safety and efficacy when transplanted to humans. It is their use as *in vitro* cellular models that is more likely to pioneer novel clinical applications of the specialized human cell types that can be derived from stem cells. These cells, including cultured neurons, cardiomyocytes, kidney cells, lung cells, and numerous others, will imminently provide a novel platform technology for drug discovery and testing. The applications of such cellular models are likely to be extensive, leading to development of new medicines for a myriad of human health problems. The wide availability of these specialized human cells will also provide an opportunity to evaluate the stability and function of stem cell progeny in the Petri dish well before they are used in transplantation. Finally, we should not overlook the importance of stem cells and their progeny as models for understanding human developmental processes. While we cannot foresee the impact of a profound understanding of human cellular differentiation, it has the potential of transcending even the most remarkable applications that we can imagine involving transplantation.

Despite the links of stem cell research to other fields of biology and to established technologies, growing and differentiating stem cells systematically in culture requires specific skills and knowledge beyond basic cell and developmental biology techniques. These *Methods in Enzymology* volumes include the most current techniques in the stem cell field, written by leading scientists with hands-on expertise in methods they have developed or in which they are recognized as experts. Each chapter is written as a short review of outcomes from the particular method, with an easy-to-follow set of protocols that should enable less experienced researchers to successfully establish the method in their laboratories. Together, the three volumes cover the spectrum of both embryonic and adult stem cells and provide tools for extending the uses of stem cells to tissue engineering. It is hoped that the availability and wide dissemination of these methods will provide wider access to the stem cell field, thereby accelerating acquisition of the knowledge needed to apply stem cell research in novel ways to improve our understanding of human biology and health.

ROGER A. PEDERSEN, PH.D.
PROFESSOR OF REGENERATIVE MEDICINE
UNIVERSITY OF CAMBRIDGE

METHODS IN ENZYMOLOGY

VOLUME XXXV. Lipids (Part B)
Edited by JOHN M. LOWENSTEIN

VOLUME XXXVI. Hormone Action (Part A: Steroid Hormones)
Edited by BERT W. O'MALLEY AND JOEL G. HARDMAN

VOLUME XXXVII. Hormone Action (Part B: Peptide Hormones)
Edited by BERT W. O'MALLEY AND JOEL G. HARDMAN

VOLUME XXXVIII. Hormone Action (Part C: Cyclic Nucleotides)
Edited by JOEL G. HARDMAN AND BERT W. O'MALLEY

VOLUME XXXIX. Hormone Action (Part D: Isolated Cells, Tissues,
and Organ Systems)
Edited by JOEL G. HARDMAN AND BERT W. O'MALLEY

VOLUME XL. Hormone Action (Part E: Nuclear Structure and Function)
Edited by BERT W. O'MALLEY AND JOEL G. HARDMAN

VOLUME XLI. Carbohydrate Metabolism (Part B)
Edited by W. A. WOOD

VOLUME XLII. Carbohydrate Metabolism (Part C)
Edited by W. A. WOOD

VOLUME XLIII. Antibiotics
Edited by JOHN H. HASH

VOLUME XLIV. Immobilized Enzymes
Edited by KLAUS MOSBACH

VOLUME XLV. Proteolytic Enzymes (Part B)
Edited by LASZLO LORAND

VOLUME XLVI. Affinity Labeling
Edited by WILLIAM B. JAKOBY AND MEIR WILCHEK

VOLUME XLVII. Enzyme Structure (Part E)
Edited by C. H. W. HIRS AND SERGE N. TIMASHEFF

VOLUME XLVIII. Enzyme Structure (Part F)
Edited by C. H. W. HIRS AND SERGE N. TIMASHEFF

VOLUME XLIX. Enzyme Structure (Part G)
Edited by C. H. W. HIRS AND SERGE N. TIMASHEFF

VOLUME L. Complex Carbohydrates (Part C)
Edited by VICTOR GINSBURG

VOLUME LI. Purine and Pyrimidine Nucleotide Metabolism
Edited by PATRICIA A. HOFFEE AND MARY ELLEN JONES

VOLUME LII. Biomembranes (Part C: Biological Oxidations)
Edited by SIDNEY FLEISCHER AND LESTER PACKER

VOLUME LIII. Biomembranes (Part D: Biological Oxidations)
Edited by SIDNEY FLEISCHER AND LESTER PACKER

VOLUME LIV. Biomembranes (Part E: Biological Oxidations)
Edited by SIDNEY FLEISCHER AND LESTER PACKER

VOLUME LV. Biomembranes (Part F: Bioenergetics)
Edited by SIDNEY FLEISCHER AND LESTER PACKER

VOLUME LVI. Biomembranes (Part G: Bioenergetics)
Edited by SIDNEY FLEISCHER AND LESTER PACKER

VOLUME LVII. Bioluminescence and Chemiluminescence
Edited by MARLENE A. DELUCA

VOLUME LVIII. Cell Culture
Edited by WILLIAM B. JAKOBY AND IRA PASTAN

VOLUME LIX. Nucleic Acids and Protein Synthesis (Part G)
Edited by KIVIE MOLDAVE AND LAWRENCE GROSSMAN

VOLUME LX. Nucleic Acids and Protein Synthesis (Part H)
Edited by KIVIE MOLDAVE AND LAWRENCE GROSSMAN

VOLUME 61. Enzyme Structure (Part H)
Edited by C. H. W. HIRS AND SERGE N. TIMASHEFF

VOLUME 62. Vitamins and Coenzymes (Part D)
Edited by DONALD B. MCCORMICK AND LEMUEL D. WRIGHT

VOLUME 63. Enzyme Kinetics and Mechanism (Part A: Initial Rate and
Inhibitor Methods)
Edited by DANIEL L. PURICH

VOLUME 64. Enzyme Kinetics and Mechanism
(Part B: Isotopic Probes and Complex Enzyme Systems)
Edited by DANIEL L. PURICH

VOLUME 65. Nucleic Acids (Part I)
Edited by LAWRENCE GROSSMAN AND KIVIE MOLDAVE

VOLUME 66. Vitamins and Coenzymes (Part E)
Edited by DONALD B. MCCORMICK AND LEMUEL D. WRIGHT

VOLUME 67. Vitamins and Coenzymes (Part F)
Edited by DONALD B. MCCORMICK AND LEMUEL D. WRIGHT

VOLUME 68. Recombinant DNA
Edited by RAY WU

VOLUME 69. Photosynthesis and Nitrogen Fixation (Part C)
Edited by ANTHONY SAN PIETRO

VOLUME 70. Immunochemical Techniques (Part A)
Edited by HELEN VAN VUNAKIS AND JOHN J. LANGONE

VOLUME 71. Lipids (Part C)
Edited by JOHN M. LOWENSTEIN

VOLUME 72. Lipids (Part D)
Edited by JOHN M. LOWENSTEIN

VOLUME 73. Immunochemical Techniques (Part B)
Edited by JOHN J. LANGONE AND HELEN VAN VUNAKIS

VOLUME 74. Immunochemical Techniques (Part C)
Edited by JOHN J. LANGONE AND HELEN VAN VUNAKIS

VOLUME 75. Cumulative Subject Index Volumes XXXI, XXXII, XXXIV–LX
Edited by EDWARD A. DENNIS AND MARTHA G. DENNIS

VOLUME 76. Hemoglobins
Edited by ERALDO ANTONINI, LUIGI ROSSI-BERNARDI, AND EMILIA CHIANCONE

VOLUME 77. Detoxication and Drug Metabolism
Edited by WILLIAM B. JAKOBY

VOLUME 78. Interferons (Part A)
Edited by SIDNEY PESTKA

VOLUME 79. Interferons (Part B)
Edited by SIDNEY PESTKA

VOLUME 80. Proteolytic Enzymes (Part C)
Edited by LASZLO LORAND

VOLUME 81. Biomembranes (Part H: Visual Pigments and Purple Membranes, I)
Edited by LESTER PACKER

VOLUME 82. Structural and Contractile Proteins (Part A: Extracellular Matrix)
Edited by LEON W. CUNNINGHAM AND DIXIE W. FREDERIKSEN

VOLUME 83. Complex Carbohydrates (Part D)
Edited by VICTOR GINSBURG

VOLUME 84. Immunochemical Techniques (Part D: Selected Immunoassays)
Edited by JOHN J. LANGONE AND HELEN VAN VUNAKIS

VOLUME 85. Structural and Contractile Proteins (Part B: The Contractile Apparatus and the Cytoskeleton)
Edited by DIXIE W. FREDERIKSEN AND LEON W. CUNNINGHAM

VOLUME 86. Prostaglandins and Arachidonate Metabolites
Edited by WILLIAM E. M. LANDS AND WILLIAM L. SMITH

VOLUME 87. Enzyme Kinetics and Mechanism (Part C: Intermediates, Stereo-chemistry, and Rate Studies)
Edited by DANIEL L. PURICH

VOLUME 88. Biomembranes (Part I: Visual Pigments and Purple Membranes, II)
Edited by LESTER PACKER

VOLUME 89. Carbohydrate Metabolism (Part D)
Edited by WILLIS A. WOOD

VOLUME 244. Proteolytic Enzymes: Serine and Cysteine Peptidases
Edited by ALAN J. BARRETT

VOLUME 245. Extracellular Matrix Components
Edited by E. RUOSLAHTI AND E. ENGVALL

VOLUME 246. Biochemical Spectroscopy
Edited by KENNETH SAUER

VOLUME 247. Neoglycoconjugates (Part B: Biomedical Applications)
Edited by Y. C. LEE AND REIKO T. LEE

VOLUME 248. Proteolytic Enzymes: Aspartic and Metallo Peptidases
Edited by ALAN J. BARRETT

VOLUME 249. Enzyme Kinetics and Mechanism (Part D: Developments in Enzyme Dynamics)
Edited by DANIEL L. PURICH

VOLUME 250. Lipid Modifications of Proteins
Edited by PATRICK J. CASEY AND JANICE E. BUSS

VOLUME 251. Biothiols (Part A: Monothiols and Dithiols, Protein Thiols, and Thiyl Radicals)
Edited by LESTER PACKER

VOLUME 252. Biothiols (Part B: Glutathione and Thioredoxin; Thiols in Signal Transduction and Gene Regulation)
Edited by LESTER PACKER

VOLUME 253. Adhesion of Microbial Pathogens
Edited by RON J. DOYLE AND ITZHAK OFEK

VOLUME 254. Oncogene Techniques
Edited by PETER K. VOGT AND INDER M. VERMA

VOLUME 255. Small GTPases and Their Regulators (Part A: Ras Family)
Edited by W. E. BALCH, CHANNING J. DER, AND ALAN HALL

VOLUME 256. Small GTPases and Their Regulators (Part B: Rho Family)
Edited by W. E. BALCH, CHANNING J. DER, AND ALAN HALL

VOLUME 257. Small GTPases and Their Regulators (Part C: Proteins Involved in Transport)
Edited by W. E. BALCH, CHANNING J. DER, AND ALAN HALL

VOLUME 258. Redox-Active Amino Acids in Biology
Edited by JUDITH P. KLINMAN

VOLUME 259. Energetics of Biological Macromolecules
Edited by MICHAEL L. JOHNSON AND GARY K. ACKERS

VOLUME 260. Mitochondrial Biogenesis and Genetics (Part A)
Edited by GIUSEPPE M. ATTARDI AND ANNE CHOMYN

VOLUME 261. Nuclear Magnetic Resonance and Nucleic Acids
Edited by THOMAS L. JAMES

Section I

Ectoderm

[1] Neural Stem Cell Isolation and Characterization

By RODNEY L. RIETZE and BRENT A. REYNOLDS

Abstract

Throughout the process of development and continuing into adulthood, stem cells function as a reservoir of undifferentiated cell types, whose role is to underpin cell genesis in a variety of tissues and organs. In the adult, they play an essential homeostatic role by replacing differentiated tissue cells "worn off" by physiological turnover or lost to injury or disease. As such, the discovery of such cells in the adult mammalian central nervous system (CNS), an organ traditionally thought to have little or no regenerative capacity, was most unexpected. Nonetheless, by employing a novel serum-free culture system termed the neurosphere assay, Reynolds and Weiss demonstrated the presence of neural stem cells in both the adult (Reynolds and Weiss, 1992) and embryonic mouse brain (Reynolds et al., 1992). Here we describe how to generate, serially passage, and differentiate neurospheres derived from both the developing and adult brain, and provide more technical details that will enable one to achieve reproducible cultures, which can be passaged over an extended period of time.

Introduction

Although originally debated, it is now clear that neurogenesis continues in at least two regions of the adult mammalian brain, namely, the olfactory bulb and hippocampal formation (Gross, 2000). This continuous and robust generation of new cells strongly argues for the existence of a founder cell with the ability to proliferate, self-renew, and ultimately generate a large number of differentiated progeny, that is, a stem cell (Potten and Loeffler, 1990). One of the difficulties in identifying and studying stem cells is their poorly defined physical nature, which affects our ability to directly measure their presence and to monitor their activity. This problem has been overcome by defining stem cells on the basis of a functional criterion such that stem cells, in general, are defined by what they do, not by what they look like. This creates a number of problems, both conceptual and practical, with the most obvious being that a stem cell must first be forced to act in order to determine its presence; hence, questions are raised concerning whether the action of imposing an action accurately reflects the original or

METHODS IN ENZYMOLOGY, VOL. 419
Copyright 2006, Elsevier Inc. All rights reserved.
0076-6879/06 $35.00
DOI: 10.1016/S0076-6879(06)19001-1

true nature of the cell in question. Clearly what is needed is a specific selective positive marker that will allow us to definitely identify stem cells both *in vivo* and *in vitro*. In this review we discuss and detail a culture methodology that allows for the isolation, propagation, and identification of stem cells from the mammalian brain and provide practical advice on the use of flow cytometry to isolate a relatively pure population of putative stem cells.

Although their presence was suggested indirectly in a number of previous studies, elucidation of the appropriate culture conditions that permitted the functional attributes of a stem cell to be demonstrated enabled the unequivocal demonstration of a neural stem cell, for the first time, in 1992. To isolate and expand the putative stem cell from the adult brain, Reynolds and Weiss employed a serum-free culture system known as the neurosphere assay (NSA), wherein most of the primary differentiated CNS cells harvested would not be able to survive. Although this system caused the death of the majority of cell types harvested from the periventricular region within 3 days of culture, it allowed a small population ($<0.1\%$) of epidermal growth factor (EGF)-responsive stem cells to enter a period of active proliferation, even at low cell densities (Reynolds and Weiss, 1992). By using such a system, Reynolds and Weiss were able to demonstrate that a single adult CNS cell could proliferate to form a ball of undifferentiated cells they called a neurosphere, which in turn could (1) be dissociated to form more numerous secondary spheres, or (2) be induced to differentiate, generating the three major cell types of the CNS. In doing so, they showed that the cell they had isolated exhibited the stem cell attributes of proliferation, self-renewal, and the ability to give rise to a number of differentiated, functional progeny (Hall and Watt, 1989; Potten and Loeffler, 1990). Subsequent studies have since demonstrated that by following a well-defined protocol, and by using EGF, basic fibroblast growth factor (bFGF), or both as mitogens, it was possible to produce a consistent, renewable source of undifferentiated CNS precursors (a portion of which are stem cells), which could be expanded as neurospheres, or reliably differentiated into defined proportions of neurons, astrocytes, and oligodendrocytes (Gritti *et al.*, 1995, 1996, 1999; Reynolds and Weiss, 1996; Reynolds *et al.*, 1992; Weiss *et al.*, 1996a,b).

The more than 1000 citations to date that have employed the NSA attest to the robust and reliable nature of the assay, and its value in studying developmental processes and elucidating the role of genetic and epigenetic factors in triggering the potential of CNS stem cells and in determining CNS phenotypes. Although the methodology seems relatively simple to carry out, strict adherence to the procedures described here is required to achieve reliable and consistent results. Here we describe in detail the protocols for the isolation and culture of neural stem cells harvested from various regions of the embryonic and adult murine brain. These protocols assume a basic knowledge of murine brain anatomy. The reader is referred to O'Connor *et al.* (1998) for information on this topic, which is essential to perform the procedures for culturing murine neural stem cells outlined in this chapter.

Reagents and Instrumentation

Dissection Equipment

Large scissors
Small fine scissors
Ultrafine spring microscissors (cat. no. 15396-01; Fine Science Tools, Vancouver, BC, Canada)
Small forceps (cat. no. 11050-10; Fine Science Tools)
Small fine forceps (cat. no. 11272-30; Fine Science Tools)
Ultrafine curved forceps (cat. no. 11251-35; Fine Science Tools)
Bead sterilizer (cat. no. 250; Fine Science Tools)
Dissection microscope

Tissue Culture Equipment

Flasks
    25 cm^2, 0.2-μm vented filter cap [cat. no. 9026; Techno Plastic Products (TPP), Trasadingen, Switzerland]
    75 cm^2, 0.2-μm vented filter cap (cat. no. 90076; TPP)
    175 cm^2, 0.2-μm vented filter cap (cat. no. 90151; TPP)
Tubes
    17 × 100 mm polystyrene, sterile (cat. no. 91015; TPP)
    50-ml polypropylene, sterile (cat. no. 91050; TPP)
    Fluorescence-activated cell-sorting (FACS), sterile (Falcon, cat. no. 352054; BD Biosciences Discovery Labware, Bedford, MA)
Petri dishes: 100 and 35 mm (Nunc cat. nos. 351029 and 174926, respectively; Nalge Nunc International, Rochester, NY)
Tissue sieve, 70-μm pore size (Falcon, cat. no. 352350; BD Biosciences Discovery Labware)
Tissue culture (TC) plates: 6, 24, and 96 well (Falcon, cat. nos. 353046, 353047, and 353072, respectively; BD Biosciences Discovery Labware)
8-well coated chamber slides
    Poly-D-lysine/laminin (BD BioCoat, cat. no. 35-4688; BD Biosciences Discovery Labware)

Human fibronectin (BD BioCoat, cat. no. 35-4631; BD Biosciences Discovery Labware)

Growth Factors

EGF (human recombinant, cat. no. 02633; StemCell Technologies, Vancouver, BC, Canada): For a stock solution of 10 μg/ml add 10 ml of hormone-supplemented neural culture medium to each vial of EGF. Store as 100-μl aliquots at $-20°$

bFGF (human recombinant, cat. no. 02634; StemCell Technologies): For a stock solution of 10 μg/ml add 999 μl of hormone-supplemented neural culture medium and 1 μl of bovine serum albumin (BSA) to each vial of bFGF. Store as 100-μl aliquots at $-20°$

0.2% heparin: Mix 100 mg of heparin (cat. no. H-3149; Sigma, St. Louis, MO) in 50 ml of water. Filter sterilize. Store at $4°$

Medium Solutions

These cultures are extremely sensitive to contaminants present in water or glassware. If medium is being made in the laboratory, use only tissue culture-grade components. We strongly suggest that as many components as possible be purchased, as this will minimize batch-to-batch inconsistencies and provide greater consistency of results overall. Optimized reagents for the culture and differentiation of neurospheres are available from StemCell Technologies (www.stemcell.com).

Commercial Medium Components

Phosphate-buffered saline (PBS, cat. no. 37350; StemCell Technologies)

Basal medium (NeuroCult NSC basal medium, cat. no. 05700; Stem-Cell Technologies)

10× hormone mix (NeuroCult NSC proliferation supplement, cat. no. 05701; StemCell Technologies)

Differentiation medium (NeuroCult differentiation supplement, cat. no. 05703; StemCell Technologies)

Preparation of complete NSC medium is thoroughly described at www.stemcell.com/technical/manuals/asp.

As with the in-laboratory preparation of medium components described in the next section, combining 450 ml of NeuroCult NSC basal medium with 50 ml of NeuroCult NSC proliferation supplement will produce the hormone-supplemented growth medium described in Stock

Solutions (item 4). As described in Stock Solutions, complete NSC growth medium is produced by the addition of EGF and/or bFGF.

Medium Preparation Components

For the in-laboratory preparation of tissue culture media and hormone mix, a set of glassware to be used only for tissue cultures should be prepared. Bottles, cylinders, beakers, and so on, should be accurately rinsed several times with distilled water before being sterilized in an autoclave that is used for tissue culture purposes only. We strongly suggest that all media and stock solutions be prepared only in sterile disposable tubes and/or bottles, thereby avoiding contamination caused by cleaning solution residue or poor autoclaving techniques. Whenever possible, commercial stock solutions should also be employed.

30% glucose (cat. no. G-7021; Sigma): Mix 30 g of glucose in 100 ml of distilled water. Filter sterilize and store at $4°$

7.5% sodium bicarbonate (cat. no. S-5761; Sigma): Mix 7.5 g of $NaHCO_3$ in 100 ml of water. Filter sterilize and store at $4°$

1 M HEPES (cat. no. H-0887; Sigma): Dissolve 238.3 g of HEPES in 1 liter of distilled water. Store at $4°$

3 mM sodium selenite (cat. no. S-9133; Sigma): Add 1.93 ml of distilled water to a 1-mg vial of sodium selenite. Mix, aliquot into sterile tubes, and store at $-20°$

2 mM progesterone (cat. no. P-6149; Sigma): Add 1.59 ml of 95% ethanol to a 1-mg vial of progesterone. Mix, aliquot into sterile tubes, and store at $-20°$

200 mM L-glutamine (cat. no. 25030-024; Invitrogen GIBCO, Grand Island, NY)

Apotransferrin (cat. no. 820056-1; Serologicals, Atlanta, GA): Dissolve 400 mg of apotransferrin directly into 10× hormone mix solution

Insulin (cat. no. 977420; Roche, Indianapolis, IN): Dissolve 100 mg of bovine insulin in 4 ml of sterile 0.1 N HCl, and then add 36 ml of distilled water to this solution. Transfer the entire volume to 10× hormone mix

Putrescine (cat. no. P-7505; Sigma): Dissolve 38.6 mg of putrescine in 40 ml of distilled water. Transfer the entire volume to 10× hormone mix

0.1% DNase I (cat. no. 704159; Roche, Mannheim, Germany): Dissolve 100 mg of DNase I in 100 ml of Hanks' Eagle medium (HEM). Mix thoroughly, filter sterilize, aliquot into sterile tubes (1 ml/aliquot), and then store $-20°$

Propidium iodide (cat. no. P-4170; Sigma)

Trypsin (Calbiochem, cat. no. 6502; EMD Biosciences, San Diego, CA)
Trypsin inhibitor (T-6522; Sigma): Combine 14 mg of trypsin inhibitor,
1 ml of 0.1% DNase I, and 99 ml of HEM. Mix well, filter sterilize,
and store at 4° for a maximum of 14 days
Minimal essential medium (cat. no. 41500-018; Invitrogen GIBCO)

Stock Solutions

Preparation of 10× Dulbecco's modified Eagle's medium (DMEM)–
F12: Combine five 1-liter packages of DMEM (cat. no. 12100-046;
Invitrogen GIBCO) and five 1-liter packages of F12 powder (cat.
no. 21700-075; Invitrogen GIBCO) in 1 liter of water under gentle
continuous stirring. Filter sterilize and store at 4°
Preparation of 10× hormone mix: Combine individual components in
the following order: (1) 300 ml of ultrapure distilled water, (2) 40 ml
of 10× DMEM–F12, (3) 8 ml of 30% glucose, (4) 6 ml of 7.5%
$NaHCO_3$, and (5) 2.5 ml of 1 M HEPES. Mix well, and then add
(1) 400 mg of apotransferrin, (2) 40 ml of 2.5-mg/ml insulin stock,
(3) 40 ml of 10-mg/ml putrescine stock, (4) 40 μl of 3 mM sodium
selenite, and (5) 40 μl of 2 mM progesterone. Mix all components
thoroughly, filter sterilize, and then aliquot into 10- or 25-ml
volumes in sterile tubes and store at $-20°$
Preparation of basal medium (for 450 ml): Combine individual
components in the following order: (1) 375 ml of ultrapure distilled
water, (2) 50 ml of 10× DMEM–F12 stock, (3) 10 ml of 30%
glucose, (4) 7.5 ml of 7.5% $NaHCO_3$, (5) 2.5 ml of 1 M HEPES, and
(6) 5 ml of 20 nM L-glutamine. Mix thoroughly, filter sterilize, and
store at 4° for a maximum of 3 months
Preparation of hormone-supplemented growth medium (for 500 ml):
Combine 50 ml of 10× hormone mix with 450 ml of basal medium,
mix thoroughly, and store at 4° for a maximum of 1 week. Add 1 ml
of 0.2% heparin, 20 μl of EGF, and/or 10 μl of bFGF stock (final
concentrations: EGF, 20 ng/ml; bFGF, 10 ng/ml)
Preparation of complete NSC medium: Add 2 μl of EGF for every 1 ml of
hormone-supplemented growth medium and/or 1 μl of bFGF and 1 μl
of heparin for every 1 ml of hormone-supplemented growth medium
Preparation of tissue dissociation medium (for 200 ml): Add 476 mg
of HEPES, 40 mg of EDTA, 50 mg of trypsin, and 1 ml of 0.1%
DNase I to 200 ml of Ca^{2+}/Mg^{2+} Hanks' balanced salt solution
(HBSS). Mix well, filter sterilize, and then aliquot (3 ml/aliquot)
and store at $-20°$

Preparation of Hanks' Eagle medium (HEM): (for 8.75 liters): Add the contents of one 10-liter packet of minimal essential medium to 3 liters of distilled water in a 5-liter flask. Combine 160 ml of 1 M HEPES and 175 ml of penicillin–streptomycin (1:50 dilution) in a separate flask containing 3 liters of distilled water. Combine and adjust to pH 7.2 with 10 M NaOH. Filter sterilize and aliquot into 100-ml portions. Store at 4° for a maximum of 3 months

Miscellaneous

10× PBS: Without calcium, without magnesium (cat. no. 14200-067; Invitrogen, Carlsbad, CA)
Penicillin–streptomycin (cat. no. 15140-114; Invitrogen)
Trypsin–EDTA (cat. no. E-6511; Sigma)
Matrigel (growth factor reduced, cat. no. 40230; BD Biosciences Discovery Labware)
Laminin (cat. no. 1243217; Roche)
Poly-L-ornithine (cat. no. P-3655; Sigma)
Fetal bovine serum (cat. no. 10106-151; Invitrogen)

Methods

Establishment of Primary Embryonic Neurosphere Cultures

Neurospheres have been generated from various regions of the embryonic CNS and from numerous strains of mice. As such, the protocol that we describe here has been made sufficiently broad so as to increase its applicability, yet most accurately reflects the methodology required to generate neurospheres from the lateral and medial ganglionic eminences of embryonic day 14 (E14) mice, as originally described by Reynolds *et al.* (1992).

Dissection of Embryonic Tissue

Mice (e.g., CD1 albino) are typically mated overnight and then separated the next morning and checked for the presence of a gestational plug. This will count as embryonic day 0 (E0). Alternatively, time-pregnant animals can be purchased from specialized animal care facilities. For the establishment of embryonic neurosphere cultures we typically harvest pups at E14 to E15 (note that dissection of embryonic CNS is much easier on E15), killing the mother in accordance with rules dictated by the animal ethics committee. Perform the dissection as quickly as possible (within 2 h), as tissue becomes soft and sticky over time and may be difficult to dissect. If it is estimated that more than 2 h is required, remove and dissect 8–10 brains at a time, keeping the remaining embryos at 4°.

Setup

1. Add cold sterile HEM to two 100-mm sterile plastic Petri dishes.
2. Sterilize dissection tools immediately before use by using a glass bead sterilizer, or well in advance by autoclaving (120° for 20 min). Tools needed for the gross dissection include the following: large scissors, small pointed scissors, larger forceps, and small curved forceps. Ultrafine forceps and scissors will be used for the microdissection of CNS tissue.
3. Place gauze on the bottom of a small glass beaker and then fill the beaker with 70% ethanol. This is where forceps and scissors are stored during the dissection so as to reduce contamination.
4. Prepare a gross dissection area on a laboratory bench by laying several absorbent towels flat, and then soaking the towels with 70% ethanol. Place gross dissection tools to the side.
5. Arrange the dissecting microscope, two Petri dishes containing HEM, and the ultrafine dissection tools within a laminar flow hood. As a precaution, keep some sterile Petri dishes and HEM ready at hand.
6. Warm culture medium to 37° in a thermostatic water bath.

Harvesting of Embryonic Brain Tissue

1. Anesthetize the pregnant mother by an intraperitoneal injection of pentobarbital (120 mg/kg) and, on deep anesthesia, sacrifice the mother by cervical dislocation.

2. Lay the pregnant mother on its back on the absorbent towels, and then liberally rinse the abdomen with 70% ethanol so as to sterilize the area.

3. Grasp the skin above the genitalia, using large forceps, and then cut through the skin and fascia with large scissors so as to expose the peritoneal cavity sufficiently to view the uteri.

4. Remove the uteri with small forceps and scissors and transfer them into a 100-mm dish containing HEM. A litter size of 8–12 pups is typical; however, only 2 or 3 are needed to establish a bulk culture. Ensure that tools are rinsed frequently in ethanol, so as to exclude fur. On completion of the dissection, dispose of the carcass immediately.

5. Transfer uterine tissues to the laminar flow hood and then rinse once or twice by placing them in 100-mm Petri dishes containing fresh sterile HEM.

6. Cut open the uterine horns and transfer the pups to a new 100-mm dish containing HEM, using small forceps. At this point, check the age of

the pups and discard those that appear malformed, or too small with respect to gestational age.

7. Separate the head(s) of the pup(s) at the level just below the cervical spinal cord, discarding the skulls.

8. Transfer each tissue culture dish to a dissecting microscope and, under ×10 magnification, begin to remove the brain by positioning the head side up. Hold it from the caudal side at the ears, using fine curved forceps. Use microscissors to cut a horizontal opening above the eyes and tease the brain out of the opening by gently pushing on the head from the side opposite to the cut.

9. After removing all of the brain, increase the magnification (×25) and dissect out the desired brain region(s) to be used for establishing the culture. Typically the lateral and medial ganglionic eminences are removed, but refer to a rodent brain atlas for details on how to dissect the specific areas.

10. Transfer harvested brain regions to a 15-ml Falcon tube containing 2 ml of ice-cold HEM.

Establishing Primary Embryonic Cultures

1. Several methods may be used to mechanically dissociate the dissected tissue, including a fire-polished glass pipette or 200-μl plastic tips together with a Gilson pipette (we routinely use a Pipetman P200; Gilson, Middleton, WI). In either case wet the plastic tip or glass pipette by sucking (and discarding) a small amount of sterile medium, and then proceed to triturate the tissue approximately 10 times until a milky single-cell suspension is achieved. Make sure to avoid generating air bubbles, as this reduces the number of viable cells and makes for inefficient trituration. Also, the expulsion of cells during the trituration should not be too vigorous, as this will also significantly reduce viability.

2. If undissociated pieces of tissue are still present in the suspension after the initial trituration, wait 2 min, which will allow the undissociated cells and tissue to settle, and then transfer the majority of the supernatant containing single cells into a fresh tube, leaving the undissociated tissue behind. Add an appropriate volume of complete NSC medium to the undissociated cells so as to bring the total volume to 0.5–2 ml (depending on the volume of tissue and method of dissociation). Repeat step 1.

3. Pool the two suspensions that have been created and then centrifuge the resulting suspension at 800 rpm (110g) for 5 min. Aspirate the supernatant and gently resuspend the cells to achieve a final 2-ml volume of complete NSC medium.

4. Combine a 10-μl aliquot of the cell suspension with 90 μl of trypan blue in a microcentrifuge tube, mix, and then transfer 10 μl to a

hemacytometer so as to determine the number of viable cells in the suspension.

5. For primary cultures, seed cells at a density of 2×10^6 cells per 10 ml (25-cm^2 flask) or 8×10^6 cells in 40 ml of medium (175-cm^2 flask), in complete NSC medium. Please note that the cell density for plating primary cells harvested directly from the E14 CNS is higher than that prescribed for subsequent subculturing conditions.

General Comments

- On plating primary cells, individual cells will become hypertrophic and adhere to the substrate, while the majority of cells will either die or differentiate. After 2–3 days in culture, proliferative cells will lift off the base of the tissue culture vessels. Aggregates of cells resembling neurospheres will most likely be observed within the first 48 h of culture. These should not be mistaken for primary spheres. The prevalence of aggregates is directly related to the amounts of debris and/or dead cells in the cultures. Typically, these pseudo-spheres are quite large, but are composed of unusually small, phase-dark, and irregularly shaped cells.

- Bona fide neurospheres will appear phase bright and exhibit a somewhat spherical form to begin with, becoming more spherical as size increases. As shown in Fig. 1, small microspikes should be apparent on the outer surface of viable spheres by day 3.

- Primary neurospheres are often associated with cellular debris; however, subculturing will effectively select for proliferating precursor cells and remove cell aggregates, debris, and dead cells.

Establishment of Primary Adult Neurosphere Cultures

De novo neurogenesis has been reported to occur within discrete areas of the adult brain, namely the olfactory bulb, hippocampus, and cortex. Here, we describe how to isolate adult murine neural stem cells and to establish continuous stem cell lines by means of growth factor stimulation. This protocol may also be applied to rats, and implies the use of enzymatic predigestion before mechanical dissociation. Note that although stem cells isolated from many different mouse strains display similar general features, differences regarding their growth rate and differentiation capacity may also be observed.

Setup

Killing of animals, and removal and dissection of brain and/or spinal cord, are performed outside the laminar flow hood. Particular caution should be exercised to avoid contamination. Have all the materials and instrumentation ready before starting the dissection procedure.

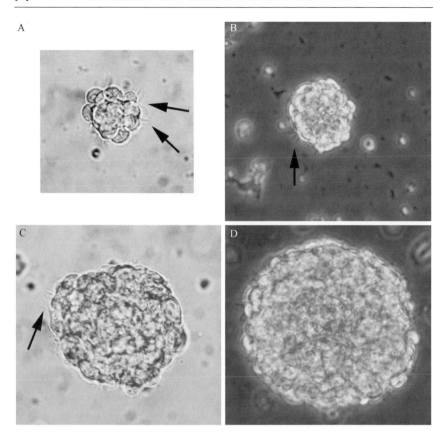

FIG. 1. EGF-responsive murine neural stem cells, isolated from E14 striatum, were grown for 7 days in culture and then passaged. Small clusters of cells may be identified 2 days after passaging (A). The shape and opacity of the sphere, along with the presence of microspikes (arrows), assist in identifying a young, healthy neurosphere. Microspikes are still present in neurospheres after 3 days *in vitro* (DIV) (B) and 4 DIV (C). By 6 DIV the neurosphere is ready to be passaged. Original magnification: ×200.

1. Add cold HEM to sterile plastic Petri dishes: one or two 100-mm dishes to hold tissue, several 60-mm dishes to wash tissues, and some 35-mm dishes to hold dissected tissues.
2. Dissection tools may be sterilized in a hot bead sterilizer, in a preheated oven (250° for 2 h), or by autoclaving (120° for 20 min).
3. Select the tools needed to remove the brain and spinal cord (large scissors, small pointed scissors, large forceps, small curved forceps, and a small spatula) or for tissue dissection (small forceps, curved fine forceps, small scissors, curved fine scissors, and scalpel). Immerse the

two sets of tools in 70% ethanol in two beakers with gauze at the bottom, to avoid spoiling the tips of the microforceps and scissors.

4. Warm the culture medium and tissue dissociation medium to 37° in a thermostatic water bath.

5. Begin the dissection.

Dissection of Adult Periventricular Region

1. Anesthetize mice by intraperitoneal injection of pentobarbital (120 mg/kg) and kill them by cervical dislocation. Tissues from two or three mice (age, from 2 to 8 months) are generally pooled to start a culture.

2. Using large scissors, cut off the head just above the cervical spinal cord region. Rinse the head with 70% ethanol.

3. Using small pointed scissors, make a medial caudal–rostral cut and part the skin of the head to expose the skull. Rinse the skull with sterile HEM.

4. Using the skin to hold the head in place, place each blade of a pair of small scissors in the orbital bone, so as to make a coronal cut between the orbits of the eyes.

5. Using the coronal cut as an entry point, make a longitudinal cut through the skull along the sagittal suture. Be careful not to damage the brain; make small cuts, ensuring the angle of the blades is as shallow as possible. Cut the entire length of the skull to the foramen magnum.

6. Using curved, pointed forceps, grasp and peel the skull of each hemisphere outward to expose the brain. Using a small wetted curved spatula, scoop the brain into a Petri dish containing HEM.

7. Repeat steps 1–6 until all the brains have been harvested.

8. Wash the brains twice by subsequently transferring them to new Petri dishes containing PBS.

9. To dissect the forebrain subventricular region, place each dish containing a brain under the dissecting microscope (×10 magnification). Position the brain flat on its ventral surface and hold it from the caudal side, using fine curved forceps placed on either side of the cerebellum. Use a scalpel to make a coronal cut just behind the olfactory bulbs.

10. After removal of the olfactory bulbs, rotate the brain to expose the ventral aspect. Make a coronal cut at the level of the optic chiasm (Fig. 2A), discarding the caudal aspect of the brain.

11. Repeat steps 8–10 until all the brains are sectioned.

12. Shift to ×25 magnification. Rotate the rostral aspect of the brain with the presumptive olfactory bulb facing downward. Using fine curved microscissors, first remove the septum and discard, and then cut the thin layer of tissue surrounding the ventricles, excluding the striatal parenchyma and the corpus callosum (Fig. 2B). Pool the dissected tissue in a newly labeled 35-mm Petri dish.

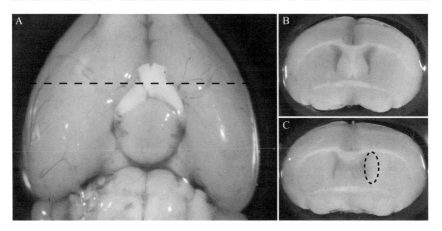

FIG. 2. (A) Ventral view of an adult C57BL/6 mouse brain, illustrating the rostral–caudal coordinate (dotted line) to section the brain coronally in order to harvest the rostral periventricular region of the lateral ventricles. (B) Resulting coronal section when brain is sliced along the dotted line in (A). (C) Dotted line highlights the periventricular region that is harvested in a typical dissection after removal of the septum.

13. On harvesting the periventricular regions from all brains, transfer the dish to the tissue culture laminar flow hood. Continue to use strict sterile technique.

Dissociation Protocol

1. Using a scalpel blade, mince tissue for ~1 min until only small pieces remain.

2. Using a filter-tipped glass pipette and a total volume of 3 ml of tissue dissociation medium, transfer all the minced tissues into the base of a 15-ml tube.

3. Incubate the tube for 7 min in a 37° water bath. Greater incubation times may be required, depending on the amount of tissue and on the overall size of the particles (larger pieces may be present due to inadequate mincing of the tissue).

4. At the end of the enzymatic incubation, return the tube to the hood and add an equal volume of trypsin inhibitor (3 ml).

5. Avoiding the generation of air bubbles, mix well and then pellet the tissue suspension by centrifugation at 110g for 7 min.

6. Discard virtually all the supernatant overlaying the pellet, and then add an appropriate volume of HEM so as to attain a final volume of 1 ml. Using a Gilson P1000 pipette (or similar) and a wetted 1000-μl filter tip, begin to dissociate by triturating once or twice, and then place the tip at

the bottom of the tube so as to restrict the flow of cells by $\sim$50% and continue triturating five to seven times until the cell suspension takes on a milky or smooth appearance. Let the suspension settle for 3–4 min.

7. If many undissociated pieces of tissue are left, move the cell suspension to a clean, labeled tube, leaving about 100 μl behind. To the latter, add 900 μl of HEM and triturate again five to seven times, until almost no undissociated pieces are left. Let the suspension settle for 3–4 min. Transfer all but 100 μl from this tube to the labeled tube, thus pooling the cells from both trituration steps.

8. Bring the resulting cell suspension to a total volume of 14 ml by adding fresh HEM, and then pass the suspension through a 70-μm pore size sieve into a 15-ml tube, so as to remove debris or undissociated pieces, and then pellet the cells by centrifugation at 110g for 7 min.

9. Remove virtually all the supernatant, and resuspend the pellet in complete NSC culture medium so as to bring the total volume of the resulting cell suspension to 0.5 ml.

10. Combine a 10-μl aliquot from the cell suspension with 90 μl of trypan blue in a microcentrifuge tube, mix, and then transfer 10 μl to a hemacytometer so as to perform a cell count.

11. Seed cells at a density of 3500 viable cells/cm^2 in complete culture medium in untreated 6-well tissue culture dishes (volume, 3 ml) or 25 cm^2-tissue culture flasks (volume, 5 ml).

12. Incubate at 37°, 5% CO_2 in a humidified incubator.

13. Cells should proliferate to form spherical clusters that eventually lift off as they grow larger. These primary spheres should be ready for subculturing 7–10 days after plating, depending on the growth factors used.

Comments

• The 3-ml volume of tissue dissociation solution is sufficient for good digestion of tissue from up to eight mice. In the case of cell sorting, where 8–16 mice are used, use a single 15-ml tube containing 3 ml of tissue dissociation solution for every 8 brains.

• In primary cultures from adult brain significant debris is normally present, particularly in spinal cord cultures, together with adherent cells. To reduce debris, rinse the tissue more frequently (steps 8 and 9). In general, debris and adherent cells are eliminated after about two passages.

• Counting cells is sometimes difficult, because of the presence of debris and the large number of blood-derived cells, and because of the small number of CNS cells that can be isolated. In our experience this protocol should yield about 5 × 10^4 cells from the subventricular region of one brain. Accurate quantification based on low cell counts of the CNS-derived cells with a hemacytometer can be misleading. Thus, if quantification of the primary neural cell number is not to be carried out, a cell suspension derived from two mice may be plated in four dishes of a 6-well tissue culture dish, yielding an approximate final cell density of about 3500 cells/cm^2, or in one 25-cm^2 tissue culture flask, obtaining a final density of about 4000 cells/cm^2. Once competent with this procedure, it is usual to generate 400–600 neurospheres per mouse.

Passaging Neurosphere Cultures

As a rule of thumb, embryonic primary and passaged neurospheres should be ready for subculture between 4 and 5 days after plating, whereas adult primary and passaged neurosphere cultures should be ready for subculture 7–10 and 5–7 days after plating, respectively. However, it is desirable to monitor the cultures each day to ensure that neurospheres are not allowed to grow too large. Typically, a variety of diameters is apparent in a bulk culture. To determine whether spheres are ready to passage, the majority of neurospheres should be 150 μm in diameter. If neurospheres are allowed to grow too large, they become difficult to dissociate and eventually begin to differentiate *in situ*.

1. Observe the neurosphere cultures under a microscope to determine whether the NSCs are ready for passaging. The average size of neurospheres across the culture should be ~150 μm. If neurospheres are attached to the culture substrate, forcefully strike the side of the tissue culture flask (attempting to minimize vessel movement by applying an equal force with the opposing hand).
2. Remove medium with suspended cells and place it in an appropriately sized sterile tissue culture tube. If some cells remain attached to the substrate, detach them by shooting a stream of medium across the attached cells. Spin at 400 rpm ($75g$) for 5 min.
3. Remove essentially 100% of the supernatant and resuspend the cells in 1 ml of trypsin–EDTA, incubating them at room temperature for 2 min (this volume allows for the most efficient trituration manipulations and is recommended for 75-cm^2 flasks). If more than one tube was used to harvest cultures, resuspend each pellet in 1 ml of trypsin–EDTA. If a 175-cm^2 flask is used, increase the volume of trypsin–EDTA to 3 ml and incubate for 7 min.
4. Add an equal volume of trypsin inhibitor (as compared with trypsin–EDTA) to each tube, mix well, and then centrifuge the cell suspension(s) at 800 rpm ($110g$) for 5 min.
5. Remove essentially 100% of the supernatant and resuspend the cells by the addition of ~950 μl of complete NSC medium so as to produce a total volume of 1 ml. Using a Gilson P1000 pipette (or similar) and a wetted 1000-μl

filter tip, begin to dissociate by triturating once or twice, and then place the tip at the bottom of the tube so as to restrict the flow of cells by ~50% and continue triturating five to seven times until the cell suspension takes on a milky or smooth appearance.

6. Combine a 10-μl aliquot from the cell suspension with 90 μl of trypan blue in a microcentrifuge tube, mix, and then transfer 10 μl to a hemacytometer so as to perform a cell count. If whole spheres appear, triturate the cell suspension two or three times and recount.

7. Seed cells for the next culture passage in complete NSC medium at a density of 7.5×10^5 cells/ml.

Differentiation of Neurosphere Cultures

When cultured in the presence of EGF and/or bFGF, neural stem cells and progenitor cells proliferate to form neurospheres that, when harvested at the appropriate time point and using appropriate methods as described here, can be passaged practically indefinitely. However, on removal of the growth factors and addition of a small amount of serum, neurosphere-derived cells are induced to differentiate into neurons, astrocytes, and oligodendrocytes (see Fig. 3). Overall, two methods have been described for the differentiation of neurospheres: as whole spheres cultured at low density (typically used to demonstrate individual spheres are multipotent) or as dissociated cells at high density (typically used to determine the relative percentage of differentiated cell types generated). The techniques for both methods are provided here.

Differentiation of Whole Neurospheres

If poly-L-ornithine-coated coverslips are to be used, precoat glass slides by adding a sufficient volume of poly-L-ornithine (15 mg/ml) to completely cover the glass coverslip for a period of 2 h at 37°. Alternatively, 96-well plates can be precoated with poly-L-ornithine. Aspirate poly-L-ornithine and immediately rinse three times (10 min each) with sterile PBS (do not allow the coverslips or plate to dry). Remove the PBS immediately before the addition of neurospheres and differentiation medium.

1. Once primary or passaged neurospheres reach 150 μm (typically after 7–8 days *in vitro*), use percussion to remove adherent spheres, and then transfer the contents of the flask to an appropriately sized sterile tissue culture tube. Spin at 400 rpm (75g) for 5 min.

2. Aspirate essentially 100% of the growth medium and then gently resuspend the neurospheres (so as not to dissociate any) with an appropriate volume of basal medium–1% sterile fetal calf serum. *Note*: An equal volume of commercially available NSC differentiation medium can also be used

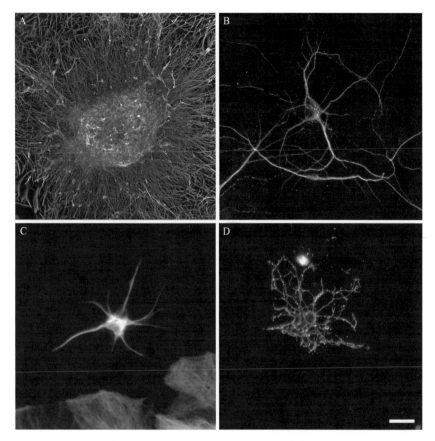

FIG. 3. When transferred to differentiating conditions for 7 DIV, neurospheres will lose their spherical shape and flatten to form essentially a monolayer. The greatest concentration of cells will remain in the center of the neurosphere [4′,6-diamidine-2-phenylindole (DAPI)-positive cells, blue], with astrocytes apparent throughout the sphere [glial fibrillary acidic protein (GFAP), green], and neurons (β-tubulin, red) surrounding the core of the sphere, lying on top of the astrocytes (A). Neurons are identified with a fluorescently labeled antibody raised against β-tubulin, a neuron-specific antigen found in cell bodies and processes (B). Both protoplasmic and stellate astrocytes are identified with a fluorescently tagged antibody against the astrocyte-specific protein GFAP (C). Oligodendrocytes are identified with an antibody against myelin basic protein (MBP) (D). Scale bar (B–D): 20 μm. (See color insert.)

here (NeuroCult differentiation supplement, cat. no. 05703; StemCell Technologies).

3. Transfer the neurosphere suspension to a 60-mm dish (or other sized vessel) to enable the harvesting/plucking of individual neurospheres with a disposable plastic pipette.

4. Transfer approximately 10 neurospheres, using a sterile disposable plastic pipette or a Gilson P1000 pipette, to NSC differentiation medium in individual wells of a 24- or 96-well tissue culture plate with a poly-L-ornithine-coated surface. Alternatively, commercially available, precoated chamber slides can be employed here.

5. After 6–8 days *in vitro*, individual neurospheres should have attached to the substrate and dispersed in such a manner so as to appear as a flattened monolayer of cells.

6. Proceed to fix the cells by the addition of 4% paraformaldehyde (in PBS, pH 7.2) for 10 min at room temperature and then process the adherent cells for immunocytochemistry as required.

Differentiation of Dissociated Cells

1. Once primary or passaged neurospheres reach 150 μm (typically after 7–8 days *in vitro*), apply percussion to remove adherent spheres and then transfer the contents of the flask to an appropriately sized sterile tissue culture tube. Spin at 400 rpm (75g) for 5 min.

2. Remove essentially 100% of the supernatant and resuspend the cells in 1 ml of trypsin–EDTA, incubating at room temperature for 2 min (this volume allows for the most efficient trituration manipulations). If more than one tube was used to harvest cultures, resuspend each pellet in 1 ml of trypsin–EDTA.

3. Add 1 ml of trypsin inhibitor to each tube, mix well, and then centrifuge the cell suspension(s) at 800 rpm (110g) for 5 min.

4. Remove essentially 100% of the supernatant and resuspend the cells by the addition of 1 ml of basal medium–1% sterile fetal calf serum. *Note*: An equal volume of commercially available NSC differentiation medium can also be used here (NeuroCult differentiation supplement, cat. no. 05703; StemCell Technologies). Triturate the cells until the suspension appears milky and no spheres can be seen (about five to seven times).

5. Combine a 10-μl aliquot from the cell suspension with 90 μl of trypan blue in a microcentrifuge tube, mix, and then transfer 10 μl to a hemacytometer so as to perform a cell count.

6. Prepare the appropriate cell suspension in 1 ml of complete NSC differentiation medium so as to seed individual wells of 24-well tissue culture plate containing a poly-L-ornithine-coated glass coverslip with 5 × 10^5 cells. Alternatively, commercially available, precoated chamber slides can be employed here, seeding wells at the same density.

7. After 4–6 days *in vitro*, neurosphere-derived cells will have differentiated sufficiently. Proceed to fix the cells by the addition of 4% paraformaldehyde (in PBS, pH 7.2) for 10 min at room temperature and then process the adherent cells for immunocytochemistry as required.

Flow Cytometric Enrichment of Adult Neural Stem Cells

Although approximately 1:300 cells harvested from the periventricular region of the adult mouse brain has the ability to form neurospheres, we have previously described a negative selection flow cytometric method by which neural stem cells can be greatly enriched (Rietze *et al.*, 2001). This protocol essentially begins with the addition of peanut agglutinin (PNA) and heat-stable antigen (HSA, or mCD24a) to a single-cell suspension of adult cells, the preparation of which is described previously (see Establishment of Primary Adult Neurosphere Cultures). This protocol has been established for CBA mice, but has been found to be applicable to many different mouse strains.

1. Harvest the periventricular region from 16 adult mice, processing as 2 separate samples (8 brains each), bringing both to a single-cell suspension as described above (see Establishment of Primary Adult Neurosphere Cultures). When combined, the total volume of the suspension should equal 400 μl.

2. Add 175 μl of complete NSC medium and 25 μl of the adult cell suspension to a total of four FACS tubes [labeled (a) cells alone, (b) PI, (c) PNA–FITC, and (d) HSA–PE]; these will serve as controls. Transfer the remaining 300 μl to a single FACS tube labeled "sort sample."

3. Add 2 μl of PNA–FITC (fluorescein isothiocyanate) to control tube c, and 1 μl of HSA–PE (phycoerythrin) to control tube d. Add 3 μl of PNA–FITC and 1.5 μl of HSA–PE to the sort sample tube. Cap the tubes and incubate on ice in the dark for 15 min.

4. Add 2.5 ml of NSC medium to tubes a, c, and d, whereas tube b receives 2.5 ml of propidium iodide (PI) rinsing solution. Add 5 ml of PI rinsing solution to the sort sample tube. Mix the contents of each tube with a pipette, and then centrifuge at 110g for 7 min.

5. Remove essentially 100% of the supernatant and resuspend each control pellet with 300 μl of complete medium, and the sort sample pellet with 2 ml of complete medium.

6. Bring the FACS tubes to a cytometer, using each of the control tubes to set the appropriate voltage and compensation. Voltages should be adjusted so that the forward-versus-side scatter pattern appears essentially as in Fig. 4A, and FITC/PE detectors as in Fig. 4C.

7. A triangle gate should be set first as shown in Fig. 4A, and then a second gate should be set so as to exclude dead (PI-positive) cells from those included within the triangle gate (Fig. 4B).

8. Neural stem cells are greatly enriched by selecting for the PNA^{lo}HSA^{lo} population, as shown in Fig. 4C. Sorted cells should be collected in a 96-well plate containing 200 μl of complete NSC medium in each well. Given the low frequency of stem cells, a maximum of 20 wells is typically required to collect all the PNA^{lo}HSA^{lo} population from the sort tube.

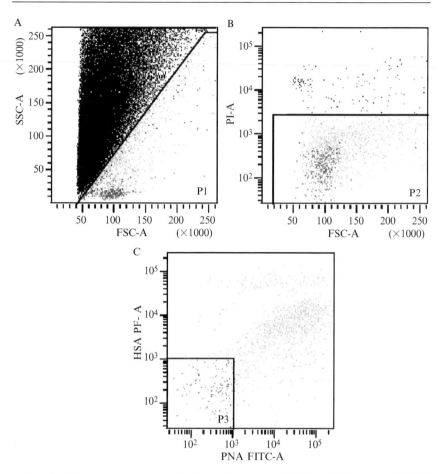

Fɪɢ. 4. (A) Dot plot comparing the forward scatter (FSC-A) and side scatter (SSC-A) attributes of periventricular cells harvested from the rostral periventricular region. Selecting cells in population 1 (P1) excludes the majority of cellular debris without affecting the number of neurospheres generated. (B) Viable cells are distinguished from those cells contained within P1 in (A), by comparing FSC-A and propidium iodide intensity, and then gating for those cells within the propidium iodide-negative population (P2). (C) Dot plot of viable periventricular cells comparing PNA and HSA staining intensities. Harvesting cells in the PNA^{lo}HSA^{lo} population (P3) will greatly enrich for stem cell activity. (See color insert.)

Acknowledgments

The authors thank Dr. Preethi Eldi, Ms. Kristin Hatherley, and Dr. Dan Blackmore for assistance in the preparation of this chapter.

References

Gritti, A., Cova, L., Parati, E. A., Galli, R., and Vescovi, A. L. (1995). Basic fibroblast growth factor supports the proliferation of epidermal growth factor-generated neuronal precursor cells of the adult mouse CNS. *Neurosci. Lett.* **185,** 151–154.

Gritti, A., Parati, E. A., Cova, L., Frolichsthal, P., Galli, R., Wanke, E., Faravelli, L., Morassutti, D. J., Roisen, F., Nickel, D. D., and Vescovi, A. L. (1996). Multipotential stem cells from the adult mouse brain proliferate and self-renew in response to basic fibroblast growth factor. *J. Neurosci.* **16,** 1091–1100.

Gritti, A., Frolichsthal-Schoeller, P., Galli, R., Parati, E., Cova, L., Pagano, S., Bjornson, C., and Vescovi, A. (1999). Epidermal and fibroblast growth factors behave as mitogenic regulators for a single multipotent stem cell-like population from the subventricular region of the adult mouse forebrain. *J. Neurosci.* **19,** 3287–3297.

Gross, C. G. (2000). Neurogenesis in the adult brain: Death of a dogma. *Nat. Rev. Neurosci.* **1,** 67–73.

Hall, P. A., and Watt, F. M. (1989). Stem cells: The generation and maintenance of cellular diversity. *Development* **106,** 619–633.

O'Connor, T. J., Vescovi, A. L., and Reynolds, B. A. (1998). "Isolation and Propagation of Stem Cells from Various Regions of the Embryonic Mammalian Central Nervous System." Academic Press, London.

Potten, C. S., and Loeffler, M. (1990). Stem cells: Attributes, cycles, spirals, pitfalls and uncertainties—lessons for and from the crypt. *Development* **110,** 1001–1020.

Reynolds, B. A., and Weiss, S. (1992). Generation of neurons and astrocytes from isolated cells of the adult mammalian central nervous system. *Science* **255,** 1707–1710.

Reynolds, B. A., and Weiss, S. (1996). Clonal and population analyses demonstrate that an EGF-responsive mammalian embryonic cns precursor is a stem cell. *Dev. Biol.* **175,** 1–13.

Reynolds, B. A., Tetzlaff, W., and Weiss, S. (1992). A multipotent EGF-responsive striatal embryonic progenitor cell produces neurons and astrocytes. *J. Neurosci.* **12,** 4565–4574.

Rietze, R. L., Valcanis, H., Brooker, G. F., Thomas, T., Voss, A. K., and Bartlett, P. F. (2001). Purification of a pluripotent neural stem cell from the adult mouse brain. *Nature* **412,** 736–739.

Weiss, S., Reynolds, B. A., Vescovi, A. L., Morshead, C., Craig, C. G., and van der Kooy, D. (1996a). Is there a neural stem cell in the mammalian forebrain? *Trends Neurosci.* **19,** 387–393.

Weiss, S., Dunne, C., Hewson, J., Wohl, C., Wheatley, M., Peterson, A., and Reynolds, B. (1996b). Multipotent CNS stem cells are present in the adult mammalian spinal cord and ventricular neuroaxis. *J. Neurosci.* **16,** 7599–7609.

[2] Neural Stem Cells and Their Manipulation

By PRITHI RAJAN and EVAN SNYDER

Abstract

Extracellular signals dictate the biological processes of neural stem cells (NSCs) both *in vivo* and *in vitro*. The intracellular response elicited by these signals is dependent on the context in which the signal is received, which in turn is decided by previous and concurrent signals impinging on

METHODS IN ENZYMOLOGY, VOL. 419
Copyright 2006, Elsevier Inc. All rights reserved.
0076-6879/06 $35.00
DOI: 10.1016/S0076-6879(06)19002-3

the cell. A synthesis of signaling pathways that control proliferation, survival, and differentiation of NSCs *in vivo* and *in vitro* will lead to a better understanding of their biology, and will also permit more precise and reproducible manipulation of these cells to particular end points. In this review we summarize the known signals that cause proliferation, survival, and differentiation in mammalian NSCs.

Introduction

Neural stem cells (NSCs) may be isolated from embryonic and adult brains, and are defined by their dual properties of self-renewal and their capacity to differentiate into the fates characteristic of the adult nervous system. The resident population of NSCs in the developing brain peaks before embryonic days 12–14 (E12–E14) in the rat, and gradually diminishes because of differentiation into neurons initially, followed by astrocytes and then oligodendrocytes. Neurogenesis is maximal around E14, followed by gliogenesis, which peaks around E19 (Caviness *et al.*, 1995; Frederiksen and McKay, 1988). Neuronal architecture and glial differentiation continue to occur postnatally, especially synaptic pruning and myelination. NSCs are isolated with ease from all areas of the embryonic central nervous system, including cerebral cortex, hippocampus, striatum, mid-brain including the substantia nigra, cerebellum, and spinal cord. In the adult there are structural zones to which these cells are restricted; this is discussed in relative detail below. NSCs have also been isolated from the neural crest and retina, and have been used as biological platforms to study the mechanisms of regulation of proliferation, survival, and differentiation. Neural crest stem cells (NCSCs) are isolated from chick, mouse, or rat neural tube explants and give rise to tissue derivatives of the neural crest including neurons and glia from the peripheral nervous system and smooth muscle (Shah and Anderson, 1997). Retinal stem cells have been isolated from the ciliary margin zone of the adult eyes of amphibians and fish, and have also been cultured from pigmented ciliary margin of mouse retinas (Moshiri *et al.*, 2004; Tropepe *et al.*, 2000).

The most obvious practical advantage of NSCs is their potential to be an unrestricted source of neurons for replacement therapies. In addition to these transplantation therapies other important uses of stem cells lie in the creation of platforms for drug discovery (Rajan *et al.*, 2006). To make these processes more efficient by the successful manipulation of these cells, it is necessary to study the biology of NSCs in *in vitro* culture systems that are used to maintain and propagate them, and to determine the signaling pathways that control the proliferation, differentiation, and survival of these cells in culture. Although the population of NSCs in the adult brain is appreciably less than in the embryonic brain, NSCs exist in localized regions called niches. The biology of the adult NSCs within their niches also needs to be studied in detail for two reasons: in addition to replacement therapies in which cells from allogeneic donors may be transplanted, resident stem cells may also be mobilized in order to harness their therapeutic potential. The use of fetal and adult NSCs for transplantation therapies may also benefit by knowledge of the *in vivo* survival and differentiation requirements of these cells.

In this review we attempt to summarize the signals controlling some aspects of neural stem cell biology including the *in vivo* niches in which neural stem cells reside in the adult and during development in the embryo, and the signals that are known to orchestrate the proliferation, survival, and differentiation of NSCs *in vitro*. Inclusion of the details of each pathway is beyond the scope of this review, but may be found in several excellent reviews focusing on individual pathways (Bieberich, 2004; Blaise *et al.*, 2005; Cross and Templeton, 2004; Kleber and Sommer, 2004; Louvi and Artavanis-Tsakonas, 2006; McMahon, 2000; Polster and Fiskum, 2004; Stupack, 2005; Sela-Donenfeld and Wilkinson, 2005). Depending on the laboratory in which these studies were performed, NSC cultures have been derived from rodent or human tissues, and cultured under conditions that vary in three-dimensional structure (monolayer or neurosphere), and culture additives [serum, B27, epidermal growth factor (EGF), or leukemia inhibitory factor (LIF)]. Admittedly, these parameters confer significant differences to the NSC cultures generated. For example, neurospheres, which are NSC cultures maintained in suspended balls of cells, generate cultures of vastly different local density and extracellular matrix composition when compared with NSCs generated in monolayers. In addition, NSC cultures prepared from tissue derived from identical brain regions but from different ages of embryo differ in their responses to growth factors, attesting to the idea that the response elicited from a cell by an impinging ligand is affected by the context in which the signal is received, which in turn is dictated by cell-intrinsic and extracellular cues. For these contextual reasons, only those results that have been obtained in mammalian, and preferably neural-related, systems are considered here. Finally, some conclusions have been drawn by overexpression of signaling proteins and transcription factors, which may lead to spurious effects. However, in this review we consider data generated using all these paradigms in addition to *in vivo* studies to generate a heuristic model that will undoubtedly undergo refinements as our understanding of stem cell biology progresses.

Adult Niches for Stem Cells *In Vivo*

The adult mammalian brain has two prominent zones wherein stem cells reside. Stem cells are known to exist in the subventricular zone (SVZ), which extends anatomically around the ventricle in the cerebral cortex.

These stem cells participate in the rostral migratory stream (RMS) in rodents, generating interneurons in the olfactory bulb; however, there is little evidence for the presence of a migrating stream similar to the RMS in humans (Sanai *et al.*, 2004). The second concentration of stem cells in the adult occurs in the subgranular zone (SGZ) in the hippocampus, from which neurons are generated in the dentate gyrus (Gage, 2000). Although the stem cells that reside in the SVZ were originally thought to comprise the layer of ependymal cells that line the ventricle (Johansson *et al.*, 1999), the current consensus suggests instead that astrocytes serve as functional stem cells (Alvarez-Buylla *et al.*, 2002), whereas the ependymal cells participate in regulating the niche that these cells occupy (Lim *et al.*, 2000). The current model suggests that the niche comprises three types of cell: the stem cell/ astrocyte, which contacts the basal lamina and is capable of self-renewal (B cell); the preneuronal cell (C cell); and the young migrating neuron (A cell). B cells express glial fibrillary acidic protein (GFAP), a mature astrocytic marker, whereas stage-specific embryonic antigen 1 (SSEA1), which is considered characteristic of rodent embryonic stem cells, is present on a subset of astrocytes *in vivo* (Alvarez-Buylla and Lim, 2004). C cells may also be considered stem cells and respond to EGF as a mitogen *in vitro* to give rise to NSC cultures. A similar hierarchy of cells is present in the SGZ in the adult hippocampus. In this case GFAP-positive astrocytes give rise to neuronal precursors, which mature into granule cells that populate the dentate gyrus.

In addition, there is some evidence for the presence of restricted progenitors, distributed in the white matter of the cerebral cortex, which have been designated white matter precursor cells (WMPCs) (Goldman and Sim, 2005). These are largely glial progenitors and less restricted multipotential progenitors that are scattered in the SVZ and throughout the parenchyma of the brain. WMPCs express platelet-derived growth factor receptor α(PDGFRα) and the A2B5 epitope and thus appear to be oligodendrocyte precursors (Scolding *et al.*, 1998, 1999), but have the capacity to generate all neural phenotypes when cultured *in vitro* (Nunes *et al.*, 2003). When sorted WMPCs were analyzed for their gene expression profiles they exhibited some characteristics of neural progenitor cells such as HES1, musashi, doublecortin, and MASH1 (Goldman and Sim, 2005). Surprisingly, about 4% of the dissociated white matter from human surgical biopsies was shown to comprise these A2B5-positive cells.

Although the biology by which the WMPCs are maintained in the brain remains to be elucidated, a picture is emerging of the SVZ and SGZ niches, mostly by *in vivo* transgenic and gene ablation studies. The participation of extracellular matrix (ECM), basal lamina, and blood vessels and the paracrine effects of the cells themselves are important for the maintenance of these niches *in vivo*. The ligands that are important for the maintenance, proliferation, differentiation, and motility of these cells include Notch/ jagged, bone morphogenetic protein (BMP)/noggin, transforming growth factor α (TGF-α), vascular endothelial growth factor 2 (VEGF2), Ephrins/ Ephs, sonic hedgehog (shh), and soluble amyloid precursor protein (sAPP), slit1, and slit2 (Alvarez-Buylla and Lim, 2004; Conti and Cattaneo, 2005; Hu *et al.*, 1996; Wu *et al.*, 1999). Most interestingly, noggin appears to be expressed by the ependymal cells, which inhibits BMP-mediated astrocyte differentiation, thus maintaining the neurogenic niche. In the SVZ the secreted fragment of β-amyloid precursor protein (sAPP) has been shown to be a mitogen. This is an interesting function for this protein, which is involved in the pathogenesis of Alzheimer's disease (Caille *et al.*, 2004). The polycomb transcription factor bmi1 and the forebrain-restricted orphan nuclear receptor tlx are also thought to participate in the maintenance of the stem cell population in the developing, postnatal, and adult brain (Leung *et al.*, 2004; Roy *et al.*, 2004; Shi *et al.*, 2004; Zencak *et al.*, 2005).

Not surprisingly, it is largely the same players that appear in studies of cultured NSCs. NSCs have been successfully cultured from various brain regions and ages. It is hoped that these cultures will be sufficiently reproducible and predictable across researchers that one may arrive at a desired end point, be it the large-scale expansion of NSCs or a precursor, or the complete maturation of a particular subset of neurons or glia.

In Vitro Manipulation of NSCs

Proliferation

The successful culture of any cell type is a result of the processes of proliferation and survival. Proliferation may be further qualified as symmetric or asymmetric cell division, and the regulation of apoptosis may be included in cell survival (Fig. 1). Manipulation of the speed and length of the cell cycle appears to be critical for the nature of cells that result. Cell divisions that have an abridged G_1 phase result in the proliferation of undifferentiated stem cells, whereas a progressive increase in the length of G_1 facilitates the onset of differentiation (Calegari and Huttner, 2003).

Proliferation of NSCs in culture is most commonly controlled by the addition of basic fibroblast growth factor (bFGF) and EGF. Implicit in the culture conditions is also the presence of extracellular matrix molecules such as fibronectin, laminin, collagen, or vitronectin if serum is used, and ECM molecules that are secreted by the cell themselves. The self-secreted ECM is especially relevant in the case of neurospheres, to which

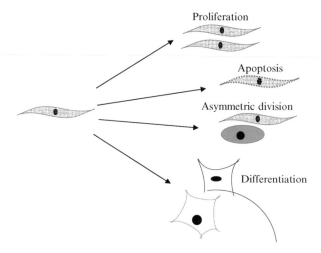

FIG. 1. Processes that are regulated in neural stem cell (NSC) biology. The processes of proliferation, survival/apoptosis, and differentiation comprise the major biological processes of an NSC. Proliferation may lead to asymmetric division, which may result in the generation of two similar or different progeny, or in the death of one of the daughter cells. Proliferation and the signals involved are described in Figs. 2 and 3, whereas differentiation and its signals are described in Figs. 4 and 5.

no other extraneous ECM molecules are added. The involvement of FGF and EGF normally alludes to activation of the mitogen-activated protein kinase (MAPK) pathway. However, as seen in Fig. 2, there is evidence for the involvement of other protein and lipid signal mediators. In addition to MAPK activation, bFGF is also known to increase levels of active β-catenin, which is an important mediator of what is commonly called the canonical signaling pathway of the wnt ligand, and appears to maintain NSCs in a proliferative state when cultured as neurospheres (Israsena *et al.*, 2004; Viti *et al.*, 2003). Wnt, a signal upstream of β-catenin, has been shown to regulate the proliferation of a subset of neural progenitor cells in mouse forebrain and chick spinal cord, where the expression of Wnt-1 or stabilized β-catenin results in increased numbers of neural progenitors along with decreased neuronal differentiation (Chenn and Walsh, 2002; Megason and McMahon, 2002; Sommer, 2004). Along with wnt, notch and shh also regulate NSC proliferation. The response elicited by notch signaling appears to be dependent on the levels of expression of notch and its ligands within the tissue. The mammalian ligands of notch include delta and jagged (Louvi and Artavanis-Tsakonas, 2006). The effects mediated by notch in response to its ligands are regulated by the relative expression of the ligand

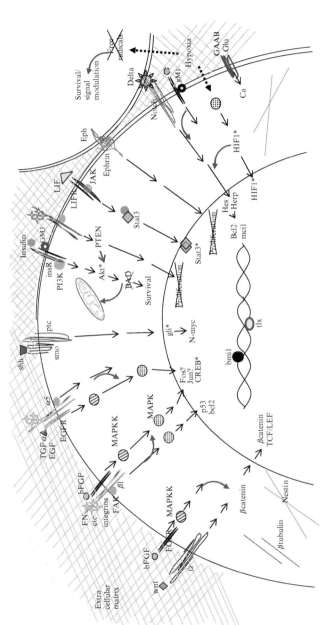

Fɪɢ. 2. Signals regulating proliferation of NSCs. Extracellular signals that regulate the proliferation of NSCs may be soluble ligands, extracellular matrix (ECM) molecules, and immobilized ligands present on neighboring cells. Interactions occur between intracellular signals generated by the growth factor receptor tyrosine kinases and the integrin receptors that respond to ECM molecules. Interactions also occur between the mitogen-activated protein (MAP) kinase pathway and wnt signals, and between hypoxia-induced signals and notch. Stat3 signals are activated by leukemia inhibitory factor (LIF), whereas sonic hedgehog possibly acts through the activation of myc in the nucleus. The epigenetic modifiers bmi1 and tlx are also involved in proliferation of NSCs. In addition to protein signaling intermediates, lipid signals are also involved, especially at the level of receptors at the cell surface [fibroblast growth factor (FGF) and insulin receptors]. It is interesting to note that almost all the transcription factors that are involved [T cell factor/lymphoid enhancing factor (TCF/LEF), Stat3, N-myc, p53, and bmi1], and proteins that are upregulated as a result of the signals (bcl2 and mci1), are associated with oncogenesis. *, activated transcription factor. See text for details and references. (See color insert.)

in adjacent cells. When there are high levels of delta in all cells, equal intensities of signaling between all the cells cause proliferation (Akai *et al.*, 2005; Jessell and Sanes, 2000). However, an imbalance in notch signaling in adjacent cells, for example, by modulation of levels of delta expression in adjacent cells by intrinsic or extrinsic signals, leads to an instructive notch signal (Artavanis-Tsakonas *et al.*, 1999). FGF causes induction of delta expression through the action of CASH in experiments performed in the chick (Akai *et al.*, 2005), and thus precipitates notch function through Hes1, Herp, and similar transcriptional regulators. Lipid signaling is also an integral modulator of growth factor signaling and the ceramide GM1 is a cofactor of the FGF receptor (Rusnati *et al.*, 2002). Although FGF appears to be an integral regulator of proliferation in culture, it is not frequently associated with proliferation *in vivo*, and may preferentially regulate the proliferation of these cells in culture. Similarly, EGF has been shown to be a mitogen *in vitro* for the C cells present in the SVZ (described in the previous section) (Doetsch *et al.*, 2002). Although there is no evidence for the presence of EGF or bFGF *in vivo*, TGF-α is present and it functions through the EGF receptor. It is possible that it is a ligand that mediates the *in vivo* proliferation of NSCs. In fact, TGF-α null mice do show the expected deficit of reduction of proliferation of NSCs in the SVZ (Tropepe *et al.*, 1997).

Shh is thought to affect proliferation of precursors, and in keeping with this observation is mutated frequently in medulloblastomas and gliomas (Sanai *et al.*, 2005). Shh functions by activating Gli-based transcriptional *trans*-activation through a complex series of "double-negative" interactions with its receptor smoothened/patched (McMahon, 2000; Sanai *et al.*, 2005). It relieves transcriptional repression on Gli targets imposed by patched and causes activation of the same genes. Shh affects the proliferation of precursors by inhibition of Rb, the retinoblastoma protein, and induction of N-myc expression (Kenney and Rowitch, 2000; Sjostrom *et al.*, 2005). Leukemia inhibitory factor (LIF) is also considered a mitogen in some NSC neurosphere culture systems, including NSCs derived from early mouse neurectoderm and human mesencephalon (Hitoshi *et al.*, 2004; Jin *et al.*, 2005). Surprisingly, it has been shown by several groups to cause astrocytic differentiation in rodent NSCs cultured in monolayers, reiterating the fact that the biological effects elicited by factors are context, and species, dependent. EphB1–3 and EphA4 are present on cells of the SVZ and ephrinB ligands are associated with SVZ astrocytes. The Eph receptors signal via tyrosine kinase domains in their intracellular domains, and the ephrins themselves can signal through the tyrosine residues present in their intracellular domains (Holland *et al.*, 1996). This bidirectional signaling may be important for the demarcation of boundaries (Mellitzer *et al.*, 1999). Infusion of truncated EphB2 or ephrinB2 into the lateral ventricle results in an increase in cell proliferation, particularly astrocytes, and blocks migration of cells from the SVZ (Conover *et al.*, 2000). Alternatively, ephrinA/EphA receptor regulates apoptosis in the developing brain and thus brain size and cell number (Depaepe *et al.*, 2005). EphrinA2 reverse signaling induced by EphA7 negatively regulates proliferation in the adult mouse brain; a disruption of this interaction causes increased proliferation of progenitors and increased neurogenesis (Holmberg *et al.*, 2005).

In addition to the soluble and immobile factors discussed above, neurotransmitters are also mitogenic in the earlier stages of CNS development. γ-Aminobutyric acid (GABA) and glutamine appear before the onset of synapse formation during CNS development, and regulate proliferation of the precursor cell population depending on the stage of development, and the paradigm studied (Haydar *et al.*, 2000). GABA has been shown to inhibit the proliferative effect of bFGF on cortical progenitor cells (Antonopoulos *et al.*, 1997). These neurotransmitters modulate DNA synthesis and calcium concentration within the cytoplasm of NSCs (Haydar *et al.*, 2000; LoTurco *et al.*, 1995; Owens *et al.*, 2000). Although initially used in hematopoietic stem cell cultures, the biological basis for the culture of various stem cells including NSCs under low-oxygen conditions is being increasingly understood. Hypoxia causes a general decrease in free radicals in the cell and thus decreases the amount of apoptosis that may occur in the culture (Cross and Templeton, 2004). It also causes the induction of a heterodimeric transcription factor, hypoxia-inducible factor 1 (HIF1), which causes an increase in the proliferation of undifferentiated cells (Gustafsson *et al.*, 2005). HIF1 interacts with the notch pathway to cause transcription of notch-regulated genes, which are involved in the maintenance of the undifferentiated state.

Survival/Apoptosis

Apoptosis plays a central role in NSC biology. *In vivo*, the development of the mammalian CNS undergoes phases of regulated cell death including one occurring in early embryonic development that is thought to regulate the number of neural stem cells (De Zio *et al.*, 2005). Mechanistically, reduction in NSC apoptosis was evidenced by knocking out the expression of Apaf1, caspases 3–9, and EphA7 (Depaepe *et al.*, 2005; Kuan *et al.*, 2000; Kuida *et al.*, 1996). In addition, experiments with null mice suggest that the "intrinsic pathway" of apoptosis is involved in cell death during synaptic development in the CNS (Chang *et al.*, 2002; Polster and Fiskum, 2004). This involves the formation of active caspase 9 in the presence of Apaf1 and cytochrome *c*, which then leads to activation of caspase 3. Although it has not been shown in stem cells per se, insulin mediates the survival of cells in culture via Akt, which phosphorylates and causes degradation of the proapoptotic protein BAD (bcl2-associated death promoter) (Datta *et al.*, 1997, 1999). Akt also regulates the activity of the forkhead transcription factor and prevents it from transcribing genes that induce apoptosis, including the FAS (FS-7 cell-associated cell surface antigen) ligand (Brunet *et al.*, 1999). Interestingly, the GM3 ceramide is involved in downregulating signals from the insulin receptor (Yamashita *et al.*, 2003). In addition, by extrapolation from experiments performed in related systems, it is possible that the MAPK pathway functions in NSC survival via the regulation of BAD (Bonni *et al.*, 1999). Notch is also involved in regulating survival of NSCs and appears to work by increasing the expression of the prosurvival genes *bcl-2* and *mcl-1* (Oishi *et al.*, 2004).

Integrins are also major effectors of cell survival and are included among the "dependence receptors," which, generally speaking, mediate survival in the presence of their respective ligand and mediate apoptosis in its absence. The family includes the receptors DCC (deleted in colon cancer) and RET (rearranged during transfection), and the integrin receptors (Stupack, 2005). There are 18 α subunits and 8 β subunits, which form at least 24 heterodimeric integrin receptors that mediate signals from a limited set of ECM molecules that form their cognate ligands (Hynes, 2002). The integrin receptor itself mediates MAPK signals via kinases such as focal adhesion kinase (FAK), but also mediates pathways related to cell survival including Akt activation, p53 activation, and bcl2 expression (Matter and Ruoslahti, 2001; Stromblad *et al.*, 1996; Zhang *et al.*, 1995). In addition to eliciting these signals, the integrin receptors are also complexed to growth receptors in the cell, thus making EGF and PDGF signaling, among others, integrin dependent (Giancotti, 1997; Schneller *et al.*, 1997). The effects of integrin signaling are thus far reaching, and may not be limited to survival alone. The β_1-intergrin subunit has been shown to be essential for NSC proliferation and survival (Leone *et al.*, 2005), possibly in combination with α_5 and α_6 subunits, respectively (Jacques *et al.*, 1998). Interestingly, it has been shown that PTEN (phosphatase and tensin homolog), a phosphatase that regulates Akt and functions downstream of integrins, negatively regulates NSC proliferation by affecting cell cycle entry at G_0–G_1 (Groszer *et al.*, 2006), again demonstrating the interaction between pathways related to cell proliferation and survival.

Symmetric and Asymmetric Cell Division

Cell division in stem cells can result in two identical daughter cells that arise because of symmetric division, or two disparate progeny may be arrived at by asymmetric cell division. Symmetric division may give rise to two stem cells that are identical to the progenitor, or could give rise to two differentiated cells that are identical. Asymmetric division may give rise to one daughter cell that is differentiated while the other remains a multipotent stem cell, or may lead to the death of one cell, thus maintaining total cell number within the environment. One of the first demonstrations of asymmetric cell division was by Chenn and McConnell, who elegantly demonstrated that notch is sequestered to cells of the dividing SVZ during development of the CNS, while the other daughter cell becomes a neuron (Chenn and McConnell, 1995). Numb and Numb-like are PTB domain proteins that are thought to negatively regulate notch signaling by causing receptor turnover of the notch protein. Loss of these proteins may cause an increase in the neuronal progenitor population because of an increase in symmetric cell divisions (Castaneda-Castellanos and Kriegstein, 2004; Li et al., 2003).

NSCs are derived from the neuroepithelial layer of the developing and developed neural tube, and are thus epithelial in nature. This implies the presence of a basement membrane and an apical surface for all cells. The most obvious manner of asymmetric division in such cells would entail a horizontal axis for cytokinesis, in which one daughter cell inherits the basement membrane and all its associated characteristics, while the other daughter inherits the apical membrane. However, a large proportion of asymmetric divisions appear to undergo cytokinesis in the vertical axis, which is not as perfectly perpendicular to the axis of the basement membrane of the cell, thus conferring one daughter cell with a larger portion of the apical membrane than the other (Huttner and Kosodo, 2005) (Fig. 3). This appears to be a method for the specification of radial glial fate (cells that inherit the apical membrane) versus neuronal fate in the mammalian CNS (Kosodo et al., 2004). The biochemical signals that are sequestered in the apical membrane are not clear; however, it is possible that lipid rafts are involved in anchoring intercellular components (Bieberich, 2004) in a manner similar to numb and prospero, as shown in Drosophila (Clevers, 2005). In addition to these physical parameters, some genes have also been identified that are involved in the specification of polarity axes and positioning of the spindle. The mammalian homolog of the Drosophila lethal giant larvae gene (Lgl1) is essential for the localization of numb and lack of function leads to hyperproliferation of the neuroepithelium and radial glial cells, and tissue disorganization in vivo (Klezovitch et al., 2004). Similarly, Lkb1 is a protein kinase that was originally isolated as a tumor suppressor gene and is known to regulate polarity in mammals (Baas et al., 2004). Nde1, a centrosome protein, and ASPM (abnormal spindle-like microcephaly associated) are thought to function in the positioning of the mitotic spindle (Bond et al., 2003; Feng and Walsh, 2004). Loss of their respective functions leads to increased horizontal/oblique axes of cell divisions, causing a depletion of NSCs and a concomitant increase in early neurogenesis.

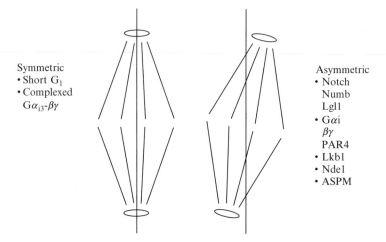

Symmetric
• Short G_1
• Complexed
  $G\alpha_{i3}$-$\beta\gamma$

Asymmetric
• Notch
  Numb
  Lgl1
• $G\alpha i$
  $\beta\gamma$
  PAR4
• Lkb1
• Nde1
• ASPM

FIG. 3. Regulation of symmetric and asymmetric division of NSCs. A schematic representation of the spindle apparatus of a dividing cell and its orientation relative to the longitudinal axis of the cells denotes the phenomena of symmetric and asymmetric cell division. Genes that have been shown to be involved in these processes in NSCs are listed.

Ceramide is also involved in signals that orchestrate asymmetric cell division, and the death of one of the progeny. Whereas ceramide itself partitions normally in the dividing neural progenitor cells, PAR4 segregates to only one of the progeny, PAR4 (prostate apoptosis response protein 4) being a protein kinase C ζ(PKCζ) inhibitor that sensitizes cells to ceramide-induced apoptosis, possibly through a p53-mediated mechanism (Bieberich *et al.*, 2003; Roussigne *et al.*, 2003; Wang *et al.*, 2005). Lipid signaling may also be involved in other aspects of interaction between nuclear and cytoskeletal proteins. Tubulin and other intermediary filaments such as nestin also localize asymmetrically during cell division, as shown in embryoid body-derived stem cells, and it is possible that interaction with ceramide is involved in these interactions (Bieberich, 2004). Because ceramide is present in membranes, it is likely that the intracellular membranes are in contact with the mitotic spindle and intermediary filaments, thus implying a tertiary level of interactions during the cell cycle that needs further investigation. G protein signals are also intertwined with the lipid-mediated signals described and may be involved in the positioning of the mitotic spindle. When $G\alpha_{i3}$-$\beta\gamma$ is present as a heterotrimer the majority of the cleavage planes are vertical, whereas nonvertical cleavage planes result when $\beta\gamma$ is free to interact with downstream effectors (Sanada and Tsai, 2005). On the other hand, $G\alpha_i$, in a manner similar to ceramide, binds tubulin via LGN and NUMA (nuclear mitotic apparatus) (Du and

Macara, 2004). The exact mechanism by which these entities regulate the cleavage plane remains to be elucidated.

The length of the cell cycle is another parameter that is manipulated for manifestation of asymmetric cell division and the resulting differentiation. Cells that have a rapid pace of cell division, that is, a short G_1 phase, will not undergo asymmetric cell divisions even if they have inherited cell components asymmetrically (Huttner and Kosodo, 2005). Lengthening of the cell cycle is required for the specification of progenitors and definitive neurons from NSCs (Cai et al., 2002; Calegari and Huttner, 2003). One mechanism that has been proposed for the lengthening of G_1 is the accumulation of ceramide in the cell during G_0–G_1. Ceramide accumulation may lead to quiescence or apoptosis in the extreme case, possibly involving the retinoblastoma protein, but its levels may also be regulated to cause just enough increase in the length of G_1 to enable cell fate decisions (Bieberich, 2004).

Differentiation

NSCs can be differentiated into the three major cell types that constitute the adult CNS—namely, neurons, astrocytes, and oligodendrocytes. NSCs may also be differentiated into cells of the neural crest lineages such as smooth muscle and neurons and glia of the peripheral nervous system (Rajan et al., 2003; Sailer et al., 2005) (Fig. 4). Fate choice decisions are orchestrated by the response of the cell to extracellular ligands, which include soluble ligands and those that are immobilized by incorporation into either the extracellular matrix or the surface membrane of neighboring cells. As mentioned previously, the end point of differentiation is a result of the sum total of all signals impinging on the cell, and the biochemical state of the cell that is receiving the signal, that is, the qualitative result of any signal is entirely context dependent. This is illustrated by the effect elicited by wnt-7a on NSC cultures: in cultures derived from embryos of age E13.5 wnt causes neuronal differentiation, whereas it causes proliferation in cultures derived from younger animals (Hirabayashi et al., 2004), and BMP differentiation of NSCs into smooth muscle cells or astrocytes depending on basal levels of Stat3 (signal transducer and activator of transcription 3) activity (Rajan et al., 2003).

Growth factors have been used extensively for NSC differentiation. The neuropoietic cytokines ciliary neurotrophic factor and leukemia inhibitory factor (CNTF and LIF) are robust inducers of astrocytic fate (Johe et al., 1996), and do so by activation of signal transducer and activator of transcription (Stat) proteins, particularly Stat1 and Stat3 (Bonni et al., 1997; Rajan and McKay, 1998). The LIF receptor (which serves as the signaling

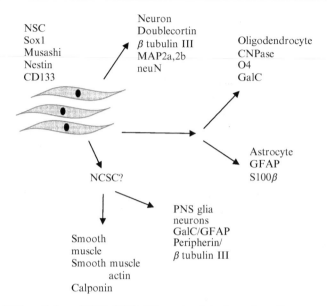

FIG. 4. Differentiation of NSCs. NSCs differentiate into the major cell types present in the adult nervous system, including the various subtypes of neurons, astrocytes, and oligodendrocytes. It is possible that NSCs can differentiate into glial progenitors that then give rise to astrocytes and oligodendrocytes, and a neural crest-like stem cell, which can then give rise to neural crest fates such as smooth muscle and glia of the peripheral nervous system (PNS). The markers generally used to identify these fates are listed by each cell type.

moiety for both CNTF and LIF) activates the janus kinase/tyrosine kinase (JAK/TYK) family of kinases and also MAPK in the cytoplasm, which results in Stat activation. The inhibition of MAPK delays initiation of CNTF-mediated differentiation, suggesting that CNTF-mediated MAPK activation is required for optimal differentiation (Rajan and McKay, 1998). BMPs also cause glial differentiation by the activation of Stat3, but do so under conditions that involve the FKBP12 rapamycin-associated protein (FRAP) and dense cultures that have high levels of basal Stat3 activation. Thus in this case the high basal levels of active Stat3 become a context in which the active Stat signal generated by BMP via FRAP yields glia (Rajan et al., 2003). The thyroid hormone triiodothyronine (T_3) causes oligoden-drocyte differentiation; however, the details of the mechanism remain to be studied (Johe et al., 1996). Shh appears to specify oligodendrocyte fate, but this effect is inhibited by BMPs, which promote astrocytic fate (Samanta and Kessler, 2004). This reflects the in vivo condition in which inhibition of BMPs by noggin in the SVZ niche causes inhibition of astrocytic fate and, in that case, promotes neuronal differentiation. BMPs also cause neuronal differentiation in both NSCs and NCSCs (Anderson *et al.*, 1997; Mabie *et al.*, 1999). Brain-derived neurotrophic factor (BDNF) has been shown to promote neuronal differentiation in NSC cultures (Caviness *et al.*, 1995; Vicario-Abejon *et al.*, 1995), whereas PDGF is though to be mitogenic for neuronal precursors (Johe *et al.*, 1996). Immobilized ligands are also integrally involved in differentiation. Notch was initially identified in *Drosophila* by its neuronal inhibition phenotype. In mammalian stem cells it causes glial differentiation by either one of two processes: lateral inhibition or inductive signaling (Gaiano and Fishell, 2002). Notch actions include activation of the transcription factors Hes and Herp. Astrocytic differentiation by notch appears to be a two-step process in which the first step blocks neuronal induction, and the second step promotes astrocytic fate (Grandbarbe *et al.*, 2003). The wnt proteins are implicated in neuronal differentiation and neural crest induction (Lee *et al.*, 2004; Sommer, 2004). Wnt-1 and -3 are required for the specification of spinal interneurons and promote neuronal differentiation in NSC cultures (Muroyama *et al.*, 2002, 2004), whereas wnt-7a causes neuronal differentiation in NSC cultures by regulating the expression of neurogenin via activation of the transcription factor TCF/LEF (T cell factor/lymphoid enhancing factor) (Hirabayashi *et al.*, 2004) (Fig. 5).

It has been known for more than a decade that manipulation of oxygen affects stem cell cultures (Cipolleschi *et al.*, 1993). Oxygen levels in developing tissues are regulated, and thus it is no surprise that oxygen levels have specific effects in NSCs in culture and *in vivo*. The generation of free radicals inside cells is dependent on the ambient oxygen tension, and thus the extent of apoptosis generated by these harmful free radicals (Cross and Templeton, 2004). However, the activation of HIF1 adds a new dimension to the regulation of cellular physiology in proliferating and differentiating NSCs. HIF1 is activated by MAPK in cells and when translocated to the nucleus causes the activation of about 50 genes, which include some obvious candidates such as erythropoietin, FGF, and transferrin and its receptor. Other survival-related genes such as FGF and stem cell factor (SCF) appear to be upregulated by signals other than HIF1 in response to hypoxia (Zhu *et al.*, 2005). Hypoxia has been differentially shown to induce differentiation and proliferation of undifferentiated NSCs and Notch signaling is thought to be required for the maintenance of the undifferentiated state of NSCs under hypoxic conditions (Gustafsson *et al.*, 2005). Nitric oxide (NO) also regulates cellular homeostasis in tissues. Although the effect of the gas is thought to extend only over several cell diameters, modification of hemoglobin by NO causes its effects to be felt over greater distances. NO is synthesized by NO synthetase (NOS), of which there are three isoforms in mammals (Mungrue *et al.*, 2003). The modes of NO action include induction of G protein signaling by direct

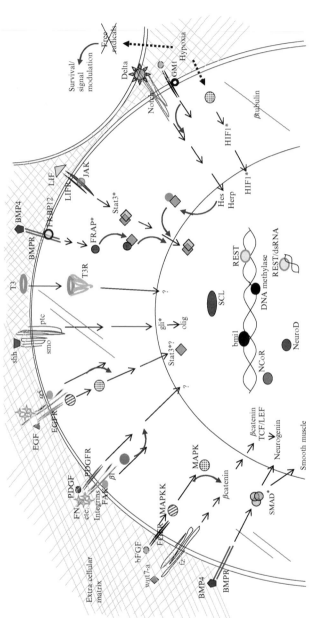

Fig. 5. Signals regulating differentiation of NSCs. As seen in the case of proliferative signals, extracellular signals that regulate the differentiation of NSCs may be soluble ligands, extracellular matrix molecules, and ligands present immobilized in neighboring cells. Almost all the ligands and receptor systems that were involved in proliferation are also involved in the regulation of differentiation, conspicuous in its absence is the Eph/ephrin system. Also, the transcription factors activated by ligands during the induction of differentiation is distinct from the nuclear effects that are elicited during proliferation. There are interactions of the receptor tyrosine kinases with the integrin receptors; however, in this case the signals lead to differentiation. Wnt and BMP activation leads to neuronal differentiation, possibly through the activation of neurogenin. Platelet-derived growth factor (PDGF) appears to lead to the proliferation of a neuronal precursor. Interaction of the notch and bone morphogenetic protein (BMP) pathways leads to astrocytic differentiation, and so does leukemia inhibitory factor (LIF) receptor activation by

binding to its receptor, and direct diffusion through the cell membrane. The effects of NO include the S-nitrosylation of thiol groups in proteins, which could be an important means of posttranslational modification of proteins; G protein signaling, resulting in intracellular signaling cascades; and generation of oxidant moieties and reactive NO species (RNOS), which result in apoptosis (Blaise *et al.*, 2005). NO appears to promote cell survival or cell death depending on the system in question. There is an extensive literature relating NO to synaptic function in the mature nervous system; however, a clear function for it in NSC biology, although likely, remains to be described.

As can be seen from the above description, these interactions are complex and not completely defined. If one were to work one's way out from the nucleus to the cell membrane, it emerges from a parsimonious analysis that the Stat proteins specify astrocytic fate; the basic helix–loop–helix (bHLH) transcription factors, including NeuroD and neurogenin, appear to specify the neuronal fate; and, similarly, olig1 and olig2 (oligodendrocyte lineage transcription factors 1 and 2) specify the oligodendrocyte fate. Whereas the cytoplasmic activation pathway for Stat protein appears to involve the JAK/TYK kinases, MAPK, and FRAP, the cytoplasmic pathway for activation of the neuronal and oligodendrocyte-inducing transcription factors is less defined. Shh causes oligodendrocyte differentiation, wnt causes activation of neurogenin via activation of TCF/LEF, and notch causes astrocytic differentiation presumably by Hes and its targets. At the next level one moves on to the "cross-talk" or "signal integration" scenarios, which may be illustrated by the mechanisms of Stat activation to yield glia from NSCs. There is a density-mediated signal that causes the activation of Stat3 in NSCs, which was first documented by Rajan *et al.* (2003). Kamakura *et al.* subsequently showed that activation of notch (which could be density mediated) causes complex formation between JAK and Stat3, which is facilitated by Hes causing Stat activation (Kamakura *et al.*, 2004). Thus it is the timing and summation of the effects of two or more antagonistic or complementary signals that will finally dictate the outcome.

LIF or ciliary neurotrophic factor (CNTF). Stat3 activation appears to be integrally involved in astrocytic differentiation. shh signals lead to oligodendrocyte differentiation, and so also do signals originating from the triiodothyronine (T_3) receptor. While hypoxia activates hypoxia inducible factor 1 (HIF1), which leads to specific differentiation effects, it also regulates free radical levels in the cell, thus modulating signals in general. The nuclear component of differentiation is more complex than proliferation, and there are several more epigenetic and cofactor molecules involved. In addition, helix–loop–helix proteins including neurogenin, stem cell leukemia protein (SCL), and NeuroD are involved in fate choice specification. The complexity of possible transcription events during differentiation may perhaps be explained by the number of fates into which an NSC can differentiate. *, activated transcription factor. See text for details and references. (See color insert.)

In addition, there is extensive interaction of these transcription factors among themselves and with other constant or induced transcription factors, and the higher-order transcription machinery. For instance, SMADs, which are the major group of transcription factors activated by BMPs, are thought to interact with Stat3 on the GFAP promoter via p300, which serves as transcriptional coactivator and that, in addition to binding the transcription factors themselves, also possesses histone acetyltransferase activity that facilitates transcription (Nakashima et al., 1999). Neurogenin appears to inhibit this interaction when it is bound to DNA by sequestering the p300/SMAD complex (Sun et al., 2001). The other higher-order transcriptional proteins that have been implicated are NcoR (nuclear receptor corepressor) in glial differentiation (Hermanson et al., 2002), and REST (RE-1-silencing transcription factor) in glial maintenance, which act by blocking neuronal genes in nonneuronal cells in most cases (Lunyak and Rosenfeld, 2005). In an added level of regulation REST function is modulated by a double-stranded RNA, which appears to relieve repression of neuronally related genes that are blocked by REST (Kuwabara et al., 2004). Transcription of the GFAP promoter is influenced by methylation of a cytosine residue in the promoter (Takizawa et al., 2001). It is thus possible that methylases are controlled on the basis of the stage of proliferation/ differentiation of NSCs, which provides a further "context" to the cell receiving the signal. There is also increasing evidence that bmi1, which is one of the polycomb group of genes, an "oncogene" that regulates cell cycle-related genes such as INK4a and p16ARF, is involved in the maintenance, proliferation, and differentiation states of neural stem cells. bmi1 forms part of a complex called the polycomb repressive complex 1 (PRC1), thought to stably maintain gene expression by regulation of epigenetic chromatin modifications (Valk-Lingbeek et al., 2004). Mice deficient in bmi1 display deficits in neural stem cells and cerebellar neurons (Leung et al., 2004). Interestingly, they also show an increase in astrocytes (Zencak et al., 2005). Finally, the bHLH protein SCL (stem cell leukemia protein) is thought to be involved in glial specification in the developing embryo (Muroyama et al., 2005).

Differentiation may be brought about by inhibition of the cell cycle, or by the induction of differentiation per se without cell cycle inhibition. In the former case inhibition of the cell cycle is brought about by the regulation of cdk inhibitors such as p27, which is described in Edlund and Jessell (1999). Other cdk inhibitors that function in a similar manner are p21 and p15; these interactions result mainly in inhibition of phosphorylation of the retinoblastoma protein (Rb), which would permit cells to reenter the cell cycle (Weinberg, 1995). In the second case, regulation of differentiation occurs by proteins such as Id and Hes, which are directly involved in the induction of differentiation regimens (Norton, 2000; Ohtsuka *et al.*, 2001).

Conclusions and Projections

Complex signaling networks regulate the life and death of an NSC. In this review we have created a model for some of the permutations of the signals that are associated with NSC proliferation and differentiation, albeit a simplistic one. Our model is based largely on physical interaction of signaling entities and chemical modifications of enzymes and substrates, and implicit throughout the discussion is the assumption that temporal and spatial factors affect signal intensity and generation. "Context" is provided by prior or simultaneously occurring chemical, spatial, and temporal events, and is crucial for the outcome of signaling events; a balance of signaling events will decide whether a cell will divide, differentiate, or die. Computer modeling could be a further approach by which networks of signals arising from the ligand–receptor interaction at the cell surface to transcription factor activation in the nucleus may be described, along with regulatory motifs along the cellular networks (Ma'ayan *et al.*, 2005). The models provided in Figs. 2 and 5 depict pathways at their most basic level. An inclusion of all the "points of regulation" with detailed depictions of pathways and in four dimensions will exponentially increase the complexity of the interactions.

A comparison of Figs. 2 and 5 shows that there are several apparently common signals involved in proliferation and differentiation. However, their interactions and the resulting end points are substantially different. Some striking examples are shh, which regulates proliferation by the activation of N-myc, and differentiation by the activation of the olig transcription factor. The delta/notch system regulates proliferation, possibly by genes that are activated by its responding transcription factors Hes/Herp, but activation of Stat proteins may be a signal that is involved in mediating the complex differentiation phenotype of notch. Another example is MAPK, which although usually associated with proliferation can mediate differentiation by augmenting signals such as those mediated by wnt and CNTF. Integrins can modulate signals from several receptor tyrosine kinases (RTKs) including FGF, EGF, and PDGF.

Signals that regulate survival/apoptosis are more apparent and integral to proliferation than to differentiation. Interestingly, several of the genes that are involved in these processes were identified as oncogenes or suppressor genes. These include signals that regulate bcl2, mci1, and p53. Among the transcriptional regulators of NSC proliferation are Stat3, N-myc, and bmi1. Stat3 seems to be required for the proliferation of retinal progenitors; N-myc is responsible for proliferation in related systems, and it functions downstream of shh; bmi1, one of the polycomb group of proteins, is involved in epigenetic regulation of the genome and, along with tlx1, appears to be involved in NSC renewal. The Eph/ephrin proteins appear to have important roles in the regulation of NSCs *in vivo*. It is possible that genes identified as oncogenes and tumor suppressor genes have as yet undefined roles in NSC proliferation and maintenance.

The number and complexity of nuclear signals associated with differentiation far exceed those responsible for proliferation of NSCs. This is not surprising because there are several more outcomes that can occur because of differentiation, and possibly many avenues by which these fates can be arrived at, depending on the context in which the differentiation event is occurring. Determination of the cytoplasmic signals that regulate these nuclear proteins will enhance our understanding of the regulation of differentiation of NSCs, and the mechanisms by which cytoplasmic interactions occur. Our current knowledge indicates that BMPs, CNTF/LIF, notch, wnt, shh, and T_3 are inducers of specific differentiation events. Robust inducers like CNTF can induce differentiation independently, whereas others such as BMPs require a specific context to bring about a particular fate, and the RTKs appear to have a supporting role.

As mentioned previously, some factors elicit different responses from NSCs derived from different species; for example, LIF causes differentiation of NSCs derived from E14 rat embryos and supports proliferation of NSCs derived from younger embryos, and in human NSCs. Because of such context considerations only those results that have been obtained in mammalian, and preferably neural-related, systems have been considered here. However, some extrapolations have been made from related systems, such as the case of insulin and integrins. The assumption of the involvement of Akt/BAD and integrin signaling in NSC cultures is not unreasonable. Insulin is included in most cell culture systems, including NSC cultures. Adherent monolayer cultures are dependent on a coating of matrix molecule such as fibronectin or laminin on the culture dish for survival and proliferation, whereas neurosphere cultures synthesize their own extracellular matrix. Ideas and practical methods to manipulate cell surface and cytoplasmic signals independently and in concert will permit appreciable control over NSCs and may even allow us to minimize differences that arise because of culture protocols. This may involve concerted inhibition signals by inhibition of enzymes, manipulation of density, extracellular matrix, temporally regulated coactivation of pathways, and compartmentalized activation of signals especially for immobile ligands, and the configuration of intersecting gradients for immobile and soluble ligands.

Sample Protocols for the Culture and Characterization of NSCs

NSCs from various sources are capable of continued proliferation as undifferentiated cells and of then differentiating *in vitro* into either mixed populations of clonally related neurons, astrocytes, or oligodendrocytes or populations relatively enriched for one neural cell type versus another. Rodent NSCs are preferentially maintained in serum-free medium supplemented with bFGF as mitogen. Differentiation toward enriched populations of various neural lineages is routinely accomplished by allowing the cells to exit the cell cycle by removing mitogens from the medium while simultaneously exposing them to various inducing agents including NT-3 for neurons, CNTF and serum for astrocytes, and T_3 and PDGF for oligodendrocytes (Johe *et al.*, 1996). These cultures are characterized by immunocytochemistry (ICC) and/or reverse transcriptase-polymerase chain reaction (RT-PCR) with stem cell markers. Differentiated cultures are characterized with ICC markers, PCR, and whole-cell patch-clamp to confirm the functionality of neurons. Here we describe one of several existing methods used to culture NSCs from mammalian tissue.

Culture of Human NSCs

We have been successful in isolating human NSCs from the telencephalic ventricular zone of a 13-week gestation human fetal cadaver.

1. Briefly, we first establish a primary dissociated, stable serum-containing monolayer culture of fetal ventricular zone. Cultures that do not grow well or do not continue to proliferate are no longer pursued. A promising culture is then subjected to a 6- to 8-week sequential growth factor selection process based on growth parameters rather than on markers (as per Flax *et al.*, 1998). In fact, cells that form clusters that are greater than 10 cell diameters and cannot be readily disaggregated are excluded.

2. Cells grown in serum are switched to serum-free conditions containing EGF (human recombinant, 20 ng/ml; EMD Biosciences, San Diego, CA), bFGF (human recombinant, 20 ng/ml; EMD Biosciences), and LIF (human recombinant, 10 ng/ml; Chemicon International, Temecula, CA). They are passaged once per week for 2 weeks.

3. Cells that are successfully passaged are then switched to bFGF alone. They are similarly passaged once per week for 2 weeks.

4. Cell that are successfully passaged in bFGF are then switched to EGF alone. They are similarly passaged once per week for 2 weeks.

5. Cells that are successfully passaged are then switched back to bFGF and the 2-week selection process is continued.

6. Cells that are successfully passaged in bFGF are then switched to bFGF plus LIF.

7. Cultures that have been successfully passaged over the previous 6–8 weeks and continue to maintain stem-like growth after this selection process are then subjected to an *in vitro* and *in vivo* functional screen. *In vitro*, the cells must be able to express undifferentiated markers yet, in response to induction, differentiate in a dish. The *in vivo* functional screen entails continuing to use only those cells that have the ability to engraft, migrate, and differentiate *in vivo* after implantation into the ventricles and cerebella of newborn (P0) mice and yield olfactory bulb neurons or cerebellar granule neurons, respectively. After 3–4 weeks, the mice are killed to determine which hNSCs yielded neurons in the olfactory bulb, glia in the cortex, and granule neurons in the cerebellum.

8. On the basis of this protracted screen, only a few lines are ultimately selected for further use. These are stored as stable lines to be used for multiple experiments. They are not grown as floating clusters ("spheres") but rather as monolayers in serum-free medium in bFGF plus LIF with the addition (1:1) of medium conditioned by hNSCs (i.e., 50% "self-conditioned medium"). No genetic manipulations are imposed on these cultures. Monolayers and any clusters are disaggregated frequently with trypsin or Accutase (Innovative Cell Technologies, San Diego, CA). Karyotypes have remained stable. Most flask cultures are grown with bFGF and LIF, with the addition (1:1) of medium self-conditioned by hNSCs and stored for future use. The hNSCs are predominantly adherent and grow as a monolayer, although there is a moderate percentage of floating cells. Tapping the flask and/or triturating easily dislodges the adherent cells to permit passaging and to preclude terminal differentiation. The cells tend to form clusters both when floating and when attached. Frequent passaging excludes those cells that are not actively proliferative and hence allows their cell cycle and differentiation state to be more or less synchronized.

9. hNSCs are grown in 25-cm^2 flasks (Falcon, tissue culture-treated; BD Biosciences Discovery Labware, Bedford, MA) containing 8 ml of medium, fed, and/or split 1:2 once per week. Concentrated splits do better than more dilute splits. hNSCs are dissociated into single-cell suspensions when any floating or adherent cellular clusters grow to such a point that they can no longer readily be dissociated mechanically by simple trituration, or when they are too adherent to be dislodged by simple agitation, or when they are greater than 10 cell diameters in width—and before all transplantations. This generally coincides with our once-per-week passaging regimen. Again, when passaging, we split only 1:2 each time. All cells, both floating and adherent, are passaged. For routine passaging, Accutase is used as the dissociating agent. Accutase can be inactivated by simply diluting with fresh medium.

Immunocytochemistry of Markers to Identify Stem Cells and
Differentiation Products

1. Fix the cells with PAF solution [4% paraformaldehyde, 0.12 *M* sucrose, in phosphate-buffered saline (PBS, pH 7.2)]. The cells are washed once, quickly and gently, with PBS (room temperature) and fixed by adding PAF (0.5 ml/well for 24 wells, 0.1 ml/well for 96 wells) for 20 min at room temperature.

2. Wash for 10 min with PBS. Repeat twice.

3. Saturate nonspecific sites and permeabilize with PBS containing 0.3% Triton and 10% normal goat serum (NGS) for 40 min at room temperature. Wash for 5 min with PBS. Repeat twice.

4. Incubate the cells with primary antibodies for 1 h at room temperature: Dilute the antibodies in PBS containing 5% NGS. Wash for 5 min with PBS. Repeat twice.

5. Add the secondary antibody for 1 h at room temperature. Dilute the antibody in PBS containing 5% NGS. Wash for 5 min with PBS. Repeat twice.

6. Permeabilize the nuclei with 100% methanol (−20°) for 5 min. Wash for 5 min with PBS. Repeat twice.

7. Incubate the cells with bisbenzimide for 10 min at room temperature: Wash the cells three times with ultrapure water and mount them with Aqua-Poly/Mount (Polysciences, Warrington, PA). Let it dry and store at 4°.

8. Some antibody sources: mouse anti-MAP-2 (diluted 1:250; Sigma, St. Louis, MO), mouse anti-β-tubulin III (diluted 1:600; Covance Research Products, Denver, PA), neurofilament M (NF-m, diluted 1:500; Chemicon International), GFAP (diluted 1:500; Chemicon International), glial cell line-derived neurotrophic factor (GDNF, diluted 1:1000; Chemicon International), nestin (diluted 1:100; Chemicon International), and Alexa-conjugated secondary antibodies (diluted 1:500; Invitrogen, Carlsbad, CA).

Acknowledgments

The authors acknowledge the financial support of the A-T Children's Project to P.R. and of the NIH to E.Y.S., and to Dr. J. F. Loring for suggestions on the manuscript.

References

Akai, J., Halley, P. A., and Storey, K. G. (2005). FGF-dependent Notch signaling maintains the spinal cord stem zone. *Genes Dev.* **19,** 2877–2887.

Alvarez-Buylla, A., and Lim, D. A. (2004). For the long run: Maintaining germinal niches in the adult brain. *Neuron* **41,** 683–686.

Alvarez-Buylla, A., Seri, B., and Doetsch, F. (2002). Identification of neural stem cells in the adult vertebrate brain. *Brain Res. Bull.* **57,** 751–758.

Anderson, D. J., Groves, A., Lo, L., Ma, Q., Rao, M., Shah, N. M., and Sommer, L. (1997). Cell lineage determination and the control of neuronal identity in the neural crest. *Cold Spring Harb. Symp. Quant. Biol.* **62,** 493–504.

Antonopoulos, J., Pappas, I. S., and Parnavelas, J. G. (1997). Activation of the GABAA receptor inhibits the proliferative effects of bFGF in cortical progenitor cells. *Eur. J. Neurosci.* **9,** 291–298.

Artavanis-Tsakonas, S., Rand, M. D., and Lake, R. J. (1999). Notch signaling: Cell fate control and signal integration in development. *Science* **284,** 770–776.

Baas, A. F., Smit, L., and Clevers, H. (2004). LKB1 tumor suppressor protein: PARtaker in cell polarity. *Trends Cell Biol.* **14,** 312–319.

Bieberich, E. (2004). Integration of glycosphingolipid metabolism and cell-fate decisions in cancer and stem cells: Review and hypothesis. *Glycoconj. J.* **21,** 315–327.

Bieberich, E., MacKinnon, S., Silva, J., Noggle, S., and Condie, B. G. (2003). Regulation of cell death in mitotic neural progenitor cells by asymmetric distribution of prostate apoptosis response 4 (PAR-4) and simultaneous elevation of endogenous ceramide. *J. Cell Biol.* **162,** 469–479.

Blaise, G. A., Gauvin, D., Gangal, M., and Authier, S. (2005). Nitric oxide, cell signaling and cell death. *Toxicology* **208,** 177–192.

Bond, J., Scott, S., Hampshire, D. J., Springell, K., Corry, P., Abramowicz, M. J., Mochida, G. H., Hennekam, R. C., Maher, E. R., Fryns, J. P., Alswaid, A., Jafri, H., Rashid, Y., Mubaidin, A., Walsh, C. A., Roberts, E., and Woods, C. G. (2003). Protein-truncating mutations in ASPM cause variable reduction in brain size. *Am. J. Hum. Genet.* **73,** 1170–1177.

Bonni, A., Sun, Y., Nadal-Vicens, M., Bhatt, A., Frank, D. A., Rozovsky, I., Stahl, N., Yancopoulos, G. D., and Greenberg, M. E. (1997). Regulation of gliogenesis in the central nervous system by the JAK–STAT signaling pathway. *Science* **278,** 477–483.

Bonni, A., Brunet, A., West, A. E., Datta, S. R., Takasu, M. A., and Greenberg, M. E. (1999). Cell survival promoted by the Ras–MAPK signaling pathway by transcription-dependent and -independent mechanisms. *Science* **286,** 1358–1362.

Brunet, A., Bonni, A., Zigmond, M. J., Lin, M. Z., Juo, P., Hu, L. S., Anderson, M. J., Arden, K. C., Blenis, J., and Greenberg, M. E. (1999). Akt promotes cell survival by phosphorylating and inhibiting a Forkhead transcription factor. *Cell* **96,** 857–868.

Cai, L., Hayes, N. L., Takahashi, T., Caviness, V. S., Jr., and Nowakowski, R. S. (2002). Size distribution of retrovirally marked lineages matches prediction from population measurements of cell cycle behavior. *J. Neurosci. Res.* **69,** 731–744.

Caille, I., Allinquant, B., Dupont, E., Bouillot, C., Langer, A., Muller, U., and Prochiantz, A. (2004). Soluble form of amyloid precursor protein regulates proliferation of progenitors in the adult subventricular zone. *Development* **131,** 2173–2181.

Calegari, F., and Huttner, W. B. (2003). An inhibition of cyclin-dependent kinases that lengthens, but does not arrest, neuroepithelial cell cycle induces premature neurogenesis. *J. Cell Sci.* **116,** 4947–4955.

Castaneda-Castellanos, D. R., and Kriegstein, A. R. (2004). Controlling neuron number: Does Numb do the math? *Nat. Neurosci.* **7,** 793–794.

Caviness, V. S., Jr., Takahashi, T., and Nowakowski, R. S. (1995). Numbers, time and neocortical neuronogenesis: A general developmental and evolutionary model. *Trends Neurosci.* **18,** 379–383.

Chang, L. K., Putcha, G. V., Deshmukh, M., and Johnson, E. M., Jr. (2002). Mitochondrial involvement in the point of no return in neuronal apoptosis. *Biochimie* **84,** 223–231.

Chenn, A., and McConnell, S. K. (1995). Cleavage orientation and the asymmetric inheritance of Notch1 immunoreactivity in mammalian neurogenesis. *Cell* **82,** 631–641.

Chenn, A., and Walsh, C. A. (2002). Regulation of cerebral cortical size by control of cell cycle exit in neural precursors. *Science* **297,** 365–369.

Cipolleschi, M. G., Dello Sbarba, P., and Olivotto, M. (1993). The role of hypoxia in the maintenance of hematopoietic stem cells. *Blood* **82,** 2031–2037.

Clevers, H. (2005). Stem cells, asymmetric division and cancer. *Nat. Genet.* **37,** 1027–1028.

Conover, J. C., Doetsch, F., Garcia-Verdugo, J. M., Gale, N. W., Yancopoulos, G. D., and Alvarez-Buylla, A. (2000). Disruption of Eph/ephrin signaling affects migration and proliferation in the adult subventricular zone. *Nat. Neurosci.* **3,** 1091–1097.

Conti, L., and Cattaneo, E. (2005). Controlling neural stem cell division within the adult subventricular zone: An APPealing job. *Trends Neurosci.* **28,** 57–59.

Cross, J. V., and Templeton, D. J. (2004). Thiol oxidation of cell signaling proteins: Controlling an apoptotic equilibrium. *J. Cell. Biochem.* **93,** 104–111.

Datta, S. R., Dudek, H., Tao, X., Masters, S., Fu, H., Gotoh, Y., and Greenberg, M. E. (1997). Akt phosphorylation of BAD couples survival signals to the cell-intrinsic death machinery. *Cell* **91,** 231–241.

Datta, S. R., Brunet, A., and Greenberg, M. E. (1999). Cellular survival: A play in three Akts. *Genes Dev.* **13,** 2905–2927.

Depaepe, V., Suarez-Gonzalez, N., Dufour, A., Passante, L., Gorski, J. A., Jones, K. R., Ledent, C., and Vanderhaeghen, P. (2005). Ephrin signalling controls brain size by regulating apoptosis of neural progenitors. *Nature* **435,** 1244–1250.

De Zio, D., Giunta, L., Corvaro, M., Ferraro, E., and Cecconi, F. (2005). Expanding roles of programmed cell death in mammalian neurodevelopment. *Semin. Cell Dev. Biol.* **16,** 281–294.

Doetsch, F., Petreanu, L., Caille, I., Garcia-Verdugo, J. M., and Alvarez-Buylla, A. (2002). EGF converts transit-amplifying neurogenic precursors in the adult brain into multipotent stem cells. *Neuron* **36,** 1021–1034.

Du, Q., and Macara, I. G. (2004). Mammalian Pins is a conformational switch that links NuMA to heterotrimeric G proteins. *Cell* **119,** 503–516.

Edlund, T., and Jessell, T. M. (1999). Progression from extrinsic to intrinsic signaling in cell fate specification: A view from the nervous system. *Cell* **96,** 211–224.

Feng, Y., and Walsh, C. A. (2004). Mitotic spindle regulation by Nde1 controls cerebral cortical size. *Neuron* **44,** 279–293.

Flax, J. D., Aurora, S., Yang, C., Simonin, C., Wills, A. M., Billinghurst, L. L., Jendoubi, M., Sidman, R. L., Wolfe, J. H., Kim, S. U., and Snyder, E. Y. (1998). Engraftable human neural stem cells respond to developmental cues, replace neurons, and express foreign genes. *Nat. Biotechnol.* **16,** 1033–1039.

Frederiksen, K., and McKay, R. D. (1988). Proliferation and differentiation of rat neuroepithelial precursor cells *in vivo. J. Neurosci.* **8,** 1144–1151.

Gage, F. H. (2000). Mammalian neural stem cells. *Science* **287,** 1433–1438.

Gaiano, N., and Fishell, G. (2002). The role of notch in promoting glial and neural stem cell fates. *Annu. Rev. Neurosci.* **25,** 471–490.

Giancotti, F. G. (1997). Integrin signaling: Specificity and control of cell survival and cell cycle progression. *Curr. Opin. Cell Biol.* **9,** 691–700.

Goldman, S. A., and Sim, F. (2005). Neural progenitor cells of the adult brain. *Novartis Found. Symp.* **265,** 66–80; discussion 82–97.

Grandbarbe, L., Bouissac, J., Rand, M., Hrabe de Angelis, M., Artavanis-Tsakonas, S., and Mohier, E. (2003). Delta-Notch signaling controls the generation of neurons/glia from neural stem cells in a stepwise process. *Development* **130,** 1391–1402.

Groszer, M., Erickson, R., Scripture-Adams, D. D., Dougherty, J. D., Le Belle, J., Zack, J. A., Geschwind, D. H., Liu, X., Kornblum, H. I., and Wu, H. (2006). PTEN negatively regulates neural stem cell self-renewal by modulating G_0–G_1 cell cycle entry. *Proc. Natl. Acad. Sci. USA* **103**, 111–116.

Gustafsson, M. V., Zheng, X., Pereira, T., Gradin, K., Jin, S., Lundkvist, J., Ruas, J. L., Poellinger, L., Lendahl, U., and Bondesson, M. (2005). Hypoxia requires notch signaling to maintain the undifferentiated cell state. *Dev. Cell* **9**, 617–628.

Haydar, T. F., Wang, F., Schwartz, M. L., and Rakic, P. (2000). Differential modulation of proliferation in the neocortical ventricular and subventricular zones. *J. Neurosci.* **20**, 5764–5774.

Hermanson, O., Jepsen, K., and Rosenfeld, M. G. (2002). N-CoR controls differentiation of neural stem cells into astrocytes. *Nature* **419**, 934–939.

Hirabayashi, Y., Itoh, Y., Tabata, H., Nakajima, K., Akiyama, T., Masuyama, N., and Gotoh, Y. (2004). The Wnt/β-catenin pathway directs neuronal differentiation of cortical neural precursor cells. *Development* **131**, 2791–2801.

Hitoshi, S., Seaberg, R. M., Koscik, C., Alexson, T., Kusunoki, S., Kanazawa, I., Tsuji, S., and van der Kooy, D. (2004). Primitive neural stem cells from the mammalian epiblast differentiate to definitive neural stem cells under the control of Notch signaling. *Genes Dev.* **18**, 1806–1811.

Holland, S. J., Gale, N. W., Mbamalu, G., Yancopoulos, G. D., Henkemeyer, M., and Pawson, T. (1996). Bidirectional signalling through the EPH-family receptor Nuk and its transmembrane ligands. *Nature* **383**, 722–725.

Holmberg, J., Armulik, A., Senti, K. A., Edoff, K., Spalding, K., Momma, S., Cassidy, R., Flanagan, J. G., and Frisen, J. (2005). Ephrin-A2 reverse signaling negatively regulates neural progenitor proliferation and neurogenesis. *Genes Dev.* **19**, 462–471.

Hu, H., Tomasiewicz, H., Magnuson, T., and Rutishauser, U. (1996). The role of polysialic acid in migration of olfactory bulb interneuron precursors in the subventricular zone. *Neuron* **16**, 735–743.

Huttner, W. B., and Kosodo, Y. (2005). Symmetric versus asymmetric cell division during neurogenesis in the developing vertebrate central nervous system. *Curr. Opin. Cell Biol.* **17**, 648–657.

Hynes, R. O. (2002). Integrins: Bidirectional, allosteric signaling machines. *Cell* **110**, 673–687.

Israsena, N., Hu, M., Fu, W., Kan, L., and Kessler, J. A. (2004). The presence of FGF2 signaling determines whether β-catenin exerts effects on proliferation or neuronal differentiation of neural stem cells. *Dev. Biol.* **268**, 220–231.

Jacques, T. S., Relvas, J. B., Nishimura, S., Pytela, R., Edwards, G. M., Streuli, C. H., and Ffrench-Constant, C. (1998). Neural precursor cell chain migration and division are regulated through different β_1 integrins. *Development* **125**, 3167–3177.

Jessell, T. M., and Sanes, J. R. (2000). Development: The decade of the developing brain. *Curr. Opin. Neurobiol.* **10**, 599–611.

Jin, G., Tan, X., Tian, M., Qin, J., Zhu, H., Huang, Z., and Xu, H. (2005). The controlled differentiation of human neural stem cells into TH-immunoreactive (ir) neurons *in vitro*. *Neurosci. Lett.* **386**, 105–110.

Johansson, C. B., Momma, S., Clarke, D. L., Risling, M., Lendahl, U., and Frisen, J. (1999). Identification of a neural stem cell in the adult mammalian central nervous system. *Cell* **96**, 25–34.

Johe, K. K., Hazel, T. G., Muller, T., Dugich-Djordjevic, M. M., and McKay, R. D. (1996). Single factors direct the differentiation of stem cells from the fetal and adult central nervous system. *Genes Dev.* **10**, 3129–3140.

Kamakura, S., Oishi, K., Yoshimatsu, T., Nakafuku, M., Masuyama, N., and Gotoh, Y. (2004). Hes binding to STAT3 mediates crosstalk between Notch and JAK-STAT signalling. *Nat. Cell Biol.* **6**, 547–554.

Kenney, A. M., and Rowitch, D. H. (2000). Sonic hedgehog promotes G_1 cyclin expression and sustained cell cycle progression in mammalian neuronal precursors. *Mol. Cell. Biol.* **20,** 9055–9067.

Kleber, M., and Sommer, L. (2004). Wnt signaling and the regulation of stem cell function. *Curr. Opin. Cell Biol.* **16,** 681–687.

Klezovitch, O., Fernandez, T. E., Tapscott, S. J., and Vasioukhin, V. (2004). Loss of cell polarity causes severe brain dysplasia in Lgl1 knockout mice. *Genes Dev.* **18,** 559–571.

Kosodo, Y., Roper, K., Haubensak, W., Marzesco, A. M., Corbeil, D., and Huttner, W. B. (2004). Asymmetric distribution of the apical plasma membrane during neurogenic divisions of mammalian neuroepithelial cells. *EMBO J.* **23,** 2314–2324.

Kuan, C. Y., Roth, K. A., Flavell, R. A., and Rakic, P. (2000). Mechanisms of programmed cell death in the developing brain. *Trends Neurosci.* **23,** 291–297.

Kuida, K., Zheng, T. S., Na, S., Kuan, C., Yang, D., Karasuyama, H., Rakic, P., and Flavell, R. A. (1996). Decreased apoptosis in the brain and premature lethality in CPP32-deficient mice. *Nature* **384,** 368–372.

Kuwabara, T., Hsieh, J., Nakashima, K., Taira, K., and Gage, F. H. (2004). A small modulatory dsRNA specifies the fate of adult neural stem cells. *Cell* **116,** 779–793.

Lee, H. Y., Kleber, M., Hari, L., Brault, V., Suter, U., Taketo, M. M., Kemler, R., and Sommer, L. (2004). Instructive role of Wnt/β-catenin in sensory fate specification in neural crest stem cells. *Science* **303,** 1020–1023.

Leone, D. P., Relvas, J. B., Campos, L. S., Hemmi, S., Brakebusch, C., Fassler, R., Ffrench-Constant, C., and Suter, U. (2005). Regulation of neural progenitor proliferation and survival by β_1 integrins. *J. Cell Sci.* **118,** 2589–2599.

Leung, C., Lingbeek, M., Shakhova, O., Liu, J., Tanger, E., Saremaslani, P., Van Lohuizen, M., and Marino, S. (2004). Bmi1 is essential for cerebellar development and is overexpressed in human medulloblastomas. *Nature* **428,** 337–341.

Li, H. S., Wang, D., Shen, Q., Schonemann, M. D., Gorski, J. A., Jones, K. R., Temple, S., Jan, L. Y., and Jan, Y. N. (2003). Inactivation of Numb and Numblike in embryonic dorsal forebrain impairs neurogenesis and disrupts cortical morphogenesis. *Neuron* **40,** 1105–1118.

Lim, D. A., Tramontin, A. D., Trevejo, J. M., Herrera, D. G., Garcia-Verdugo, J. M., and Alvarez-Buylla, A. (2000). Noggin antagonizes BMP signaling to create a niche for adult neurogenesis. *Neuron* **28,** 713–726.

LoTurco, J. J., Owens, D. F., Heath, M. J., Davis, M. B., and Kriegstein, A. R. (1995). GABA and glutamate depolarize cortical progenitor cells and inhibit DNA synthesis. *Neuron* **15,** 1287–1298.

Louvi, A., and Artavanis-Tsakonas, S. (2006). Notch signalling in vertebrate neural development. *Nat. Rev. Neurosci.* **7,** 93–102.

Lunyak, V. V., and Rosenfeld, M. G. (2005). No rest for REST: REST/NRSF regulation of neurogenesis. *Cell* **121,** 499–501.

Ma'ayan, A., Jenkins, S. L., Neves, S., Hasseldine, A., Grace, E., Dubin-Thaler, B., Eungdamrong, N. J., Weng, G., Ram, P. T., Rice, J. J., Kershenbaum, A., Stolovitzky, G. A., Blitzer, R. D., and Iyengar, R. (2005). Formation of regulatory patterns during signal propagation in a mammalian cellular network. *Science* **309,** 1078–1083.

Mabie, P. C., Mehler, M. F., and Kessler, J. A. (1999). Multiple roles of bone morphogenetic protein signaling in the regulation of cortical cell number and phenotype. *J. Neurosci.* **19,** 7077–7088.

Matter, M. L., and Ruoslahti, E. (2001). A signaling pathway from the $\alpha_5\beta_1$ and $\alpha_v\beta_3$ integrins that elevates bcl-2 transcription. *J. Biol. Chem.* **276,** 27757–27763.

McMahon, A. P. (2000). More surprises in the Hedgehog signaling pathway. *Cell* **100**, 185–188.

Megason, S. G., and McMahon, A. P. (2002). A mitogen gradient of dorsal midline Wnts organizes growth in the CNS. *Development* **129**, 2087–2098.

Mellitzer, G., Xu, Q., and Wilkinson, D. G. (1999). Eph receptors and ephrins restrict cell intermingling and communication. *Nature* **400**, 77–81.

Moshiri, A., Close, J., and Reh, T. A. (2004). Retinal stem cells and regeneration. *Int. J. Dev. Biol.* **48**, 1003–1014.

Mungrue, I. N., Stewart, D. J., and Husain, M. (2003). The Janus faces of iNOS. *Circ. Res.* **93**, e74.

Muroyama, Y., Fujihara, M., Ikeya, M., Kondoh, H., and Takada, S. (2002). Wnt signaling plays an essential role in neuronal specification of the dorsal spinal cord. *Genes Dev.* **16**, 548–553.

Muroyama, Y., Kondoh, H., and Takada, S. (2004). Wnt proteins promote neuronal differentiation in neural stem cell culture. *Biochem. Biophys. Res. Commun.* **313**, 915–921.

Muroyama, Y., Fujiwara, Y., Orkin, S. H., and Rowitch, D. H. (2005). Specification of astrocytes by bHLH protein SCL in a restricted region of the neural tube. *Nature* **438**, 360–363.

Nakashima, K., Yanagisawa, M., Arakawa, H., Kimura, N., Hisatsune, T., Kawabata, M., Miyazono, K., and Taga, T. (1999). Synergistic signaling in fetal brain by STAT3–Smad1 complex bridged by p300. *Science* **284**, 479–482.

Norton, J. D. (2000). ID helix–loop–helix proteins in cell growth, differentiation and tumorigenesis. *J. Cell Sci.* **113**, 3897–3905.

Nunes, M. C., Roy, N. S., Keyoung, H. M., Goodman, R. R., McKhann, G., II, Jiang, L., Kang, J., Nedergaard, M., and Goldman, S. A. (2003). Identification and isolation of multipotential neural progenitor cells from the subcortical white matter of the adult human brain. *Nat. Med.* **9**, 439–447.

Ohtsuka, T., Sakamoto, M., Guillemot, F., and Kageyama, R. (2001). Roles of the basic helix–loop–helix genes Hes1 and Hes5 in expansion of neural stem cells of the developing brain. *J. Biol. Chem.* **276**, 30467–30474.

Oishi, K., Kamakura, S., Isazawa, Y., Yoshimatsu, T., Kuida, K., Nakafuku, M., Masuyama, N., and Gotoh, Y. (2004). Notch promotes survival of neural precursor cells via mechanisms distinct from those regulating neurogenesis. *Dev. Biol.* **276**, 172–184.

Owens, D. F., Flint, A. C., Dammerman, R. S., and Kriegstein, A. R. (2000). Calcium dynamics of neocortical ventricular zone cells. *Dev. Neurosci.* **22**, 25–33.

Polster, B. M., and Fiskum, G. (2004). Mitochondrial mechanisms of neural cell apoptosis. *J. Neurochem.* **90**, 1281–1289.

Rajan, P., and McKay, R. D. (1998). Multiple routes to astrocytic differentiation in the CNS. *J. Neurosci.* **18**, 3620–3629.

Rajan, P., Panchision, D. M., Newell, L. F., and McKay, R. D. (2003). BMPs signal alternately through a SMAD or FRAP–STAT pathway to regulate fate choice in CNS stem cells. *J. Cell Biol.* **161**, 911–921.

Rajan, P., Park, K.-I., Ourednik, V., Lee, J.-P., Imitola, J., Mueller, F.-J., Teng, Y. D., and Snyder, E. (2006). Stem cell research and applications for human therapies. *In* "Drug Discovery Research in the Post Genomics Era." John Wiley & Sons, New York. (In press).

Roussigne, M., Cayrol, C., Clouaire, T., Amalric, F., and Girard, J. P. (2003). THAP1 is a nuclear proapoptotic factor that links prostate-apoptosis-response-4 (Par-4) to PML nuclear bodies. *Oncogene* **22**, 2432–2442.

Roy, K., Kuznicki, K., Wu, Q., Sun, Z., Bock, D., Schutz, G., Vranich, N., and Monaghan, A. P. (2004). The Tlx gene regulates the timing of neurogenesis in the cortex. *J. Neurosci.* **24**, 8333–8345.

Rusnati, M., Urbinati, C., Tanghetti, E., Dell'Era, P., Lortat-Jacob, H., and Presta, M. (2002). Cell membrane GM1 ganglioside is a functional coreceptor for fibroblast growth factor 2. *Proc. Natl. Acad. Sci. USA* **99,** 4367–4372.

Sailer, M. H., Hazel, T. G., Panchision, D. M., Hoeppner, D. J., Schwab, M. E., and McKay, R. D. (2005). BMP2 and FGF2 cooperate to induce neural-crest-like fates from fetal and adult CNS stem cells. *J. Cell Sci.* **118,** 5849–5860.

Samanta, J., and Kessler, J. A. (2004). Interactions between ID and OLIG proteins mediate the inhibitory effects of BMP4 on oligodendroglial differentiation. *Development* **131,** 4131–4142.

Sanada, K., and Tsai, L. H. (2005). G protein $\beta\gamma$ subunits and AGS3 control spindle orientation and asymmetric cell fate of cerebral cortical progenitors. *Cell* **122,** 119–131.

Sanai, N., Tramontin, A. D., Quinones-Hinojosa, A., Barbaro, N. M., Gupta, N., Kunwar, S., Lawton, M. T., McDermott, M. W., Parsa, A. T., Manuel-Garcia Verdugo, J., Berger, M. S., and Alvarez-Buylla, A. (2004). Unique astrocyte ribbon in adult human brain contains neural stem cells but lacks chain migration. *Nature* **427,** 740–744.

Sanai, N., Alvarez-Buylla, A., and Berger, M. S. (2005). Neural stem cells and the origin of gliomas. *N. Engl. J. Med.* **353,** 811–822.

Schneller, M., Vuori, K., and Ruoslahti, E. (1997). $\alpha_v\beta_3$ Integrin associates with activated insulin and PDGFβ receptors and potentiates the biological activity of PDGF. *EMBO J.* **16,** 5600–5607.

Scolding, N., Franklin, R., Stevens, S., Heldin, C. H., Compston, A., and Newcombe, J. (1998). Oligodendrocyte progenitors are present in the normal adult human CNS and in the lesions of multiple sclerosis. *Brain* **121,** 2221–2228.

Scolding, N. J., Rayner, P. J., and Compston, D. A. (1999). Identification of A2B5-positive putative oligodendrocyte progenitor cells and A2B5-positive astrocytes in adult human white matter. *Neuroscience* **89,** 1–4.

Sela-Donenfeld, D., and Wilkinson, D. G. (2005). Eph receptors: Two ways to sharpen boundaries. *Curr. Biol.* **15,** R210–R212.

Shah, N. M., and Anderson, D. J. (1997). Integration of multiple instructive cues by neural crest stem cells reveals cell-intrinsic biases in relative growth factor responsiveness. *Proc. Natl. Acad. Sci. USA* **94,** 11369–11374.

Shi, Y., Chichung Lie, D., Taupin, P., Nakashima, K., Ray, J., Yu, R. T., Gage, F. H., and Evans, R. M. (2004). Expression and function of orphan nuclear receptor TLX in adult neural stem cells. *Nature* **427,** 78–83.

Sjostrom, S. K., Finn, G., Hahn, W. C., Rowitch, D. H., and Kenney, A. M. (2005). The Cdk1 complex plays a prime role in regulating N-myc phosphorylation and turnover in neural precursors. *Dev. Cell* **9,** 327–338.

Sommer, L. (2004). Multiple roles of canonical Wnt signaling in cell cycle progression and cell lineage specification in neural development. *Cell Cycle* **3,** 701–703.

Stromblad, S., Becker, J. C., Yebra, M., Brooks, P. C., and Cheresh, D. A. (1996). Suppression of p53 activity and p21WAF1/CIP1 expression by vascular cell integrin $\alpha_v\beta_3$ during angiogenesis. *J. Clin. Invest.* **98,** 426–433.

Stupack, D. G. (2005). Integrins as a distinct subtype of dependence receptors. *Cell Death Differ.* **12,** 1021–1030.

Sun, Y., Nadal-Vicens, M., Misono, S., Lin, M. Z., Zubiaga, A., Hua, X., Fan, G., and Greenberg, M. E. (2001). Neurogenin promotes neurogenesis and inhibits glial differentiation by independent mechanisms. *Cell* **104,** 365–376.

Takizawa, T., Nakashima, K., Namihira, M., Ochiai, W., Uemura, A., Yanagisawa, M., Fujita, N., Nakao, M., and Taga, T. (2001). DNA methylation is a critical cell-intrinsic determinant of astrocyte differentiation in the fetal brain. *Dev. Cell* **1,** 749–758.

Tropepe, V., Craig, C. G., Morshead, C. M., and van der Kooy, D. (1997). Transforming growth factor-α null and senescent mice show decreased neural progenitor cell proliferation in the forebrain subependyma. *J. Neurosci.* **17,** 7850–7859.

Tropepe, V., Coles, B. L., Chiasson, B. J., Horsford, D. J., Elia, A. J., McInnes, R. R., and van der Kooy, D. (2000). Retinal stem cells in the adult mammalian eye. *Science* **287,** 2032–2036.

Valk-Lingbeek, M. E., Bruggeman, S. W., and van Lohuizen, M. (2004). Stem cells and cancer: The polycomb connection. *Cell* **118,** 409–418.

Vicario-Abejon, C., Johe, K. K., Hazel, T. G., Collazo, D., and McKay, R. D. (1995). Functions of basic fibroblast growth factor and neurotrophins in the differentiation of hippocampal neurons. *Neuron* **15,** 105–114.

Viti, J., Gulacsi, A., and Lillien, L. (2003). Wnt regulation of progenitor maturation in the cortex depends on Shh or fibroblast growth factor 2. *J. Neurosci.* **23,** 5919–5927.

Wang, G., Silva, J., Krishnamurthy, K., Tran, E., Condie, B. G., and Bieberich, E. (2005). Direct binding to ceramide activates protein kinase Czeta before the formation of a pro-apoptotic complex with PAR-4 in differentiating stem cells. *J. Biol. Chem.* **280,** 26415–26424.

Weinberg, R. A. (1995). The retinoblastoma protein and cell cycle control. *Cell* **81,** 323–330.

Wu, W., Wong, K., Chen, J., Jiang, Z., Dupuis, S., Wu, J. Y., and Rao, Y. (1999). Directional guidance of neuronal migration in the olfactory system by the protein Slit. *Nature* **400,** 331–336.

Yamashita, T., Hashiramoto, A., Haluzik, M., Mizukami, H., Beck, S., Norton, A., Kono, M., Tsuji, S., Daniotti, J. L., Werth, N., Sandhoff, R., Sandhoff, K., and Proia, R. L. (2003). Enhanced insulin sensitivity in mice lacking ganglioside GM3. *Proc. Natl. Acad. Sci. USA* **100,** 3445–3449.

Zencak, D., Lingbeek, M., Kostic, C., Tekaya, M., Tanger, E., Hornfeld, D., Jaquet, M., Munier, F. L., Schorderet, D. F., van Lohuizen, M., and Arsenijevic, Y. (2005). Bmi1 loss produces an increase in astroglial cells and a decrease in neural stem cell population and proliferation. *J. Neurosci.* **25,** 5774–5783.

Zhang, Z., Vuori, K., Reed, J. C., and Ruoslahti, E. (1995). The $\alpha_5\beta_1$ integrin supports survival of cells on fibronectin and up-regulates Bcl-2 expression. *Proc. Natl. Acad. Sci. USA* **92,** 6161–6165.

Zhu, L. L., Wu, L. Y., Yew, D. T., and Fan, M. (2005). Effects of hypoxia on the proliferation and differentiation of NSCs. *Mol. Neurobiol.* **31,** 231–242.

[3]   Retinal Stem Cells

By Thomas A. Reh and Andy J. Fischer

Abstract

During the embryonic development of the eye, a group of founder cells in the optic vesicle gives rise to multipotent progenitor cells that generate all the neurons and the Müller glia of the mature retina. In most vertebrates, a small group of retinal stem cells persists at the margin of the retina, near the junction with the ciliary epithelium. In fish and amphibians, the retinal stem

METHODS IN ENZYMOLOGY, VOL. 419   0076-6879/06 $35.00
Copyright 2006, Elsevier Inc. All rights reserved.
DOI: 10.1016/S0076-6879(06)19003-5

cells continue to produce progenitors throughout life, adding new retina to the periphery of the existing retina as the eye grows. In birds the new retinal addition is more limited, and it is absent in those mammals that have been analyzed. Nevertheless, cells from the retinal periphery and ciliary body of mammals can be isolated and grown *in vitro* for extended periods. Methods for the study of both embryonic progenitors and adult retinal stem cells *in vitro* and *in vivo* have led to a better understanding of retinal development, allowed for the screening of factors important in retinal growth and differentiation, and enabled the development of methods to direct stem and progenitor cells to specific fates. These methods may ultimately lead to the development of strategies for retinal repair.

Introduction

Retinal Stem/Progenitor Cells During Development

The vertebrate retina arises during development as an evagination, called the optic vesicle, of the diencephalon of the neural tube. The cells in the early optic vesicle express a unique complement of transcription factors, termed eye-field transcription factors (EFTFs), including Pax6, Rx, Six3, and Chx10 (see Zuber *et al.*, 2003, for review). On the basis of this combination of transcription factors, the cells of the optic vesicle can be distinguished from the other cells of the neural tube. The optic vesicle cells undergo extensive divisions to generate all the neurons and the Müller glia of the adult retina. Analysis of the lineages of the proliferating cells from the early stages of eye development indicates that they undergo both symmetric and asymmetric divisions, and that many form clones containing large numbers of cells (2795 cells; Fekete *et al.*, 1994). The majority of the clones obtained by labeling the dividing cells at early stages of retinal development contain multiple types of retinal neurons, as well as Müller glia. Thus, these cells are known as multipotent retinal progenitors.

Although the clonal analysis of retinal progenitor cells demonstrates a wide variety of clone size and composition, some patterns have emerged. First, ever since the first birth-dating studies of Sidman (1961), it has been consistently found that the different types of retinal neurons are generated in a sequence (Fig. 1), with ganglion cells, cone photoreceptors, and horizontal cells generated during early stages of development, and most amacrine rod photoreceptors, bipolar cells, and Müller glia generated in the latter half of the period of retinogenesis. This has led to the hypothesis that retinal progenitor cells undergo a progressive change during development that constrains them to a smaller range of fates (Reh and Kljavin, 1989). However, an alternative model is that a changing environment directs the

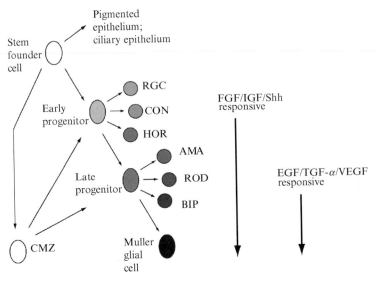

FIG. 1. Diagram showing the possible relationships among the various types of retinal progenitor cells and retinal stem cells. The early optic vesicle is composed of cells that ultimately give rise to all the retinal cells, as well as the pigmented epithelium and ciliary epithelium. Progenitor cells in the retina generate the neurons and ultimately the Müller glia and in nonmammalian vertebrates there is a specialized zone of proliferating cells at the junction between the retina and the ciliary epithelium, called the CMZ. At least some cells of the CMZ or adjacent ciliary epithelium must be true retinal stem cells, because these cells are capable of generating all of the various types of retinal neurons and Müller glia and persist throughout the life of the animal. AMA, amacrine cells; BIP, bipolar cells; CMZ, ciliary marginal zone; CON, cone photoreceptors; FGF, fibroblast growth factor; HOR, horizontal cells; IGF, insulin-like growth factor; RGC, retinal ganglion cells; ROD, rod photoreceptors; Shh, Sonic hedgehog; TGF-α, transforming growth factor α; VEGF, vascular endothelial growth factor.

cells to progressively later fates, but the progenitor cells themselves remain competent to generate all retinal cell types throughout the period of retinogenesis (James *et al.*, 2003) There is experimental support for both models, and both the environment and intrinsic state of the cell are likely to be important factors in determining its ultimate fate [see Reh and Cagan (1994) and Livesey and Cepko (2001) for review].

The first evidence that progenitor cells at a given stage of retina development are not all identical came from an analysis of their proneural gene expression. Retinal progenitor cells express at least one of the following members of this class: Ascl1, Ngn2, and NeuroD1. The basic helix–loop–helix (bHLH) transcription factor Ascl1 (formerly known as Mash1 or Cash1) is expressed in only a subset of retinal progenitors (Jasoni and Reh, 1996; Jasoni *et al.*, 1994), and the proneural gene for the remaining progenitors appears to be either Ngn2 or NeuroD1 (Akagi *et al.*, 2004). More recently, another class of transcription factors, the Fox class, also reveals heterogeneity in the progenitor pool. Foxn4 is expressed in a subset of progenitors and specifically biases them to generate either amacrine or horizontal cells (Li *et al.*, 2004). Retinal progenitors can also be distinguished by their response to growth factors. Progenitors isolated from the early embryonic retina are stimulated to proliferate by fibroblast growth factor (FGF), but are only minimally responsive to epidermal growth factor (EGF) or transforming growth factor α (TGF-α) (Anchan *et al.*, 1991; Lillien and Cepko, 1992). By embryonic day 18 in the rat, the progenitors have now acquired a robust response to EGF (Anchan *et al.*, 1991). In addition, progenitor cells differ in their response to changes in intracellular cAMP. Postnatal progenitors are induced to differentiate by treatments that raise cAMP, whereas embryonic progenitors are not (Taylor and Reh, 1990). The evidence for progenitor heterogeneity is somewhat at variance with the indeterminate lineages of single progenitor cells. One possibility is that the different types of progenitor cells can interconvert. For example, loss of Ascl1 causes an expansion of the Ngn2-expressing progenitors (Akagi *et al.*, 2004). There is also evidence from other regions of the developing CNS that FGF-responsive neural stem cells can be converted to EGF-responsive stem cells (Ciccolini and Svendsen, 1998).

In summary, the majority of mitotically active cells in the embryonic retina are competent to generate multiple different types of retinal neurons, as well as Müller glia. Although these cells are typically referred to as multipotent progenitors, those isolated at early stages of retinogenesis could also be considered retinal stem cells on the basis of their potential to generate all retinal cell types. Moreover, the fact that they can generate large clones indicates that many of their divisions are symmetric. In addition, several groups have shown that early progenitors retain the capacity to proliferate *in vitro* for extended periods of time, and can be cultured as "neurospheres," a capacity that neural stem cells are known to possess (see, e.g., Klassen *et al.*, 2004a–c). As described below, the adult retina of some vertebrates continues to add new neurons and glia at the peripheral margin, and thus true retinal stem cells exist. Presumably these cells were derived from a population of similar cells in the developing retina, but at this point there is no definitive way to distinguish the stem cells from the progenitors during retinogenesis.

Retinal Stem Cells and the Ciliary Marginal Zone

In many vertebrates, the development of the retina is not complete after the embryonic or neonatal period, but rather continues throughout life. This is most dramatically observed in teleosts; from the time of their hatching to when they reach their mature size, the eye of a teleost fish can grow 100-fold. The cellular mechanism that enables new retinal neurons to be generated throughout the lifetime of the fish is found at the peripheral margin of the retina, where it joins with the ciliary epithelium. In this region there is a small cluster of cells that forms a ring around the ciliary margin of the retina, the so-called ciliary marginal zone (or CMZ; Hollyfield, 1968).

The cells of the ciliary marginal zone in fish and frogs resembles the early progenitor cells of the eye, and possibly even the "founder" cells of the optic vesicle. Lineage-tracing studies of CMZ cells have shown that they can give rise to clones that contain all types of retinal neurons, including those that are generated both early and late in embryonic development (Wetts and Fraser, 1988). Most of the CMZ cells express the transcription factor profile of retinal progenitors, including the paired-class transcription factors, such as Rx, Chx10, and Pax6 (Perron et al., 1998). The CMZ cells also express proneural transcription factors, such as Ngn2 and Ascl1 (Harris and Perron, 1998), and at least some of them respond to mitogenic growth factors like their counterparts in the embryo (Mack and Fernald, 1993). Because these cells are capable of generating most of the retina of the mature frog (Reh and Constantine-Paton, 1983) or fish, and they continue to generate new retina throughout the lifetime of the animal, it is likely that this region contains a population of true retinal stem cells. This zone of cells is extremely productive in fish and larval frogs, but in other vertebrates it is greatly reduced or absent. Although it is relatively easy to identify the CMZ cells in fish and amphibians, it is not known how many cells in this zone represent true retinal stem cells and what proportion of them are progenitors.

In the eyes of amphibians and teleost fish, the retina continues to grow in parallel to the overall growth of the eye, whereas in birds most of the retina is generated in ovo with at least 90% of the retinal cells generated more than 1 week before hatching (Prada et al., 1991). At the time of hatching most birds have a fully functional retina, and it was generally assumed that the retina of postembryonic birds lacked the CMZ. However, a study published in 1976 first described the addition of newly generated cells to the peripheral edge of the postnatal chicken retina (Morris et al., 1976). This work had gone largely unnoticed until we demonstrated that new retinal neurons are generated at the peripheral edge of the retina in chickens up to 1 month of age (Fischer and Reh, 2000). This CMZ-like zone has also been identified in the adult quail eye (Kubota et al., 2002), and therefore may be a common feature of the avian eye. In addition to their potential to generate new retinal neurons, chicken CMZ cells express a number of different genes that are also expressed by embryonic retinal progenitor cells. These genes include Pax6, Chx10, PCNA

(Fischer and Reh, 2000), Notch1, cHairy (Fischer, 2005), transitin, the avian homolog of mammalian nestin (Fischer and Omar, 2005), Gli1, Gli3 (Moshiri *et al.*, 2005), and Cath5 when stimulated by the combination of insulin and FGF-2 (Fischer *et al.*, 2002).

Although the zone of proliferating cells in the postnatal chicken eye is reminiscent of the CMZ of fish and amphibians, the CMZ of chickens appears to generate only bipolar and amacrine cells (Fischer and Reh, 2000). We fail to find evidence for the production of photoreceptor, horizontal, or ganglion cells in the untreated chicken CMZ, suggesting that progenitors in the CMZ may be limited to producing particular neuronal cell types. However, we found that the combination of insulin and FGF-2 stimulated the production of ganglion cells, suggesting that the types of cells produced in the avian CMZ are limited by local microenvironment rather than by cell-intrinsic limitations that restrict the multipotency of CMZ progenitors (Fischer *et al.*, 2002). Unlike the CMZ progenitors in cold-blooded vertebrates, the CMZ progenitors in birds do not regenerate the retina when it has been damaged (see Moshiri *et al.*, 2004). Toxic doses of *N*-methyl-D-aspartate (NMDA) or kainic acid, which destroy numerous retinal neurons, do not stimulate the proliferation of CMZ progenitors (Fischer, 2005; Fischer and Reh, 2000).

The proliferation of cells in the normal posthatch chicken CMZ is relatively modest, but can be increased as much as 10-fold by intraocular delivery of growth factors (Fischer and Reh, 2000). Growth factors that stimulate the proliferation of CMZ progenitors include insulin, insulin-like growth factor I (IGF-I), EGF (Fischer and Reh, 2000, 2002), and sonic hedgehog (Shh) (Moshiri *et al.*, 2005), but not FGF (Fischer and Reh, 2000). Because the levels of proliferation are low in the untreated CMZ, it is possible that the factors that stimulate proliferation and neuronal differentiation are present in limiting quantities in the postnatal retina (Fischer and Reh, 2000). Alternatively, the proliferation of progenitors in the CMZ may be suppressed by factors produced by mature retinal neurons. For example, an unusual type of glucagon-expressing neuron within the retina produces neurites that project into and densely ramify within the CMZ and glucagon acts to suppress the proliferation of CMZ progenitors (Fischer *et al.*, 2005). In addition to exogenous growth factors, the proliferation of CMZ progenitors can be increased by experimentally increasing rates of ocular growth. Visual deprivation nearly doubles the number of cells that are added to the edge of the retina (Fischer and Reh, 2000). The mechanisms that link postnatal ocular growth and the addition of cells to the peripheral edge of the retina may involve glucagon-expressing retinal neurons that are known to respond to growth-guiding visual cues (Fischer *et al.*, 1999) and influence the proliferation of CMZ progenitors (Fischer *et al.*, 2005).

Other Sources of Retinal Stem Cells

Ciliary Epithelium

The ciliary epithelium of the ciliary body, like the retina, is derived from the optic vesicle during embryonic development and has been shown to contain cells with neurogenic potential. For example, the nonpigmented epithelium (NPE) of the ciliary body is capable of producing neurons in the intact chicken eye (Fischer and Reh, 2003). Intraocular injection of growth factors (insulin, FGF-2, and EGF) stimulates the proliferation and neuronal differentiation of NPE cells within the ciliary body (Fischer and Reh, 2003). Like the CMZ, NPE cells express Chx10 and Pax6, but are found up to 3 mm anterior to the peripheral edge of the retina. This region of Pax6/Chx10-expressing cells in the NPE of the ciliary body coincides with the region where proliferating and newly generated neurons appear in response to growth factor treatments. Newly generated neurons in the NPE express markers for amacrine cells, ganglion cells, and Müller glia, but not for bipolar cells or photoreceptors. The potential for the ciliary epithelium and adjacent iris to generate neurons may extend to the mammalian retina; forced expression of the paired-class homeodomain transcription factor Crx induces the expression of photoreceptor genes in cells derived from the rat iris (Haruta et al., 2001). Neuron-like cells have been identified in the NPE of adult nonhuman primates (Fischer et al., 2001). Furthermore, a report from Zhao et al. (2002) has indicated that the blockade of bone morphogenetic protein (BMP) signaling interferes with the normal formation of the NPE of the ciliary body and promotes ectopic neural differentiation in the developing NPE of the rodent eye. Several groups have also reported that extended culture of both pigmented and nonpigmented cells of the ciliary epithelium results in progenitor-like cells. These cells form neurospheres and can be passaged at least once. Moreover, the cultured cells forming the pigmented ciliary epithelium express many of the markers of retinal progenitors and their progeny express proteins normally present in subtypes of retinal neurons. As a result of these characteristics, the cells have been termed retinal stem cells (Ahmad et al., 2000; Coles et al., 2004; Tropepe et al., 2000). The relationship between sphere-forming pigmented cells and true retinal stem cells present in the CMZ of fish and frogs is not clear, because the latter are not thought to be pigmented. In addition, it is not clear how either of these cell types relate to the "founder" cells of the optic vesicle that produce all the progenitors of the retina. These issues may eventually be resolved by developing better markers that discriminate among the different types of "retinal stem cells."

Pigmented Epithelium

It has been known for more than half a century that the pigmented cells of the eye are capable of acting as a source of retinal regeneration (or neural stem cells) (Coulombre and Coulombre, 1965; Orts-Llorca and Genis-Galvez, 1960; Reh et al., 1987; Stone, 1950; Stone and Steinitz, 1957). The RPE is a well-known source of retinal stem cells in neotenic amphibians, larval anurans, and embryonic chicks (see Moshiri et al., 2004, for review). Neurogenesis from RPE cells requires dedifferentiation, loss of pigmentation, and cell division (Stone, 1950; Stroeva and Mitashov, 1983). Collectively, this process has been named transdifferentiation (Okada, 1980). RPE cells that have been stimulated to transdifferentiate produce new neurons in a manner that resembles normal retinal histogenesis (Reh et al., 1987; Sakaguchi et al., 1997). In the embryonic chick and rodent, the ability of RPE cells to become retinal stem cells is lost during early stages of development (Park and Hollenberg, 1989; Pittack et al., 1991; Zhao et al., 1995). During the dedifferentiation process, the pigmented epithelial cells acquire a gene expression profile that resembles that of retinal progenitor cells (Sakami et al., 2005). It is possible that these cells go through a stage in which they resemble stem or "founder" cells, because the RPE cells can regenerate the entire retina in some species, up to four complete times (Stone and Steinitz, 1957).

Müller Glia

Mature Müller glia, the major type of support cell in the retina, are capable of dedifferentiating into proliferating progenitor-like cells in the retinas of chickens (Fischer and Reh, 2001; Fischer et al., 2002), zebrafish (Yurco and Cameron, 2005), and rat (Ooto et al., 2004). Under normal conditions, Müller glia are the predominant type of support cell in retina, providing structural, nutritive, and metabolic support to retinal neurons. In response to sufficient retinal damage, or on exposure to a combination of insulin and FGF-2 without damage (Fischer et al., 2002), Müller glia re-enter the cell cycle and express transcription factors found in embryonic retinal progenitors. These transcription factors include Ascl1, Pax6, Chx10 (Fischer and Reh, 2001), and Six3 (Fischer, 2005). Although Müller glia undergo only one round of division in vivo, these cells continue to proliferate and produce some new neurons when dissociated from the intact retina and grown in culture (Fischer and Reh, 2001). In vivo, the majority (about 80%) of cells that are generated by proliferating Müller glia remain as undifferentiated progenitor-like cells, whereas some differentiate into Müller glia and a few differentiate into amacrine or bipolar neurons.

Destruction of ganglion cells, combined with insulin/FGF-2 treatment, stimulated the regeneration of a few ganglion cells (Fischer and Reh, 2002).

In the adult zebrafish, Yurco and Cameron (2005) have reported that acute lesions to the retina result in the reentry of Müller glia into the cell cycle. This study indicated that the proliferating Müller glia become progenitor-like and suggested that the glia may regenerate neurons in the damaged teleost retina.

In the rat retina Müller glia have been shown to be a potential source of retinal regeneration (Ooto *et al.*, 2004). Similar to studies in the chicken retina, Ooto and colleagues used NMDA to induce excitotoxicity and damage the adult rat retina. In response to damage, Müller glia were stimulated to proliferate and produce new neurons. Although these newly produced neurons were limited in number, some regenerated neurons were increased by treatment with retinoic acid or the misexpression of bHLH and homeobox genes. These findings suggest the Müller glial cells are a potential source of neural regeneration in the adult mammalian retina, but may require stimulation (drugs and/or gene therapy) to regenerate significant numbers of neurons to treat retinal degenerative diseases.

Materials and Methods

Primary Cell Culture of Retinal Stem/Progenitor Cells

To establish and maintain cell cultures of retinal progenitor/stem cells, the optimal source is embryonic retina from either rodent or chick. The following methods work well for chick embryos from stages 25–35 (embryonic days 4–8), and for either rat or mouse from embryonic days 14.5 to birth and up to postnatal day 7 (Fig. 2) (Anchan *et al.*, 1991; Kelley *et al.*, 1994; Levine *et al.*, 1997; Reh and Kljavin, 1989). After postnatal day 7 in the rodent there are few progenitor/stem cells, and the vast majority of cells that proliferate *in vitro* after this stage are likely Müller glia (Close *et al.*, 2005). We have also used the same methods for fetal human retina, up to day 70 postconception (Kelley *et al.*, 1995).

Harvest and Culture of Retinal Stem/Progenitor Cells

The embryos are harvested or the pup is killed in accordance with approved protocols for the institution. The eyes are removed and placed in a sterile Petri dish containing cold (4°) Hanks' balanced salt solution (HBSS) with 3% D-glucose and 0.01 M HEPES, pH 7.4 (HBSS+). The retinas are dissected from the extraocular tissue in this solution. The pigmented epithelium, lens, and scleral tissue are easily removed, and the retina is then transferred to a new sterile Petri dish containing

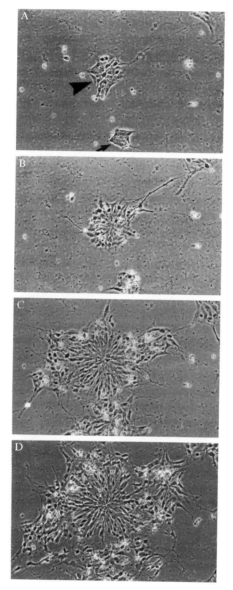

FIG. 2. Stem/progenitor cells grown *in vitro* on adherent substrates and supplemented with TGF-α. (A–D) The same cluster of cells was continuously monitored by time-lapse microscopy over several days. The arrowhead in (A) points to a small group of cells that eventually grows to a large rosette, containing both dividing cells and differentiated neurons (as determined by subsequent immunolabeling for retinal neuron-specific antigens). Reprinted from Anchan *et al.* (1991).

calcium/magnesium-free (CMF)-HBSS, using forceps or a Pasteur pipette. The retinas are then transferred to a 15-ml centrifuge tube containing 0.25% trypsin, in HBSS-CMF (stock at 2.5%; 0.5 ml plus 4.5 ml of HBSS-CMF). The centrifuge tube containing the retinas is then put on a rocking platform in a 37° incubator and gently rocked for 5–15 min. As a guideline, embryonic chick retinas typically require only 5 min, whereas postnatal mouse retina can require up to 10 min for thorough dissociation. In all cases, however, care must be taken not to overtreat the retinas with trypsin, because this will result in low yields of viable progenitor cells. The trypsin treatment is complete when the retinas are broken up into small pieces, but not yet into single cells. Add 0.5 ml of fetal bovine serum (FBS) to the tube to inactivate the trypsin. Centrifuge the tube at 1500 rpm (approximately 750g) for 10 min to pellet the cells. Carefully remove the supernatant (leaving a small amount of solution at the bottom, so that the pellet is not disturbed). Resuspend the pelleted cells in 2 ml of medium (see below) by gentle trituration with a (fire-polished) Pasteur pipette. Determine the number of cells with a hemocytometer; trypan blue can be used to estimate the percentage viability. Plate cells between 50,000 (low density) and 500,000 (high density) cells per well (for a 24-well plate) onto poly-D-lysine/Matrigel-coated coverslips (see below). Cultures are maintained at 37° and 5% CO_2.

The culture medium contains Dulbecco's modified Eagle's medium (DMEM)–F12 (without glutamate or aspartate), insulin (25 μg/ml), transferrin (100 μg/ml), 60 μM putrescine, 30 nM selenium, 20 nM progesterone, penicillin (100 U/ml), streptomycin (100 μg/ml), 0.05 M HEPES, and 1% FBS (Invitrogen, Carlsbad, CA). We prefilter all the stock solutions so there is no need to filter the final medium; however, if contamination is suspected, we have found that the medium can be filtered once without loss of potency. Once the retinal cells are established *in vitro*, one-half the medium is replaced every other day. The final medium is effective for approximately 1 week when stored at 4°. The progenitor cells of chick or rodent retina can be maintained in this medium for up to 1 week and they retain their ability to generate neurons, as demonstrated by double labeling for bromodeoxyuridine (BrdU)/ thymidine and neuron-specific markers (see below). We typically use a serum concentration of 1%, but this can be reduced to 0.1% with little reduction in proliferation. The medium can also be supplemented with growth factors to stimulate the proliferation of the progenitor cells. We have also used the same medium for serum-free cultures, but under these conditions the progenitor cells are more likely to differentiate.

Substrate for Adherent Culture of Retinal Stem/Progenitor Cells

Retinal progenitor/stem cells proliferate in adherent cultures on glass coverslips coated with poly-D-lysine and Matrigel. We have also maintained these cells on laminin-coated coverslips and our more recent experience indicates this is as effective and much easier. However, the method we have used in previous publications is given here. To prepare the coverslips, use high molecular weight poly-D-lysine (MW 30,000–70,000). The poly-D-lysine is dissolved in sterile water at a concentration of 0.5 mg/ml and aliquoted into sterile 15-ml conical tubes for storage at –20°. Before use, a tube is thawed, and 9 ml of sterile water is added for a final concentration of 50 μg/ml. The coverslips used most frequently are circular (diameter, 12 mm). They are placed into a small vial for sterilization with an autoclave. Typically, the retinal cells from a litter of embryonic day 18 (E18) rats or from one E5–E7 chick embryo are plated in the well of a 24-well plate, and therefore 25–30 coverslips are placed in a large Petri dish for coating. The poly-D-lysine solution is added to the dish, taking care that all coverslips are immersed in the solution, and the coverslips are then incubated at 37° for 15–30 min. The poly-D-lysine solution is removed and the coverslips are washed in sterile water three times for 5 min each. Care is taken to wash the coverslips well, mixing the coverslips with flamed forceps, because poly-D-lysine in solution can be toxic to cells.

The coverslips can be dried and stored for up to 2 weeks in the Petri dish at 4°, or used immediately. When ready to use, put one coverslip in each well of a 24-well plate, using flamed forceps. Then proceed to coat them with Matrigel. Matrigel is supplied by the manufacturer as a frozen solution. Thaw the bottle slowly on ice (for several hours) to prevent gel formation. Small (200-μl) aliquots are distributed to precooled tubes (15 ml) on ice, using a prechilled pipette. *Note:* If the Matrigel warms during the aliquoting, it will gel and not be effective for the cell cultures. The aliquots are stored at –20° for up to 6 months. To coat the coverslips, remove one aliquot of Matrigel from the –20° freezer and place it on ice for 15–30 min to thaw (200 μl is used for one 24-well plate). Add 10 ml of ice-cold HBSS+ to the 15-ml tube containing 200 μl of thawed Matrigel. Mix gently. Immediately, put 0.5 ml of the dilute Matrigel solution into each well of a 24-well plate, in which polylysine-coated coverslips have already been placed, and place the plate in the incubator for 30 min at 37°. Remove the plate from the incubator and, under the sterile hood, remove nearly all the liquid from the wells. A small amount of the dilute Matrigel solution is left in the well, just enough to cover the coverslips. Let the plate dry in the hood uncovered for 15–30 min. The Matrigel will dry into a thin coating. Plate cells onto the Matrigel, or store the plate at 4° for no more than 2 days before plating the cells.

Analysis of Culture of Retinal Stem/Progenitor Cells

One of the primary methods used for the characterization of retinal stem/progenitor cells and their progeny is immunofluorescence labeling.

With this technique, the overall number of stem/progenitor cells in the cultures can be estimated, and their potential for generation of the different types of retinal neurons can be evaluated. The techniques for labeling the dissociated retinal stem/progenitor cell cultures on coverslips are presented, because this is typically how we do the analysis; however, we have successfully used the same procedures when the cells have been directly cultured in the tissue culture wells without coverslips.

To label the cultures for stem/progenitor cell-specific antigens or retinal neuron-specific antigens, we use the following technique. The coverslips are fixed in 4% paraformaldehyde (PFA) in PBS (0.05 M sodium phosphate, 195 mM NaCl, pH 7.4) for 30 min. The fixing solution is removed and replaced with PBS, and the plates are stored in the cold room for up to 1 week before staining. In preparation for immunofluorescence labeling, the coverslips are incubated with a "blocking solution" to prevent nonspecific binding of the antibody and to reduce background fluorescence. The blocking solution is PBS with 0.3% Triton X-100 and 5% serum (goat serum for non-goat antibodies; FBS for goat antibodies). The primary antibody is then diluted in blocking solution and the coverslips are incubated with the primary antibody overnight. We typically do this at room temperature; carrying out this incubation at 4° can reduce the background labeling for some antibodies. The primary antibody solution is then removed and the coverslips are washed three times with 10- to 15-min incubations in PBS. The coverslips are next incubated in a secondary antibody solution. We have used many different types of secondary antibodies, but we currently prefer 1:500 dilutions of Alexa Fluor-conjugated antibodies. These can be purchased in a variety of different fluorescence emission wavelengths for simultaneous labeling of multiple different antigens in the same cultures. The secondary antibodies are diluted in PBS and 0.3% Triton X-100 and incubated for 1.5 h in the dark at room temperature. The coverslips are then rinsed three times for 10 to 15 min in PBS, and then finally in water to remove residual salt. They are then dried at room temperature and mounted, cell side down, on slides, with a drop of Fluoromount (Electron Microscopy Sciences, Hatfield, PA). We frequently use 4′,6-diamidino-2-phenylindole (DAPI) to label all cell nuclei, but because this emits a short wavelength, blue light, it can be used only with secondary antibodies that emit at longer wavelengths (e.g., Alexa Fluor 488 or 568; and see below). To double label for two or more different antibodies, we incubate the coverslips with multiple primary antibodies simultaneously, as long as they have been raised in different species (e.g., one raised in rabbit and one raised in mouse or rat). Multiple secondary antibodies are also added at the secondary antibody incubation step.

There are many different primary antibodies that can be used to label retinal stem/progenitor cells, although none is definitive. Studies of the developing retina have shown that Pax6, Chx10, Sox2, Prox1, Six3, nestin [transitin in the chick (monoclonal antibody 7b36); J. Weston, Institute of Neuroscience, University of Oregon, Eugene, OR], mushashi, vimentin, Mash1, and Ngn2 are expressed by retinal stem/progenitor cells during development and regeneration (Anchan *et al.*, 1991; Belecky-Adams *et al.*, 1997; Burmeister *et al.*, 1996; Fischer and Omar, 2005; Fischer and Reh, 2000; Jasoni and Reh, 1996; Mathers *et al.*, 1997; Oliver *et al.*, 1995). This is by no means an exhaustive list, however; antibodies raised against these antigens have been used to label cells in various species, and double-labeling with BrdU has demonstrated that these proteins are expressed by stem/progenitor cells. One problem with these markers is that they are not only expressed by stem/progenitor cells, but most are also expressed in one or more types of differentiated retinal neurons. Another problem is that many are also expressed in Müller glia (albeit at lower levels; Blackshaw *et al.*, 2003), and so they cannot distinguish between stem/progenitor cells and glia (although Mash1 and Ngn2 may be the exception and are not expressed in Müller glia). Last, none of these markers has been used to discriminate between a stem cell and a progenitor cell. With these caveats in mind, the use of these markers can be informative; however, the only definitive way to demonstrate that there are stem/progenitor cells in the culture is to show that the cells can generate neurons and glia of the retina.

Immunofluorescence analysis of stem/progenitors has been used to demonstrate that neurons are generated *in vitro*, when a neuron-specific antibody is used in conjunction with BrdU. In this method, BrdU is added to the culture, and the mitotically active stem/progenitor cells incorporate the nucleotide into their DNA during S phase. After several days *in vitro*, the stem/progenitor cells give rise to new neurons, and by double-labeling with both an antibody raised against BrdU and a neuron-specific antibody, one can definitively demonstrate that neurogenesis is occurring *in vitro* (see, e.g., Anchan *et al.*, 1991; Fischer and Reh, 2001; Reh, 1992; Kelley *et al.*, 1994; Levine *et al.*, 1997). There are many good retinal neuron-specific antibodies that give reliable labeling when used in conjunction with BrdU. For rod photoreceptors, we have used monoclonal antibodies raised against rhodopsin (3A6 and 4D2 from R. S. Molday, Department of Biochemistry and Molecular Biology, University of British Columbia, Vancouver, BC, Canada). To label all photoreceptors (and a few bipolar cells), we have used anti-recoverin antibodies (J. Hurley, Department of Biochemistry, University of Washington, Seattle, WA). Recoverin has the advantage of being expressed more quickly after the final mitotic division in the new rods and cones, whereas rhodopsin is not expressed for several days after the progenitor cells have generated the new rod. More recently, antibodies generated against photoreceptor-specific transcription factors have been generated and are likely to be ideal for *in vitro* studies, because of their nuclear localization. The use of multiple colocalized markers, along with BrdU, is the best way to ensure that the specific cell type is generated in the cultures, and that indeed retinal stem/progenitor cells are present. Other retinal cell types can also be identified *in vitro*, using antibodies that are frequently used in other regions of the nervous system. TuJ1, an antibody directed against neuron-specific tubulin, will label most inner retinal neurons, including ganglion cells, amacrine cells, horizontal cells, and bipolar cells (although less well). We have also used commercially available antibodies raised against Hu, NeuN, Brn3, calbindin, and calretinin. The reader is referred to earlier publications for details of the different antibodies used to characterize the various types of retinal neurons.

To label for proliferating cells, we add BrdU (final concentration, 1–10 μg/ml) to the cultures before fixation. To identify the cells that have incorporated the BrdU, we pretreat the coverslips with 4 N HCl for 7–8 min, rinse three times for 5 min each with PBS, and block nonspecific labeling with 5% goat serum in ~0.1% Triton X-100 in PBS. Next we dilute rat anti-BrdU 1:250 in 5% goat serum in ~0.2% Triton X-100 in PBS and incubate overnight. The coverslips are then rinsed three times for 5 min each in PBS, and the appropriate secondary antibody is applied as described previously. As described above, the coverslips to be labeled can be incubated with two or more primary antibodies simultaneously. When labeling for BrdU alone it is important to wash the cells with Triton X-100 before the acid treatment. The acid treatment is far less effective if the cells have not been permeabilized. Also, double-labeling for BrdU and neuronal markers often requires that the neuronal marker and appropriate secondary be applied first, followed by a brief fixation (2% PFA for 15 min), and then the acid treatment. Most antigens lose their antigenicity with the acid treatment.

Although many different secondary antibodies are available and work well, in general Alexa Fluor-conjugated secondary antibodies (Invitrogen Molecular Probes, Eugene, OR) provide the brightest and most photostable choices. Alexa Fluor-conjugated secondary antibodies include anti-rabbit, anti-rat, and anti-mouse diluted to 1:1000 in PBS plus 0.2% Triton X-100. For triple-labeling, provided that the microscope is equipped with the appropriate filter sets and a camera capable of detecting far red wavelengths, it is best to combine Alexa Fluor 488 and Alexa Fluor 568 with Alexa Fluor 647 (far red), versus Alexa Fluor 350 (blue), because Alexa Fluor 647 is brighter and far more photostable than Alexa Fluor 350.

In Vivo *Methods for the Study of Retinal/Stem Progenitor Cells*

Intraocular Injections

We have developed methods to study retinal progenitor cells *in vivo*, both at the retinal margin, that is, the ciliary epithelium, and those derived from Müller glia. The method described here details the intraocular injections and analysis of the tissue common to all these studies. The reader is referred to specific relevant publications for additional details relevant to specific experimental paradigms. Before intraocular injections, posthatch chicks must be anesthetized. The simplest way to anesthetize a postnatal chicken is to place about 1 ml of a 4:1 mixture of mineral oil to halothane or isoflurane into the bottom of a large glass jar. Inhalation of the halothane or isoflurane vapors will render the chicks unconscious within 1 min. Once removed from the jar, the animals will regain consciousness within 2 min. To keep the chickens unconscious for longer periods of time and to more precisely deliver anesthetic, machines equipped with the appropriate halothane/isoflurane vaporizer can be used. Once the chickens are unconscious, the eyelids are swabbed with 70% ethanol in water or povidone–iodine (Betadine; Purdue Pharma, Stamford, CT) solution to sterilize the area where needle will enter the eye. Intravitreal injections can be made with standard 26-gauge needles (Hamilton, Reno, NV). The standard 26-gauge needle is 55 mm long and flexible, making it difficult to control during insertion into the eye and difficult to estimate whether the tip of the needle is in the desired location within the liquid vitreous. However, if cost is not a concern, custom 22-mm 26-gauge needles can be obtained. The maximum volume per injection should not exceed 30 μl. Larger injection volumes will result in back-flow from the injection site. Puncturing the eye through the pars plana does not influence the proliferation of NPE cells or cells in the CMZ (Fischer and Reh, 2003). Doses of growth factors can vary from 10 to 2000 ng, but growth factors such as IGF-I, EGF, BMP4, and FGF-2 influence progenitors in the CMZ at 100 ng per dose delivered over 2 to 3 consecutive days. Most of the growth factors we have used in our studies have been obtained commercially and the reader is referred to the relevant publications for details (Fig. 3).

Dissection and Fixation of Tissues

Our studies have relied heavily on the immunolabeling of the chicken retina prepared as cryosections or whole mounts. The eyes of postnatal chickens are large and easy to dissect. The ease of dissection accommodates expeditious isolation of tissues and obviates the need for perfusion. To prepare tissues for immunolabeling, brief (less than 30 min) exposure to

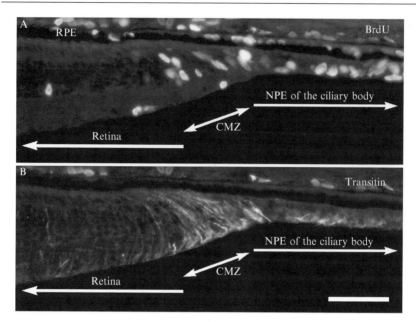

FIG. 3. Photomicrographs of the CMZ of a posthatch chicken retina, showing (A) the mitotically active, BrdU-labeled cells in this zone after an intraocular injection of BrdU. Notice that cells in the ciliary nonpigmented epithelium (NPE) and the pigmented epithelium (RPE) are also labeled with BrdU. (B) The CMZ cells also label with transitin, a homolog to nestin, a neural stem and progenitor marker.

fixation is preferable. Although longer exposure to fixative may result in high-quality sections and better preservation of gross tissue morphology, the modification of basic side chains in antigens with PFA often prevents interactions with antibodies. Enucleated eyes are hemisected equatorially with a fresh razor blade and the gel vitreous is removed with forceps from the posterior eye cup. Eye cups are fixed (4% paraformaldehyde plus 3% sucrose in 0.1 M phosphate buffer, pH 7.4; 30 min at 20°), washed three times in PBS (0.05 M sodium phosphate, 195 mM NaCl, pH 7.4), cryoprotected in PBS plus 30% sucrose, immersed in embedding medium (Tissue-Tek O.C.T. compound; Sakura Finetek, Torrance, CA), and freeze-mounted onto sectioning blocks. Vertical sections, nominally 12 μm thick, are thaw-mounted onto SuperFrost Plus slides (Fisher Scientific, Pittsburgh, PA), air dried at 37°, and stored at –20° until use. Depending on the quality of the fixation, the sections can be safely stored for several months without compromising immunoreactivity within the tissue.

Retinal sections are thawed, ringed with rubber cement, washed three times in PBS, covered with primary antibody solution (200 μl of antiserum diluted in PBS plus 5% normal goat serum, 0.2% Triton X-100, and 0.01% NaN$_3$), and incubated for 24 h at room temperature in a humidified chamber. The slides are washed three times in PBS, covered with second-ary antibody solution, and incubated for at least 1 h at room temperature in a humidified chamber. Finally, samples are washed three times in PBS, the rubber cement is removed from the slides, and a coverglass is mounted in glycerol–water (4:1, v/v).

For whole mount preparations of the chicken retina, fixed retinas are dissected away from the pigmented epithelium, choroid, and sclera, cryo-protected in 30% sucrose in PBS and taken through three cycles of freezing (at $-80°$) and thawing (on a 37° hot plate) (Fischer and Stell, 1997). Samples are washed three times in PBS, placed in 250 μl of primary antibody, and incubated at room temperature on an oscillating shaker for 24–48 h. A small amount (0.01%) of sodium azide (NaN$_3$) should be added to the antibody diluent to prevent the growth of bacteria or fungus. At the end of incubation the primary antibody is aspirated (and can be saved at 4° and reused), samples are washed three times in PBS, 250 μl of secondary antibody solution is added, and tissues are incubated at room temperature on an oscillating shaker for 24 h. Finally, samples are washed three times in PBS and each is mounted under a coverglass in glycerol–water (4:1, v/v) .

The primary antibodies described in the previous section on *in vitro* immunofluorescence will also label the cells *in vivo*. In addition, the sec-ondary antibodies used for labeling the sections or flatmount preparations are similar to those used for the cells in culture.

Overview

The development of *in vitro* and *in vivo* methods for the study of retinal progenitor cells and stem cells has allowed rapid progress in our under-standing of the factors that regulate the generation of retinal neurons and glia. The continued application of these methods, as well as the develop-ment of more efficient gene transfer methods *in vivo*, will enable the next level of analysis.

References

Ahmad, I., Tang, L., and Pham, H. (2000). Identification of neural progenitors in the adult mammalian eye. *Biochem. Biophys. Res. Commun.* **270**, 517–521.
Akagi, T., Inoue, T., Miyoshi, G., Bessho, Y., Takahashi, M., Lee, J. E., Guillemot, F., and Kageyama, R. (2004). Requirement of multiple basic helix–loop–helix genes for retinal neuronal subtype specification. *J. Biol. Chem.* **279**, 28492–28498.

Anchan, R. M., Reh, T. A., Angello, J., Balliet, A., and Walker, M. (1991). EGF and TGF-α stimulate retinal neuroepithelial cell proliferation *in vitro*. *Neuron* **6,** 923–936.

Belecky-Adams, T., Tomarev, S., Li, H. S., Ploder, L., McInnes, R. R., Sundin, O., and Adler, R. (1997). Pax-6, Prox 1, and Chx10 homeobox gene expression correlates with phenotypic fate of retinal precursor cells. *Invest. Ophthalmol. Vis. Sci.* **38,** 1293–1303.

Blackshaw, S., Kuo, W. P., Park, P. J., Tsujikawa, M., Gunnersen, J. M., Scott, H. S., Boon, W. M., Tan, S. S., and Cepko, C. L. (2003). MicroSAGE is highly representative and reproducible but reveals major differences in gene expression among samples obtained from similar tissues. *Genome Biol.* **4,** R17.

Burmeister, M., Novak, J., Liang, M. Y., Basu, S., Ploder, L., Hawes, N. L., Vidgen, D., Hoover, F., Goldman, D., Kalnins, V. I., Roderick, T. H., Taylor, B. A., Hankin, M. H., and McInnes, R. R. (1996). Ocular retardation mouse caused by Chx10 homeobox null allele: Impaired retinal progenitor proliferation and bipolar cell differentiation. *Nat. Genet.* **12,** 376–384.

Ciccolini, F., and Svendsen, C. N. (1998). Fibroblast growth factor 2 (FGF-2) promotes acquisition of epidermal growth factor (EGF) responsiveness in mouse striatal precursor cells: Identification of neural precursors responding to both EGF and FGF-2. *J. Neurosci.* **18,** 7869–7880.

Close, J. L., Gumuscu, B., and Reh, T. A. (2005). Retinal neurons regulate proliferation of postnatal progenitors and Muller glia in the rat retina via TGFβ signaling. *Development* **132,** 3015–3026.

Coles, B. L., Angenieux, B., Inoue, T., Del Rio-Tsonis, T., Spence, J. R., McInnes, R. R., Arsenijevic, Y., and van der Kooy, D. (2004). Facile isolation and the characterization of human retinal stem cells. *Proc. Natl. Acad. Sci. USA* **101,** 15772–15777.

Coulombre, J. L., and Coulombre, A. J. (1965). Regeneration of neural retina from the pigmented epithelium in the chick embryo. *Dev. Biol.* **12,** 79–92.

Fekete, D. M., Perez-Miguelsanz, J., Ryder, E. F., and Cepko, C. L. (1994). Clonal analysis in the chicken retina reveals tangential dispersion of clonally related cells. *Dev. Biol.* **166,** 666–682.

Fischer, A. J. (2005). Neural regeneration in the chick retina. *Prog. Retin. Eye Res.* **24,** 161–182.

Fischer, A. J., and Omar, G. (2005). Transitin, a nestin-related intermediate filament, is expressed by neural progenitors and can be induced in Muller glia in the chicken retina. *J. Comp. Neurol.* **484,** 1–14.

Fischer, A. J., and Reh, T. A. (2000). Identification of a proliferating marginal zone of retinal progenitors in postnatal chickens. *Dev. Biol.* **220,** 197–210.

Fischer, A. J., and Reh, T. A. (2001). Muller glia are a potential source of neural regeneration in the postnatal chicken retina. *Nat. Neurosci.* **4,** 247–252.

Fischer, A. J., and Reh, T. A. (2002). Exogenous growth factors stimulate the regeneration of ganglion cells in the chicken retina. *Dev. Biol.* **251,** 367–379.

Fischer, A. J., and Reh, T. A. (2003). Growth factors induce neurogenesis in the ciliary body. *Dev. Biol.* **259,** 225–240.

Fischer, A. J., and Stell, W. K. (1997). Light-modulated release of RFamide-like neuropeptides from nervus terminalis axon terminals in the retina of goldfish. *Neuroscience* **77,** 585–597.

Fischer, A. J., McGuire, J. J., Schaeffel, F., and Stell, W. K. (1999). Light- and focus-dependent expression of the transcription factor ZENK in the chick retina. *Nat. Neurosci.* **2,** 706–712.

Fischer, A. J., Hendrickson, A., and Reh, T. A. (2001). Immunocytochemical characterization of cysts in the peripheral retina and pars plana of the adult primate. *Invest. Ophthalmol. Vis. Sci.* **42,** 3256–3263.

Fischer, A. J., Dierks, B. D., and Reh, T. A. (2002). Exogenous growth factors induce the production of ganglion cells at the retinal margin. *Development* **129**, 2283–2291.

Fischer, A. J., Omar, G., Walton, N. A., Verrill, T. A., and Unson, C. G. (2005). Glucagon-expressing neurons within the retina regulate the proliferation of neural progenitors in the circumferential marginal zone of the avian eye. *J. Neurosci.* **25**, 10157–10166.

Harris, W. A., and Perron, M. (1998). Molecular recapitulation: The growth of the vertebrate retina. *Int. J. Dev. Biol.* **42**, 299–304.

Haruta, M., Kosaka, M., Kanegae, Y., Saito, I., Inoue, T., Kageyama, R., Nishida, A., Honda, Y., and Takahashi, M. (2001). Induction of photoreceptor-specific phenotypes in adult mammalian iris tissue. *Nat. Neurosci.* **4**, 1163–1164.

Hollyfield, J. G. (1968). Differential addition of cells to the retina in *Rana pipiens* tadpoles. *Dev. Biol.* **18**, 163–179.

James, J., Das, A. V., Bhattacharya, S., Chacko, D. M., Zhao, X., and Ahmad, I. (2003). *In vitro* generation of early-born neurons from late retinal progenitors. *J. Neurosci.* **23**, 8193–8203.

Jasoni, C. L., and Reh, T. A. (1996). Temporal and spatial pattern of MASH-I expression in the developing rat retina demonstrates progenitor cell heterogeneity. *J. Comp. Neurol.* **369**, 319–327.

Jasoni, C. L., Walker, M. B., Morris, M. D., and Reh, T. A. (1994). A chicken achaete-scute homolog (CASH-1) is expressed in a temporally and spatially discrete manner in the developing nervous system. *Development* **120**, 769–783.

Kelley, M. W., Turner, J. K., and Reh, T. A. (1994). Retinoic acid promotes differentiation of photoreceptors *in vitro*. *Development* **120**, 2091–2102.

Kelley, M. W., Turner, J. K., and Reh, T. A. (1995). Regulation of proliferation and photoreceptor differentiation in fetal human retinal cell cultures. *Invest. Ophthalmol. Vis. Sci.* **36**, 1280–1289.

Klassen, H., Ziaeian, B., Kirov, I. I., Young, M. J., and Schwartz, P. H. (2004a). Isolation of retinal progenitor cells from post-mortem human tissue and comparison with autologous brain progenitors. *J. Neurosci. Res.* **77**, 334–343.

Klassen, H., Sakaguchi, D. S., and Young, M. J. (2004b). Stem cells and retinal repair. *Prog. Retin. Eye Res.* **23**, 149–181.

Klassen, H. J., Ng, T. F., Kurimoto, Y., Kirov, I., Shatos, M., Coffey, P., and Young, M. J. (2004c). Multipotent retinal progenitors express developmental markers, differentiate into retinal neurons, and preserve light-mediated behavior. *Invest. Ophthalmol. Vis. Sci.* **45**, 4167–4173.

Kubota, R., Hokoc, J. N., Moshiri, A., McGuire, C., and Reh, T. A. (2002). A comparative study of neurogenesis in the retinal ciliary marginal zone of homeothermic vertebrates. *Brain Res. Dev. Brain Res.* **134**, 31–41.

Levine, E. M., Roelink, H., Turner, J., and Reh, T. A. (1997). Sonic hedgehog promotes rod photoreceptor differentiation in mammalian retinal cells *in vitro*. *J. Neurosci.* **17**, 6277–6288.

Li, S., Mo, Z., Yang, X., Price, S. M., Shen, M. M., and Xiang, M. (2004). Foxn4 controls the genesis of amacrine and horizontal cells by retinal progenitors. *Neuron* **43**, 795–807.

Lillien, L., and Cepko, C. (1992). Control of proliferation in the retina: Temporal changes in responsiveness to FGF and TGF-α. *Development* **115**, 253–266.

Livesey, R., and Cepko, C. (2001). Neurobiology: Developing order. *Nature* **413**, 471–473.

Mack, A. F., and Fernald, R. D. (1993). Regulation of cell division and rod differentiation in the teleost retina. *Brain Res. Dev. Brain Res.* **76**, 183–187.

Mathers, P. H., Grinberg, A., Mahon, K. A., and Jamrich, M. (1997). The Rx homeobox gene is essential for vertebrate eye development. *Nature* **387**, 603–607.

Morris, V. B., Wylie, C. C., and Miles, V. J. (1976). The growth of the chick retina after hatching. *Anat. Rec.* **184,** 111–113.

Moshiri, A., and Reh, T. A. (2004). Persistent progenitors at the retinal margin of ptc$^{+/-}$ mice. *J. Neurosci.* **24,** 229–237.

Moshiri, A., Close, J., and Reh, T. A. (2004). Retinal stem cells and regeneration. *Int. J. Dev. Biol.* **48,** 1003–1014.

Moshiri, A., McGuire, C. R., and Reh, T. A. (2005). Sonic hedgehog regulates proliferation of the retinal ciliary marginal zone in posthatch chicks. *Dev Dyn.* **233,** 66–75.

Okada, T. S. (1980). Cellular metaplasia or transdifferentiation as a model for retinal cell differentiation. *Curr. Top. Dev. Biol.* **16,** 349–380.

Oliver, G., Mailhos, A., Wehr, R., Copeland, N. G., Jenkins, N. A., and Gruss, P. (1995). Six3, a murine homologue of the sine oculis gene, demarcates the most anterior border of the developing neural plate and is expressed during eye development. *Development* **121,** 4045–4055.

Ooto, S., Akagi, T., Kageyama, R., Akita, J., Mandai, M., Honda, Y., and Takahashi, M. (2004). Potential for neural regeneration after neurotoxic injury in the adult mammalian retina. *Proc. Natl. Acad. Sci. USA* **101,** 13654–13659.

Orts-Llorca, F., and Genis-Galvez, J. M. (1960). Experimental production of retinal septa in the chick embryo: Differentiation of pigment epithelium into neural retina. *Acta Anat. (Basel)* **42,** 31–70.

Park, C. M., and Hollenberg, M. J. (1989). Basic fibroblast growth factor induces retinal regeneration *in vivo. Dev. Biol.* **134,** 201–205.

Perron, M., Kanekar, S., Vetter, M. L., and Harris, W. A. (1998). The genetic sequence of retinal development in the ciliary margin of the *Xenopus* eye. *Dev. Biol.* **199,** 185–200.

Pittack, C., Jones, M., and Reh, T. A. (1991). Basic fibroblast growth factor induces retinal pigment epithelium to generate neural retina *in vitro. Development* **113,** 577–588.

Prada, C., Puga, J., Perez-Mendez, L., Lopez, R., and Ramirez, G. (1991). Spatial and temporal patterns of neurogenesis in the chick retina. *Eur. J. Neurosci.* **3,** 559–569.

Reh, T. A. (1992). Cellular interactions determine neuronal phenotypes in rodent retinal cultures. *J. Neurobiol.* **23,** 1067–1083.

Reh, T. A., and Cagan, R. L. (1994). Intrinsic and extrinsic signals in the developing vertebrate and fly eyes: Viewing vertebrate and invertebrate eyes in the same light. *Perspect. Dev. Neurobiol.* **2,** 183–190.

Reh, T. A., and Constantine-Paton, M. (1983). Qualitative and quantitative measures of plasticity during the normal development of the Rana pipiens retinotectal projection. *Brain Res.* **312,** 187–200.

Reh, T. A., and Kljavin, I. J. (1989). Age of differentiation determines rat retinal germinal cell phenotype: Induction of differentiation by dissociation. *J. Neurosci.* **9,** 4179–4189.

Reh, T. A., Nagy, T., and Gretton, H. (1987). Retinal pigmented epithelial cells induced to transdifferentiate to neurons by laminin. *Nature* **330,** 68–71.

Sakaguchi, D. S., Janick, L. M., and Reh, T. A. (1997). Basic fibroblast growth factor (FGF-2) induced transdifferentiation of retinal pigment epithelium: Generation of retinal neurons and glia. *Dev. Dyn.* **209,** 387–398.

Sakami, S., Hisatomi, O., Sakakibara, S., Liu, J., Reh, T. A., and Tokunaga, F. (2005). Downregulation of Otx2 in the dedifferentiated RPE cells of regenerating newt retina. *Brain Res. Dev. Brain Res.* **155,** 49–59.

Sidman, R. L. (1961). Histogenesis of mouse retina studied with ^{3}H-thymidine. *In* "The Structure of the Eye" (G. Smelser, ed.), pp. 487–506. New York: Academic Press.

Stone, L. S. (1950). Neural retina degeneration followed by regeneration from surviving retinal pigment cells in grafted adult salamander eyes. *Anat. Rec.* **106,** 89–109.

Stone, L. S., and Steinitz, H. (1957). Regeneration of neural retina and lens form retina pigment cell grafts in adult newts. *J. Exp. Zool.* **135,** 301–317.

Stroeva, O. G., and Mitashov, V. I. (1983). Retinal pigment epithelium: Proliferation and differentiation during development and regeneration. *Int. Rev. Cytol.* **83,** 221–293.

Taylor, M., and Reh, T. A. (1990). Induction of differentiation of rat retinal, germinal, neuroepithelial cells by dbcAMP. *J. Neurobiol.* **21,** 470–481.

Tropepe, V., Coles, B. L., Chiasson, B. J., Horsford, D. J., Elia, A. J., McInnes, R. R., and van der Kooy, D. (2000). Retinal stem cells in the adult mammalian eye. *Science* **287,** 2032–2036.

Wetts, R., and Fraser, S. E. (1988). Multipotent precursors can give rise to all major cell types of the frog retina. *Science* **239,** 1142–1145.

Yurco, P., and Cameron, D. A. (2005). Responses of Muller glia to retinal injury in adult zebrafish. *Vision Res.* **45,** 991–1002.

Zhao, S., Thornquist, S. C., and Barnstable, C. J. (1995). *In vitro* transdifferentiation of embryonic rat retinal pigment epithelium to neural retina. *Brain Res.* **677,** 300–310.

Zhao, S., Chen, Q., Hung, F. C., and Overbeek, P. A. (2002). BMP signaling is required for development of the ciliary body. *Development* **129,** 4435–4442.

Zuber, M. E., Gestri, G., Viczian, A. S., Barsacchi, G., and Harris, W. A. (2003). Specification of the vertebrate eye by a network of eye field transcription factors. *Development* **130,** 5155–5167.

[4]   Epithelial Skin Stem Cells

By TUDORITA TUMBAR

Abstract

Major progress in understanding epithelial skin stem cells has been accomplished. This has been possible by developing new methods for labeling, tracking, isolating, and characterizing enriched populations of stem cells. This chapter summarizes *in vivo* and *in vitro* assays that are currently employed to analyze skin epithelial stem cells. Despite progress, the definition of a stem cell is currently a functional one. Unambiguous identification of a stem cell in intact tissue is still not possible. These limitations hamper molecular studies aimed at unraveling the cellular mechanisms operating in the stem cell compartment. This chapter emphasizes current methods for analyzing hair follicle stem cells, as opposed to other epithelial compartments, because the hair follicle has been most intensively studied up to date.

Introduction to Skin Epithelial Stem Cells

The main function of the skin is to provide a body cover in order to prevent dehydration and to protect against environmental insults (Fuchs, 1990). Skin is composed of two major compartments: a mesenchymal inner

METHODS IN ENZYMOLOGY, VOL. 419   0076-6879/06 $35.00
Copyright 2006, Elsevier Inc. All rights reserved.   DOI: 10.1016/S0076-6879(06)19004-7

compartment (dermis) and an epithelial outer compartment (epidermis). Mesenchymal skin is made largely of fibroblast cells, loosely distributed in a dense mass of extracellular matrix. Epithelial skin is made of epidermis and its appendages: the hair follicle, the sebaceous gland, and the sweat gland. Sweat gland biology is poorly understood, and is not considered further in this chapter (Niemann and Watt, 2002; Odland, 1991; Watt, 2004). Skin epithelial cells are called keratinocytes, and are specialized cells that express a large variety of intermediate filament keratins, cytoskeleton protein fibers thought to confer tissue resilience (Fuchs, 1995).

Stem cells (SCs) are defined as cells that can self-renew and differentiate during the life of the animal. They constitute an unlimited source of cells that contribute to tissue morphogenesis, homeostasis, and injury repair (Melton and Cowan, 2004). It is generally thought that each of the epithelial compartments (the interfollicular epidermis, the hair follicle, and the sebaceous gland) has its own specialized stem cells capable of sustaining tissue growth independently. It has been demonstrated that at times of high need, such as rapid growth or injury, hair follicle stem cells can contribute not only to the hair follicle, but also to the epidermis and the sebaceous gland (Bickenbach and Grinnell, 2004; Lavker et al., 2003; Taylor et al., 2000; Tumbar and Fuchs, 2004; Watt and Hogan, 2000). This led to the hypothesis that the hair follicle is the residence of multipotent stem cells in the epithelial skin (Fuchs et al., 2001). However, this did not explain how hairless skin maintained its epidermis for extended periods of time. It has been directly demonstrated that the hair follicle does not contribute cells to normal epidermis homeostasis (Ito et al., 2005; Levy et al., 2005); nonetheless, significant but transient contribution occurs during injury (Claudinot et al., 2005; Ito et al., 2005). An alternative to the hypothesis that hair follicle stem cells are the common epithelial skin stem cells is that epithelial stem cells from each compartment are equivalent (Fig. 1) (Niemann and Watt, 2002; Watt, 2004). Thus, each class would be capable of making any of the other two epithelial compartments, if provided with the right environment. Although it has been demonstrated that hair follicle stem cells were capable of such contributions, it has been less apparent that stem cells from the sebaceous gland and the interfollicular epidermis had similar abilities. Moreover, studies suggest that hair follicle stem cells contribute only transiently to the interfollicular epidermis compartment, and thus they are not capable of replenishing this stem cell compartment (Ito et al., 2005; Levy et al., 2005). This interchangeability hypothesis is supported by studies showing that interfollicular epidermis can produce hair follicles de novo, when induced to overexpress a constitutively active form of β-catenin (Gat et al., 1998; Silva-Vargas et al., 2005). This process takes place independent of the hair follicle stem cells (Silva-Vargas et al., 2005). Although

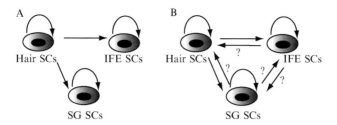

FIG. 1. Models for stem cell activity in the epithelial skin. (A) Hair follicle stem cells give rise to self, and all the differentiated hair follicle cells, but can also contribute to epidermis and sebaceous gland (SG) under special circumstances. In contrast, sebaceous gland and interfollicular epidermis (IFE) stem cells are restricted to make only self and cells of their own lineage, and cannot contribute to the other two compartments. (B) All three classes of epithelial stem cells have equal potency, and can be used interchangeably in the skin.

attractive, the interchangeability model requires more substantiation, especially with respect to the potency of sebaceous gland stem cells, which are poorly understood.

Several important features of the epithelial skin make it an ideal model system to study stem cell behavior. First, the skin epithelial tissue has the ability to regenerate. This process relies on the self-renewal of epithelial stem cells, which undergo cycles of activation to generate a constant flux of differentiating cells (Fuchs and Raghavan, 2002). These cells withdraw from the cell cycle and migrate upward toward the skin surface to make the outermost layer of the epidermis or cornified envelope, and the innermost layer of the follicle, the hair shaft. Second, skin is a large and easily accessible organ, providing sufficient numbers of stem cells for *in vitro* experiments and analyses, such as fluorescence-activated cell-sorting (FACS) profiles and cell cycle, cell growth, cell graft, and gene expression studies including reverse transcription-polymerase chain reaction (RT-PCR) and microarray. Third, because skin tissue is highly structured spatially and temporally it is possible to determine the state of stem cell activation solely on the basis of tissue morphology. Moreover, the synchrony of hair follicle development in young mouse skin allows the tracking of stem cell progression through their activation and dormancy cycles. These features make it possible to determine in intact skin whether a particular mutation induces perturbations in the stem cell activation process. *In vivo* tracking assays are extremely valuable, because they avoid perturbation of the normal biology of the cells inherent to cell isolation approaches. This chapter emphasizes methods of studying characteristics of the hair epithelial stem cells, because this is an area of major progress (Fuchs *et al.*, 2004; Gambardella and Barrandon, 2003) in skin biology.

Epithelial Skin Organization and the Hair Cycle

Epidermis is made of several layers of keratinocytes, of which the innermost basal layer (BL) of mitotically active keratinocytes expresses keratins K5 and K14 (Fuchs, 1995). The interfollicular epidermal stem cells reside in this basal layer, interspersed among rapidly dividing transient amplifying (TA) cells. The BL cells divide and move outward toward the skin surface, withdraw from the cell cycle, and differentiate to form the suprabasal layers of the epidermis. The suprabasal layer of cells located right above the basal layer is made of spinous cells, which turn off the expression of K5 and K14 and express differentiation-specific keratin K1 and K10. These differentiating cells are continuously moving outward, forming the granular layer and finally the highly keratinized cornified envelope (Candi *et al.*, 2005).

Epidermal appendages, including the hair follicle and the sebaceous gland (SG) (Fig. 2), are inserted deep into the dermis. The hair follicle is a complex structure made of at least eight different cell types. The hair shaft (hs), or the hair per se, is centrally located and grows outward, toward the skin surface. Concentric layers of cells surrounding the shaft form the outer and inner root sheaths (ORS and IRS, respectively). The basal layer of the epidermis is contiguous with the ORS of the follicle. The BL and ORS

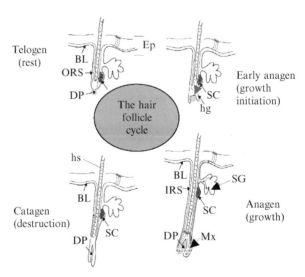

FIG. 2. The hair follicle cycle. BL, basal layer; ORS, outer root sheath; IRS, inner root sheath; hg, hair germ; DP, dermal papillae; hs, hair shaft; Ep, epidermis; SG, sebaceous gland; Mx, matrix; SC, stem cells.

contain the epidermal and hair follicle SCs, respectively. Hair follicle SCs are localized in a specialized permanent structure of the follicle, the bulge. Below the bulge is the temporary portion of the follicle, which contains the matrix (Mx), a highly proliferating pocket of bulge-derived cells. The matrix cells give rise to all the inner differentiated layers of the hair follicles. The matrix encloses a mesenchymal pocket of cells, the dermal papillae (DP), which has signaling properties important for hair follicle growth (Hardy, 1992; Jahoda and Reynolds, 1993; Rogers, 2004).

Postnatal hair follicles are in a dynamic state of perpetual cycles of growth (anagen), regression (catagen), and rest (telogen) (Fig. 2). For the early hair cycles mouse follicles grow in relative synchrony. At around 17 days of postnatal life, anagen ceases and massive cell death occurs in the bulb of the hair follicle, beginning the catagen phase. Follicle cells below the bulge area are destroyed, except for the DP, which is dragged upward by the shrinking basement membrane that separates the epithelial and mesenchymal compartments. The DP comes to rest just beneath the bulge and the hair follicles enter their resting, or telogen, phase. The new anagen is reinitiated around postnatal day 21 (P21), when bulge cells are activated to proliferate and give rise to different cell types. Throughout life, the hair follicles repeatedly undergo this precise sequence of changes. Signals from the surrounding environment and from the DP are thought to be necessary to reactivate bulge SCs to restart the process of follicle morphogenesis (Hardy, 1992). The hair follicle cycle has provided an excellent model system for studying SC activation by natural tissue growth signals.

Characteristics of Skin Epithelial Stem Cells

Skin epithelial stem cells are currently attributed several characteristics: (1) infrequent division and/or localization in stem cell niches (bulge), (2) expression of a combination of putative stem cell "markers," (3) contribution to tissue morphogenesis and wound repair *in vivo*, (4) high proliferative potential in tissue culture, (5) self-renewal and differentiation potential *in vitro*, and (5) long-term potential to regenerate their own tissue of residence. A more refined characteristic of stem cells is their potency, the ability of a single stem cell to give rise by division to (1) self and a single type of differentiated cell (unipotent SCs) or (2) self and multiple types of differentiated cell (multipotent SCs). Individually, each of these criteria has particular strengths and caveats; their use in combination is recommended and provides a more complete assessment of the stem cell phenotype. The following sections discuss each of these characteristics and up-to-date methods developed to describe them.

Epithelial SCs and the Frequency of Their Division

Tissue stem cells are thought to proliferate, self-renew, and differentiate throughout the entire life of an animal. They participate with cell progeny in tissue homeostasis and repair (Fuchs *et al.*, 2004). If these long-lived cells did not develop mechanisms to prevent the accumulation of replication errors, repeated divisions would lead to a high rate of tissue abnormalities (Cairns, 1975). One long-standing hypothesis explains this apparent paradox by suggesting that tissue SCs are relatively quiescent cells, each of which divides infrequently and gives rise to another SC daughter and to a rapidly proliferating, transient amplifying (TA) daughter cell (Bickenbach and Grinnell, 2004; Falciatori *et al.*, 2004; Hudson *et al.*, 2000; Jensen *et al.*, 2004; Webb *et al.*, 2004). TA cells, unlike SCs, have limited growth potential. Yet, for a short period of time they divide repeatedly to make a large number of progeny, and fulfill the need for cells of the tissue of origin. For several decades identification of slowly cycling or infrequently dividing cells within tissues has been equated with searching for SCs. Slowly cycling cells are found by a repeated "pulse" with 5-bromo-2′-deoxyuridine (BrdU) or [^{3}H]thymidine to label proliferating cells, followed by a "chase" period. The rapidly dividing TA cells divide and dilute the label, whereas infrequently dividing cells retain the label (label-retaining cells or LRCs). Thus, label retention is circumstantial and reflects the growth history of dividing cells within tissues. Pulse–chase labeling schemes can be highly variable because they are dependent on the growth kinetics of the tissue in study. In the hair follicle, pulse–chase experiments have identified two classes of cells: one is a recognized class of transient amplifying cells residing in the matrix; the other, composed of infrequently dividing LRCs, the putative stem cells, resides in the hair bulge (Cotsarelis *et al.*, 1990). The hair bulge has been shown by independent methods to contain multipotent stem cells (Gambardella and Barrandon, 2003). Although normally quiescent, LRCs proliferate in adult life in response to tissue growth or repair signals. Other stimuli, such as phorbol ester (Braun *et al.*, 2003) and growth hormone (Ohlsson *et al.*, 1992), induce LRCs to proliferate. When in culture, LRCs show higher proliferative potential relative to other keratinocytes, consistent with their predicted long-term self-renewal capacity (Bickenbach and Chism, 1998; Morris and Potten, 1994; Tumbar *et al.*, 2004). Whereas LRCs express SC markers and other markers specific to undifferentiated cells of their tissue of residence, they do not express differentiation markers (Tumbar *et al.*, 2004). Selective photoinduced killing of LRCs impairs normal tissue growth (Kameda *et al.*, 2002). Several targeted mutations of specific genes perturbed normal LRC behavior, a phenomenon attributed to SC misregulation (Arnold and Watt, 2001; Benitah *et al.*, 2005; Doetsch *et al.*, 2002; Flores *et al.*, 2005; Gandarillas and Watt, 1997; Sarin *et al.*, 2005).

To label infrequently dividing cells, mice are injected subcutaneously with a solution of BrdU dissolved in phosphate-buffered saline (PBS), at a dose of 50 μg/g of body mass. The most common injection schemes are every 8–12 h, for three consecutive days on postnatal days 3, 4, and 5 (Cotsarelis *et al.*, 1990; Taylor *et al.*, 2000), or for two consecutive days on postnatal days 10 and 11 (Braun *et al.*, 2003). To detect LRCs, a chase period of 4 weeks is sufficient for the interfollicular epidermis (IFE) LRCs, which divide more frequently than bulge LRCs (Bickenbach and Dunnwald, 2000; Bickenbach and Mackenzie, 1984; Cotsarelis *et al.*, 1999). Chases longer than 4 weeks result in the disappearance of interfollicular epidermal LRCs, presumably through division, with distinct label retention only in cells from the bulge area of the hair follicle (Cotsarelis *et al.*, 1990; Taylor *et al.*, 2000). The bulge area is easily recognized as the follicle segment located underneath the sebaceous gland. Microscopy analysis is typically performed on thin skin sections (5–10 μm), which capture approximately one-fifth of the entire follicle. A whole mount technique to visualize entire hair follicles has been developed, and is described in detail elsewhere (Braun *et al.*, 2003; Braun and Watt, 2004).

We have developed a second method for sensitive *in vivo* green fluorescent protein (GFP) detection of infrequently dividing cells, which has allowed for the first time isolation of live LRCs (Tumbar *et al.*, 2004). In our first application of this method we were able to purify LRCs from the hair follicle SC niche, and to demonstrate that they contain cells with some SC phenotypes.

Briefly, the system is composed of two transgenic mouse lines: (1) keratin 5-driven tetracycline repressor mice (K5-tTA) (Diamond *et al.*, 2000) and (2) tetracycline response element-driven histone H2B-GFP transgenic mice (pTRE-H2B-GFP) (Tumbar *et al.*, 2004). The H2B-GFP mice can now be purchased from Jackson Laboratory (Bar Harbor, ME). We maintain each of the two transgenic lines as individual, heterozygous mouse lines. We cross animals from the individual lines to obtain double-transgenic keratin 5-dependent inducible H2B-GFP-expressing mice. We turn off the H2B-GFP expression by feeding the animals doxycyline (1 mg/g) in their food (Bio-Serv, Frenchtown, NJ) starting at 4 weeks of age, after separation from the parent animals. This allows for efficient use of our parental breeding pairs, which never receive the doxycycline. If doxycycline must be administered to pups during lactation, approximately 1 mo of drug-free diet must be allowed for it to clear off from the mother's circulatory system, before initiating a new breeding cycle. We monitor GFP expression in live animals by using an ultraviolet (UV)-based portable lamp, equipped with GFP filters (BioLabs, Budapest, Hungary). Approximately 30% of bulge cells retain bright GFP fluorescence and could be detected with a fluorescence microscope (Fig. 3), or

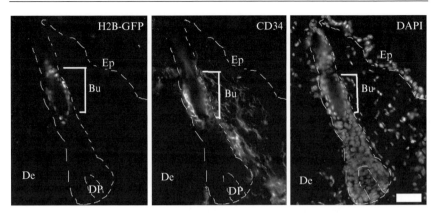

FIG. 3. Skin section from double-transgenic K5-TTA/pTRE-H2B-GFP animals at 8.5 weeks of life and 4.5 weeks of chase shows anagen follicle and epidermis. The H2B-GFP LRCs are found in the bulge (Bu) area, the zone coincidental with CD34 expression, as shown by immunofluorescence staining with antibodies specific to CD34. DNA-specific 4', 6-diamino-2-phenylindole dihydrochloride (DAPI) stain is also shown. De, dermis; Ep, epidermis; Bu, bulge; DP, dermal papillae. Scale bar: 50 μm.

with a flow cytometer equipped with appropriate lasers and GFP filters. Several concerns need to be taken into consideration when using the BrdU and H2B-GFP labeling systems. It is clear that some of the BrdU and H2B-GFP LRCs are cells that contribute to hair follicle growth and/or wound repair, and that last for a long time in the bulge. However, it is less clear that all bulge LRCs are capable of such contribution. Moreover, both BrdU and H2B-GFP pulse–chase schemes are likely to miss some of the bulge SCs, if cells that divide and dilute the label still retain their stemness (Oshima *et al.*, 2001). In addition, BrdU labeling could be incomplete, if the most quiescent cells of the bulge did not divide during the 2–3 days of the pulse-labeling period. In contrast, the H2B-GFP labeling is more inclusive, but both labeling methods could include extremely quiescent cells that might proliferate only in response to high levels of stress. Despite the caveats, detecting stem cells on the basis of their potential label-retaining property is currently the method of choice when analyzing hair follicle stem cell phenotypes in mutant animals.

Protocol for Labeling and Detection of BrdU LRCs in Whole Mount Epidermis

See Braun *et al.* (2003) for details.

1. Ten-day-old mice are injected with a 50-mg/kg body weight concentration of BrdU (20 μl of BrdU at 12.5 mg/ml) every 12 h for a total of four injections to label dividing cells. The mice are healthy and the

only evidence that they are affected by BrdU is a transient loss of dorsal hair before the first postnatal hair cycle.

2. Normally, mice are maintained for a minimum of 70 days after the final BrdU injection in order to detect LRCs. The localization of cells containing the BrdU label can be assessed after a chase period of less than 70 days.

3. To prepare whole mounts of mouse tail epidermis, a scalpel is used to slit the tail lengthways. Skin is peeled from the tail, cut into pieces (0.5 × 0.5 cm²), and incubated in 5 mM EDTA in PBS at 37° for 4 h. Forceps are used to gently peel the intact sheet of epidermis away from the dermis.

4. Fix the epidermal tissue in 4% formal saline (Sigma, St. Louis, MO) for 2 h at room temperature. Store in PBS containing 0.2% sodium azide at 4° for up to 8 weeks before labeling.

5. To prepare whole mounts from mouse back skin, the skin surface is shaved with electric clippers and treated topically with hair removal cream (Immac; Veet UK, Clevedon, UK) for 5 min. Remaining hair is scraped from the dorsal surface and the skin is removed from the mouse. Fat and most of the connective tissue are separated with scissors, and the skin is cut into pieces (0.5 × 0.5 cm²). Skin is incubated in 0.25% trypsin at 4° for 24 to 48 h; the epidermis is removed as soon as it can be separated from the dermis as an intact sheet. Fixation and storage conditions are the same as described above.

6. To detect BrdU, monoclonal antibodies are used [Oxford Biotechnology (Kidlington, UK) or BD Biosciences Pharmingen (San Diego, CA)]. Secondary antibodies are conjugated to Alexa Fluor 488 or Alexa Fluor 594 (Invitrogen Molecular Probes, Eugene, OR).

7. Epidermal sheets are blocked and permeabilized by incubation in PB buffer for 30 min. PB buffer consists of 0.5% skim milk powder, 0.25% fish skin gelatin (Sigma), and 0.5% Triton X-100 in TBS (0.9% NaCl, 20 mM HEPES, pH 7.2).

8. Epidermal sheets are incubated for 20–30 min in 2 M HCl at 37° to denature the DNA and allow for antibody penetration.

9. Primary anti-BrdU antibodies are diluted in PB buffer and tissue is incubated overnight at room temperature with gentle agitation.

10. Epidermal whole mounts are then washed for at least 4 h in PBS containing 0.2% Tween 20, changing the buffer several times.

11. Incubation with secondary antibodies is performed in the same way.

12. Samples are rinsed in distilled water and mounted in Gelvatol (Monsanto, St. Louis, MO) containing 0.5% 1,4-diazabicyclo[2.2.2]octane (DABCO; Sigma).

13. A dissecting microscope is used to isolate individual hair follicles from labeled epidermal sheets. The epidermis is placed in PBS with the basal layer facing upward. Fine forceps are used to grasp a hair follicle at the infundibulum and the follicle is pulled gently until it separates from the epidermal sheet.

14. Images are acquired with a confocal microscope and a long-working distance lens that allows deep penetration into the thick sample (hair follicle diameter, $\sim$50–100 μm).

Putative Stem Cell Markers: Genes Preferentially Expressed by Bulge Cells

A long-sought characteristic that would uniquely identify stem cells in a skin section or by FACS analysis is the potential expression of a set of specific biochemical markers (Akiyama *et al.*, 2000; Watt, 1998). Studies have unraveled a plethora of new bulge-preferred markers. These would make it possible to carry out many functional studies to investigate in more depth the molecular mechanisms that operate in epithelial hair stem cells (Ito *et al.*, 2004, 2005; Kizawa and Ito, 2005; Lyle *et al.*, 1998; Michel *et al.*, 1996; Morris *et al.*, 2004; Trempus *et al.*, 2003; Tumbar *et al.*, 2004). Despite the usefulness of these markers for defining and isolating bulge cells, there is no definitive biochemical marker for a hair stem cell, and it is difficult to unambiguously identify stem cells in a skin section (Gambardella and Barrandon, 2003). Even though it is now generally agreed that the bulge is the main location of epithelial stem cells, it is thought that only a fraction of the bulge cells are stem cells (Claudinot *et al.*, 2005; Gambardella and Barrandon, 2003). The distinguishing stem cell microscopic features, and their precise location within the niche, are unclear. Future experiments aimed at refining our current understanding of the genetic make-up of bulge cells and the architecture of the stem cell niche will address these questions.

This section summarizes the most useful biochemical bulge markers that have been described to date. The most traditional epithelial stem cell markers are integrins (Watt, 2002) β_1 and α_6, which are expressed at increased levels in the bulge area, relative to the lower segments of the anagen phase follicle (Jones and Watt, 1993; Tani *et al.*, 2000). β_1-Integrin is not differentially regulated in the mouse hair follicle ORS relative to the bulge, but shows clear differences in human follicles and epidermis. Integrin β_6 is upregulated in the bulge and ORS during early anagen, and is downregulated in telogen (Tumbar *et al.*, 2004). Upregulation of integrins in the stem cell compartment has been attributed to the need to keep stem cells tight within the niche by adherence to each other and to the basement membrane (Watt, 1998, 2002). When stem cells migrate out of the niche in anagen, integrins required for anchorage, such as β_1 and α_6, are downregulated, especially below the bulge, whereas those involved in cell migration, such as integrin β_6, are upregulated. However, integrins also regulate cell proliferation, and it is possible that cells with high proliferative potential have elevated integrin levels (Brakebusch *et al.*, 2000; Raghavan *et al.*, 2000). Integrins are thought to play a major role in the skin, in regulating epidermal adhesion, growth, and differentiation

(Brakebusch *et al.*, 2000; De Arcangelis and Georges-Labouesse, 2000; Watt, 2002).

Several important studies have suggested a number of bulge stem cell markers: keratin 15 (Lyle *et al.*, 1998) and keratin 19 (Michel *et al.*, 1996), both also expressed in a patchy pattern by basal layer cells; CD71^{low} (Tani *et al.*, 2000); S100 proteins (Ito *et al.*, 2004, 2005; Kizawa and Ito, 2005); E-cadherin^{low} (Akiyama *et al.*, 2000); p63 (Pellegrini *et al.*, 2001); CD34, shown in Fig. 3 (Trempus *et al.*, 2003); and Sox9 (Michel *et al.*, 1996). Of these, CD34, S100A6, S100A4, and Sox9 were all upregulated at the level of mRNA expression in the bulge (Tumbar *et al.*, 2004).

Other markers have been described, when transcriptional profiles obtained from isolated infrequently cycling H2B-GFP bulge cells were compared with progeny cells in the basal layer of the epidermis and ORS of the hair follicle (Tumbar *et al.*, 2004). Probing one-third of the mouse genome resulted in identification of ~100 mRNAs that were expressed at levels 2-fold or greater in the bulge LRCs relative to the BL/ORS. Preliminary analyses by immunofluorescence microscopy demonstrated that a number of the proteins encoded by these mRNAs are expressed within the bulge (Tumbar *et al.*, 2004). Some of the factors were more specific to cells of the bulge, whereas others were expressed in other cell types within the skin (Tumbar *et al.*, 2004). Interestingly, these markers were not specific to the brightest LRCs of the bulge but were, rather, expressed by the cells within the entire bulge compartment. In a separate study, Cotsarelis and colleagues have isolated bulge cells from a transgenic animal expressing GFP driven by the keratin 15 promoter, and demonstrated extensive overlap in the expression profile of all-bulge cells and that of the H2B-GFP LRCs (Morris *et al.*, 2004).

Gene Circuitry in the Stem Cell Niche

Targeted loss and gain of function of bulge-specific genes is likely to unravel some of the molecular characteristics governing stem cell function in the hair follicle. Full knockouts have been used with some success, but often can lead to embryonic lethality at stages before skin and hair follicle development. Epithelial conditional gene knockout can be accomplished by using the Cre-*loxP* system, a bacteriophage P1-derived site-specific recombination system in which the Cre recombinase catalyzes recombination between *loxP* recognition sequences (Gu *et al.*, 1993; Sauer and Henderson, 1988). Through conventional ES cell-based homologous recombination, two *loxP* sites can be inserted into the mouse genome flanking a particular exon within the gene locus of interest ("floxed gene"). This gene region is usually part of a functional domain, which is essential for the gene function and results in a frame shift that would render the truncated protein inactive. The knockout is achieved by mating mice carrying the floxed gene with transgenic mice with keratinocyte-specific and/or inducible Cre expression. For epithelial specific knockouts there are at least four available transgenic lines expressing Cre from the (1) keratin 14 (K14) promoter (Vassar *et al.*, 1989); (2) keratin 5 (K5) promoter (Byrne and Fuchs, 1993); K14 promoter driving an inducible Cre-ER transgene (Vasioukhin *et al.*, 1999); (3) K15 promoter driving an inducible Cre-PR1 gene (Ito *et al.*, 2005; Liu *et al.*, 2003); and (4) sonic hedgehog (Shh) promoter driving Cre-GFP (Harfe *et al.*, 2004). Of these, the K15-driven Cre-PR1 seems to be specifically expressed in the bulge in adult stages of hair follicle development (Liu *et al.*, 2003), whereas the Shh-driven Cre is restricted to a subset of the hair follicle cells during early development. The K14 promoter is active in the BL of the epidermis and the ORS of the hair follicle. Some of these transgenic lines are available for purchase from Jackson Laboratory.

Searching for target genes regulated downstream of a particular mutation specifically within the bulge is now possible by sorting $CD34^+/\alpha_6$ integrin$^+$ skin cells from mutant and wild-type animals, and performing microarray analyses. An important issue is to focus the analysis on the earliest possible stage of the phenotype onset. This will avoid identification of genes that appear changed as a secondary consequence of the mutation, many steps removed from the initial effect on skin physiology. Typically $\sim$100,000 double-positive cells can be isolated from one adult animal, and $\sim$200 ng of RNA can be obtained by employing conventional RNA isolation techniques. We use an RNA Miniprep kit (Stratagene, La Jolla, CA), or a combination of TRIzol (Invitrogen, Carlsbad, CA) and Qiagen kits (Qiagen, Valencia, CA), and obtain high-quality RNA in sufficient amounts to perform RT-semiquantitative/quantitative PCR, and microarrays. The quality of RNA needs to be assayed with a Bioanalyzer (Agilent, Foster City, CA) that can evaluate a minimum of 200 pg of RNA, and provide useful parameters such as 28S:18S ribosomal RNA ratios and RNA integrity information. mRNAs are subsequently converted into cDNAs and linearly amplified to provide sufficient amounts of labeled probes for microarray analyses. Several T7 polymerase-based amplification methods have been developed, and we have successfully employed the Arcturus kit (Arcturus, Mountain View, CA). Various platforms for hybridizations can be used to determine gene profiles, and have been described in detail in the literature.

Protocol for Detecting Gene Expression of Putative Bulge-Expressed Genes

See Tumbar *et al.* (2004) and Blanpain *et al.* (2004) for details.

1. Adult mice (21 days of age or older) are killed and their skin is removed. If available, use a transgenic mouse that expresses GFP from the keratin 5, 15, or 14 promoter for more specific labeling of keratinocytes.

If not available, cell populations sorted on the basis of α_6-integrin will still be highly enriched in keratinocytes (>90%).

2. Fix the skin with the hypodermis facing up on a foam support, using fine needles.

3. Scrape away the subcutaneous fat and muscle, using a sharp scalpel.

4. Incubate in 0.25% trypsin and 5 mM EDTA in PBS at 4° overnight.

5. Use scalpels to scrape off the epidermis. Chop off into small pieces and incubate at 37° for 10 min.

6. Neutralize in 15% serum. Pipette up and down to obtain a single-cell suspension and filter through a cell strainer.

7. Immunostain cells with biotin-conjugated rat anti-CD34 antibodies (RAM34) and phycoerythrin-conjugated anti-α_6-integrin antibody (GoH3), followed by streptavidin–allophycocyanin conjugate (BD Biosciences Pharmingen).

8. Label with propidium iodide (PI, 5 μg/ml) to stain dead cells.

9. Resuspend cells at a density of 10 million cells/ml.

10. Use a FACS to sort live cells (PI negative) that are $\alpha_6{}^+$/GFP$^+$/CD34$^-$ (basal layer and outer root cells outside the bulge) and $\alpha_6{}^+$/GFP$^+$/CD34$^+$ (bulge-only cells).

11. Collect cells directly into RNA lysis buffer (Stratagene), and vortex every 10 min to disrupt the cell membrane and to release the RNA, which will be protected from degradation. Samples can be stored at $-80°$ until ready to prepare.

12. Use a Stratagene kit to prepare RNA and concentrate with a speed-vacuum pump.

13. Use an Agilent Bioanalyzer to determine quality and to estimate concentration.

14. Prepare cDNA with a reverse transcriptase-based kit.

15. Design specific primers to your gene of interest and perform RT-PCR with increasing numbers of PCR cycles, and use ethidium bromide–agarose gels to detect levels of expression or use quantitative RT PCR.

Contribution of Epithelial SCs to Tissue Morphogenesis and Wound Repair In Vivo

The main function of an adult stem cell is to contribute cells necessary to maintain normal tissue homeostasis or to repair injury. The epidermis self-regenerates every 3–4 weeks, the length of time necessary for a cell to transit from the undifferentiated, proliferating basal layer to the terminally differentiated, anuclear cornified envelope. The hair follicle goes through progressive cycles of growth (anagen), regression (catagen), and rest (telogen), which take place in relative synchrony during morphogenesis and the first hair cycle, and take approximately 3 weeks to complete. Later hair cycles are asynchronous, and the duration of the SC dormant state (telogen) increases.

The activation cycles of the interfollicular epidermis stem cells are not well understood. Hair follicle stem cells can be activated by two mechanisms. First, stem cell activation can occur spontaneously by hair growth stimuli coming from the dermal papillae and the surrounding environment. These stimuli cause stem cells to exit their dormancy, migrate, proliferate, and give rise to newly growing anagen follicles (Jahoda and Reynolds, 1993). Second, stem cell activation can be triggered by skin injury: epidermal wounding or hair plucking (Ito *et al.*, 2004; Taylor *et al.*, 2000; Tumbar *et al.*, 2004). Hair is ideal for studying stem cell activation because of the synchrony of the hair follicle transition from telogen to anagen for the first two cycles, a transition that takes place at defined adult stages. During hair growth, stem cells exit the bulge and proliferate to contribute to the hair germ, the ORS, and the matrix (Taylor *et al.*, 2000; Tumbar *et al.*, 2004). It has been proposed that matrix cells can be defined as three different types of committed progenitor cell: pre-IRS, precuticle, and premedulla. These progenitor cells subsequently give rise to all the inner, differentiated layers of the hair follicle, and they undergo apoptosis at the end of each hair follicle cycle (Paus *et al.*, 1999). The telogen follicle is a good starting point for monitoring the contribution of bulge cells to the hair follicle. This is because in telogen the hair follicle is composed solely of bulge cells surrounding the club hair, and a small pocket of germ cells located right below the bulge, which stays in close contact with the dermal papillae. When hairs transition into anagen, a hair bulb emanates from the bulge and gives rise to a new hair shaft. The hair germ has been an object of debate. It has been proposed to be a station area of bulge stem cells that migrated and homed there in the previous hair cycle (Panteleyev *et al.*, 2001). These cells have been identified as ORS cells at the bottom of the bulge that had collapsed around the hair club at the end of catagen (Ito *et al.*, 2004). Initiation of anagen begins with the stimulation of the hair germ cells to proliferate. This is either followed by or coincides with migration and proliferation of at least some newly activated bulge cells, which contribute to the growing germ cells and finally to the matrix.

Monitoring the contribution of bulge cells to hair morphogenesis is possible by pulse–chase labeling of bulge cells, followed by microscopy analyses of either H2B-GFP epifluorescence or BrdU LRC antibody staining, at progressive time points during the initiation of the second hair follicle cycle [roughly from postnatal day 49 (P49) to postnatal day 60 (P60)]. Double labeling for LRCs and for proliferation markers Ki67, P-histone H3, BrdU, or [^{3}H]thymidine allows capture of the activated bulge cells.

Some of the activated bulge cells are H2B-GFP LRCs, sufficiently bright to allow tracking their progeny over several population doublings. Newly arising populations of dim GFP cells, deriving from the bulge H2B-GFP

LRCs, form the matrix and contribute to the differentiated layers of the hair follicle (Tumbar *et al.*, 2004). Similar data can be obtained by involving BrdU as a means of labeling LRCs (Taylor *et al.*, 2000). In our experience, the BrdU LRCs retain challengingly small amounts of label, and tracking them over cycles of stem cell activation requires extremely sensitive detection methods. In contrast, the H2B-GFP LRCs are extremely bright and easy to track, but the disadvantage of using this labeling system, relative to the BrdU labeling system, is the long time required to introduce two transgenes (tTA and pTRE-H2B-GFP) into the genetic background of interest. A second path for LRC activation is by wounding (Tumbar *et al.*, 2004), when LRC migration and contribution to wound repair can be detected within approximately 24 h. Apparently, the bulge contribution to the epidermis is only transient. The bulge progeny cells fail to propagate in the epidermis beyond several weeks (Claudinot *et al.*, 2005; Ito *et al.*, 2005). The contribution of bulge cells to hair morphogenesis or wound repair can be elegantly studied by using a powerful system developed by Cotsarelis and colleagues (Ito *et al.*, 2005). A cloned fragment of the keratin 15 (K15) promoter drives specific LacZ, GFP, or Cre-PR1 expression in the hair follicle bulge in adult animals (Liu *et al.*, 2003). By inducing K15-driven Cre-PR1 activation during adulthood in a mouse containing the *loxP*–Stop–*loxP* Rosa 26 reporter locus, one can easily track progeny of bulge cells by virtue of their β-galactosidase expression during hair follicle growth or wound repair (Ito *et al.*, 2005). Once again, two gene loci need to be introduced into the genetic background of interest for labeling purposes, but this system is superior to other tracking methods in that the fate of the migrating bulge cells could be determined over unlimited divisions and for extended periods of time. A similar system has been reported that exploits the hair follicle-specific expression of sonic hedgehog early in skin development (Harfe *et al.*, 2004; Levy *et al.*, 2005).

Protocol for Immunofluorescence Staining of Skin Sections

See Tumbar *et al.* (2004) for details.

1. Collect skin tissue from an animal. Press it firmly on a piece of thin paper towel, avoiding folds.
2. Cut a skin piece 1.5 cm length along the anteroposterior axis by 0.5 cm wide along the dorsoventral axis of the mouse. Mount in O.C.T. resin (Sakura Finetek, Torrance, CA) on dry ice, with the long dimension facing down into the block (the short dimension should be perpendicular to the bottom). Freeze on dry ice.
3. Cut 10- to 30-μm sections, and pick up on microscope slides.
4. Fix for 10 min in 4% formaldehyde in PBS.

5. Wash three times for 5 min each in PBS.
6. Neutralize three times for 5 min each in 20 mM glycine in PBS.
7. Block in 5% normal goat serum (NGS) and 5% normal donkey serum (NDS) in 0.1% Triton X-100 in PBS for 1 h at room temperature.
8. Incubate in primary antibody for 2 h at room temperature, or overnight at 4°.
9. Wash in PBS, three times for 5 min each.
10. Incubate in secondary antibody as described above. Wash in PBS.
11. Stain the DNA with Hoescht dye (5 μg/ml; Invitrogen Molecular Probes) for 5 min.
12. Wash and mount in 80% glycerol or antifade mounting medium (Invitrogen Molecular Probes).

Epithelial Stem Cells in Culture

Whether they originate in the IFE or the bulge, epithelial stem cells are thought to be clonogenic cells, meaning those cells that form large colonies when keratinocytes isolated from epithelial skin are placed in culture (Gambardella and Barrandon, 2003; Watt, 2004; Watt and Hogan, 2000). Keratinocytes isolated from human skin not only form large colonies that can be highly expanded in culture, but also generate cultured skin explants used to replace the damaged epidermis of burn victims (Rochat and Barrandon, 2004). This artificial skin sustains itself on the patient for extended periods of time, and the extent of its survival is thought to reflect the ability to capture the highly proliferative putative stem cells from the tissue. Oshima *et al.* showed that the bulge is the hair follicle segment that contains the highest frequency of clonogenic cells (Oshima *et al.*, 2001). The bulge is the area of the hair follicle that contains (1) LRCs, (2) clonogenic cells, and (3) cells with morphogenetic abilities *in vitro* and *in vivo*. Morris and Potten have shown that BrdU LRCs are the source of large keratinocyte colonies (Morris and Potten, 1994), and Barrandon and colleagues have shown that bulge colony-forming cells contribute to skin morphogenesis and long-term homeostasis when grafted (Claudinot *et al.*, 2005). The colony formation efficiency of mouse cells is low, relative to rat or human cells, even when the cell populations in study are enriched in bulge cells. However, colonies formed by bulge populations of H2B-GFP LRCs (Tumbar *et al.*, 2004), CD34/α_6 cells (Trempus *et al.*, 2003), or K15-driven GFP-expressing cells (Liu *et al.*, 2003) are larger and more numerous than those made by other keratinocytes in the skin. Primary mouse keratinocytes grow in culture on fibroblast feeder layers, in specially formulated media (Barrandon and Green, 1987; Redvers and Kaur, 2005; Rochat and Barrandon, 2004). The feeder layers are NIH3T3 or 3T3J2 fibroblast cell lines treated with mitomycin C or lethally irradiated to prevent cell division. In our experience, large colonies made of several hundred cells arise from the initial primary keratinocytes after 1 or 2 weeks of culture. Keratinocytes can than be transferred multiple times directly onto plastic dishes, or on covered feeder dishes for extended periods of time (Redvers and Kaur, 2005).

The colony formation efficiency (seeding), the size of the colony, and the extent of cell expansion in culture depend on several factors: (1) the proliferation potential of a cell, that is, how many divisions a cell could undergo before it senesces, (2) the adhesion capacity of a cell—lack of substratum attachment will result in cell loss, and lack of colonies, and (3) the ability of a particular SC to respond to growth-promoting factors present in any particular cell culture medium, that is, the ability of stem cells to be "activated" by the growth medium. Care should be taken when performing and interpreting cell culture studies. When analyzing the phenotype of various gene mutations all the factors enumerated above might be perturbed, and appropriate tests must be done to address the different possibilities independently. Raghavan et al. have shown that in the absence of integrin β_1, keratinocytes can grow under modified culturing conditions that overcome the original impairment in cell adhesion, whereas under regular conditions the knockout cells cannot grow (Raghavan et al., 2000, 2003). This clearly demonstrated how lack of proper adhesion alone results in perturbation of cell growth. When starting culturing experiments several basic questions must be taken into consideration. Does the same number of small colonies appear early after plating, but then many fail to expand over time? Or simply, primary mutant keratinocytes float above the feeders, never attach, and therefore never form any colonies? Do various substrate and/or culturing conditions modify the cell culture potential? A study by Kaur and colleagues has clearly demonstrated the importance of slight variations in culture conditions to the outcome of the assay (Li et al., 2004). When carefully performed, and interpreted in context, this assay has proved seminal in gaining insight into stem cell function (Gambardella and Barrandon, 2003).

Cultured cells have been used in several assays to characterize putative stem cell populations. To assess multipotency and differentiation capabilities, cultured cells can be grafted onto the back of immunocompromised mice, an assay described in more detail in a subsequent section of this chapter. Alternatively, keratinocytes can be induced to differentiate by increasing the calcium concentration in the culture medium (Hennings et al., 1980a,b), or by allowing them to form organotypic cultured skin (Li et al., 2004). To assess "proliferative potential," researchers perform clonal analyses to determine the frequency of colony seeding as well as the type of colonies formed by any particular population enriched in stem cells (Barrandon and Green, 1987;

Rochat and Barrandon, 2004). Colonies are classified on the basis of (1) their size or total number of cells after a defined time period in culture (typically 1 week) and (2) the potential of progeny cells from individual colonies to expand further. For example, late-generation telomerase knockout animals give rise to colonies at a frequency comparable to that of wild-type animals, but the number of total divisions these cells can undergo is tremendously reduced (Flores *et al.*, 2005). In general, this phenotype could result from a similar number of stem cells, which were defective (i.e., lack of telomerase) in long-term proliferative potential. This phenotype can be attributed to early stem cell senescence. In other instances, improper adhesion and cell spreading can also lead to similar small colony phenotypes (Raghavan *et al.*, 2003). Other common cell biology assays pertain to the kinetics of cell proliferation, and involve determining growth curves over time and analyses of cell cycle profiles. Having recognized the caveats of taking the cells out of their normal context (niche), culture studies have proven enormously useful in studying epithelial stem cell biology.

Self-Renewal and Differentiation of Epithelial SCs

The defining characteristic of a stem cell is thought to be its ability to self-renew and differentiate for the entire life of the animal. This ability is difficult to demonstrate *in vivo*, in the absence of major perturbation of the normal biology of an SC. An ideal tracking system would be one in which a single cell is genetically marked to express a specific reporter gene. The expression of this reporter gene should be maintained in all progeny cells derived from this single cell. Division of a stem cell is thought to be asymmetric, and would result in another identical stem cell, and a more differentiated transient amplifying (TA) cell (Fuchs *et al.*, 2001; Watt and Hogan, 2000). The reporter gene should be induced rapidly to detectable levels of expression in the single stem cell alone. Ideally, a second reporter gene would turn on specifically in the newly formed TA cell, and would be expressed in all its downstream cell progeny. Fate mapping based on tracking the two individual reporter genes during relevant developmental stages (i.e., various hair cycle stages) would allow us to identify both the self-maintaining population of single-labeled stem cells and the newly emerging double-labeled population of progeny cells. Marking single cells with accuracy, and without perturbing the system by manipulation and surgery, is technically challenging. The closest systems available so far are those in which a mosaic animal is created either by the virtue of chimera generation (Kopan *et al.*, 2002), by retroviral infection (Ghazizadeh and Taichman, 2001), or by using inefficient or mosaic expression of the Cre recombinase, coupled with the Rosa26–*loxP* system (Legue and

Nicolas, 2005). In all these cases some, but not all, cells of a certain kind are marked. However, the precise number, the exact tissue origin, and the type of marked stem or progenitor cells are unknown.

At present, the self-renewal and differentiation capacity of a cell is best addressed *in vitro*. Even though *in vitro* approaches are extremely powerful, they might introduce major modification to biological systems (Gabay *et al.*, 2003; Potten, 2004; Potten and Booth, 2002). Data should be interpreted with care when *in vitro* results are used to explain *in vivo* biological processes. Nevertheless, *in vitro* studies are extremely important for clinical application.

Several studies that address the self-renewal of bulge cells have been performed. These studies can be used as starting points in setting up similar new experiments. Figure 4 presents a step-wise "protocol" synthesizing the various approaches described below. Barrandon and colleagues reported in 2001 the first studies on the self-renewal potential of bulge cells (Oshima *et al.*, 2001). Entire bulges transplanted onto the back of immunocompromised mice contributed to all the different epithelial lineages, including all layers of the hair follicle, sebaceous glands, and epidermis. Morris and colleagues (2004) followed with grafting of bulge cells sorted by their GFP expression driven by the K15 promoter (Liu *et al.*, 2003). Blanpain *et al.* attempted first to take this approach to the clonal level (Blanpain *et al.*, 2004), using a keratinocyte-grafting method developed in the Yuspa laboratory (Lichti *et al.*, 1993). Bulge cells isolated by fluorescence-activated cell sorting from 2-month-old mouse skin by virtue of their CD34 and α_6 integrin expression were cultured onto 3T3J2 fibroblast feeders (Blanpain *et al.*, 2004). Large colonies, presumably coming from single cells, were subsequently picked up from the plate and expanded to large numbers of $\sim 10^6$ total cells. These cells were then mixed with 10^7 dermal fraction newborn mouse skin cells (containing dermal fibroblast cells, hair follicle cells, sebaceous gland cells, etc.), and subsequently grafted onto the backs of nude animals (Blanpain *et al.*, 2004). Genetically marked cells originating in the single-bulge cell could than be detected in the graft as contributors to all the epithelial lineages. This experiment has 3-fold relevance. First, it shows that extensive culture conditions do not perturb the morphogenetic abilities of bulge epithelial stem cells. Second, it provides evidence that the bulge epithelial cells retain their stem cell potential over many cell divisions; in other words, they self-renew. Finally, it shows that single epithelial stem cells are multipotent, one cell being capable of differentiating into many cell types of the hair follicle, epidermis, and sebaceous gland (all epithelial).

These short-term analyses do not distinguish between a multipotent short-lived transient amplifying cell and a stem cell, and cell fate tracking of the transplanted cells must be accomplished for prolonged periods of time. This has been performed by Barrandon and colleagues (Claudinot

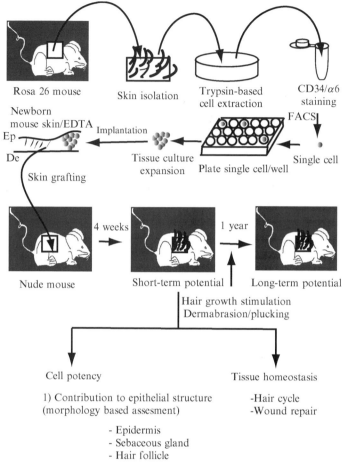

Rosa 26 mouse   Skin isolation   Trypsin-based   CD34/α6
Newborn   cell extraction   staining
mouse skin/EDTA   Implantation   FACS
Ep
De   Tissue culture   Single cell
        Skin grafting   expansion   Plate single cell/well

        4 weeks   1 year

Nude mouse   Short-term potential   Long-term potential
                                    Hair growth stimulation
                                    Dermabrasion/plucking

Cell potency   Tissue homeostasis

1) Contribution to epithelial structure   -Hair cycle
(morphology based assesment)   -Wound repair

                - Epidermis
                - Sebaceous gland
                - Hair follicle

2) Contribution to specific cell types
(immunostaining for specific markers)

        - BL/ORS: Keratin 5, Keratin 14, Integrin β1, Integrin β4, Integrin α6
        - Suprabasal epidermal layers: Keratin 1&10, Loricrin, Filagrin, Involucrin
        - Bulge: CD34, Sox 9, S100 A4, S100 A6, LTBP1, Tektin, TCF3;
        - IRS: Gata3, AE15
        - Matrix: Ephrin B1, Msx2, Shh
        - Pre-cortex/cortex: Lef1, AE13
        - Medulla: AE15, Keratin 6
        - ORS companion layer: Keratin 6

FIG. 4. Long-term self-renewal and differentiation potential of hair epithelial stem cells. A genetically marked reporter mouse such as Rosa 26 (Soriano, 1999) is used to isolate skin cells, using a trypsin-based skin digestion procedure (Trempus *et al.*, 2003). Cell suspensions are subsequently subjected to immunofluorescence staining with antibodies against cell surface

et al., 2005), who clearly demonstrated that a single clonogenic cell can be highly expanded to a population of cells that can subsequently reconstitute all the hair lineages, including making more stem cells. The graft was allowed to develop for approximately 1 year, a significant fraction of a mouse's life. These authors proposed that any clonogenic cell is a multipotent stem cell *in vitro*, capable of tissue morphogenesis and prolonged tissue homeostasis, and paved the way for many more studies. The technique used by Barrandon and colleagues involved plating single bulge cells isolated from dissected rat whisker bulges. After cell culture expansion, 5×10^5 cells are implanted in a pocket created between the epidermis and the dermis of newborn mouse skin, and transplanted onto an acceptor mouse. To determine the long-term potential of bulge stem cells, follicles are repeatedly stimulated to grow by skin dermabrasion. A second round of single-cell culturing and transplantation shows that the self-renewal ability of bulge cells has not diminished over time.

In future it will be important to transplant transient amplifying cells in parallel with stem cells, and to monitor their fate for extended periods of time. The current difficulty consists in the apparent incapacity of cells other than stem cells to be propagated in culture, even though some reports suggest that transient amplifying cells and stem cells can both be propagated and can contribute to skin at least over a 10-week time period (Li *et al.*, 2004). It is possible that in that study there was overlap between the transient amplifying cells and stem cell populations, or that the transient amplifying cells undergo dedifferentiation during the culturing process. Some of these issues could be resolved by transplanting single, freshly isolated stem cells from the bulge, and transient amplifying cells from the

---

markers CD34 and α_6-integrin. Double-positive bulge cells isolated by fluorescence-activated cell sorting (FACS) can be directly collected in 96-well plates containing irradiated feeders as single bulge cells per well. One to 2 weeks of culture results in cell colony formation. Cells can be transferred and expanded further to 5×10^5 cells. During this time presumably stem cells undergo multiple divisions, and thus they must self-renew, if any stem cells are to be found in culture at the end of this extensive expansion. Newborn mouse skin is treated with EDTA to loosen the epidermal–dermal junction, and the cultured keratinocytes are implanted in a pocket created at this junction. The newborn skin is now grafted onto an immunocompromised mouse to prevent rejection, and the fate of the genetically marked cells is monitored over time. Analyses of the graft after a short period of time ($\sim$4 weeks) allow tracking the contribution of the cultured cells to various epithelial lineages, including the hair follicle, the sebaceous gland, and the epidermis. In addition, it could be established whether the transplanted cells are able to respond to injury and sustain a normal hair follicle cycle. To determine the long-term potential of the cells, stem cells are repeatedly activated to proliferate by induced hair growth (dermabrasion or plucking), during extended periods of time ($\sim$1 year). For a more stringent approach, a second round of single-cell isolation, cell culture expansion, and grafting on a secondary acceptor mouse can be employed (Claudinot *et al.*, 2005). BL/ORS, basal layer/outer root sheath; IRS, inner root sheath.

bulge or matrix. This approach has been perfected for the hematopoietic system, and should be applied in future to the skin and the hair follicle.

It has been suggested that *in vitro* approaches are more a reflection of stem cell response to stress or injury, rather than of normal tissue homeostasis (Potten, 2004; Potten and Booth, 2002). Despite the inherent uncertainty when studying stem cells outside their environment, the approach is extremely valuable for clinical application, which is the ultimate goal of stem cell research, and can provide useful insight into the basic biology of epithelial stem cells (Claudinot *et al.*, 2005; Gambardella and Barrandon, 2003; Oshima *et al.*, 2001).

SCs: Life as Long as That of the Animal?

Cell senescence, skin aging, and epithelial stem cells must be tightly linked together. The effect of telomere length and telomerase activity on the lifetime of skin epithelial stem cells has been addressed (Flores *et al.*, 2005; Sarin *et al.*, 2005). Even though in principle adult stem cells have been considered immortal because of their postulated expression of telomerase, this concept has been challenged by findings that melanocyte stem cells might be progressively lost with aging, a process that could lead to hair graying (Nishimura *et al.*, 2005). The epithelial skin structure changes in aging animals. Thinning of the epidermis, loss of hair cycle synchrony, and prolonged dormancy of hair follicle stem cells (telogen) are apparent in older mouse skin. How these processes are related to stem cell activity is not clear. The growth potential of human keratinocytes progressively decreases with age (Rochat and Barrandon, 2004), and mouse keratinocytes show similar, though less well documented, declines. The residence of stem cells in the permanent portion of the hair follicle (the bulge) suggests that these cells live for the entire life of the animal. Some bulge cells can be traced for extended periods of time (Cotsarelis *et al.*, 1990; Morris and Potten, 1999) by virtue of their long-term label retention. However, it is not clear whether the most quiescent bulge cells are, in fact, the same cells that normally contribute to hair growth and tissue repair (Morris and Potten, 1999). As we learn more about the molecular mechanisms governing epithelial skin stem cell function, we might be able in future to manipulate stem cell behavior to overcome the effects of currently irreversible skin conditions, such as aging.

References

Akiyama, M., Smith, L. T., and Shimizu, H. (2000). Changing patterns of localization of putative stem cells in developing human hair follicles. *J. Invest. Dermatol.* **114,** 321–327.

Arnold, I., and Watt, F. M. (2001). c-Myc activation in transgenic mouse epidermis results in mobilization of stem cells and differentiation of their progeny. *Curr. Biol.* **11,** 558–568.

Barrandon, Y., and Green, H. (1987). Three clonal types of keratinocyte with different capacities for multiplication. *Proc. Natl. Acad. Sci. USA* **84,** 2302–2306.

Benitah, S. A., Frye, M., Glogauer, M., and Watt, F. M. (2005). Stem cell depletion through epidermal deletion of Rac1. *Science* **309,** 933–935.

Bickenbach, J. R., and Chism, E. (1998). Selection and extended growth of murine epidermal stem cells in culture. *Exp. Cell Res.* **244,** 184–195.

Bickenbach, J. R., and Dunnwald, M. (2000). Epidermal stem cells: Characteristics and use in tissue engineering and gene therapy. *Adv. Dermatol.* **16,** 159–183; discussion 184.

Bickenbach, J. R., and Grinnell, K. L. (2004). Epidermal stem cells: Interactions in developmental environments. *Differentiation* **72,** 371–380.

Bickenbach, J. R., and Mackenzie, I. C. (1984). Identification and localization of label-retaining cells in hamster epithelia. *J. Invest. Dermatol.* **82,** 618–622.

Blanpain, C., Lowry, W. E., Geoghegan, A., Polak, L., and Fuchs, E. (2004). Self-renewal, multipotency, and the existence of two cell populations within an epithelial stem cell niche. *Cell* **118,** 635–648.

Brakebusch, C., Grose, R., Quondamatteo, F., Ramirez, A., Jorcano, J. L., Pirro, A., Svensson, M., Herken, R., Sasaki, T., Timpl, R., *et al.* (2000). Skin and hair follicle integrity is crucially dependent on β_1 integrin expression on keratinocytes. *EMBO J.* **19,** 3990–4003.

Braun, K. M., and Watt, F. M. (2004). Epidermal label-retaining cells: Background and recent applications. *J. Investig. Dermatol. Symp. Proc.* **9,** 196–201.

Braun, K. M., Niemann, C., Jensen, U. B., Sundberg, J. P., Silva-Vargas, V., and Watt, F. M. (2003). Manipulation of stem cell proliferation and lineage commitment: Visualisation of label-retaining cells in wholemounts of mouse epidermis. *Development* **130,** 5241–5255.

Byrne, C., and Fuchs, E. (1993). Probing keratinocyte and differentiation specificity of the human K5 promoter *in vitro* and in transgenic mice. *Mol. Cell. Biol.* **13,** 3176–3190.

Cairns, J. (1975). Mutation selection and the natural history of cancer. *Nature* **255,** 197–200.

Candi, E., Schmidt, R., and Melino, G. (2005). The cornified envelope: A model of cell death in the skin. *Nat. Rev. Mol. Cell. Biol.* **6,** 328–340.

Claudinot, S., Nicolas, M., Oshima, H., Rochat, A., and Barrandon, Y. (2005). Long-term renewal of hair follicles from clonogenic multipotent stem cells. *Proc. Natl. Acad. Sci. USA* **102,** 14677–14682.

Cotsarelis, G., Sun, T. T., and Lavker, R. M. (1990). Label-retaining cells reside in the bulge area of pilosebaceous unit: Implications for follicular stem cells, hair cycle, and skin carcinogenesis. *Cell* **61,** 1329–1337.

Cotsarelis, G., Kaur, P., Dhouailly, D., Hengge, U., and Bickenbach, J. (1999). Epithelial stem cells in the skin: Definition, markers, localization and functions. *Exp. Dermatol.* **8,** 80–88.

De Arcangelis, A., and Georges-Labouesse, E. (2000). Integrin and ECM functions: Roles in vertebrate development. *Trends Genet.* **16,** 389–395.

Diamond, I., Owolabi, T., Marco, M., Lam, C., and Glick, A. (2000). Conditional gene expression in the epidermis of transgenic mice using the tetracycline-regulated transactivators tTA and rTA linked to the keratin 5 promoter. *J. Invest. Dermatol.* **115,** 788–794.

Doetsch, F., Verdugo, J. M., Caille, I., Alvarez-Buylla, A., Chao, M. V., and Casaccia-Bonnefil, P. (2002). Lack of the cell-cycle inhibitor p27Kip1 results in selective increase of transit-amplifying cells for adult neurogenesis. *J. Neurosci.* **22,** 2255–2264.

Falciatori, I., Borsellino, G., Haliassos, N., Boitani, C., Corallini, S., Battistini, L., Bernardi, G., Stefanini, M., and Vicini, E. (2004). Identification and enrichment of spermatogonial stem cells displaying side-population phenotype in immature mouse testis. *FASEB J.* **18,** 376–378.

Flores, I., Cayuela, M. L., and Blasco, M. A. (2005). Effects of telomerase and telomere length on epidermal stem cell behavior. *Science* **309,** 1253–1256.

Fuchs, E. (1990). Epidermal differentiation. *Curr. Opin. Cell Biol.* **2,** 1028–1035.

Fuchs, E. (1995). Keratins and the skin. *Annu. Rev. Cell. Dev. Biol.* **11,** 123–153.

Fuchs, E., and Raghavan, S. (2002). Getting under the skin of epidermal morphogenesis. *Nat. Rev. Genet.* **3,** 199–209.

Fuchs, E., Merrill, B. J., Jamora, C., and DasGupta, R. (2001). At the roots of a never-ending cycle. *Dev. Cell* **1,** 13–25.

Fuchs, E., Tumbar, T., and Guasch, G. (2004). Socializing with the neighbors: Stem cells and their niche. *Cell* **116,** 769–778.

Gabay, L., Lowell, S., Rubin, L. L., and Anderson, D. J. (2003). Deregulation of dorsoventral patterning by FGF confers trilineage differentiation capacity on CNS stem cells *in vitro. Neuron* **40,** 485–499.

Gambardella, L., and Barrandon, Y. (2003). The multifaceted adult epidermal stem cell. *Curr. Opin. Cell Biol.* **15,** 771–777.

Gandarillas, A., and Watt, F. M. (1997). c-Myc promotes differentiation of human epidermal stem cells. *Genes Dev.* **11,** 2869–2882.

Gat, U., DasGupta, R., Degenstein, L., and Fuchs, E. (1998). *De novo* hair follicle morphogenesis and hair tumors in mice expressing a truncated β-catenin in skin. *Cell* **95,** 605–614.

Ghazizadeh, S., and Taichman, L. B. (2001). Multiple classes of stem cells in cutaneous epithelium: A lineage analysis of adult mouse skin. *EMBO J.* **20,** 1215–1222.

Gu, H., Zou, Y. R., and Rajewsky, K. (1993). Independent control of immunoglobulin switch recombination at individual switch regions evidenced through Cre–*loxP*-mediated gene targeting. *Cell* **73,** 1155–1164.

Hardy, M. H. (1992). The secret life of the hair follicle. *Trends Genet.* **8,** 55–61.

Harfe, B. D., Scherz, P. J., Nissim, S., Tian, H., McMahon, A. P., and Tabin, C. J. (2004). Evidence for an expansion-based temporal Shh gradient in specifying vertebrate digit identities. *Cell* **118,** 517–528.

Hennings, H., Holbrook, K., Steinert, P., and Yuspa, S. (1980a). Growth and differentiation of mouse epidermal cells in culture: Effects of extracellular calcium. *Curr. Probl. Dermatol.* **10,** 3–25.

Hennings, H., Michael, D., Cheng, C., Steinert, P., Holbrook, K., and Yuspa, S. H. (1980b). Calcium regulation of growth and differentiation of mouse epidermal cells in culture. *Cell* **19,** 245–254.

Hudson, D. L., O'Hare, M., Watt, F. M., and Masters, J. R. (2000). Proliferative heterogeneity in the human prostate: Evidence for epithelial stem cells. *Lab. Invest.* **80,** 1243–1250.

Ito, M., Kizawa, K., Hamada, K., and Cotsarelis, G. (2004). Hair follicle stem cells in the lower bulge form the secondary germ, a biochemically distinct but functionally equivalent progenitor cell population, at the termination of catagen. *Differentiation* **72,** 548–557.

Ito, M., Liu, Y., Yang, Z., Nguyen, J., Liang, F., Morris, R. J., and Cotsarelis, G. (2005). Stem cells in the hair follicle bulge contribute to wound repair but not to homeostasis of the epidermis. *Nat. Med.* **11,** 1351–1354.

Jahoda, C. A., and Reynolds, A. J. (1993). Dermal–epidermal interactions: Follicle-derived cell populations in the study of hair-growth mechanisms. *J. Invest. Dermatol.* **101,** 33S–38S.

Jensen, C. H., Jauho, E. I., Santoni-Rugiu, E., Holmskov, U., Teisner, B., Tygstrup, N., and Bisgaard, H. C. (2004). Transit-amplifying ductular (oval) cells and their hepatocytic progeny are characterized by a novel and distinctive expression of delta-like protein/preadipocyte factor 1/fetal antigen 1. *Am. J. Pathol.* **164,** 1347–1359.

Jones, P. H., and Watt, F. M. (1993). Separation of human epidermal stem cells from transit amplifying cells on the basis of differences in integrin function and expression. *Cell* **73,** 713–724.

Kameda, T., Hatakeyama, S., Ma, Y. Z., Kawarada, Y., Kawamata, M., Terada, K., and Sugiyama, T. (2002). Targeted elimination of the follicular label-retaining cells by photo-induced cell killing caused a defect on follicular renewal on mice. *Genes Cells* **7,** 923–931.

Kizawa, K., and Ito, M. (2005). Characterization of epithelial cells in the hair follicle with S100 proteins. *Methods Mol. Biol.* **289,** 209–222.

Kopan, R., Lee, J., Lin, M. H., Syder, A. J., Kesterson, J., Crutchfield, N., Li, C. R., Wu, W., Books, J., and Gordon, J. I. (2002). Genetic mosaic analysis indicates that the bulb region of coat hair follicles contains a resident population of several active multipotent epithelial lineage progenitors. *Dev. Biol.* **242,** 44–57.

Lavker, R. M., Sun, T. T., Oshima, H., Barrandon, Y., Akiyama, M., Ferraris, C., Chevalier, G., Favier, B., Jahoda, C. A., Dhouailly, D., *et al.* (2003). Hair follicle stem cells. *J. Investig. Dermatol. Symp. Proc.* **8,** 28–38.

Legue, E., and Nicolas, J. F. (2005). Hair follicle renewal: Organization of stem cells in the matrix and the role of stereotyped lineages and behaviors. *Development* **132,** 4143–4154.

Levy, V., Lindon, C., Harfe, B. D., and Morgan, B. A. (2005). Distinct stem cell populations regenerate the follicle and interfollicular epidermis. *Dev. Cell* **9,** 855–861.

Li, A., Pouliot, N., Redvers, R., and Kaur, P. (2004). Extensive tissue-regenerative capacity of neonatal human keratinocyte stem cells and their progeny. *J. Clin. Invest.* **113,** 390–400.

Lichti, U., Weinberg, W. C., Goodman, L., Ledbetter, S., Dooley, T., Morgan, D., and Yuspa, S. H. (1993). *In vivo* regulation of murine hair growth: Insights from grafting defined cell populations onto nude mice. *J. Invest. Dermatol.* **101,** 124S–129S.

Liu, Y., Lyle, S., Yang, Z., and Cotsarelis, G. (2003). Keratin 15 promoter targets putative epithelial stem cells in the hair follicle bulge. *J. Invest. Dermatol.* **121,** 963–968.

Lyle, S., Christofidou-Solomidou, M., Liu, Y., Elder, D. E., Albelda, S., and Cotsarelis, G. (1998). The C8/144B monoclonal antibody recognizes cytokeratin 15 and defines the location of human hair follicle stem cells. *J. Cell Sci.* **111,** 3179–3188.

Melton, D., and Cowan, C. (2004). "Stemness": Definition, criteria, and standards. *In* "Handbook of Stem Cells" (R. Lanza, H. Blau, J. Gearhart, B. Hogan, D. Melton, M. Moore, R. Pedersen, E. Donnall Thomas, J. Thomson, C. Verfaillie, I. Weissman, and M. West, eds.), Vol. 1, pp. xxv–xxxi. Academic Press, Orlando, FL.

Michel, M., Torok, N., Godbout, M. J., Lussier, M., Gaudreau, P., Royal, A., and Germain, L. (1996). Keratin 19 as a biochemical marker of skin stem cells *in vivo* and *in vitro*: Keratin 19 expressing cells are differentially localized in function of anatomic sites, and their number varies with donor age and culture stage. *J. Cell Sci.* **109,** 1017–1028.

Morris, R. J., and Potten, C. S. (1994). Slowly cycling (label-retaining) epidermal cells behave like clonogenic stem cells *in vitro*. *Cell Prolif.* **27,** 279–289.

Morris, R. J., and Potten, C. S. (1999). Highly persistent label-retaining cells in the hair follicles of mice and their fate following induction of anagen. *J. Invest. Dermatol.* **112,** 470–475.

Morris, R. J., Liu, Y., Marles, L., Yang, Z., Trempus, C., Li, S., Lin, J. S., Sawicki, J. A., and Cotsarelis, G. (2004). Capturing and profiling adult hair follicle stem cells. *Nat. Biotechnol.* **22,** 411–417.

Niemann, C., and Watt, F. M. (2002). Designer skin: Lineage commitment in postnatal epidermis. *Trends Cell Biol.* **12,** 185–192.

Nishimura, E. K., Granter, S. R., and Fisher, D. E. (2005). Mechanisms of hair graying: Incomplete melanocyte stem cell maintenance in the niche. *Science* **307,** 720–724.

Odland, G. F. (1991). Structure of the skin. *In* "Physiology, Biochemistry, and Molecular Biology of the Skin," 2nd Ed., pp. 3–62. Oxford University Press, New York.

Ohlsson, C., Nilsson, A., Isaksson, O., and Lindahl, A. (1992). Growth hormone induces multiplication of the slowly cycling germinal cells of the rat tibial growth plate. *Proc. Natl. Acad. Sci. USA* **89,** 9826–9830.

Oshima, H., Rochat, A., Kedzia, C., Kobayashi, K., and Barrandon, Y. (2001). Morphogenesis and renewal of hair follicles from adult multipotent stem cells. *Cell* **104,** 233–245.

Panteleyev, A. A., Jahoda, C. A., and Christiano, A. M. (2001). Hair follicle predetermination. *J. Cell Sci.* **114,** 3419–3431.

Paus, R., Muller-Rover, S., Van Der Veen, C., Maurer, M., Eichmuller, S., Ling, G., Hofmann, U., Foitzik, K., Mecklenburg, L., and Handjiski, B. (1999). A comprehensive guide for the recognition and classification of distinct stages of hair follicle morphogenesis. *J. Invest. Dermatol.* **113,** 523–532.

Pellegrini, G., Dellambra, E., Golisano, O., Martinelli, E., Fantozzi, I., Bondanza, S., Ponzin, D., McKeon, F., and De Luca, M. (2001). p63 identifies keratinocyte stem cells. *Proc. Natl. Acad. Sci. USA* **98,** 3156–3161.

Potten, C. S. (2004). Keratinocyte stem cells, label-retaining cells and possible genome protection mechanisms. *J. Investig. Dermatol. Symp. Proc.* **9,** 183–195.

Potten, C. S., and Booth, C. (2002). Keratinocyte stem cells: A commentary. *J. Invest. Dermatol.* **119,** 888–899.

Raghavan, S., Bauer, C., Mundschau, G., Li, Q., and Fuchs, E. (2000). Conditional ablation of β_1 integrin in skin: Severe defects in epidermal proliferation, basement membrane formation, and hair follicle invagination. *J. Cell Biol.* **150,** 1149–1160.

Raghavan, S., Vaezi, A., and Fuchs, E. (2003). A role for $\alpha\beta_1$ integrins in focal adhesion function and polarized cytoskeletal dynamics. *Dev. Cell* **5,** 415–427.

Redvers, R. P., and Kaur, P. (2005). Serial cultivation of primary adult murine keratinocytes. *Methods Mol. Biol.* **289,** 15–22.

Rochat, A., and Barrandon, Y. (2004). Regeneration of epidermis from adult keratinocyte stem cells. *In* "Handbook of Stem Cells" (R. Lanza, H. Blau, J. Gearhart, B. Hogan, D. Melton, M. Moore, R. Pedersen, E. Donnall Thomas, J. Thomson, C. Verfaillie, I. Weissman, and M. West, eds.), Vol. 2, pp. 763–772. Academic Press, Orlando, FL.

Rogers, G. E. (2004). Hair follicle differentiation and regulation. *Int. J. Dev. Biol.* **48,** 163–170.

Sarin, K. Y., Cheung, P., Gilison, D., Lee, E., Tennen, R. I., Wang, E., Artandi, M. K., Oro, A. E., and Artandi, S. E. (2005). Conditional telomerase induction causes proliferation of hair follicle stem cells. *Nature* **436,** 1048–1052.

Sauer, B., and Henderson, N. (1988). Site-specific DNA recombination in mammalian cells by the Cre recombinase of bacteriophage P1. *Proc. Natl. Acad. Sci. USA* **85,** 5166–5170.

Silva-Vargas, V., Lo Celso, C., Giangreco, A., Ofstad, T., Prowse, D. M., Braun, K. M., and Watt, F. M. (2005). β-Catenin and Hedgehog signal strength can specify number and location of hair follicles in adult epidermis without recruitment of bulge stem cells. *Dev. Cell* **9,** 121–131.

Soriano, P. (1999). Generalized lacZ expression with the ROSA26 Cre reporter strain. *Nat. Genet.* **21,** 70–71.

Tani, H., Morris, R. J., and Kaur, P. (2000). Enrichment for murine keratinocyte stem cells based on cell surface phenotype. *Proc. Natl. Acad. Sci. USA* **97,** 10960–10965.

Taylor, G., Lehrer, M. S., Jensen, P. J., Sun, T. T., and Lavker, R. M. (2000). Involvement of follicular stem cells in forming not only the follicle but also the epidermis. *Cell* **102,** 451–461.

Trempus, C. S., Morris, R. J., Bortner, C. D., Cotsarelis, G., Faircloth, R. S., Reece, J. M., and Tennant, R. W. (2003). Enrichment for living murine keratinocytes from the hair follicle bulge with the cell surface marker CD34. *J. Invest. Dermatol.* **120,** 501–511.

Tumbar, T., and Fuchs, E. (2004). "Epithelial Skin Stem Cells." Academic Press, Orlando, FL.

Tumbar, T., Guasch, G., Greco, V., Blanpain, C., Lowry, W. E., Rendl, M., and Fuchs, E. (2004). Defining the epithelial stem cell niche in skin. *Science* **303,** 359–363.

Vasioukhin, V., Degenstein, L., Wise, B., and Fuchs, E. (1999). The magical touch: Genome targeting in epidermal stem cells induced by tamoxifen application to mouse skin. *Proc. Natl. Acad. Sci. USA* **96,** 8551–8556.

Vassar, R., Rosenberg, M., Ross, S., Tyner, A., and Fuchs, E. (1989). Tissue-specific and differentiation-specific expression of a human K14 keratin gene in transgenic mice. *Proc. Natl. Acad. Sci. USA* **86,** 1563–1567.

Watt, F. M. (1998). Epidermal stem cells: Markers, patterning and the control of stem cell fate. *Philos. Trans. R. Soc. Lond. B Biol. Sci.* **353,** 831–837.

Watt, F. M. (2002). Role of integrins in regulating epidermal adhesion, growth and differentiation. *EMBO J.* **21,** 3919–3926.

Watt, F. M. (2004). "Human Epidermal Stem Cells." Academic Press, Orlando, FL.

Watt, F. M., and Hogan, B. L. (2000). Out of Eden: Stem cells and their niches. *Science* **287,** 1427–1430.

Webb, A., Li, A., and Kaur, P. (2004). Location and phenotype of human adult keratinocyte stem cells of the skin. *Differentiation* **72,** 387–395.

[5] Dental Pulp Stem Cells

By He Liu, Stan Gronthos, and Songtao Shi

Abstract

Postnatal stem cells have been isolated from a variety of tissues. These stem cells are thought to possess great therapeutic potential for repairing damaged and/or defective tissues. Clinically, hematopoietic stem cells have been successfully used for decades in the treatment of various diseases and disorders. However, the therapeutic potential of other postnatal stem cell populations has yet to be realized, because of the lack of detailed understanding of their stem cell characteristics at the cellular and molecular levels. Furthermore, there is limited knowledge of their therapeutic value at the preclinical level. Therefore, it is necessary to develop optimal strategies and approaches to overcome the substantial challenges currently faced by researchers examining the clinical efficacy of different postnatal stem cell populations. In this review, we introduce methodologies for isolating postnatal stem cells from human dental pulp and discuss their potential role in tissue regeneration.

Introduction

A tooth can be grossly described as having two basic components: the crown and the root. The crown is gradually exposed in the oral cavity after the tooth eruption process, whereas the root is permanently embedded in

METHODS IN ENZYMOLOGY, VOL. 419
Copyright 2006, Elsevier Inc. All rights reserved.
0076-6879/06 $35.00
DOI: 10.1016/S0076-6879(06)19005-9

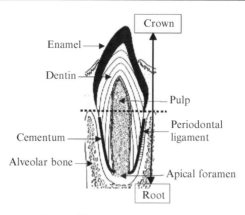

FIG. 1. Diagram of tooth structure.

the alveolar sockets and may only be exposed to the oral cavity under pathological conditions, such as periodontal tissue resorption caused by disease. A mature tooth is composed of the hard tissues enamel, dentin, and cementum, as well as a soft tissue core known as dental pulp (Fig. 1). Enamel, the hardest mineralized tissue in the human body, covers the tooth crown, whereas cementum, a thin layer of orofacial bone-like tissue, covers the root surface (Fig. 1). Underlying the enamel and cementum is the dentin, which is a specialized and vital mineralized matrix that harbors odontoblastic processes and nerve fibers and is directly linked to the pulp core by odontoblasts. Furthermore, blood vessels and nerve bundles enter the pulp through the apical foramen to provide nutrition and sensation for responding to external stimuli. The dental pulp contains connective tissue, mesenchymal cells, neural fibers, blood vessels, and lymphatics. The tooth is surrounded by condensed connective tissue called the periodontal ligament, which contains Sharpey's fibers embedded in the cementum on one side and the alveolar bone on the other to attach and suspend the tooth inside the alveolar socket (Fig. 1). Alveolar bone is in the base of the alveolar socket and maintains teeth *in situ* and supports them for masticatory function.

Enamel is formed by epithelial cell-derived ameloblasts in a process involving reciprocal induction with odontoblasts (Slavkin and Diekwisch, 1997; Thesleff and Aberg, 1999). After the completion of crown formation, ameloblasts undergo programmed cell death and totally lose the potential to repair enamel *in vivo*. In contrast, after mutual induction with ameloblasts, odontoblasts form primary dentin and subsequently line the inner surface of newly formed dentin to maintain a long cytoplasmic process inside the dentinal tubules (Mjor *et al.*, 2001; Ruch, 1998; Thesleff, 2003).

Although odontoblasts are postmitotic cells that are not able to divide and repair damaged dentin, the progenitor or stem cells of odontoblasts are capable of migrating into the dentin surface and differentiating into odonto-blasts to form reparative or tertiary dentin (About *et al.*, 2001; Batouli *et al.*, 2003; Murray *et al.*, 2000, 2001; Ruch, 1998). Unlike the primary dentin, this reparative dentin is poorly organized, with irregular dentinal tubules embedded in the dentin matrix, offering a protective barrier to the dental pulp and facilitating maintenance of the integrity and vitality of dental pulp tissue.

Dentin is composed of collagens and noncollagenous proteins (NCPs). Collagen accounts for most of the matrix and is primarily type I collagen. It is believed that collagen provides a scaffold for mineralization, whereas the NCPs initiate and regulate the process of mineralization (Butler and Ritchie, 1995; Butler *et al.*, 1997). Dentin phosphoprotein (DPP) and dentin sialoprotein (DSP) are two important NCPs involved in dentin formation and mineralization (D'Souza *et al.*, 1997; Feng *et al.*, 1998; MacDougall *et al.*, 1997). In fact, they are specific cleavage products of a single gene named dentin sialophosphoprotein (*DSPP*). Initially, DPP and DSP were considered to be dentin-specific proteins (Ritchie *et al.*, 1994), but later it was found that other tissues such as bone, periodontium, and kidney tissue also express DSP, but at a much lower level in comparison with the expression in dentin and odontoblasts (Baba *et al.*, 2004; Ogbureke and Fisher, 2005; Qin *et al.*, 2003). Previously, DSP has been used as a marker to differentiate odontoblasts from their stem progenitors and osteoblastic cells, because the latter two cell populations express undetectable levels of DSP based on immunohistochemical staining (Gronthos *et al.*, 2002).

Identification of Dental Pulp Stem Cells

Tooth regeneration is one of the ultimate goals of restoring the loss of natural teeth. Studies have indicated that cell-based strategies show promising potential for regenerating the whole tooth structure in rodents (Chai and Slavkin, 2003; Duailibi *et al.*, 2004; Ohazama *et al.*, 2004; Yen and Sharpe, 2006). Moreover, stem cell-based regeneration of human tooth structures has been achieved in immunocompromised mouse models (Gronthos *et al.*, 2000, 2002; Seo *et al.*, 2004). So far, several human tooth-associated stem cell populations, including dental pulp stem cells (DPSCs), periodontal ligament stem cells (PDLSCs), and stem cells from human exfoliated deciduous teeth (SHED), have been isolated from dental pulp and periodontal ligament tissues (Gronthos *et al.*, 2000; Miura *et al.*, 2003; Seo *et al.*, 2004). More importantly, DPSCs and PDLSCs are capable of forming a dentin–pulp complex and cementum–periodontal ligament, respectively, when transplanted subcutaneously into immunocompromised mice (Gronthos *et al.*, 2000; Seo *et al.*, 2004), demonstrating their potential for regenerating human dental tissues *in vivo*. In addition, SHED are able to form significant amounts of bone *in vivo*, providing an alternative population of postnatal stem cells for alveolar and orofacial bone regeneration (Miura *et al.*, 2003). However, to achieve the goal of functional biotooth regeneration, a more thorough study of the intricate processes of tooth growth and development is necessary, in order to fully understand the processes of mutual induction between odontoblasts and ameloblasts and the complicated multiple structures of tooth composition (Slavkin and Diekwisch, 1997; Thesleff and Aberg, 1999).

Researchers have recognized that dental pulp contains *ex vivo*-expandable cells called dental pulp cells. These cells express osteogenic markers, such as alkaline phosphatase, type I collagen, bone sialoprotein, osteocalcin, osteopontin, transforming growth factor β(TGF-β), and bone morphogenetic proteins (BMPs) (Chen *et al.*, 2005; Kuo *et al.*, 1992; Nakashima *et al.*, 1994; Pavasant *et al.*, 2003; Shiba *et al.*, 1998). They also respond to the induction of BMP, fibroblast growth factor 2 (FGF-2), matrix extracellular phosphoglycoprotein (MEPE), and TGF-β by undergoing osteogenic differentiation (Alliot-Licht *et al.*, 2005; Dobie *et al.*, 2002; Iohara *et al.*, 2004; Liu *et al.*, 2005; Nakao *et al.*, 2004; Saito *et al.*, 2004; Unda *et al.*, 2000). Collectively, this experimental evidence suggests that dental pulp cells might be similar to osteoblast-like cells in terms of expressing bone markers and forming mineralized nodules when cultured under osteoinductive conditions. It was discovered that dental pulp cells might represent a distinct population of cells because of their unique ability to form specific crystalline structures in mineralized nodules, similar to physiological dentin but different from bone structures (About *et al.*, 2000a). In addition, dental pulp cells express high levels of DSP, which is usually expressed at a low level in osteogenic cells (Batouli *et al.*, 2003; Gronthos *et al.*, 2002). On the basis of microarray analysis, multiple genes are differentially expressed between dental pulp cells and osteogenic cells (Shi *et al.*, 2001). For instance, cyclin-dependent kinase 6 and D-cyclins are highly expressed in dental pulp cells; these molecules are able to promote cell proliferation. In contrast, *IGFBP7*, an inhibitor gene of cell growth, is highly expressed in osteogenic cells (Shi *et al.*, 2001). This evidence may account for the fact that dental pulp cells show a high proliferation capacity compared with osteogenic cells (Gronthos *et al.*, 2000). Given what we know about the unique characteristics of dental pulp cells and their capacity to form reparative dentin *in vivo*, it is reasonable to presume that dental pulp may contain stem progenitors. Moreover, it may be possible to isolate these stem cells by strategies analogous to those for retrieving mesenchymal stem cells from bone marrow aspirates (Gronthos *et al.*, 2000; Shi and Gronthos, 2003).

Human DPSCs were initially identified on the basis of their traits of forming single colonies in culture, self-renewal *in vivo*, and multidifferentiation *in vitro* (Gronthos *et al.*, 2000). One of the most important advances in DPSC research was the revelation of their stem cell niche in the perivascular region (Shi and Gronthos, 2003; Shi *et al.*, 2005). Extensive immunophenotyping of *ex vivo*-expanded DPSCs demonstrated their expression of various markers associated with endothelial and/or smooth muscle cells, such as stromal-derived factor 1 (STRO-1), vascular cell adhesion molecule 1 (VCAM-1), melanoma-associated antigen/mucin 18 (MUC-18), and smooth muscle α-actin (Gronthos *et al.*, 2000). In addition, smooth muscle α-actin-positive cells have also been detected close to mineralized deposits in human dental pulp cultures (Alliot-Licht *et al.*, 2001). It is hoped that further characterization of DPSCs by current molecular technology will provide novel markers that will be useful in their identification *in situ*, and isolation and purification for *ex vivo* expansion.

Isolation of DPSCs

Dental pulp tissues can be easily collected from clinically extracted human teeth and the individual cells can be enzymatically released from the pulp tissue (Gronthos *et al.*, 2000). After plating at a low density, DPSCs are able to attach to the culture dish and form a single-colony cluster after 10–14 days of culture (Gronthos *et al.*, 2000, 2002). The colonies contain fibroblast-like cells similar to colony-forming unit-fibroblasts (CFU-F) formed by bone marrow mesenchymal stem cells (BMMSCs) (Castro-Malaspina *et al.*, 1980; Friedenstein, 1976). At plating densities of $0.1–2.5 \times 10^4$ cells, the frequency of colony-forming cells derived from dental pulp tissue is 22–70 colonies/10^4 cells (Gronthos *et al.*, 2000). Single-colony-derived cells can be selected with a cloning ring and subjected to secondary *ex vivo* culture to expand cell numbers for further experimental requirements. Interestingly, individual colonies demonstrated marked differences in their proliferation rates according to a bromodeoxyuridine (BrdU) uptake assay. Only 20% of colonies are able to proliferate beyond 20 population doublings under regular culture conditions (Gronthos *et al.*, 2002). This result implies that the progeny of this small percentage of highly proliferative colonies will finally dominate the cell population in multicolony-derived cells. These data, combined with immunohistochemistry studies showing the uneven expression of markers between subsets of cells, support the concept that DPSCs are a heterogeneous population of postnatal stem cells akin to BMMSCs.

STRO-1 has been shown to be an early marker of various mesenchymal stem cell populations (Gronthos *et al.*, 1994, 2003). At present, on the basis of the high expression of STRO-1 in combination with the expression of CD146 (MUC-18) and VCAM-1, selected BMMSCs show characteristics of high-purity stem cells (Gronthos *et al.*, 2003; Shi and Gronthos, 2003). Because of the limited cell numbers in dental pulp, several pulp tissues obtained from three or four different third molars must be pooled. Subsequently, immunomagnetic bead selection is carried out to assess whether STRO-1 or CD146 antigens can be used to select for the highly purified DPSC populations. Results show that the majority of colony-forming cells (82%) are represented in the minor STRO-1$^+$ cell fraction, 6-fold greater than in unfractionated pulp cells, whereas a high proportion (96%) of CFU-F cells is present in the CD146$^+$ population, 7-fold greater than in unfractionated pulp cells (Shi and Gronthos, 2003). Purified DPSC populations have been subsequently expanded *in vitro* and then transplanted into immunocompromised mice and show a similar capacity for regenerating the dentin–pulp complex compared with multicolony-derived DPSCs.

Protocol for DPSC Isolation

1. Collect extracted normal human teeth and place into a sterile container containing saline. Third molars are the most common resource for dental stem cell isolation in dental clinics because of the widespread extraction of wisdom teeth. As the third molar is the last tooth developed in humans, it is usually at an earlier developmental stage and is capable of providing an optimal amount of dental pulp tissue for DPSC isolation.

2. Scrape the gingival and periodontal tissue, if there is any, from the tooth surface. Clean the tooth surfaces with iodine and 70% ethanol to prevent contamination from oral bacteria. Wash five times with 1× phosphate-buffered saline (PBS) to remove iodine and ethanol.

3. Use sterilized high-speed dental fissure burrs to cut around the cementum–enamel junction to separate the crown and root and expose the pulp tissue in the pulp chamber. Alternatively, it may be possible to wrap the tooth in sterile gauze and squeeze it in a vise until cracked, to expose the pulp tissue.

4. Pick up the pulp tissue with sterile tweezers and mince it, using a scalpel blade under moist conditions (α-MEM). Digest the minced pulp tissue in a solution containing collagenase type I (3 mg/ml) and dispase (4 mg/ml) for 30–45 min at 37° to totally digest the pulp tissue. It is important to always keep the pulp pellet moist during the DPSC isolation procedures.

5. After digestion, add 5 volumes of culture medium (α-MEM) containing 10% serum. After centrifugation at 1200 rpm for 10 min, suspend the cell pellet in regular culture medium and pass the cells through a 70-μm pore size strainer to obtain single-cell suspensions.

6. Seed the cells into culture plates, dishes, or flasks with α-MEM supplemented with 10% fetal calf serum (FCS), 100 μM L-ascorbic acid 2-phosphate, 2 mM L-glutamine, penicillin (100 U/ml), and streptomycin (100 mg/ml), and then incubate the cells at 37° with 5% CO_2.

7. Cells grow slowly in the initial stage, with attachment to the culture dish. Single colonies can be identified after 10–14 days of culture if DPSCs are plated at low density.

8. To select purified DPSCs, several pulp tissues obtained from three or four third molars should be pooled to obtain a sufficient number of cells for magnetic bead sorting. After collagenase–dispase digestion the pulp cells should be washed twice and then incubated with 0.5 ml of either STRO-1 (mouse IgM, anti-BMSSC), 3G5 (mouse IgM, anti-pericyte), or CC9 (mouse IgG2a, anti-CD146) (Shi and Gronthos, 2003) for 1 h on ice. Supernatants are used undiluted, whereas purified antibodies are used at 20 μg/ml. After washing twice with PBS–1% bovine serum albumin, the cells are incubated with either sheep anti-mouse IgG-conjugated or rat anti-mouse IgM-conjugated magnetic Dynabeads (Invitrogen, Carlsbad, CA) for 1 h on a rotary mixer at 4°. Separation of the bead-positive cells is performed with a Dynal MPC-1 magnetic particle concentrator (Invitrogen). The magnetic wash step is repeated three times to reduce contamination with bead-negative cells. After immunomagnetic selection, single-cell suspensions of bead-positive dental pulp cells are cultured as described above.

Differentiation of DPSCs

One of the most important characteristics of stem cells is their capacity to differentiate into multiple cell lineages. To prove that DPSCs satisfy this stem cell property, a variety of inductive culture conditions can be used to assess whether DPSCs can differentiate into multiple types of cells *in vitro*.

Previous studies have showed that cells derived from dental pulp are capable of forming mineralized nodules *in vitro* in the presence of inductive media containing ascorbic acid, dexamethasone, and an excess of phosphate (Hanks *et al.*, 1998; Tsukamoto *et al.*, 1992; Unda *et al.*, 2000; Yokose *et al.*, 2000). The use of methodology such as infrared microspectroscopic examination and X-ray diffraction electron microscopy has confirmed the dentin-like nature of the crystalline structures that comprise the mineralized nodules *in vitro*, which are distinct from the crystal structures of mineralized enamel and bone *in vivo* (About *et al.*, 2000a). Furthermore, we demonstrated that *ex vivo*-expanded human DPSCs are capable not only of forming mineralized nodules under osteoinductive culture conditions, but also of generating a dentin/pulp-like complex *in vivo* in conjunction with hydroxyapatite/tricalcium phosphate (HA/TCP) as a carrier

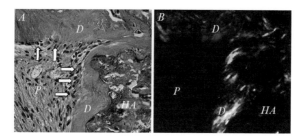

FIG. 2. Eight weeks posttransplantation, DPSCs are capable of differentiating into odontoblasts (arrows) that are responsible for dentin formation (*D*) on the surface of the hydroxyapatite/tricalcium (*HA*) carrier. Pulp-like tissue (*P*) is generated with newly formed dentin as shown by hematoxylin–eosin staining (A) and polarizing light microscopy (B). (See color insert.)

vehicle (Fig. 2) (Gronthos *et al.*, 2000, 2002). These data clearly indicated the tissue regeneration capacity of DPSCs.

Studies have explored whether DPSCs possess the ability to self-renew. To answer this question, we harvested primary DPSC implants at 2 months posttransplantation and liberated the cells by enzymatic digestion for subsequent expansion *in vitro*. Donor human cells were isolated from the cultures by fluorescence-activated cell sorting (FACS), using a human β_1-integrin-specific monoclonal antibody, and then retransplanted into immunodeficient mice for 2 months. Recovered secondary transplants yielded the same dentin/pulp-like structures as observed in the primary transplants (Gronthos *et al.*, 2002). In addition, human DSP was found to be present in the dentin matrix by immunohistochemical staining, and *in situ* hybridization studies confirmed the human origin of the odontoblast/pulp cells contained within the secondary DPSC transplants (Gronthos *et al.*, 2002).

To determine whether DPSCs represent multipotent stem cells, we cultured the DPSCs in various inductive media previously shown to promote the differentiation of adipocytes. Our data suggest that DPSCs are capable of forming adipocytes when cultured with a potent cocktail of adipogenic agents (0.5 m*M* methylisobutylxanthine, 0.5 μM hydrocortisone, 60 μM indomethacin) (Gimble *et al.*, 1995), showing the presence of oil red O-positive fat-containing adipocytes in DPSC culture after several weeks of induction (Gronthos *et al.*, 2002). This was also correlated with an upregulation of the early adipogenic master regulatory gene, *PPARγ2*, and the mature adipocyte marker, lipoprotein lipase, using reverse transcription-polymerase chain reaction (RT-PCR) (Gronthos *et al.*, 2002). These observations highlight the flexibility of the DPSC population to develop into functional stromal cell types not normally associated with dental pulp tissue.

During development, odontoblasts are presumed to originate from craniofacial neural crest cells (Chai *et al.*, 2000). Investigations have explored the possibility that DPSCs have the potential to differentiate into neural-like cells (Gronthos *et al.*, 2002; Miura *et al.*, 2003; Nosrat *et al.*, 2004). *Ex vivo*-expanded DPSCs were shown to constitutively express nestin, an early marker of neural precursor cells, and glial fibrillary acid protein (GFAP), an antigen characteristic of glial cells (Fig. 3). In accord with these findings, other investigators have identified the same markers in dental pulp tissue *in situ* (About *et al.*, 2000b; Hainfellner *et al.*, 2001). When DPSCs were cultured under defined neural-inductive conditions, there was enhanced expression of both nestin and GFAP. Morphological assessment of induced DPSCs identified long cytoplasmic processes protruding from rounded cell bodies, in contrast to their usual bipolar fibroblastic-like appearance. Moreover, dental pulp cells under neural-inductive culture conditions were

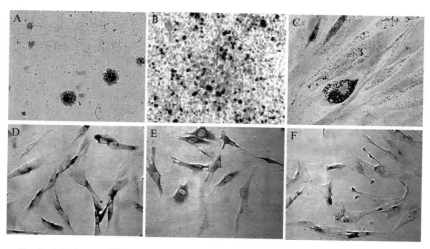

Fig. 3. Multipotent differentiation of DPSCs. DPSCs recovered from human dental pulp were capable of forming heterogeneous single-colony clusters after being plated at low density and cultured with regular culture medium for 10 days (A). DPSCs were cultured with L-ascorbate 2-phosphate, dexamethasone, and inorganic phosphate for 4 weeks. Alizarin red staining showed mineralized nodule formation (B). DPSCs were able to form oil red O-positive lipid clusters after 5 weeks of induction in the presence of 0.5 mM isobutylmethylxanthine, 0.5 μM hydrocortisone, and 60 μM indomethacin (C). Immunocytochemical staining depicts cultured DPSCs expressing nestin (D), GFAP (E), and neurofilament M (F), with culture medium containing Neurobasal A (Invitrogen GIBCO-BRL, Grand Island, NY), B27 supplement (GIBCO-BRL), 1% penicillin, epidermal growth factor (EGF, 20 ng/ml; BD Biosciences Discovery Labware, Bedford, MA), and FGF (40 ng/ml; BD Biosciences Discovery Labware). (See color insert.)

found to express the neuron-specific marker NeuN (neuronal nuclei) by immunohistochemical staining (Gronthos et al., 2002; Miura et al., 2003). These studies provide the first experimental evidence that adult human DPSCs may possess the potential to differentiate into neural-like cells, with the expression of nestin, GFAP, and NeuN in vitro.

To determine the capacity of ex vivo-expanded DPSCs to generate a functional dentin/pulp-like tissue in vivo, we used an established transplantation system previously optimized for the formation of ectopic bone by cultured BMMSCs (Krebsbach et al., 1997; Kuznetsov et al., 1997). Previously, it was demonstrated that rodent or bovine developing dental papilla tissues are capable of forming ectopic dentin in vivo (Holtgrave and Donath, 1995; Ishizeki et al., 1990; Lyaruu et al., 1999; Prime and Reade, 1980). However, similar studies using human intact developing dental papilla or adult dental pulp tissue failed to generate a mineralized dentin matrix and/or odontoblast-like cells after transplantation into immunocompromised mice (Gimble et al., 1995; Prime et al., 1982). It is likely that human mesenchymal cells require a suitable conductive carrier such as HA/TCP particles to initiate the mineralization process (Holtgrave and Donath, 1995; Krebsbach et al., 1997). HA/TCP and other biomaterials have also been clinically used, with partial success, to stimulate a pulpal proliferation response to aid in the repair of damaged dentin (Kaigler and Mooney, 2001; Levin, 1998).

Typical DPSC transplants developed areas of vascularized pulp tissue surrounded by a well-defined layer of odontoblast-like cells aligned around mineralized dentin, with their processes extending into tubular structures. The odontoblast-like cells and fibrous pulp tissue in the transplants were shown to have originated from the donor material by their reactivity to the human-specific Alu cDNA probe (Gronthos et al., 2000). In addition, orientation of the collagen fibers within the dentin was characteristic of ordered primary dentin, perpendicular to the odontoblast layer. Back-scatter electron microscopy analysis demonstrated that the dentin-like material formed in the transplants had a globular appearance consistent with the structure of dentin in situ. Moreover, the presence of human DSP detected in the transplants confirmed the ability of DPSCs to regenerate a human dentin–pulp microenvironment in vivo (Gronthos et al., 2002).

Protocol for DPSC Transplantation

1. Adherent cultures are washed once with serum-free PBS and the cells are liberated by the addition of 2 ml of 0.5% trypsin–EDTA solution per 75-cm^2 flask for 5–10 min at 37°.

2. Single-cell suspensions of DPSCs are washed in growth medium, by centrifugation, and then resuspended in 1 ml of culture medium.

3. Previous studies have shown that human dental pulp cells need a carrier to induce ectopic dentin formation. Hydroxyapatite/tricalcium phosphate (HA/TCP) particles have been identified as a suitable carrier for human DPSCs. *Ex vivo*-expanded DPSCs ($\sim$2.0–4.0 $\times$ 10^6) are mixed with 40 mg of HA/TCP ceramic particles and then incubated at 37° for 90 min under rotation at 25 rpm. After centrifugation at 1200 rpm for 15 min, the supernatant is discarded and DPSC–HA/TCP particles are suitable for transplantation.

4. Immunocompromised *bg/nu/xid* mice (8–12 weeks old) are used for the transplantation. The mice are first anesthetized by intraperitoneal injection with 2.5% tribromoethanol at 0.018 ml/g body weight. Approximately 1-cm-long midlongitudinal skin incisions are then prepared on the dorsal surface of recipient mice and blunt dissection is used to produce subcutaneous pockets for transplantation of DPSC–HA/TCP particles.

5. Animal wound clips are then used to seal the wound.

6. The transplants are recovered 8–12 weeks posttransplantation, fixed with 4% formalin, decalcified with buffered 10% EDTA (pH 8.0), and then embedded in paraffin. Sections (5 μm) are deparaffinized for histological and immunological analysis by hematoxylin–eosin and immunohistochemical staining.

7. Human-specific Alu and murine-specific pF1 sequences labeled with digoxigenin are used as probes for *in situ* hybridization as previously described (Gronthos *et al.*, 2000). The primers used were as follows. Human Alu: sense, 5'-TGGCTCACGCCTGTAATCC-3' (base numbers 90–108); antisense, 5'-TTTTTTGAGACGGAGTCTCGC-3' (base numbers 344–364) (GenBank accession number AC004024). Murine pF1: sense, 5'-CCGGGCAGTGGT GGCGCATGCCTTTAAATCCC-3' (base numbers 170–201); antisense, 5'-GTTTGGTTTTTGAGCAGGGTTCTCTGTGTAGC-3' (base numbers 275–306) (GenBank accession number X78319). The probes are prepared by PCR with 1$\times$ PCR buffer (Applied Biosystems, Foster City, CA), 0.1 mM dATP, 0.1 mM dCTP, 0.1 mM dGTP, 0.065 mM dTTP, 0.035 mM digoxigenin-11-dUTP, 10 pmol of specific primers, and 100 ng of human genomic DNA as templates. Unstained sections are deparaffinized and hybridized with the digoxigenin-labeled Alu probe, using an mRNAlocator-Hyb kit (catalog no. 1800; Ambion, Austin, TX). After hybridization, the presence of Alu or pF1 in tissue sections is detected by immunoreactivity with an anti-digoxigenin alkaline phosphatase-conjugated Fab fragment (Roche, Indianapolis, IN).

References

About, I., Bottero, M. J., de Denato, P., Camps, J., Franquin, J. C., and Mitsiadis, T. A. (2000a). Human dentin production *in vitro*. *Exp. Cell Res.* **258**, 33–41.

About, I., Laurent-Maquin, D., Lendahl, U., and Mitsiadis, T. A. (2000b). Nestin expression in embryonic and adult human teeth under normal and pathological conditions. *Am. J. Pathol.* **157**, 287–295.

About, I., Murray, P. E., Franquin, J. C., Remusat, M., and Smith, A. J. (2001). The effect of cavity restoration variables on odontoblast cell numbers and dental repair. *J. Dent.* **29**, 109–117.

Alliot-Licht, B., Hurtrel, D., and Gregoire, M. (2001). Characterization of α-smooth muscle actin positive cells in mineralized human dental pulp cultures. *Arch. Oral Biol.* **46**, 221–228.

Alliot-Licht, B., Bluteau, G., Magne, D., Lopez-Cazaux, S., Lieubeau, B., Daculsi, G., and Guicheux, J. (2005). Dexamethasone stimulates differentiation of odontoblast-like cells in human dental pulp cultures. *Cell Tissue Res.* **321**, 391–400.

Baba, O., Qin, C., Brunn, J. C., Jones, J. E., Wygant, J. N., McIntyre, B. W., and Butler, W. T. (2004). Detection of dentin sialoprotein in rat periodontium. *Eur. J. Oral Sci.* **112**, 163–170.

Batouli, S., Miura, M., Brahim, J., Tsutsui, T. W., Fisher, L. W., Gronthos, S., Robey, P. G., and Shi, S. (2003). Comparison of stem-cell-mediated osteogenesis and dentinogenesis. *J. Dent. Res.* **82**, 976–981.

Butler, W. T., and Ritchie, H. (1995). The nature and functional significance of dentin extracellular matrix proteins. *Int. J. Dev. Biol.* **39**, 169–179.

Butler, W. T., Ritchie, H. H., and Bronckers, A. L. (1997). Extracellular matrix proteins of dentine. *Ciba Found. Symp.* **205**, 107–115.

Castro-Malaspina, H., Gay, R. E., Resnick, G., Kapoor, N., Meyers, P., Chiarieri, D., McKenzie, S., Broxmeyer, H. E., and Moore, M. A. (1980). Characterization of human bone marrow fibroblast colony-forming cells (CFU-F) and their progeny. *Blood* **56**, 289–301.

Chai, Y., and Slavkin, H. C. (2003). Prospects for tooth regeneration in the 21st century: A perspective. *Microsc. Res. Tech.* **60**, 469–479.

Chai, Y., Jiang, X., Ito, Y., Bringas, P., Jr., Han, J., Rowitch, D. H., Soriano, P., McMahon, A. P., and Sucov, H. M. (2000). Fate of the mammalian cranial neural crest during tooth and mandibular morphogenesis. *Development* **127**, 1671–1679.

Chen, S., Santos, L., Wu, Y., Vuong, R., Gay, I., Schulze, J., Chuang, H. H., and MacDougall, M. (2005). Altered gene expression in human cleidocranial dysplasia dental pulp cells. *Arch. Oral Biol.* **50**, 227–236.

Dobie, K., Smith, G., Sloan, A. J., and Smith, A. J. (2002). Effects of alginate hydrogels and TGF-β_1 on human dental pulp repair *in vitro. Connect. Tissue Res.* **43**, 387–390.

D'Souza, R. N., Cavender, A., Sunavala, G., Alvarez, J., Ohshima, T., Kulkarni, A. B., and MacDougall, M. (1997). Gene expression patterns of murine dentin matrix protein 1 (Dmp1) and dentin sialophosphoprotein (DSPP) suggest distinct developmental functions *in vivo. J. Bone Miner. Res.* **12**, 2040–2049.

Duailibi, M. T., Duailibi, S. E., Young, C. S., Bartlett, J. D., Vacanti, J. P., and Yelick, P. C. (2004). Bioengineered teeth from cultured rat tooth bud cells. *J. Dent. Res.* **83**, 523–528.

Feng, J. Q., Luan, X., Wallace, J., Jing, D., Ohshima, T., Kulkarni, A. B., D'Souza, R. N., Kozak, C. A., and MacDougall, M. (1998). Genomic organization, chromosomal mapping, and promoter analysis of the mouse dentin sialophosphoprotein (Dspp) gene, which codes for both dentin sialoprotein and dentin phosphoprotein. *J. Biol. Chem.* **273**, 9457–9464.

Friedenstein, A. J. (1976). Precursor cells of mechanocytes. *Int. Rev. Cytol.* **47**, 327–359.

Gimble, J. M., Morgan, C., Kelly, K., Wu, X., Dandapani, V., Wang, C. S., and Rosen, V. (1995). Bone morphogenetic proteins inhibit adipocyte differentiation by bone marrow stromal cells. *J. Cell. Biochem.* **58**, 393–402.

Gronthos, S., Graves, S. E., Ohta, S., and Simmons, P. J. (1994). The STRO-1$^+$ fraction of adult human bone marrow contains the osteogenic precursors. *Blood* **84**, 4164–4173.

Gronthos, S., Mankani, M., Brahim, J., Robey, P. G., and Shi, S. (2000). Postnatal human dental pulp stem cells (DPSCs) *in vitro* and *in vivo. Proc. Natl. Acad. Sci. USA* **97**, 13625–13630.

Gronthos, S., Brahim, J., Li, W., Fisher, L. W., Cherman, N., Boyde, A., DenBesten, P., Robey, P. G., and Shi, S. (2002). Stem cell properties of human dental pulp stem cells. *J. Dent. Res.* **81,** 531–535.

Gronthos, S., Zannettino, A. C., Hay, S.J, Shi, S., Graves, S. E., Kortesidis, A., and Simmons, P. J. (2003). Molecular and cellular characterisation of highly purified stromal stem cells derived from human bone marrow. *J. Cell Sci.* **116,** 1827–1835.

Hainfellner, J. A., Voigtlander, T., Strobel, T., Mazal, P. R., Maddalena, A. S., Aguzzi, A., and Budka, H. (2001). Fibroblasts can express glial fibrillary acidic protein (GFAP) *in vivo. J. Neuropathol. Exp. Neurol.* **60,** 449–461.

Hanks, C. T., Sun, Z. L., Fang, D. N., Edwards, C. A., Wataha, J. C., Ritchie, H. H., and Butler, W. T. (1998). Cloned 3T6 cell line from CD-1 mouse fetal molar dental papillae. *Connect. Tissue Res.* **37,** 233–249.

Holtgrave, E. A., and Donath, K. (1995). Response of odontoblast-like cells to hydroxyapatite ceramic granules. *Biomaterials* **16,** 155–159.

Iohara, K., Nakashima, M., Ito, M., Ishikawa, M., Nakasima, A., and Akamine, A. (2004). Dentin regeneration by dental pulp stem cell therapy with recombinant human bone morphogenetic protein 2. *J. Dent. Res.* **83,** 590–595.

Ishizeki, K., Nawa, T., and Sugawara, M. (1990). Calcification capacity of dental papilla mesenchymal cells transplanted in the isogenic mouse spleen. *Anat. Rec.* **226,** 279–287.

Kaigler, D., and Mooney, D. (2001). Tissue engineering's impact on dentistry. *J. Dent. Educ.* **65,** 456–462.

Krebsbach, P. H., Kuznetsov, S. A., Satomura, K., Emmons, R. V., Rowe, D. W., and Robey, P. G. (1997). Bone formation *in vivo*: Comparison of osteogenesis by transplanted mouse and human marrow stromal fibroblasts. *Transplantation* **63,** 1059–1069.

Kuo, M. Y., Lan, W. H., Lin, S. K., Tsai, K. S., and Hahn, L. J. (1992). Collagen gene expression in human dental pulp cell cultures. *Arch. Oral Biol.* **37,** 945–952.

Kuznetsov, S. A., Krebsbach, P. H., Satomura, K., Kerr, J., Riminucci, M., Benayahu, D., and Robey, P. G. (1997). Single-colony derived strains of human marrow stromal fibroblasts form bone after transplantation *in vivo. J. Bone Miner. Res.* **12,** 1335–1347.

Levin, L. G. (1998). Pulpal regeneration. *Pract. Periodontics Aesthet. Dent.* **10,** 621–624.

Liu, H., Li, W., Shi, S., Habelitz, S., Gao, C., and Denbesten, P. (2005). MEPE is downregulated as dental pulp stem cells differentiate. *Arch. Oral Biol.* **50,** 923–928.

Lyaruu, D. M., van Croonenburg, E. J., van Duin, M. A., Bervoets, T. J., Woltgens, J. H., and de Blieck-Hogervorst, J. M. (1999). Development of transplanted pulp tissue containing epithelial sheath into a tooth-like structure. *J. Oral Pathol. Med.* **28,** 293–296.

MacDougall, M., Simmons, D., Luan, X., Nydegger, J., Feng, J., and Gu, T. T. (1997). Dentin phosphoprotein and dentin sialoprotein are cleavage products expressed from a single transcript coded by a gene on human chromosome 4: Dentin phosphoprotein DNA sequence determination. *J. Biol. Chem.* **272,** 835–842.

Miura, M., Gronthos, S., Zhao, M., Lu, B., Fisher, L. W., Robey, P. G., and Shi, S. (2003). SHED: Stem cells from human exfoliated deciduous teeth. *Proc. Natl. Acad. Sci. USA* **100,** 5807–5812.

Mjor, I. A., Sveen, O. B., and Heyeraas, K. J. (2001). Pulp–dentin biology in restorative dentistry. 1. Normal structure and physiology. *Quintessence Int.* **32,** 427–446.

Murray, P. E., About, I., Lumley, P. J., Smith, G., Franquin, J. C., and Smith, A. J. (2000). Postoperative pulpal and repair responses. *J. Am. Dent. Assoc.* **131,** 321–329.

Murray, P. E., About, I., Franquin, J. C., Remusat, M., and Smith, A. J. (2001). Restorative pulpal and repair responses. *J. Am. Dent. Assoc.* **132,** 482–491.

Nakao, K., Itoh, M., Tomita, Y., Tomooka, Y., and Tsuji, T. (2004). FGF-2 potently induces both proliferation and DSP expression in collagen type I gel cultures of adult incisor immature pulp cells. *Biochem. Biophys. Res. Commun.* **325,** 1052–1059.

Nakashima, M., Nagasawa, H., Yamada, Y., and Reddi, A. H. (1994). Regulatory role of transforming growth factor-β, bone morphogenetic protein-2, and protein-4 on gene expression of extracellular matrix proteins and differentiation of dental pulp cells. *Dev. Biol.* **162,** 18–28.

Nosrat, I. V., Smith, C. A., Mullally, P., Olson, L., and Nosrat, C. A. (2004). Dental pulp cells provide neurotrophic support for dopaminergic neurons and differentiate into neurons *in vitro*: Implications for tissue engineering and repair in the nervous system. *Eur. J. Neurosci.* **19,** 2388–2398.

Ogbureke, K. U., and Fisher, L. W. (2005). Renal expression of SIBLING proteins and their partner matrix metalloproteinases (MMPs). *Kidney Int.* **68,** 155–166.

Ohazama, A., Modino, S. A., Miletich, I., and Sharpe, P. T. (2004). Stem-cell-based tissue engineering of murine teeth. *J. Dent. Res.* **83,** 518–522.

Pavasant, P., Yongchaitrakul, T., Pattamapun, K., and Arksornnukit, M. (2003). The synergistic effect of TGF-β and 1,25-dihydroxyvitamin D_3 on SPARC synthesis and alkaline phosphatase activity in human pulp fibroblasts. *Arch. Oral Biol.* **48,** 717–722.

Prime, S. S., and Reade, P. C. (1980). Xenografts of recombined bovine odontogenic tissues and cultured cells to hypothymic mice. *Transplantation* **30,** 149–152.

Prime, S. S., Sim, F. R., and Reade, P. C. (1982). Xenografts of human ameloblastoma tissue and odontogenic mesenchyme to hypothymic mice. *Transplantation* **33,** 561–562.

Qin, C., Brunn, J. C., Cadena, E., Ridall, A., and Butler, W. T. (2003). Dentin sialoprotein in bone and dentin sialophosphoprotein gene expressed by osteoblasts. *Connect. Tissue Res.* **44**(Suppl. 1), 179–183.

Ritchie, H. H., Hou, H., Veis, A., and Butler, W. T. (1994). Cloning and sequence determination of rat dentin sialoprotein, a novel dentin protein. *J. Biol. Chem.* **269,** 3698–3702.

Ruch, J. V. (1998). Odontoblast commitment and differentiation. *Biochem. Cell Biol.* **76,** 923–938.

Saito, T., Ogawa, M., Hata, Y., and Bessho, K. (2004). Acceleration effect of human recombinant bone morphogenetic protein-2 on differentiation of human pulp cells into odontoblasts. *J Endod.* **30,** 205–208.

Seo, B. M., Miura, M., Gronthos, S., Bartold, P. M., Batouli, S., Brahim, J., Young, M., Robey, P. G., Wang, C. Y., and Shi, S. (2004). Investigation of multipotent postnatal stem cells from human periodontal ligament. *Lancet* **364,** 149–155.

Shi, S., and Gronthos, S. (2003). Perivascular niche of postnatal mesenchymal stem cells in human bone marrow and dental pulp. *J. Bone Miner. Res.* **18,** 696–704.

Shi, S., Robey, P. G., and Gronthos, S. (2001). Comparison of gene expression profiles between human bone marrow stromal cells and human dental pulp stem cells by microarray analysis. *Bone* **29,** 532–539.

Shi, S., Bartold, P. M., Miura, M., Seo, B. M., Robey, P., and Gronthos, S. (2005). The efficacy of mesenchymal stem cells to regenerate and repair dental structures. *Orthod. Craniofacial Res.* **8,** 191–199.

Shiba, H., Fujita, T., Doi, N., Nakamura, S., Nakanishi, K., Takemoto, T., Hino, T., Noshiro, M., Kawamoto, T., Kurihara, H., and Kato, Y. (1998). Differential effects of various growth factors and cytokines on the syntheses of DNA, type I collagen, laminin, fibronectin, osteonectin/secreted protein, acidic and rich in cysteine (SPARC), and alkaline phosphatase by human pulp cells in culture. *J. Cell. Physiol.* **174,** 194–205.

Slavkin, H. C., and Diekwisch, T. G. (1997). Molecular strategies of tooth enamel formation are highly conserved during vertebrate evolution. *Ciba Found. Symp.* **205,** 73–80.

Thesleff, I. (2003). Developmental biology and building a tooth. *Quintessence Int.* **34,** 613–620.

Thesleff, I., and Aberg, T. (1999). Molecular regulation of tooth development. *Bone* **25,** 123–125.

Tsukamoto, Y., Fukutani, S., Shin-Ike, T., Kubota, T., Sato, S., Suzuki, Y., and Mori, M. (1992). Mineralized nodule formation by cultures of human dental pulp-derived fibroblasts. *Arch. Oral Biol.* **37,** 1045–1055.

Unda, F. J., Martin, A., Hilario, E., Begue-Kirn, C., Ruch, J. V., and Arechaga, J. (2000). Dissection of the odontoblast differentiation process *in vitro* by a combination of FGF1, FGF2, and TGFβ_1. *Dev. Dyn.* **218,** 480–489.

Yen, A. H., and Sharpe, P. T. (2006). Regeneration of teeth using stem cell-based tissue engineering. *Expert Opin. Biol. Ther.* **6,** 9–16.

Yokose, S., Kadokura, H., Tajima, Y., Fujieda, K., Katayama, I., Matsuoka, T., and Katayama, T. (2000). Establishment and characterization of a culture system for enzymatically released rat dental pulp cells. *Calcif. Tissue Int.* **66,** 139–144.

Section II

Mesoderm

[6] Postnatal Skeletal Stem Cells

By Paolo Bianco, Sergei A. Kuznetsov,
Mara Riminucci, and Pamela Gehron Robey

Abstract

Postnatal skeletal stem cells are a subpopulation of the bone marrow stromal cell network. To date, the most straightforward way of assessing the activity of skeletal stem cells within the bone marrow stromal cell (BMSC) population is via analysis of the rapidly adherent, colony-forming unit-fibroblast (CFU-F), and their progeny, BMSCs. Several *in vitro* methods are employed to determine the differentiation capacity of BMSCs, using osteogenic and adipogenic "cocktails" and staining protocols, and pellet cell culture for chondrogenic differentiation. However, true differentiation potential is best determined by *in vivo* transplantation in either closed or open systems. By *in vivo* transplantation, ~10% of the clonal strains are able to form bone, stroma, and marrow adipocytes, and are true skeletal stem cells. Furthermore, when derived from patients or animal models with abnormalities in gene expression, they recapitulate the disease phenotype on *in vivo* transplantation. Although *ex vivo* expansion of BMSCs inevitably dilutes the skeletal stem cells, when used *en masse*, they are attractive candidates for reconstruction of segmental bone defects, and as targets for gene therapy.

Introduction

The existence of multipotent nonhematopoietic stem cells within postnatal bone marrow stroma was brought to light by the work of Friedenstein, and later Owen and coworkers (Friedenstein *et al.*, 1966, 1968; Owen and Friedenstein, 1988). As early as in the late 1960s, and only a few years after the first direct experimental evidence for the existence of hematopoietic stem cells in the postnatal mouse bone marrow, these studies were the first to reveal the existence, in the postnatal bone marrow, of multipotent progenitors giving rise to multiple, distinct mesodermal tissues. Therefore, these studies revolutionized our concept of the bone marrow, revealing a dual system of stem and progenitor cells that are located in the same organ and feed into a dual system of differentiated tissues—the hematopoietic tissue and blood cells downstream of the hematopoietic stem cell, and the "stromal system" (including bone, cartilage, adipocytes, fibroblasts, and myelosupportive

METHODS IN ENZYMOLOGY, VOL. 419
Copyright 2006, Elsevier Inc. All rights reserved.
0076-6879/06 $35.00
DOI: 10.1016/S0076-6879(06)19006-0

stroma) downstream of the newly recognized, postnatal common progenitor (reviewed in Bianco and Robey, 2004).

The applicative potential of the discovery of skeletal stem cells was perceived only two decades later, and then dawned to popular awareness in the late 1990s, when biotech companies emerged to exploit that potential. No doubt, the notion that the skeleton is made of tissues that can be generated *de novo* from postnatal cells that can be isolated and grown in culture, has an impact as significant in technology as in science. Not only can the repair of bone or cartilage be conceived, based on the notion of skeletal stem cells, but the entire physiology and pathology of the skeleton can be reinterpreted on the basis of the notion of stem cells, progenitors, and lineages. In a field of medicine traditionally slow in adopting advances from basic science, the discovery of skeletal stem cells (and more so its belated awareness) indeed changed the pace, the rhythm, and the theme of the basic science–clinical medicine dialogue (Bianco and Robey, 2001; Robey and Bianco, 2004). A pale reflection of this, but also a visible token, can be found in the almost universal and rapid adoption of stromal cell cultures as the pivotal model for *in vitro* studies on bone metabolism.

Skeletal Stem Cells and Mesenchymal Stem Cells

Originally called "precursors of mechanocytes" (where "mechanocyte" was an imaginative term to indicate cells of connective tissue, notoriously endowed with mechanical function related to gravity and motion) (Friedenstein, 1976), skeletal stem cells were to be recognized as stem cells later. Their origin from unidentified stromal cells in the bone marrow, and their putative "stem cell" nature, were recalled in the term "bone marrow stromal stem cells" coined and adopted by Friedenstein and Owen in the 1980s (Owen and Friedenstein, 1988). The term "mesenchymal stem cells" was later proposed to denote the bone marrow-derived nonhematopoietic stem cells. As popularly used, the term "mesenchymal stem cell" (1) vows a resemblance of the ability of these cells (i.e., to give rise to multiple mesoderm-derived tissue) to the properties of primitive embryonic mesenchyme, (2) implies that a broader range of tissues than originally indicated by Friedenstein's experiments may be generated by a common postnatal progenitor, and (3) postulates that multipotent progenitors of mesodermal tissues also exist beyond the anatomical boundaries of the bone marrow, and in multiple postnatal tissues.

However, on the basis of results of proper *in vivo* transplantation studies, bone marrow-derived stromal progenitors only give rise to a defined subset of tissues. This subset includes all tissues that are part of the developing or postnatal skeleton (cartilage, bone, adipocytes, fibrous tissue, smooth muscle cells, and myelosupportive stroma), but excludes other tissues (e.g., skeletal muscle and endothelia) (Bianco and Robey, 2004). Furthermore, the embryonic "mesenchyme" is an anatomically defined tissue, rather than a developmentally defined entity. Neither the developmental origin, nor the differentiation potential, of embryonic "mesenchymal" cells is uniform. Mesenchyme originates from two germ layers (neuroectoderm and mesoderm) and multiple specifications of mesoderm. Lateral mesoderm and somites both generate "mesenchyme." However, lateral mesoderm does not give rise to skeletal muscle, and dermatomyotome cells do not give rise to bone and cartilage. Finally, so far formal *in vivo* transplantation studies have failed to demonstrate that the differentiation potential of "mesenchymal" stem cells purportedly derived from different sources (e.g., bone marrow vs. muscle vs. fat) share in fact a similar *in vivo* differentiation potential.

These considerations should discourage the use of the term "mesenchymal stem cells." Current evidence does support the existence of a bona fide stem cell within postnatal bone marrow that is capable of reforming, *in vivo*, all the components of skeletal tissue proper. Hence the term, skeletal stem cell, which the authors of these pages favor and support.

Defining a Skeletal Stem Cell

Stromal Nature

When plating a single-cell suspension of bone marrow at low density, a subset of cells rapidly adheres and proliferates to establish a colony. These clonogenic cells were originally called the colony-forming unit-fibroblast (CFU-F) (Friedenstein *et al.*, 1976). On *ex vivo* expansion, these bone marrow stromal cells (BMSCs), when studied *en masse*, have the ability to recreate a bone/marrow organ on *in vivo* transplantation with appropriate carriers in open systems in which vasculature can readily infiltrate the developing transplant (Friedenstein *et al.*, 1968; Krebsbach *et al.*, 1997). In closed systems, when vascularization is physically prevented, primarily bone and cartilage and, to a lesser extent, fibrous tissue and fat are formed (Ashton *et al.*, 1980; Friedenstein *et al.*, 1970). In the original studies, the nonhematopoietic nature of the CFU-F was deduced by the consistent lack of functional or enzymatic properties of phagocytes. Likewise, the lack of expression of the restricted range of endothelial markers known at the time excluded their endothelial origin. On this basis, the CFU-Fs were traced to the nonhematopoietic tissue included in the intact bone marrow; that is, the bone marrow stroma on which hematopoietic cells commit, proliferate, and mature. As more refined phenotype analyses became available, the nonhematopoietic, nonendothelial, stromal nature of the CFU-F was thoroughly confirmed (reviewed in Bianco and Robey, 2004).

Clonogenicity

The ability of single stromal cells to generate a colony in low-density culture reflects directly their competence for density-insensitive growth. This distinguishes a small subset of stromal cells from the whole population of stromal cells capable of adherence and growth *in vitro* per se. Density-insensitive growth is an indirect expression of a greater proneness to growth, independent of paracrine stimuli derived from neighboring cells. Colonies formed by bone marrow CFU-Fs under proper culture conditions are clonal. This was originally shown by Friedenstein *et al.* (1974a,b), and more recently also demonstrated by other approaches. For example, colonies established by cell suspensions derived from somatic mosaics are either 100% mutant, or 100% wild-type (Bianco *et al.*, 1998).

Colony formation and clonogenicity coincide with one another only when colonies are formed at clonal density. Colonies per se also appear when single-cell suspensions are plated at nonclonal density. Under these conditions, colonies are not necessarily clonal, and the formation of colonies per se does not denote density-insensitive growth. This is especially relevant when tissues other than the hematopoietic bone marrow are subjected to colony-forming efficiency assays. In a bone marrow cell suspension, the vast majority of cells are nonadherent. In any other nonhematopoietic connective tissue (e.g., fibrotic bone marrow, periosteum, dental pulp, and adipose tissue) the vast majority of cells are adherent. Hence, the same cell number used for clonogenicity assay with hematopoietic bone marrow is not legitimate for other tissues, as it unavoidably results in the establishment of a culture at nonclonal density. Enumeration of colonies formed under these circumstances is meaningless and does not assess the actual existence and frequency of clonogenic cells. Clonogenic (density-insensitive) cells do exist in all connective tissues, but can only be revealed, in any tissue other than the hematopoietic bone marrow, by plating no more than 1.6 cells/cm^2.

Multipotency

The multipotency of an individual CFU-F is demonstrated by *in vivo* transplantation of the clonal progeny of individual CFU-Fs. The progeny of single, individual bone marrow CFU-Fs (although not of every CFU-F) is sufficient to generate a complete heterotopic ossicle, including multiple cell types (e.g., bone cells, stroma, adipocytes, and fibroblasts), on transplantation *in vivo*, or cartilage in micromass ("pellet") cultures (Fig. 1). Hence, multipotent cells are found among clonogenic stromal cells (Kuznetsov

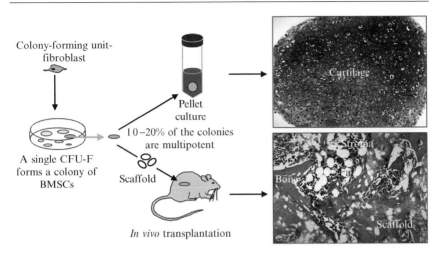

FIG. 1. Proof of the existence of a multipotential skeletal ("mesenchymal") stem cell in postnatal bone marrow. When plated at low density, the colony-forming unit-fibroblast (CFU-F) rapidly adheres, and proliferates to form a colony of bone marrow stromal cells (BMSCs). Approximately 10% of these colonies can be further expanded and, when placed in pellet cultures under relatively anaerobic conditions, they form cartilage. When attached to appropriate scaffolds and transplanted subcutaneously into immunocompromised mice, they form bone, myelosupportive stroma, and marrow adipocytes. (See color insert.)

et al., 1997). It is important to note that multipotency cannot be assessed, by any means, in nonclonal cell strains, either *in vitro* or *in vivo*. Hence, subjecting nonclonal cell strains to *in vitro* differentiation assays, as is commonplace in the literature, does not prove multipotency in any way. In addition, *in vitro* differentiation assays, albeit convenient, are empirical surrogates of *in vivo* differentiation assays, and expose cells to nonphysiological, artificial stimuli. Although a formal comparison of *in vitro* and *in vivo* differentiation assays has not yet been reported, *in vitro* assays do not predict *in vivo* differentiation.

Self-Renewal

Self-renewal is the outcome of asymmetric cell division. Asymmetric cell division has never been directly demonstrated in any postnatal human cell, but is postulated to characterize stem cells *in vivo*. *In vivo*, asymmetric cell division of a stem cell would generate one daughter cell that would clonally expand and/or differentiate, and one daughter cell that would neither expand nor differentiate. *In vitro*, all nontransformed postnatal cells (stem cells or not stem cells) grow according to asymmetric *kinetics*, which emanates from asymmetric *division*, and implies replicative senescence

(Bianco and Robey, 2004). Bone marrow stromal cell strains are no exception to this rule and, consistent with their lack of detectable telomerase activity, do undergo replicative senescence and therefore obey asymmetric kinetics. Interestingly, transduction of human bone marrow stromal cells with human telomerase reverse transcriptase (hTERT) results in greater ability for *ex vivo* expansion, and shifting from asymmetric to symmetric growth kinetics, but also in the emergence of cell transformation (Serakinci *et al.*, 2004; Simonsen *et al.*, 2002).

In the relevant literature, self-renewal is commonly mistaken for the extent of *in vitro* expansion, so that the ability of a given cell strain to undergo a relatively high number of population doublings is hastily translated into evidence for self-renewal. Considering a cell culture initiated by a single stromal CFU-F, and assuming that the culture-initiating cell (as a stem cell) undergoes asymmetric cell division in culture each and every time that it divides in culture, then the resulting cell strain includes the expanding and differentiating progeny of the original stem cell and the single original stem cell itself, with no stem cell "expansion." If, at the other extreme, one assumes that the single culture-initiating cell is constantly expanding in culture, then each cell division must be symmetrical and therefore must result in two daughter stem cells, with no differentiating or senescent progeny. Whereas all bone marrow stromal cell strains do senesce in culture, providing evidence for asymmetric kinetics, there is no direct evidence that "expansion" of skeletal ("mesenchymal") stem cells ever occurs in culture. What expands is, of course, the progeny of the culture-initiating cells. The practical implication of this is that no bone marrow stromal cell strain is ever a culture of "mesenchymal stem cells," and there is actually no evidence whatsoever that skeletal ("mesenchymal") stem cells can themselves be "cultured" or culture-expanded. It must be noted that exactly the same conclusion applies to other stem cells, notably hematopoietic stem cells.

Self-renewal of stem cells cannot be demonstrated without *in vivo* assays. The only system for which this has been formally accomplished is the hematopoietic system. Here, direct evidence for stem cell self-renewal was made possible by two conditions: (1) the definition of the surface phenotype of hematopoietic stem cells (HSCs) in the mouse; and (2) the availability of the proper *in vivo* assay (serial hematopoietic reconstitution in lethally irradiated mice). Similar studies have not been conducted with skeletal ("mesenchymal") stem cells, and are needed. At this time, self-renewal can only be indirectly surmised. Because skeletal stem cells are comprised in the bone marrow stroma, regeneration of a functional (hematopoiesis-supporting) bone marrow stroma in a heterotopic ossicle provides evidence for self-renewal at least of the anatomical compartment in which skeletal stem cells reside, and from which they are explanted. In practice, generation of a heterotopic ossicle in which hematopoiesis is maintained proves the establishment of a functional bone marrow stroma. However, it must be noted that direct evidence for the formation *in vivo* of specific stromal cell types originating from transplanted progenitors has not been given so far. Generation of heterotopic bone in the absence of hematopoiesis, instead, probes osteogenic progenitors, but falls short of proving self-renewal of stromal cells. Direct evidence for the donor origin of bone cells after *in vivo* local transplantation of stromal progenitors has been repeatedly demonstrated (Bianco *et al.*, 1998; Friedenstein *et al.*, 1968; Kuznetsov *et al.*, 1997).

Phenotype

Definition of a minimal surface phenotype of skeletal ("mesenchymal") stem cells (SSCs) would be essential for addressing their ability to self-renew, and also to address many important theoretical and applicative issues. It would allow purification of uncultured SSCs, which might have significant bearing on their applicative use for tissue engineering. It would also allow correlation of *ex vivo* properties with *in vivo* identity of specific stromal cell types, and therefore to (1) anatomically identify SSCs and (2) trace their fate after *in vivo* transplantation. Finally, it would allow definition of the biological properties of the stem cells proper, with no confounding influence of the properties of the stem cell progeny that are generated in culture, and of culture conditions themselves. These considerations justify the intense search for suitable surface markers of SSCs that has taken place, and the significant amount of published data on the topic.

Taken together, studies conducted on nonclonal cultures of bone marrow stromal cell strains have outlined a sufficiently characteristic profile, which can be summarized as follows:
$CD34^-CD45^-CD14^-$
$CD13^+CD29^+CD44^+CD49a^+CD63^+CD90^+CD105^+CD106^+CD146^+$
$CD166^+$ (Barry *et al.*, 1999; Deschaseaux and Charbord, 2000; Filshie *et al.*, 1998; Gronthos *et al.*, 1999; Stewart *et al.*, 2003; Vogel *et al.*, 2003; Zannettino *et al.*, 2003; reviewed in Bianco and Robey, 2004).

Consistent and reproducible as it may be, this is the phenotype of stromal cell cultures, not of "mesenchymal stem cells," and heavily reflects the conditions under which the strain is cultured. Virtually all proposed "markers" of "mesenchymal stem cells" are modulated in culture, and most of these markers are also expressed at comparable levels in strains of "fibroblasts" from multiple sources, which do not behave like bone marrow stromal cell strains either *in vitro* or *in vivo*.

Studies conducted on uncultured bone marrow cell suspensions, on the other hand, have aimed at identifying one or more markers suitable for isolation of clonogenic stromal cells. The typical design of these studies involves immunoselection of a marker-defined fraction of bone marrow cells, followed by assessment of clonogenic efficiency in the sorted fraction compared with the unsorted total population. These studies have provided evidence for the expression of several markers (e.g., STRO-1^{bright}, CD49a, CD63, and CD146) in the clonogenic fraction of bone marrow stromal cells (reviewed in Bianco and Robey, 2004). However, claims toward effective "purification" of genuine skeletal ("mesenchymal") stem cells by the use of any marker in immunoselection protocols appear unjustified. First, what these procedures in fact purify is the clonogenic fraction of stromal cells. Separation of the clonogenic fraction by simple adherence to the substrate in clonal cultures is in itself a purification procedure, and no study has ever shown that the "purified" population obtained by immunoselection is biologically different from the population obtained through the establishment of classical CFU-F cultures. Second, although all immunoselection procedures do result in a significant enrichment in CFU-Fs in the selected population over the unsorted population, no study has ever documented that more CFU-Fs can be recovered by immunoselection compared with selection by adherence in a given volume of bone marrow or bone marrow cell number. Third, whereas *in vitro* and *in vivo* assays do document significant biological diversity of the colonies (clones) originating from individual CFU-Fs, there is no indication of diversity of the immunophenotype of individual clones, or clone-generating CFU-Fs.

In addition to surface markers, bone marrow stromal cell strains are noted for other characteristic phenotypic traits, including expression of transcription factors and lineage-specific gene products. The master gene of osteogenic commitment, Runx2/CBFA1 (Ducy *et al.*, 1997; Komori *et al.*, 1997), is constitutively expressed in stromal cell cultures (Satomura *et al.*, 2000). It has been claimed that Runx2 may be rapidly turned on in culture, and not expressed in uncultured CFU-Fs. This would require proper confirmation, because expression of Runx2 can reasonably be taken as evidence of osteogenic commitment. Hence, it would be of theoretical importance to determine conclusively whether skeletal ("mesenchymal") stem cells are in fact truly uncommitted (Runx2 negative) or already committed to skeletogenesis (Runx2 positive). On the basis of separate evidence (such as the activity of the bone-specific promoter of the *COL1A1* gene) other studies have suggested that, indeed, assayable skeletal ("mesenchymal") stem cells in the bone marrow are in fact committed to osteogenesis (Kuznetsov *et al.*, 2004). The second pivotal transcription factor controlling osteogenic differentiation, Osterix (Nakashima *et al.*, 2002), and the chondrogenesis-controlling transcription factor, Sox9 (Akiyama *et al.*, 2002), are only erratically detected, at low levels, in bone marrow stromal strains. On the other hand, transcription factors controlling adipogenesis, CCAAT/enhancer-binding protein α (C/EBPα) and peroxisome proliferator-activated receptor γ (PPARγ), are expressed, albeit at low levels, without the use of adipogenic inducers (Gimble et al., 1996; Kuznetsov et al., 2001).

In addition to type I and III collagens, gene products characterizing the osteoblastic phenotype, such as bone sialoprotein, osteocalcin, osteonectin, and osteopontin, are variably expressed in bone marrow stromal cultures (Kuznetsov et al., 2001). α-Smooth muscle actin, considered a marker of smooth muscle cells, is regularly expressed (Kuznetsov et al., 2001), and its expression increases over time in culture, consistent with a serum effect on the serum-responsive element (SRE) in the α-smooth muscle actin promoter.

Origin and Nature of Skeletal Stem Cells

Bone and bone marrow coexist within bone as an organ, in a physical balance of volume and mass. The bone marrow stroma, in which skeletal stem cells are found, is established during the organogenesis of bone as a local adaptation of a primitive osteogenic tissue. During development, capture of space for hematopoiesis implies and involves negative regulation of the osteogenic potential of the primitive bone marrow stroma. This is best illustrated by the histological events that take place at the medullary aspect of a long bone growth plate. Here, while the bone grows in length, the trabecular bone of the primary spongiosa is resorbed and replaced by the trabecular bone of the secondary spongiosa. This, in turn, is resorbed and replaced by hematopoietic marrow instead of new bone (Bianco and Riminucci, 1998; Bianco et al., 1999). Targeted expression of a constitutively active parathyroid/parathyroid-related peptide (PTH/PTHrP) receptor in cells of osteoblastic lineage in mice disrupts this sequence, so that bone formation continues unabated for many more cycles than normal in the presumptive marrow cavities (Kuznetsov et al., 2004). The establishment of the cavity is significantly delayed in this way. Once a marrow cavity is eventually formed, the size of the assayable skeletal stem cell pool within the bone marrow that they occupy is significantly smaller compared with normal mice (Kuznetsov et al., 2004). Interestingly, the size of the hematopoietic stem cell pool within the same bone marrow is, conversely, expanded compared with normal mice (Calvi et al., 2003). Hence, the balance of bone tissue and bone marrow at skeletal maturity is not only dependent on the magnitude of bone formation events during growth, but also reflected in a balance of the two stem cell systems known to coexist in the bone marrow—hematopoietic stem cells and skeletal stem cells.

These data suggest that skeletal stem cells are established during bone development and growth from unused osteoprogenitor cells. Mechanisms regulating the retention of prenatal osteoprogenitor cells in a quiescent state within the postnatal bone marrow stroma may involve the interaction of local osteoprogenitors with additional cell types in the developing bone marrow, including hematopoietic cells and vascular endothelial cells. Interestingly, a link between vascular walls and hematopoiesis is apparent throughout development, and a link between skeletal stem cells and vascular walls is suggested by a variety of independent evidence. Attention has focused on the potential simultaneous emergence, in development, of definitive hematopoietic stem cells and primitive "mesenchymal" stem cells at the same site and time. Definitive hematopoietic stem cells appear in the floor of the dorsal aorta in the aorta–gonad–mesonephron (AGM) region (Sanchez et al., 1996). A specification of the subendothelial mesenchyme is formed at the same site, which is viewed as a primitive hematopoietic stroma, supporting the emergence of hematopoietic stem cells (Marshall et al., 1999). Different studies have shown that the AGM region may also contain multipotent progenitors of mesodermal tissues, assayable in vitro and in vivo (Minasi et al., 2002). Taken together, these data would suggest that a subset of "mesenchymal" cells able to support hematopoietic stem cells, and also able to give rise to multiple mesodermal tissues, is formed in close continuity with "hemogenic" endothelial cells at the floor of the embryonic dorsal aorta. These cells would thus resemble postnatal stromal cells in their two defining and unique properties. Whether skeletal stem cells can directly provide hematopoietic stem cell niches in the bone marrow remains to be determined. Likewise, the potential identity of skeletal stem cells as subendothelial cells in the postnatal bone marrow requires direct demonstration. Meanwhile, it is important to note that whereas some studies have suggested that bone surfaces may represent the physical location of hematopoietic stem cell "niches" (reviewed in Zhu and Emerson, 2004), other studies have identified microvascular walls as the "niche" (reviewed in Kopp et al., 2005). Vascular walls as a common thread uniting embryonic and postnatal hematopoiesis, as well as hematopoietic and mesenchymal progenitors throughout development and postnatal life, seems a fascinating avenue for present experimental work.

Plasticity of BMSCs

A flurry of studies has suggested that postnatal stem cells in different systems (hematopoietic, neural, and "mesenchymal") may exhibit a remarkable degree of germ layer commitment infidelity, commonly referred to as "plasticity." The view is commonly entertained, as a result, that "mesenchymal" stem cells can indeed differentiate into a host of cell types that either do not belong to the skeletal lineage (e.g., skeletal muscle and cardiomyocytes) or are not even related to mesoderm (e.g., neurons). Most of these studies, while fueling the expectation of wonder science, and perhaps providing ammunition to one party in the debate on embryonic stem cell research, have either failed to meet experimental reproduction by peers, or fallen short of adequate experimental controls (reviewed in Lakshmipathy and Verfaillie, 2005). For example, the fact that "mesenchymal" stem cells actually change shape when exposed to stimulators of cyclic adenosine monophosphate (cAMP) production, such as forskolin, although known for many years, reflects a reversible change in cytoskeletal organization, and not neural differentiation. Phenomena observed as a result of immortalization and chromosomal instability of murine cell lines cannot be reproduced in the genetically stable human stromal cells, or translated into evidence for natural differentiation potential. For example, whereas cell lines established from murine bone marrow stromal cells can be induced by specific cocktails (including potent demethylating agents such as $5'$-azacytidine) to differentiate into cardiomyocytes, primary human marrow stromal cells cannot.

Per se, the fact that one can alter to variable degrees the phenotype (gene expression) of any cell in culture is not surprising. All epigenetic states are reversible (Surani, 2001), or mammals could not be cloned, as they can, starting from a somatic cell nucleus. The use of empirically defined "cocktails" for *in vitro* "differentiation" of cells into any direction plays specifically on this established fact, but does so in an uncontrolled way. What is important is that under no circumstances can the expression of any set of "markers" after exposure to epigenetic regulators be taken as evidence of differentiation potential or lineage relationship. Whereas it remains of interest to determine to what extent and by what technical approach one can reproducibly reprogram the genome of postnatal cells, this line of activity will eventually generate an area of research that is as distant from stem cell research as epigenetic control of gene expression is distant from lineage relationships in development and growth.

For bone marrow stromal cells, "plasticity" has a distinct meaning. Phenotype conversion within the range of skeletal cell types is a naturally occurring phenomenon *in vivo*, under multiple circumstances of development, growth, and adaptation (Bianco and Riminucci, 1998; Bianco *et al.*, 1999). This degree of "plasticity" is not duplicated in any other known lineage tree emanating from a postnatal common progenitor. Unraveling the molecular mechanisms underlying the ability of chondrocytes to switch to an osteoblastic phenotype, or of stromal reticular cells to turn into adipocytes, will shed light onto the nature of a possibly unique kind of reversible differentiation, into which specific and evolutionarily conserved adaptive functions of connective tissues are rooted.

Role of Skeletal Stem Cells in Disease

The impact of the notion of skeletal stem cells as the unit of bone disease is greater than currently appreciated. Whereas the potential of skeletal stem cells as a tool for tissue engineering of bone and cartilage is clear to many, their significance as tools for other therapeutic approaches is less apparent, and their value as intellectual and technological tools for understanding disease has remained elusive to most.

For decades, the only biological key to interpretation (and therefore treatment) of bone diseases was provided by the bone-remodeling paradigm. Here, the modulated balance of the activities of only two differentiated cells (bone-forming osteoblasts emanating from mesenchymal progenitors, and osteoclasts derived from hematopoietic progenitors) would account for almost all that would need to be accounted for in bone biology—growth, turnover, senescence, and disease. The notion of skeletal stem and progenitor cells provides a paradigm shift (Bianco and Robey, 1999). In bone, as in other areas of human pathology, the function of differentiated cells is often insufficient to account for mechanisms of disease in organs and tissues that emanate from a system of progenitors, even when the etiology of disease is clearly established. Even knowing the disease genotype, one could not explain the pathology of thalassemia solely on the basis of the function of red blood cells. Likewise, one cannot explain certain bone diseases without considering that the disease conveys changes in lineage kinetics, and dysfunction in progenitor cells as well as in differentiated cells. In some instances, these changes are reflected even in physical changes in the bone marrow structure (e.g., marrow fibrosis), which remain uninterpretable outside the context of the lineage continuity between bone marrow stromal cells and bone cells proper.

Skeletal stem cells can be used for *in vitro* and *in vivo* modeling of bone disease. No doubt, *in vivo* modeling conveys a particularly novel dimension in the study of skeletal disease. Initially, its value was clearly apparent for genetic diseases of bone. When used with normal skeletal stem cells, the transplantation model allows for direct assessment of their (correct) organogenic potential by observation of the structural integrity of the transplant-generated "ossicle." Likewise, it was reasoned, skeletal stem cells carrying disease genotypes would generate ossicles reproducing the disease phenotype. This approach would (and does) allow for the generation of pathological human ossicles in immunocompromised mice, which could be used for a variety of clinically relevant studies, ranging from histopathology to drug testing. As applied to a particular human genetic disease, fibrous dysplasia (OMIM 174800), this approach has resulted in important insights, which have significantly improved our understanding of the disease phenotype (Bianco *et al.*, 1998). A number of other human diseases, in which interactions of abnormal nonbone cells with a normal marrow stromal environment are critical, can also be modeled by generating human ossicles in mice (e.g., hematological malignancies, and cancer metastasis to human bone).

The value of stem cell biology and stem cell-based technologies for modeling genetic diseases of the bone cell lineage, in which an intrinsic alteration of bone cell function is brought about, is intuitive. Their significance for the pathophysiology of systemic and local regulators of bone homeostasis is also becoming apparent, more slowly but not less surely. As highlighted by specific transgenic models, the effects of abnormal PTH/PTHrP signaling on bone, for example, result in a blatant bone phenotype that is rooted in changes in stem cell kinetics and function, as well as linked to changes in the hematopoietic function of bone marrow stroma (Kuznetsov et al., 2004).

Molecular Engineering

Skeletal stem cells are notoriously refractory to conventional transfection. Methods for efficient nonviral genetic engineering of skeletal stem cells require selection of stable transfectants, and fail to ensure long-term transduction (Hoelters et al., 2005). Adenoviral vectors are effective for short-term skeletal stem cell transduction, but are neither efficient nor neutral with respect to their biological properties (e.g., differentiation potential) (Conget and Minguell, 2000). Adeno-associated vectors permit stable transduction only with low efficiency (Chamberlain et al., 2004). Oncoretroviral and lentiviral vectors are highly efficient, and admittedly neutral with respect to the growth and differentiation properties of human skeletal stem cells. This makes both kinds of vectors valuable, whereas the purported ability of lentivectors to transduce quiescent cells (Naldini et al., 1996) would permit transduction of freshly explanted, uncultured, and mitotically quiescent skeletal stem cells, pending direct experimental proof.

The possibility of transducing skeletal stem cells stably and efficiently, with oncoretrovectors and lentivectors, allows for the production of stromal cell strains that overexpress genes of interest, whereas the combination of RNA interference technologies with lentivector transduction introduces the theoretical possibility to assess the effects of gene knockdown in skeletal stem cells. Data indicate that indeed, RNA polymerase III-dependent promoters can stably direct the transcription of short hairpin precursors of RNA-interfering sequences in skeletal stem cells, resulting in effective silencing of endogenous genes (Piersanti et al., 2006).

In vivo transplantation of skeletal stem cells results in a true organogenic process in which bone and bone marrow are formed. Much like genetic manipulation of embryonic stem cells permits the generation of genetically altered organisms, genetic manipulation of transplantable and organogenic

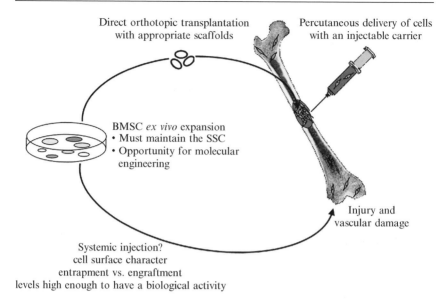

Direct orthotopic transplantation
with appropriate scaffolds

Percutaneous delivery of cells
with an injectable carrier

BMSC *ex vivo* expansion
• Must maintain the SSC
• Opportunity for molecular
  engineering

Injury and
vascular damage

Systemic injection?
cell surface character
entrapment vs. engraftment
levels high enough to have a biological activity

FIG. 2. Use of bone marrow stromal cells (BMSCs) in tissue engineering and regenerative medicine. Although *ex vivo* expansion cannot increase the numbers of skeletal stem cells contained within the BMSC population, the culture conditions must at least maintain them. Using lentiviral vectors, the *ex vivo* step also provides the opportunity to genetically modify them, to either make a deficient protein, or perhaps to silence a mutant protein by RNA interference. Cells attached to appropriate carriers can be used for direct orthotopic transplantation for bone regeneration of critical size defects through open surgery procedures, or via percutaneous delivery with an injectable carrier. Although systemic injection has been viewed as a delivery method for treatment of generalized skeletal disorders, it has yet to be demonstrated that BMSCs injected into intact animals are able to escape from the circulation; because of their cell surface characteristics they become entrapped within blood vessels. When there is vascular damage, BMSCs may be able to escape, but it is not clear that there are sufficient numbers of them to have a biological impact. (See color insert.)

postnatal stem cells permits—in principle—the *in vivo* analysis of the role of individual genes in those physiological processes that find in the *in vivo* transplantation a convenient model—bone formation and turnover, establishment of bone marrow (Fig. 2). Molecular engineering of skeletal stem cells finds here its most important experimental significance.

Potential Use of Skeletal Stem Cells in Tissue Engineering and Regenerative Medicine

The biological significance of BMSCs and skeletal stem cells in maintaining hematopoietic and skeletal homeostasis notwithstanding, their potential use in tissue engineering and regenerative medicine has clearly placed them in the public's eye. There is little doubt that after *ex vivo* expansion, BMSCs are able to regenerate segments of bone lost through trauma and focal disease by direct orthotopic placement in conjunction with appropriate scaffolds, most commonly those that contain hydroxyapatite/tricalcium phosphate (HA/TCP) (Bianco and Robey, 2001; Robey and Bianco, 2004) (Fig. 2). Numerous animal models have been developed, and treatment of bone defects in a small number of patients demonstrate efficacy (Quarto *et al.*, 2001), and larger trials are currently being planned. Furthermore, injectable carriers for delivery, percutaneously, for repair of nonunion fractures and avascular osteonecrosis, for example, without the need for open surgery would also be desirable, but currently are not commercially available. BMSCs along with appropriate carriers can be constructed around a vascular bed to produce a pedicled bone graft (with intact vasculature) (Mankani *et al.*, 2001). It can be conceived that such constructs could be grown in one part of the body, and then subsequently moved to a recipient site that would not benefit from orthotopic procedures because of morbidity arising from extensive trauma, loss of vascularity, and/or infection. A similar procedure, using freshly isolated bone marrow, was used to reconstruct a mandible in a jawless patient (Warnke *et al.*, 2004).

In addition to these applications, creation of a subcutaneous ossicle can also be viewed as a potential mode of therapy for the treatment of a disease that is characterized by lack of a particular protein in the circulation. Autologous cells could be molecularly engineered to replace the missing protein (Fig. 2), and subsequently used to generate a stable, self-renewing, subcutaneous ossicle, acting in essence as an *in vivo* bioreactor for the treatment of hemophilia A and B (lacking factor VIII and IX, respectively), or lysosomal storage diseases (lacking enzymes that degrade glycoproteins and glycolipids), as two examples.

On the basis of the success of bone marrow transplantation for treatment of hematologic disorders, there have been a number of attempts to deliver BMSCs via systemic injection for the treatment of generalized skeletal disorders. However, in an intact organism, the BMSCs are rapidly removed by the lungs, liver, and spleen by entrapment within blood vessels, and rarely escape from the circulation. In the treatment of a skeletal disorder, such as osteogenesis imperfecta, there are reports of low levels of cells incorporating into bone (Horwitz *et al.*, 1999), most likely due to vascular injury. However, it is not clear that enough cells can be incorporated into a preexisting three-dimensional structure at a level sufficient to have a biological impact (Fig. 2).

A number of studies suggest that injury may act as a homing device for BMSCs (reviewed in Barry and Murphy, 2004). Preclinical animal models suggest that direct injection of BMSCs or infusion systemically has a positive effect on the healing of myocardial infarcts. Whether the cells actually become myocardiocytes and form new blood vessels is questionable. However, the cells may be performing a "helper" function, by secreting factors that encourage endogenous repair. It has also been reported that BMSCs not only escape from immune surveillance, suggesting that nonautologous transplantation is feasible, but also that they are immunosuppressive (reviewed in Le Blanc and Ringden, 2005). However, studies in appropriate animal models have yet to verify these characteristics of BMSCs, and caution in proceeding with clinical trials in humans is warranted.

Harnessing the chondrogenic capacity of BMSCs has yet to be accomplished in reconstruction of damaged cartilage. The natural fate of most chondrocytes, including those generated by BMSCs, is to undergo hypertrophy and mineralize, setting the stage for resorption and replacement by bone. Cartilage is maintained only in certain sites in the postnatal skeleton (nasal cartilage, ear cartilage, sternal cartilage, and cartilage on articulating surfaces), and the factors that prevent its hypertrophy at these sites are not well understood. This information is needed in order to control the hypertrophic process that occurs in BMSC-generated cartilage. Furthermore, there is a need to develop scaffolds that will support cartilage formation by BMSCs, but prevent vascular ingrowth while still allowing for nutrient exchange (reviewed in Raghunath *et al.*, 2005).

Establishment of Nonclonal Populations of BMSCs *In Vitro*

Overview

Populations of BMSCs, which contain within it skeletal stem cells, can be isolated from virtually any animal species. Samples can include long bones from small animals, fragments of bone containing marrow from surgical waste, or marrow aspirates, most commonly from the iliac crest in humans. However, given the true volume of marrow spaces, only small aspirates should be used ($\sim$2.0 cm^3), because larger volumes are significantly contaminated with peripheral blood unless the position of the needle is substantially changed during harvest. The presence of peripheral blood has a substantial negative effect on the growth of BMSCs. Furthermore, the culture conditions must be fine-tuned to the species, especially regarding the percentage and lot of serum (Kuznetsov and Gehron Robey, 1996). In general, establishing BMSC cultures from rodents presents three main problems that are not a significant issue when growing human cells: (1) on occasion, rodent BMSCs, in particular murine, may expand through a limited number of passages, and cease to grow; (2) a high proportion of macrophages (sometimes even more than 25%) contaminates the primary cultures of rodent stromal cells compared with human and other species. Hence, efforts toward selection of the stromal cells proper (i.e., mesenchymal, nonhematopoietic, nonphagocytic) through simple sorting procedures (e.g., magnetic separation) may be appropriate when working with rodent cells. Detectable levels of endothelial (CD34) and hematopoietic (CD45) markers remain, even in passaged strains of rodent stromal cells, indicating the nonhomogeneous composition of the strain prepared for further studies. Nonetheless, the contaminating macrophages are negatively selected through passaging; and (3) given the high chromosomal instability of murine cells, repeated passaging of subconfluent cells may easily result in spontaneous immortalization (Tavassoli and Shall, 1988). This is an extremely important fact that must be taken into account when evaluating "unusual" differentiation properties, and high proliferating activity of murine cells derived from the bone marrow stroma. On repeated passaging, the spontaneously immortalized lines commonly acquire tumorigenic properties. In contrast, human cells do not undergo spontaneous immortalization during their growth phase. However, immortalization may rarely occur after the senescence crisis of extensively passaged human cultures (Rubio et al., 2005). This results in the emergence of a rapidly proliferative subset of cells, which, however, displays a significantly different phenotype compared with the parent population, and usually fails to demonstrate the multipotency characteristic of stromal cells (our unpublished observations).

In all species, the most straightforward way to establish a nonclonal culture of BMSCs is via the procedure established by Friedenstein et al. (1976), which relies on the rapid adherence of BMSCs under standard tissue culture conditions. Initially, these cultures represent a heterogeneous population, and include a wide variety of "stromal" cells at different stages of differentiation, and of different nature (osteoblastic, adipogenic, and fibroblastic) (Bianco et al., 1999). With multiple passages, the population appears to become more homogeneous, presumably because of the inability of more committed or differentiated cells to proliferate extensively (Bianco and Robey, 2004).

Nonclonal BMSC Culture Protocol

1. To prepare a bone marrow suspension from any animal species, bone is removed aseptically and cleaned extensively to remove associated soft connective tissues. The marrow cavities of bones from small animals are flushed with α-minimum essential medium (α-MEM) and combined. Fragments of bone from larger animals and surgical specimens from humans containing trabecular bone are scraped with a surgical blade into α-MEM, and washed extensively to remove marrow. In the case of bone marrow aspirates, $\sim$2.0 cm^3 of marrow is immediately placed into a tube

and heparin is added to a final concentration of 100 U/ml. The tube is mixed well to avoid clotting, and the contents are subsequently combined with α-MEM (20 ml) and centrifuged, and the resulting cell pellet is resuspended in fresh medium.

2. All marrow preparations are then pipetted repeatedly, and then passed through needles of decreasing gauge (down to 20 gauge) to break up aggregates. The suspension is then passed through a cell strainer (Falcon; BD Biosciences Discovery Labware, Bedford, MA) to yield a single-cell suspension. Nucleated cells are counted via a hemocytometer.

3. Cells are plated at a density of 5×10^6–5×10^7 cells per 75-cm^2 flask in nutrient medium composed of α-MEM, 2 mM glutamine, penicillin (100 U/ml), streptomycin sulfate (100 μg/ml), and 20% lot-selected fetal bovine serum (*note*: lots that are supportive of murine growth are not necessarily good for the growth of human BMSCs). Aspirates are generally plated at a higher density (up to 20×10^7) because of the dilution of marrow with peripheral blood.

4. After 1 day of incubation at 37° in a humidified atmosphere of 5% CO_2 for human cultures, and 7 days for others, the medium is replaced, and changed three times per week until the cultures become ~70% confluent (between 12 and 14 days).

5. Cells are passaged by washing extensively with Hanks' balanced salt solution followed by two treatments with 0.05% trypsin–0.53 mM EDTA for 10–15 min at room temperature. In the first passage, murine cultures may require treatment with chondroitinase ABC (20 mU/ml) for 25–30 min at 37° before trypsinization at room temperature for 25–30 min to remove matrix that is trypsin resistant. Trypsin is inhibited by the addition of fetal bovine serum (final concentration of 1%) as each fraction is collected. After the combination of fractions, cell aggregates are broken up by pipetting, collected by centrifugation, and resuspended in fresh nutrient medium.

6. BMSCs are plated at 2×10^6 cells per 75-cm^2 flask or 150-mm^2 Petri dish and fed with fresh nutrient medium three times per week until they reach 70% confluency, and are passaged as described previously.

Establishment of Clonal Populations of BMSCs *In Vitro*

Overview

Verification that populations of BMSCs contain true skeletal stem cells must rely on the isolation and characterization of clonal populations of cells. When single-cell suspensions are plated at low density, CFU-Fs adhere rapidly, and after a period of quiescence begin to proliferate to form a colony of bone marrow stromal cells (Fig. 1). These colonies display markedly different characteristics in terms of cell morphology (ranging from large flat cells to spindle-shaped fibroblastic cells), rate of proliferation and growth habit, and in phenotypic character. Approximately 50% of the colonies are alkaline phosphatase positive, of which approximately 25% multilayer and accumulate calcium as demonstrated by alizarin red S staining (the so-called bone nodule). Finally, approximately 10% contain high levels of lipid droplets as demonstrated by oil red O staining, indicative of more mature adipocytes (S. A. Kuznetsov and G. Robey, unpublished results).

Procedures for establishing clonal cell strains rely either on limiting dilution in 96-well plates, or on plating single-cell suspensions at low density and using cloning cylinders. Identical frequencies of clonogenic cells are revealed when the same sample is assayed in both ways, indicating that CFU-F estimates are not affected by the way in which clonogenic cells are assessed. As a rule of thumb, however, clones established by plating single-cell suspensions at clonal density in Petri dishes seem better able to expand further subsequent passages compared with clones that originate in 96-well plates. Whereas this phenomenon is not well understood, it is entirely possible that colonies established by limiting dilution may not exhibit the same characteristics as those established by plating at low density, because of effects of paracrine factors secreted by different colonies within a dish.

On passaging, a large majority of colonies that are generated by both methods fail to expand beyond two or three passages, most likely because of the number of population doublings reached within the clonal population. On the basis of the *in vivo* transplantation assays described later, only some clonal strains (~10–20%, depending on the animal species) display full multipotency.

Clonal BMSC Culture Protocol

1. Single-cell suspensions prepared as described previously are plated in Petri dishes at a density of $0.007–3.5 \times 10^3$ nucleated cells/cm^2 for human surgical specimens, $0.14–14.0 \times 10^3$ nucleated cells/cm^2 for aspirates, and $4.2–1 \times 10^3$ nucleated cells/cm^2 for murine cells. Cells may also be plated by limiting dilution.

2. The cultures are then vigorously washed to remove nonadherent cells with nutrient medium after 2–3 h, which is sufficient time for adherence of the CFU-Fs present in the bone marrow suspension. γ-Irradiated (6000 cGy) guinea pig bone marrow feeder cells ($0.4–0.6 \times 10^6$ nucleated cells/cm^2) are added to murine cells plated at clonal density, as it has been found that serum alone is not sufficient to support colony formation. Colony formation in rat and rabbit cell cultures has also been found to be partially dependent on irradiated feeder cells.

3. Cultures are incubated as described previously for 10–14 days, without medium change, and subsequently inspected to identify colonies that are completely round, and well separated from other colonies. After washing with HBSS, individual colonies are isolated by use of a cloning cylinder attached to the dish with sterilized high-vacuum grease.

4. Cells are released by treatment with trypsin–EDTA as described previously. Cells are transferred to individual six-well plates containing nutrient medium.

5. On reaching ~70% confluency, cells are trypsinized and plated sequentially into 25-cm^2 (passage 2) and 75-cm^2 (passage 3) flasks.

Determination of Colony-Forming Efficiency

Overview

Because of their central role in mediating postnatal skeletal metabolism, enumeration of the number of skeletal stem cells within the BMSC population is of interest with respect to the possible change in their number as a function of a disease process. The closest approximation to estimating the frequency of skeletal stem/progenitor cells derives from a colony-forming efficiency (CFE) assay, which enumerates the number of CFU-Fs, a subset of which are skeletal stem cells.

CFE in humans is generally reported to be in the range of 10–50 CFU-Fs per 10^5 nucleated marrow cells. The variability encountered in these determinations is most likely based on different methods of preparing a single-cell suspension of bone marrow (with or without density gradient centrifugation) and depends on whether a bone specimen or a marrow aspirate (which can be contaminated by peripheral blood) is used. It is thought that there is a decrease in CFE as a function of aging, with CFE as high as 80 per 10^5 in newborns, decreasing to as low as 10 per 10^5 with advanced age. However, the rate of decrease is low, and may not have a substantial impact on skeletal homeostasis. Rather, it is thought that changes in the microenvironment (endocrine, paracrine, and extracellular matrix) influence skeletal stem cell activity in maintaining homeostasis more than their slight decrease in numbers.

CFE assays are not generally performed as part of an evaluation of genetic and acquired skeletal diseases, but may provide valuable insight into mechanisms. Both in human disease and in the evaluation of animal models of disease, CFE determination provides significant information. However, it must be remembered that the estimate as applied to normal bone marrow samples is related to a cell population including a vast majority of non-adherent, hematopoietic cells. Hence, all uses of CFE determinations in disease or in experimental models must always be complemented by a direct histological assessment of the composition of tissue that is being evaluated. Fibrotic bone marrow, or any other tissue sample that does not include hematopoietic tissue, cannot be easily compared with normal bone marrow in terms of CFE.

It is also important to note that CFE assays are designed for cell suspensions made from the originally explanted tissue. As such, CFE assays estimate the frequency of cells that are mitotically quiescent in the explanted tissue, and "primed" to growth by exposure to high concentrations of serum. When applied to any cell strain that has been in culture (and therefore exposed to serum for some time, and actively growing), CFE assays provide data that are not necessarily comparable to those related to a primary cell suspension. Nonetheless, serial CFE assays can be used to longitudinally evaluate a given cell strain established in culture.

Colony-Forming Efficiency Protocol

1. Single-cell suspensions are prepared as described previously. Human cells prepared from surgical specimens are plated into 25-cm^2 flasks along with 5 ml of nutrient medium. If cells are obtained from an aspirate, or from a pathological specimen, abnormally low or high CFE may be encountered, and flasks should be plated with 1×10^4, 1×10^5, and 1×10^6 nucleated cells in the hope that one density will be appropriate for determining CFE.

2. The flasks are subsequently treated as described previously for establishment of clonal cells (step 2 in previous protocol). After washing, no more than several hundred cells should remain. Again, murine CFE assays require the use of irradiated guinea pig marrow feeder cells.

3. After incubation for 10–14 days without medium change, cultures are washed with HBSS, and then fixed with 100% methanol and stained with an aqueous solution of saturated methyl violet. Using a dissecting microscope, colonies with greater than 50 cells are counted, and the CFE is determined per 1×10^5 nucleated cells plated.

4. In some cases, it is of interest to further delineate the nature of the colonies formed by studying the distribution of colonies with different staining characteristics. In this case, after 10–14 days, medium is refreshed, and replicate cultures are incubated for another 10–14 days with medium change every 3 days. Replicate cultures are then stained for alkaline phosphatase (osteoblastic and preadipogenic marker), alizarin red S (osteoblastic marker), and oil red O (adipogenic marker). Although alcian blue staining would be indicative of chondrogenic differentiation, this is rarely seen under standard culture conditions, and is best determined by assays described later.

In Vitro Differentiation of BMSCs

Overview

There is no doubt that *in vitro* analyses are the mainstay of studies aimed at characterizing the effect of growth factors and matrices on the biological behavior of a given population of cells, and to elucidate intracellular signaling pathways that are elicited as cells are coerced to differentiate in one direction or another. There are a number of different protocols by which BMSCs can be induced to begin to differentiate *in vitro*. Although they can be informative, they can also be misleading. Given the complexity of the bone/marrow organ with respect to its structural organization and relationships between various different cell types within it, and its reliance on the vasculature, true differentiation cannot be achieved by any *in vitro* protocol, including those that incorporate the cells within a three-dimensional structure. However, some initial insight can come from *in vitro* differentiation assays (as outlined later), although caution is warranted.

Osteogenic Differentiation Protocol

1. BMSCs are plated at a density of 1.5×10^3 cells/cm^2 in nutrient medium supplemented with 10^{-4} M L-ascorbic acid 2-phosphate and 10^{-8} M dexamethasone, and either 1.8 mM potassium phosphate or 2–10 mM β-glycerol phosphate.
2. The cultures are incubated for up to 6 weeks with medium changes every 3 days.
3. Calcification is visually apparent because of its phase-bright nature (Fig. 3A), and can be stained according to two different protocols.
   a. Alizarin red S: The cultures are fixed with cold 70% ethanol for 1 h at room temperature, and incubated with alizarin red S (2% aqueous solution, pH 4.1–4.3, adjusted with ammonium hydroxide) for 30 min. Excess stain is removed by washing four times with water. Alternatively, the amount of calcium accumulated can be determined by washing the cultures first with Ca^{2+}/Mg^{2+}-free phosphate-buffered saline (PBS), and solubilizing with 0.6 N HCl. Using a commercially available kit (Sigma Diagnostics, Procedure No. 587, St. Louis, MO), the samples are reacted with *o*-cresolphthalein complexon. The color generated is compared with a standard curve for quantitation.
   b. von Kossa staining: Plates are incubated in 1% silver nitrate and then placed in natural sunlight or in ultraviolet light for 45 min. After extensive washing with distilled water, the plates are treated with 3% sodium thiosulfate for 5 min, and then washed

extensively with distilled water. Calcium deposition appears brown/black (Fig. 3A).

Adipogenic Differentiation Protocol

1. BMSCs are plated at a density of 4×10^3 cells/cm^2 in standard nutrient medium. Once they reach confluency, they are incubated in one of several cocktails that have been shown to induce adipogenesis *in vitro*: (a) nutrient medium with 0.5 mM isobutylmethylxanthine, 0.5 μM hydrocortisone, and 60 μM indomethacin, (b) nutrient medium with 10^{-4} M L-ascorbic acid 2-phosphate and 10^{-8}–10^{-7} M dexamethasone, (c) α-MEM containing

A Osteogenic
    differentiation

B Adipogenic
    differentiation

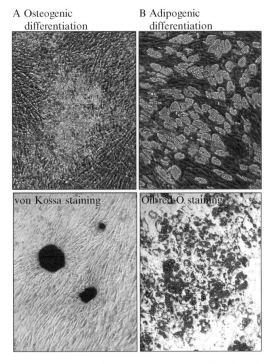

FIG. 3. *In vitro* differentiation of bone marrow stromal cells (BMSCs). (A) When incubated under osteogenic conditions (see Osteogenic Differentiation Protocol) for extended periods of time, BMSCs focally multilayer and accumulate calcium, which is phase-bright by inverted light microscopy. These condensations are readily stainable with von Kossa as shown here, or with alizarin red S. (B) Incubation of BMSCs under adipogenic conditions (see Adipogenic Differentiation Protocol) induces accumulation of fat, again readily apparent by microscopic examination, and can be further visualized by staining with oil red O. (See color insert.)

glutamine and penicillin–streptomycin, with 20% lot-selected rabbit serum, 10^{-4} L-ascorbic acid 2-phosphate, and 10^{-8} M dexamethasone, or (d) nutrient medium with a 0.1–10 μM concentration of the PPARγ ligand, rosiglitizone.

2. Cultures are incubated at 37° for up to 4 weeks with medium changes every 3 days.

3. Fat accumulation is apparent on visualization with an inverted microscope (Fig. 3B). After fixation in neutral buffered formaldehyde for 1 h, followed by 30 min with 60% isopropanol, fat can be stained with oil red O (Fig. 3B). The oil red O stock solution is prepared by dissolving 0.5 g of oil red O in 100 ml of isopropanol. The working solution is prepared fresh by diluting 30 ml of stock with 20 ml of distilled water.

Chondrogenic Differentiation Protocol

1. Passaged BMSCs are switched to Coon's modified Ham's F12 medium containing 4×10^{-6} M bovine insulin, 8×10^{-8} M human apo-transferrin, 8×10^{-8} M bovine serum albumin, 4×10^{-6} M linoleic acid, 10^{-3} M sodium pyruvate, recombinant human transforming growth factor β (rhTGF-β, 10 ng/ml), 10^{-7} M dexamethasone, and 2.5×10^{-4} M ascorbic acid.

2. Cells (2.5×10^5) are centrifuged in 15-ml polypropylene conical tubes at 500g for 5 min in 5 ml of the medium (Fig. 1).

3. The tubes are incubated with the lids partially unscrewed for 3 weeks at 37° in 5% CO_2 with medium changes three times per week.

4. Pellets are harvested and fixed for histological staining with toluidine blue (1% aqueous) (Fig. 1), alcian blue (1% alcian blue dissolved in 3% acetic acid, pH 2.5), or safranin O (0.1% aqueous) per standard procedures.

In Vivo Differentiation of BMSCs

Overview

In vivo differentiation assays can be performed in both closed and open systems. In closed systems, such as diffusion chambers filled with cells and implanted into the peritoneal cavity of mice, vascular ingrowth is prevented. Consequently, the tissue formed within the chamber is of donor origin, without the contribution of recipient cells. Narrow chambers filled with BMSCs form bone at the periphery of the chamber that is in closest contact with recipient tissue, and in wider chambers, cartilage is formed in the central portion, due most likely to the anaerobic conditions that undoubtedly exist within this part of the chamber. Fat and fibrous tissues are less often observed.

In open systems, such as underneath the renal capsule, as first performed by Friedenstein, or in a subcutaneous pocket, BMSCs are adsorbed to a number of scaffolds, hydroxyapatite/tricalcium phosphate (HA/TCP) ceramic particles (Fig. 4A), or denatured collagen sponges (murine cells; Fig. 4C). Although other types of polymeric scaffolds have been investigated, to date HA/TCP appears to be the most efficacious for murine and human BMSCs.

Evolution of the heterotopic ossicle mimics normal bone development in many ways. Initially, cells undergoing osteoblastic differentiation on the surface of the scaffold, with the separating spaces being filled with fibrous tissue, characterize the transplant. It is only after vascular invasion occurs that a myelosupportive stroma with adipocytes (of donor origin) and hematopoiesis (of recipient origin) are established. Cells of donor origin can be

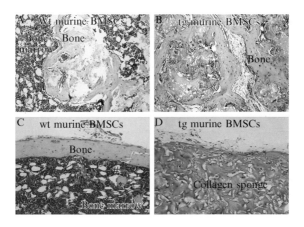

FIG. 4. *In vivo* transplantation of bone marrow stromal cells (BMSCs) and regeneration of a diseased phenotype. When *ex vivo*-expanded murine cells are transplanted with hydroxyapatite/tricalcium phosphate (HA/TCP) (A) or denatured collagen sponges (C) into subcutaneous pockets in immunocompromised mice, they completely regenerate a bone/marrow organ, with bone, hematopoietic stroma and adipocytes of donor origin, and hematopoietic cells of recipient origin. This model system can be used to determine the impact of abnormal gene expression on the development of the bone/marrow organ. As an example, BMSCs derived from transgenic mice in which a constitutively active PTH/PTHrP receptor is expressed under the control of a bone-specific promoter are capable of forming bone when transplanted with HA/TCP, but do not support formation of hematopoietic marrow (B), identical to what is observed in bone of young transgenic mice. However, when transplanted in denatured collagen sponges, transgenic BMSCs not only do not support the formation of marrow, but also do not form bone (D), indicating depletion of skeletal stem cells. Thus, these two assays are able to probe the biological activity of both committed progenitors and skeletal stem cells within the BMSC population. (See color insert.)

identified in a variety of ways by using species-specific antibodies or *in situ* hybridization, or by using cells bearing a number of markers, such as green fluorescent protein (GFP) or LacZ. Of note, hematopoiesis is never observed in the absence of bone formation, whereas exuberant bone formation is often observed, but without hematopoiesis, indicative of a population of cells that have lost multipotentiality.

Given the fact that BMSCs (and the subset of skeletal stem cells within them) are central mediators of postnatal skeletal metabolism by virtue of their contribution to hematopoietic support, bone formation, and control of bone resorption, it stands to reason that any mutation that impacts any of these functions will result in a skeletal disease. Thus, the *in vivo* transplantation assays provide a cell-based model system of disease that can directly assess their performance without the confounding effects that can arise indirectly during development. By studying the orderly progression by which the heterotopic ossicles are formed, key insights into the role of a particular gene in establishing different skeletal cell types can be obtained, for example, in fibrous dysplasia of bone (Bianco *et al.*, 1998), and in cells derived from a transgenic mouse with a bone-specific, constitutively active PTH/PTHrP receptor (Kuznetsov *et al.*, 2004). In this case, transgenic BMSCs are capable of forming bone, but do not support formation of a hematopoietic marrow, when transplanted in conjunction with HA/TCP (osteoconductive) carrier (Fig. 4B), which mirrors what occurs in the mouse at early postnatal ages. When transplanted in denatured collagen sponges (nonosteoconductive), no bone or marrow is formed, indicating depletion of the skeletal stem cell (Fig. 4). Consequently, the two different types of transplantation assays probe different biological activities of the BMSC population.

Numerous preclinical studies have demonstrated the efficacy of using *ex vivo*-expanded BMSCs for regeneration of critical size defects; that is, defects that will never completely heal on their own. Calvarial transplants in mice (described later) provide one example of a frequently used direct orthotopic assay because the defect is in non-weight-bearing bone, and does not require fixation as is needed in long bones. Development of injectable carriers will also allow the cells to be introduced into segmental defects and hold them in place. But in order to treat generalized skeletal disorders, it has long been the desire to infuse BMSCs systemically, in a fashion reminiscent of bone marrow transplantation. However, in an intact animal there is no reason to believe that cells that are extravascular in nature, or have spent their *in vitro* lives attached to a culture dish, have the proper cell surface machinery to either survive in the circulation, or to escape from it. Cells infused either by intracardiac injection or through the tail vein become entrapped within the blood vessels of the liver, spleen, and lungs, and do not survive long term, or engraft. If vascular injury is induced, it is possible that some cells may be able to exit; however, it is not clear how the cells would incorporate into a preexisting three-dimensional structure at a level sufficient to have a biological effect. Much more work is needed to begin to address these issues (Fig. 2).

Diffusion Chamber Protocol

1. A diffusion chamber is assembled from Lucite rings (external diameter, ~13 mm; internal diameter, ~9 mm; height, ~2 mm) that contain a small hole for loading the cells (commercially available from Millipore, Bedford, MA). Millipore filters with a pore size of 0.45 μm are cemented to both sides of the rings. The assembled diffusion chambers are sterilized with ethylene oxide.

2. Using aseptic technique, passaged BMSCs (2×10^5–2×10^6 cells in approximately 120 μl) are inoculated into the chambers through the hole in the ring, and the hole is sealed with a tapered plug and glue. The loaded diffusion chambers are placed in nutrient medium to prevent drying of the membranes.

3. Immunocompromised mice (*bg/bg nu/nu xid/xid*) are placed in an anesthesia chamber and anesthetized with 2% oxygen with 5% isoflurane for induction, and held at 2–2.5% isoflurane to maintain a surgical plane.

4. Mice are positioned on their backs on a heating pad, and the skin is cleaned with povidone-iodine (Betadine; Purdue Pharma, Stamford, CT) and 70% ethanol. A 2-cm midline surgical incision is made first through the skin in the lower abdomen under the rib cage. The musculoperitoneal layer is carefully lifted to avoid damage to the bowel and incised directly beneath the skin incision. The diffusion chamber is placed into the peritoneal cavity above the bowel, and the peritoneum is closed with 4.0 absorbable sutures in a continuous fashion. The skin incision is closed with two or three autoclips. The mice are given buprenorphine (0.05–0.2 mg/kg) for relief of postoperative pain.

5. The chambers are left in the mice for various time intervals up to 2 months and, after euthanasia, they are embedded noncalcified in methyl methacrylate for histological evaluation.

Subcutaneous Ossicles Formed by HA/TCP Protocol

1. HA/TCP particles (65%/35%), 100–250 μm in size, are sterilized by heating at 220° overnight and then aseptically aliquoted at 40 mg of particles per sterile 2-ml round-bottomed centrifuge tube.

2. Passaged BMSCs (2×10^6 cells resuspended in 1 ml of medium) are added to the tube with HA/TCP particles, and slowly rotated at 37° for 70–100 min to allow cells to attach.

3. The slurry is briefly centrifuged at 135g, and the supernatant is carefully removed.

4. The particles are held together by mouse fibrin, created by gently mixing 15 μl of mouse fibrinogen (3.2 mg/ml in sterile PBS) with 15 μl of mouse thrombin (25 U/ml in sterile 2% $CaCl_2$) and the particles with attached cells. The tubes are left at room temperature for 5 min in order for the clot to form, and the cap of the tube is sealed tightly to prevent drying of the construct.

5. Immunocompromised mice (*bg/bg nu/nu xid/xid*) are prepped as described previously. Mice are positioned on their stomach on a heating pad, and the skin is cleaned with Betadine and 70% ethanol. A 3-cm longitudinal incision is made in the back, and the tip of a pair of dissecting scissors is inserted under the skin and opened to create a subcutaneous pouch for the transplants.

6. A sterile spatula is used to insert the HA/TCP/fibrin constructs. Up to four can be placed into the back of each mouse.

7. Autoclips are used to close the incision, but are not removed because this procedure induces excessive bleeding in immunocompromised mice. The mice are treated for pain as described previously.

8. After euthanasia at various time points, transplants are harvested and fixed for histological evaluation.

Subcutaneous Ossicle Formed with Denatured Collagen Sponges Protocol

1. Sterile collagen sponges (Gelfoam; Pfizer, New York NY) are cut into cubes approximately $5 \times 5 \times 5$ mm, or to the desired shape and size, and placed into nutrient medium. The sponges are squeezed to remove air bubbles and to allow the sponge to regain its full size.

2. Passaged BMSCs (2×10^6 cells) in nutrient medium are pelleted by centrifugation at 135g for 10 min in 1-ml Eppendorf tubes. The supernatant is aspirated, but approximately 50 μl is left behind, and the cells are resuspended.

3. The sponges are blotted between two pieces of sterile filter paper and immediately placed into the cell suspension. As the sponges expand, the cells are drawn into the sponges. The sponges are incubated at 37° for 90 min.

4. Up to four cell-loaded sponges are transplanted into the back of immunocompromised mice prepared for surgery as described previously. Care is taken to avoid squeezing the sponges during their placement in the subcutaneous pocket. Sponges are harvested at various time points.

Calvarial Defect Protocol

1. BMSCs attached to either HA/TCP or adsorbed to collagen sponges are produced as described previously, and mice are anesthetized and cleaned as described previously and placed on their stomachs.
2. A 1-cm midline incision is made in the skin over the cranial vault, and the skin and periosteum are separated from the skull by blunt dissection.
3. Using a 5-mm trephine burr attached to a Dremell hand piece, a full-thickness defect is created, taking great care to not damage the dura mater.
4. The constructs are placed into the defect, and the skin is repositioned and closed with 5–0 Vicryl suture (Johnson&Johnson, New Brunswick, NJ). Mice are treated for pain as described previously.
5. Mice are killed at various time points, and transplants are harvested and fixed for subsequent analyses.

Acknowledgments

Work described in this chapter was supported by grants from Telethon Fondazione Onlus (grant GGP04263), MIUR and the EU (GENOSTEM) to P.B. and M.R., and the DIR, NIDCR, Intramural Research Program, NIH (S.A.K. and P.G.R.).

References

Akiyama, H., Chaboissier, M. C., Martin, J. F., Schedl, A., and de Crombrugghe, B. (2002). The transcription factor Sox9 has essential roles in successive steps of the chondrocyte differentiation pathway and is required for expression of Sox5 and Sox6. *Genes Dev.* **16,** 2813–2828.

Ashton, B. A., Allen, T. D., Howlett, C. R., Eaglesom, C. C., Hattori, A., and Owen, M. (1980). Formation of bone and cartilage by marrow stromal cells in diffusion chambers *in vivo. Clin. Orthop.* **151,** 294–307.

Barry, F. P., and Murphy, J. M. (2004). Mesenchymal stem cells: Clinical applications and biological characterization. *Int. J. Biochem. Cell Biol.* **36,** 568–584.

Barry, F. P., Boynton, R. E., Haynesworth, S., Murphy, J. M., and Zaia, J. (1999). The monoclonal antibody SH-2, raised against human mesenchymal stem cells, recognizes an epitope on endoglin (CD105). *Biochem. Biophys. Res. Commun.* **265,** 134–139.

Bianco, P., and Riminucci, M. (1998). The bone marrow stroma *in vivo*: Ontogeny, structure, cellular composition and changes in disease. *In* "Marrow Stromal Cell Culture" (J. N. Beresford and M. E. Owen, eds.), pp. 10–25. Cambridge University Press, Cambridge, UK.

Bianco, P., and Robey, P. (1999). Diseases of bone and the stromal cell lineage. *J. Bone Miner. Res.* **14,** 336–341.

Bianco, P., and Robey, P. G. (2001). Stem cells in tissue engineering. *Nature* **414,** 118–121.

Bianco, P., and Robey, P. G. (2004). Skeletal stem cells. *In* "Handbook of Adult and Fetal Stem Cells" (R. P. Lanza, ed.), pp. 415–424. Academic Press, San Diego, CA.

Bianco, P., Kuznetsov, S. A., Riminucci, M., Fisher, L. W., Spiegel, A. M., and Robey, P. G. (1998). Reproduction of human fibrous dysplasia of bone in immunocompromised mice by transplanted mosaics of normal and $G_s\alpha$-mutated skeletal progenitor cells. *J. Clin. Invest.* **101,** 1737–1744.

Bianco, P., Riminucci, M., Kuznetsov, S., and Robey, P. G. (1999). Multipotential cells in the bone marrow stroma: Regulation in the context of organ physiology. *Crit. Rev. Eukaryot. Gene Expr.* **9,** 159–173.

Calvi, L. M., Adams, G. B., Weibrecht, K. W., Weber, J. M., Olson, D. P., Knight, M. C., Martin, R. P., Schipani, E., Divieti, P., Bringhurst, F. R., Milner, L. A., Kronenberg, H. M., and Scadden, D. T. (2003). Osteoblastic cells regulate the haematopoietic stem cell niche. *Nature* **425,** 841–846.

Chamberlain, J. R., Schwarze, U., Wang, P. R., Hirata, R. K., Hankenson, K. D., Pace, J. M., Underwood, R. A., Song, K. M., Sussman, M., Byers, P. H., and Russell, D. W. (2004). Gene targeting in stem cells from individuals with osteogenesis imperfecta. *Science* **303,** 1198–1201.

Conget, P. A., and Minguell, J. J. (2000). Adenoviral-mediated gene transfer into *ex vivo* expanded human bone marrow mesenchymal progenitor cells. *Exp. Hematol.* **28,** 382–390.

Deschaseaux, F., and Charbord, P. (2000). Human marrow stromal precursors are α_1 integrin subunit-positive. *J. Cell. Physiol.* **184,** 319–325.

Ducy, P., Zhang, R., Geoffroy, V., Ridall, A. L., and Karsenty, G. (1997). Osf2/Cbfa1: A transcriptional activator of osteoblast differentiation. *Cell* **89,** 747–754.

Filshie, R. J., Zannettino, A. C., Makrynikola, V., Gronthos, S., Henniker, A. J., Bendall, L. J., Gottlieb, D. J., Simmons, P. J., and Bradstock, K. F. (1998). MUC18, a member of the immunoglobulin superfamily, is expressed on bone marrow fibroblasts and a subset of hematological malignancies. *Leukemia* **12,** 414–421.

Friedenstein, A. J. (1976). Precursor cells of mechanocytes. *Int. Rev. Cytol.* **47,** 327–359.

Friedenstein, A. J., Chailakhjan, R. K., and Lalykina, K. S. (1970). The development of fibroblast colonies in monolayer cultures of guinea-pig bone marrow and spleen cells. *Cell Tissue Kinet.* **3,** 393–403.

Friedenstein, A. J., Piatetzky-Shapiro, I. I., and Petrakova, K. V. (1966). Osteogenesis in transplants of bone marrow cells. *J. Embryol. Exp. Morphol.* **16,** 381–390.

Friedenstein, A. J., Petrakova, K. V., Kurolesova, A. I., and Frolova, G. P. (1968). Heterotopic transplants of bone marrow: Analysis of precursor cells for osteogenic and hematopoietic tissues. *Transplantation* **6,** 230–247.

Friedenstein, A. J., Chailakhyan, R. K., Latsinik, N. V., Panasyuk, A. F., and Keiliss-Borok, I. V. (1974a). Stromal cells responsible for transferring the microenvironment of the hemopoietic tissues: Cloning *in vitro* and retransplantation *in vivo. Transplantation* **17,** 331–340.

Friedenstein, A. J., Deriglasova, U. F., Kulagina, N. N., Panasuk, A. F., Rudakowa, S. F., Luria, E. A., and Ruadkow, I. A. (1974b). Precursors for fibroblasts in different populations of hematopoietic cells as detected by the *in vitro* colony assay method. *Exp. Hematol.* **2,** 83–92.

Friedenstein, A. J., Gorskaja, J. F., and Kulagina, N. N. (1976). Fibroblast precursors in normal and irradiated mouse hematopoietic organs. *Exp. Hematol.* **4,** 267–274.

Gimble, J. M., Robinson, C. E., Wu, X., and Kelly, K. A. (1996). The function of adipocytes in the bone marrow stroma: An update. *Bone* **19,** 421–428.

Gronthos, S., Zannettino, A. C., Graves, S. E., Ohta, S., Hay, S. J., and Simmons, P. J. (1999). Differential cell surface expression of the STRO-1 and alkaline phosphatase antigens on discrete developmental stages in primary cultures of human bone cells. *J. Bone Miner. Res.* **14,** 47–56.

Hoelters, J., Ciccarella, M., Drechsel, M., Geissler, C., Gulkan, H., Bocker, W., Schieker, M., Jochum, M., and Neth, P. (2005). Nonviral genetic modification mediates effective transgene expression and functional RNA interference in human mesenchymal stem cells. *J. Gene Med.* **7**, 718–728.

Horwitz, E. M., Prockop, D. J., Fitzpatrick, L. A., Koo, W. W., Gordon, P. L., Neel, M., Sussman, M., Orchard, P., Marx, J. C., Pyeritz, R. E., and Brenner, M. K. (1999). Transplantability and therapeutic effects of bone marrow-derived mesenchymal cells in children with osteogenesis imperfecta. *Nat. Med.* **5**, 309–313.

Komori, T., Yagi, H., Nomura, S., Yamaguchi, A., Sasaki, K., Deguchi, K., Shimizu, Y., Bronson, R. T., Gao, Y. H., Inada, M., Sato, M., Okamoto, R., Kitamura, Y., Yoshiki, S., and Kishimoto, T. (1997). Targeted disruption of Cbfa1 results in a complete lack of bone formation owing to maturational arrest of osteoblasts. *Cell* **89**, 755–764.

Kopp, H. G., Avecilla, S. T., Hooper, A. T., and Rafii, S. (2005). The bone marrow vascular niche: Home of HSC differentiation and mobilization. *Physiology (Bethesda)* **20**, 349–356.

Krebsbach, P. H., Kuznetsov, S. A., Satomura, K., Emmons, R. V., Rowe, D. W., and Robey, P. G. (1997). Bone formation *in vivo*: Comparison of osteogenesis by transplanted mouse and human marrow stromal fibroblasts. *Transplantation* **63**, 1059–1069.

Kuznetsov, S., and Gehron Robey, P. (1996). Species differences in growth requirements for bone marrow stromal fibroblast colony formation *in vitro*. *Calcif. Tissue Int.* **59**, 265–270.

Kuznetsov, S. A., Friedenstein, A. J., and Robey, P. G. (1997). Factors required for bone marrow stromal fibroblast colony formation *in vitro*. *Br. J. Haematol.* **97**, 561–570.

Kuznetsov, S. A., Mankani, M. H., Gronthos, S., Satomura, K., Bianco, P., and Robey, P. G. (2001). Circulating skeletal stem cells. *J. Cell Biol.* **153**, 1133–1140.

Kuznetsov, S. A., Riminucci, M., Ziran, N., Tsutsui, T. W., Corsi, A., Calvi, L., Kronenberg, H. M., Schipani, E., Robey, P. G., and Bianco, P. (2004). The interplay of osteogenesis and hematopoiesis: Expression of a constitutively active PTH/PTHrP receptor in osteogenic cells perturbs the establishment of hematopoiesis in bone and of skeletal stem cells in the bone marrow. *J. Cell Biol.* **167**, 1113–1122.

Lakshmipathy, U., and Verfaillie, C. (2005). Stem cell plasticity. *Blood Rev.* **19**, 29–38.

Le Blanc, K., and Ringden, O. (2005). Immunobiology of human mesenchymal stem cells and future use in hematopoietic stem cell transplantation. *Biol. Blood Marrow Transplant.* **11**, 321–334.

Mankani, M. H., Krebsbach, P. H., Satomura, K., Kuznetsov, S. A., Hoyt, R., and Robey, P. G. (2001). Pedicled bone flap formation using transplanted bone marrow stromal cells. *Arch. Surg.* **136**, 263–270.

Marshall, C. J., Moore, R. L., Thorogood, P., Brickell, P. M., Kinnon, C., and Thrasher, A. J. (1999). Detailed characterization of the human aorta–gonad–mesonephros region reveals morphological polarity resembling a hematopoietic stromal layer. *Dev. Dyn.* **215**, 139–147.

Minasi, M. G., Riminucci, M., De Angelis, L., Borello, U., Berarducci, B., Innocenzi, A., Caprioli, A., Sirabella, D., Baiocchi, M., De Maria, R., Boratto, R., Jaffredo, T., Broccoli, V., Bianco, P., and Cossu, G. (2002). The meso-angioblast: A multipotent, self-renewing cell that originates from the dorsal aorta and differentiates into most mesodermal tissues. *Development* **129**, 2773–2783.

Nakashima, K., Zhou, X., Kunkel, G., Zhang, Z., Deng, J. M., Behringer, R. R., and de Crombrugghe, B. (2002). The novel zinc finger-containing transcription factor osterix is required for osteoblast differentiation and bone formation. *Cell* **108**, 17–29.

Naldini, L., Blomer, U., Gallay, P., Ory, D., Mulligan, R., Gage, F. H., Verma, I. M., and Trono, D. (1996). *In vivo* gene delivery and stable transduction of nondividing cells by a lentiviral vector. *Science* **272**, 263–267.

Owen, M., and Friedenstein, A. J. (1988). Stromal stem cells: Marrow-derived osteogenic precursors. *Ciba Found. Symp.* **136,** 42–60.

Piersanti, S., Sacchetti, B., Funari, A., Di Cesare, S., Bonci, D., Cherunbini, G., Peschle, C., Riminucci, M., Bianco, P., and Saggio, I. (2006). Lentiviral transduction of human post-natal skeletal (stromal, mesenchymal) stem cells: *In vivo* transplantation and gene silencing. *Calcif. Tissue Int.* **78,** 372–384.

Quarto, R., Mastrogiacomo, M., Cancedda, R., Kutepov, S. M., Mukhachev, V., Lavroukov, A., Kon, E., and Marcacci, M. (2001). Repair of large bone defects with the use of autologous bone marrow stromal cells. *N. Engl. J. Med.* **344,** 385–386.

Raghunath, J., Salacinski, H. J., Sales, K. M., Butler, P. E., and Seifalian, A. M. (2005). Advancing cartilage tissue engineering: The application of stem cell technology. *Curr. Opin. Biotechnol.* **16,** 503–509.

Robey, P. G., and Bianco, P. (2004). Stem cells in tissue engineering. *In* "Handbook of Adult and Fetal Stem Cells" (R. P. Lanza, ed.), pp. 785–792. Academic Press, San Diego, CA.

Rubio, D., Garcia-Castro, J., Martin, M. C., de la Fuente, R., Cigudosa, J. C., Lloyd, A. C., and Bernad, A. (2005). Spontaneous human adult stem cell transformation. *Cancer Res.* **65,** 3035–3039.

Sanchez, M. J., Holmes, A., Miles, C., and Dzierzak, E. (1996). Characterization of the first definitive hematopoietic stem cells in the AGM and liver of the mouse embryo. *Immunity* **5,** 513–525.

Satomura, K., Krebsbach, P., Bianco, P., and Gehron Robey, P. (2000). Osteogenic imprinting upstream of marrow stromal cell differentiation. *J. Cell. Biochem.* **78,** 391–403.

Serakinci, N., Guldberg, P., Burns, J. S., Abdallah, B., Schrodder, H., Jensen, T., and Kassem, M. (2004). Adult human mesenchymal stem cell as a target for neoplastic transformation. *Oncogene* **23,** 5095–5098.

Simonsen, J. L., Rosada, C., Serakinci, N., Justesen, J., Stenderup, K., Rattan, S. I., Jensen, T. G., and Kassem, M. (2002). Telomerase expression extends the proliferative life-span and maintains the osteogenic potential of human bone marrow stromal cells. *Nat. Biotechnol.* **20,** 592–596.

Stewart, K., Monk, P., Walsh, S., Jefferiss, C. M., Letchford, J., and Beresford, J. N. (2003). STRO-1, HOP-26 (CD63), CD49a and SB-10 (CD166) as markers of primitive human marrow stromal cells and their more differentiated progeny: A comparative investigation *in vitro*. *Cell Tissue Res.* **313,** 281–290.

Surani, M. A. (2001). Reprogramming of genome function through epigenetic inheritance. *Nature* **414,** 122–128.

Tavassoli, M., and Shall, S. (1988). Transcription of the c-*myc* oncogene is altered in spontaneously immortalized rodent fibroblasts. *Oncogene* **2,** 337–345.

Vogel, W., Grunebach, F., Messam, C. A., Kanz, L., Brugger, W., and Buhring, H. J. (2003). Heterogeneity among human bone marrow-derived mesenchymal stem cells and neural progenitor cells. *Haematologica* **88,** 126–133.

Warnke, P. H., Springer, I. N., Wiltfang, J., Acil, Y., Eufinger, H., Wehmoller, M., Russo, P. A., Bolte, H., Sherry, E., Behrens, E., and Terheyden, H. (2004). Growth and transplantation of a custom vascularised bone graft in a man. *Lancet* **364,** 766–770.

Zannettino, A. C., Harrison, K., Joyner, C. J., Triffitt, J. T., and Simmons, P. J. (2003). Molecular cloning of the cell surface antigen identified by the osteoprogenitor-specific monoclonal antibody, HOP-26. *J. Cell. Biochem.* **89,** 56–66.

Zhu, J., and Emerson, S. G. (2004). A new bone to pick: Osteoblasts and the haematopoietic stem-cell niche. *Bioessays* **26,** 595–599.

[7]   Hematopoietic Stem Cells

By ROBERT G. HAWLEY, ALI RAMEZANI, and TERESA S. HAWLEY

Abstract

Hematopoietic stem cells (HSCs) have the capacity to self-renew and the potential to differentiate into all of the mature blood cell types. The ability to prospectively identify and isolate HSCs has been the subject of extensive investigation since the first transplantation studies implying their existence almost 50 years ago. Despite significant advances in enrichment protocols, the continuous *in vitro* propagation of human HSCs has not yet been achieved. This chapter describes current procedures used to phenotypically and functionally characterize candidate human HSCs and initial efforts to derive permanent human HSC lines.

Introduction

Hematopoietic stem cells (HSCs) are multipotent precursors that have self-renewal capacity and the ability to regenerate all the different cell types that comprise the blood-forming system (Bonnet, 2002; McCulloch and Till, 2005). Transplantation of HSCs forms the basis of consolidation therapy in cancer treatments and is used to cure or ameliorate a number of hematologic and genetic disorders (Shizuru *et al.*, 2005; Steward and Jarisch, 2005). With certain caveats (McCormack and Rabbitts, 2004), HSCs are also an attractive target cell population for gene therapies because they are readily accessible for *ex vivo* genetic modification and allow for the possibility of sustained transgene expression in circulating peripheral blood cells throughout the lifetime of an individual (Hawley, 2001; Moayeri *et al.*, 2005).

Historically, mouse HSCs were identified retrospectively by utilizing clonal *in vivo* assays wherein labeled cells (e.g., genetically tagged with reporter genes) were assessed for potential to functionally reconstitute hematopoiesis after injection into conditioned hosts, with self-renewal capacity demonstrated by serial transfer into secondary recipients (Abramson *et al.*, 1977; Capel *et al.*, 1990; Jordan and Lemischka, 1990; Keller *et al.*, 1985). Limiting dilution analysis of total bone marrow preparations allowed quantitative estimation of HSC frequencies ranging from 1 in 10,000 to 1 in 100,000 cells (Harrison, 1980; Harrison *et al.*, 1993; Szilvassy *et al.*, 1990). A major advance in the field of HSC biology was the prospective isolation of enriched populations of mouse HSCs on the basis of cell surface phenotype

METHODS IN ENZYMOLOGY, VOL. 419   0076-6879/06 $35.00
Copyright 2006, Elsevier Inc. All rights reserved.
DOI: 10.1016/S0076-6879(06)19007-2

(Spangrude *et al.*, 1988). With the exception of clinical gene-marking trials (Stewart *et al.*, 1999), analogous HSC transplantation experiments cannot be performed in humans. For this reason, xenogeneic transplant models have been developed as surrogate assays to evaluate human hematopoietic precursors for *in vivo* repopulating potential. These assays have helped to elucidate the composition of the human HSC compartment (Bhatia *et al.*, 1998; Gallacher *et al.*, 2000; Glimm *et al.*, 2001; Guenechea *et al.*, 2001; Mazurier *et al.*, 2003; Wang *et al.*, 2003; Zanjani *et al.*, 1998) and have provided paradigms for translation to clinical applications (Baum *et al.*, 1992; Civin *et al.*, 1996b; Lang *et al.*, 2004; Shizuru *et al.*, 2005; Shmelkov *et al.*, 2005; Shpall *et al.*, 1994; Yin *et al.*, 1997).

In this chapter, we discuss the phenotypic and functional characteristics of mouse and human HSCs, and describe protocols for the isolation and assay of candidate human HSCs. A procedure to derive factor-dependent human hematopoietic progenitor cell lines is also provided.

Identification and Enrichment of HSCs

Cell Surface Markers

All HSC activity in adult mouse bone marrow is contained in a population of cells characterized by expression of the c-Kit tyrosine kinase receptor (the receptor for stem cell factor, SCF), stem cell antigen-1 (Sca-1, Ly-6A/E), low levels of the Thy-1.1 cell surface antigen (Thy-1.1^{lo}), and no or low levels of expression of many cell surface antigens found on differentiated cells belonging to various lineages (referred to as lineage-negative or Lin$^-$) (Shizuru *et al.*, 2005). Mouse HSCs variably express the sialomucin CD34, depending on developmental stage and cell cycle status (Ito *et al.*, 2000; Matsuoka *et al.*, 2001; Osawa *et al.*, 1996; Sato *et al.*, 1999). Studies have identified a number of additional cell surface antigens that mark mouse HSCs, including the following: the TIE family of receptor tyrosine kinases (Arai *et al.*, 2004; Iwama *et al.*, 1993); endoglin, an ancillary transforming growth factor-β receptor (Chen *et al.*, 2002); endomucin, a CD34-like sialomucin (Matsubara *et al.*, 2005); CD150, the founding member of the SLAM family of receptors (Kiel *et al.*, 2005; Yilmaz *et al.*, 2006); CD201, the endothelial protein C receptor (Balazs *et al.*, 2006); and prion protein (Zhang *et al.*, 2006b). The receptor for thrombopoietin (TPO), c-Mpl, is also expressed on ~70% of mouse c-Kit$^+$Lin$^-$Sca-1$^+$ HSCs (Solar *et al.*, 1998).

In humans, clinical protocols involving enrichment for HSCs generally utilize cells expressing CD34 (Civin *et al.*, 1996b; Shizuru *et al.*, 2005; Shpall *et al.*, 1994), which is expressed on ~0.2–3% of the nucleated cells in cord blood, bone marrow, and mobilized peripheral blood (Civin *et al.*, 1984; Krause *et al.*, 1996; Sutherland *et al.*, 1996). Experimentally, further isolation and characterization of Lin⁻CD34⁺ subpopulations have defined more primitive precursors with hematopoietic repopulating activity that express combinations of the CD59 surface antigen related to Sca-1, the vascular endothelial growth factor receptor-2 (VEGFR2 or KDR), and low levels of c-Kit (CD117), Thy-1 (CD90), and the CD38 surface antigen (Baum *et al.*, 1992; Civin *et al.*, 1996a; Gunji *et al.*, 1993; Hill *et al.*, 1996; Kawashima *et al.*, 1996; Larochelle *et al.*, 1996; Ziegler *et al.*, 1999). As in the mouse, the TIE family of receptor tyrosine kinases and the TPO receptor c-Mpl also appear to further enrich for human HSCs, being expressed on ~80% and ~70% of CD34⁺CD38⁻ cells, respectively (Hashiyama *et al.*, 1996; Ninos *et al.*, 2006; Solar *et al.*, 1998).

It has become appreciated that the CD133 cell surface antigen is another important human HSC marker (de Wynter *et al.*, 1998; Gallacher *et al.*, 2000; Hess *et al.*, 2006; Lang *et al.*, 2004; Shmelkov *et al.*, 2005; Yin *et al.*, 1997). CD133, the human homolog of mouse Prominin-1 (Shmelkov *et al.*, 2005), was first identified as a selective human HSC surface molecule, using a monoclonal antibody recognizing a particular glycosylated form of Prominin-1 designated as AC133 (Yin *et al.*, 1997). Selection for CD133⁺ hematopoietic precursors yields >90% CD34⁺ cells containing all the human hematopoietic repopulating activity. Notably, the extremely rare CD34⁻ candidate HSCs that had previously been identified (Bhatia *et al.*, 1998; Gao *et al.*, 2001; Wang *et al.*, 2003; Zanjani *et al.*, 1998) reside within the CD133 fraction (Gallacher *et al.*, 2000).

Enriched populations of human HSCs are routinely obtained by positive selection for CD34/CD133 and/or by depletion of lineage-committed cells, using monoclonal antibodies recognizing differentiation markers (such as CD2, CD3, CD14, CD16, CD19, CD24, CD41, CD56, CD66b, and CD235a) in the context of immunomagnetic or fluorescence-activated cell-sorting methodologies. In this regard, it is important to bear in mind that physical manipulation of HSCs during the enrichment procedure may not be without effect on cell physiology (Kimura *et al.*, 2004). For example, it is conceivable that binding of antibodies to CD34/CD133 may trigger intracellular signaling pathways that could modulate HSC function. Interestingly, one study suggests that the majority of cells within the CD34⁺ CD38⁻Lin⁻ HSC compartment express the myeloid-associated lineage markers CD13, CD33, and CD123 [the low-affinity binding subunit of the interleukin (IL)-3 receptor] (Taussig *et al.*, 2005), indicating that some caution is warranted when selecting a cocktail of monoclonal antibodies for lineage marker-depletion enrichment of human HSCs.

Fluorescent Dye Staining

Hoechst 33342 and Rhodamine 123

Other strategies that have been utilized to identify and enrich for HSCs are based on the staining patterns of fluorescent dyes (Bertoncello *et al.*, 1985; Goodell *et al.*, 1996; Jones *et al.*, 1995; Leemhuis *et al.*, 1996; Storms *et al.*, 1999; Visser *et al.*, 1981; Wolf *et al.*, 1993). Rhodamine 123 (which preferentially accumulates in active mitochondria) and Hoechst 33342 (a *bis*-benzimidazole that binds to adenine–thymine-rich regions of the minor groove of DNA) are two fluorescent vital dyes that have been routinely used to characterize hematopoietic precursor populations (Bertoncello *et al.*, 1985; Leemhuis *et al.*, 1996; McAlister *et al.*, 1990; Visser *et al.*, 1981; Wolf *et al.*, 1993). Rhodamine 123 staining of mouse bone marrow cells demonstrated that HSCs with long-term repopulating potential stained dimly whereas more brightly staining hematopoietic precursors could provide only short-term repopulation (Bertoncello *et al.*, 1988, 1991; Spangrude and Johnson, 1990; Zijlmans *et al.*, 1995). Moreover, subpopulations of mouse bone marrow cells that stained most weakly with both dyes were shown to be highly enriched for long-term repopulating HSCs (Bertoncello and Williams, 2004; Leemhuis *et al.*, 1996; Wolf *et al.*, 1993). Decreased staining with these dyes generally reflects a metabolically and mitotically inactive state (Arndt-Jovin and Jovin, 1977; Johnson *et al.*, 1980; Spangrude and Johnson, 1990). However, it is now appreciated that decreased staining of HSCs with rhodamine 123 and Hoechst 33342 is also due to efflux mediated by at least two members of the ATP-binding cassette (ABC) family of transporters, ABCB1 (also referred to as MDR1 or P-glycoprotein) and ABCG2 (also referred to as BCRP, MXR, or ABCP) (Chaudhary and Roninson, 1991; Juliano and Ling, 1976; Scharenberg *et al.*, 2002; Zhou *et al.*, 2001, 2002).

Side Population Assay

A novel method that simultaneously monitors the low fluorescence intensity of Hoechst 33342 staining at ~450 nm and at >675 nm after ultraviolet excitation identifies a rare (<0.1%) subpopulation of mouse bone marrow cells, referred to as "side population" (SP) cells, which contains the vast majority of long-term hematopoietic repopulating activity (Goodell *et al.*, 1996). The ABC transporter Bcrp1 (the mouse ortholog of human ABCG2) expressed in mouse bone marrow cells is the major determinant of the mouse SP profile (Zhou *et al.*, 2001, 2002). Subsequent multiparameter flow cytometric analysis of mouse bone marrow SP cells showed that approximately one third exhibited the c-Kit$^+$Thy-1.1^{lo}Lin$^-$Sca-1$^+$ phenotype whereas approximately one half expressed CD34

(Pearce *et al.*, 2004). In another study, mouse bone marrow cells with the strongest dye efflux activity, which exhibited the highest hematopoietic repopulating activity, were shown to have a c-Kit$^+$Lin$^-$Sca-1$^+$CD34$^-$ phenotype (Matsuzaki *et al.*, 2004).

The SP assay has also been applied to human hematopoietic tissues (Eaker *et al.*, 2004; Goodell *et al.*, 1997; Naylor *et al.*, 2005; Preffer *et al.*, 2002; Scharenberg *et al.*, 2002; Storms *et al.*, 2000; Uchida *et al.*, 2001). Unlike mouse bone marrow SP cells, human hematopoietic SP cells constitute a much more phenotypically and functionally heterogeneous precursor population (Naylor *et al.*, 2005; Preffer *et al.*, 2002; Storms *et al.*, 2000; Uchida *et al.*, 2001). CD34$^-$ SP cells have been identified in several studies, but to date repopulating ability of human hematopoietic SP cells has been demonstrated only for CD34$^+$ subpopulations (Eaker *et al.*, 2004; Scharenberg *et al.*, 2002; Uchida *et al.*, 2001).

Fluorescent Substrates for Cytosolic Aldehyde Dehydrogenase Activity

Cytosolic aldehyde dehydrogenase (ALDH), an enzyme responsible for oxidizing a variety of intracellular aldehydes, is expressed at high levels in HSCs, conferring resistance to the alkylating agents cyclophosphamide and 4-hydroxyperoxycyclophosphamide (Gordon *et al.*, 1985; Kastan *et al.*, 1990; Sahovic *et al.*, 1988). Fluorescent substrates for ALDH have been developed and shown to be useful for isolating mouse and human HSCs (Fallon *et al.*, 2003; Hess *et al.*, 2004, 2006; Jones *et al.*, 1995, 1996; Storms *et al.*, 1999). In proof-of-principle studies (Jones *et al.*, 1995, 1996), dansyl-aminoacetaldehyde (DAAA) was used as an ALDH substrate. DAAA can diffuse freely across the cell membrane because it is uncharged. Cells expressing ALDH oxidize DAAA to dansyl-glycine, which is retained intracellularly by virtue of a charged carboxylate group at physiologic pH, and ALDH$^+$ cells are identified by dansyl fluorescence on excitation with ultraviolet light. More recently, a newer fluorescent substrate for ALDH, termed BODIPY-aminoacetaldehyde (BAAA), was synthesized, which uses a nontoxic visible light-excitable fluorophore BODIPY (Storms *et al.*, 1999). Similar to DAAA, BAAA is uncharged and diffuses freely across the cell membrane, becoming converted to BODIPY-aminoacetate (BAA), which is retained intracellularly because of its net negative charge in the presence of an inhibitor of the ABC transporter ABCB1 (Storms *et al.*, 1999).

Mouse hematopoietic precursors enriched for high expression of ALDH by staining with BAAA or DAAA may represent a novel class of HSCs, which express undetectable or low levels of the c-Kit, Thy-1, Sca-1, and CD34 HSC markers, but which produce long-term albeit delayed multilineage engraftment (Armstrong *et al.*, 2004; Jones *et al.*, 1996). Flow cytometric analysis of human cord blood cells stained with BAAA identified a population of cells (at a frequency of ~1%) with bright fluorescence intensity (ALDH^{br}) and low orthogonal light "side" scattering (SSC^{lo}), comprising ~74% CD34$^+$ cells and ~46% CD34$^+$CD38$^{lo/-}$ cells, which was largely depleted of cells with mature T cell, natural killer cell, myeloid, erythroid, and platelet lineage markers (Storms et al., 1999). The SSC^{lo}ALDH^{br} population still contained a small number of B cells, however (~12%) (Storms et al., 1999). In another study, Lin$^-$ depletion combined with selection for ALDH^{br} cells by BAAA staining demonstrated enrichment for hematopoietic precursors coexpressing CD133 and CD34 (~73%) and all the hematopoietic repopulating activity in human cord blood preparations (Hess et al., 2004). A follow-up study by the same group reported that prospective selection of ALDH^{br}Lin$^-$ human cord blood cells for CD133 expression yields a population of primitive precursors that are primarily CD34$^+$ (~95%) and that contains all long-term hematopoietic repopulating activity (Hess et al., 2006).

A two-step enrichment strategy for human HSCs combining positive selection for CD133$^+$ cells and an assay for high-level ALDH expression (SSC^{lo}ALDH^{br} cells) is described here.

Protocol for Isolation of Candidate Human HSCs

1. Obtain human cord blood, bone marrow, or mobilized peripheral blood cells after informed consent in conformity with a human subjects protocol approved by an institutional review board, or purchase from a commercial source [e.g., AllCells (Berkeley, CA), Cambrex (East Rutherford, NJ), or StemCell Technologies Vancouver, BC, Canada)]. For human cord blood cells, dilute anticoagulated cord blood 1:3 with phosphate-buffered saline (PBS) containing 0.6% anticoagulant citrate dextrose solution A (ACD-A, cat. no. C3821; Sigma-Aldrich, St. Louis, MO). Layer 35 ml of diluted cord blood over 12 ml of Ficoll-Paque PLUS (cat. no. 17-1440-02; GE Healthcare Life Sciences, Piscataway, NJ). Centrifuge at 375g for 30 min at room temperature (22°). Collect cells at the interface, dilute with PBS containing 0.6% ACD-A, and centrifuge at 375g for 10 min at 22°. Resuspend cells in erythrocyte lysing solution (0.15 M NH$_4$Cl, 1.0 mM KHCO$_3$, 0.1 mM EDTA, pH 7.2–7.4) and incubate for 10 min at 22°. Centrifuge at 375g for 10 min at 22° and wash once in PBS.

2. Subsequent enrichment for cells expressing CD133 can be performed with a CD133 MicroBead kit (cat. no. 130-050-801; Miltenyi Biotec, Bergisch Gladbach, Germany), utilizing superparamagnetic beads conjugated to a monoclonal mouse anti-human CD133/1 antibody and a VarioMACS separator (cat. no. 130-090-282; Miltenyi Biotec). Follow the manufacturer's recommendations and obtain ~1 × 10^6 CD133$^+$ cells/ml

with >95% purity (if necessary, repeat the enrichment with a second MACS cell separation column).

3. Prepare aliquots of ~5 × 10^4 CD133$^+$ cells/50 μl in PBS containing 2% fetal bovine serum (FBS) for staining individually with fluorochrome-conjugated anti-CD133/2 (which recognizes a different epitope than CD133/1), anti-CD34 and anti-CD38 monoclonal antibodies, as compensation controls: unstained; anti-CD133/2–phycoerythrin (PE) (cat. no. 130-090-853; Miltenyi Biotec); anti-CD34–peridinin chlorophyll protein (PerCP) (cat. no. 340430; BD Biosciences Pharmingen, San Diego, CA); and anti-CD38–allophycocyanin (APC) (cat. no. 555462; BD Biosciences Pharmingen). Incubate at 4° for 20 min. Add 2 ml of PBS containing 2% FBS. Centrifuge at 375g for 10 min at 4°. Decant the supernatant and drain. Resuspend in 300 μl of PBS containing 2% FBS. Keep on ice until analysis.

4. Identification of CD133$^+$ cells expressing high levels of ALDH activity can be facilitated by using a commercial kit for BAAA staining (ALDEFLUOR kit, cat. no. 01700; StemCell Technologies). Centrifuge 1 × 10^6 CD133$^+$ cells at 375g for 10 min at 25°. Decant the supernatant and drain. Resuspend CD133$^+$ cells in 1 ml of proprietary ALDEFLUOR assay buffer (containing an inhibitor of ABCB1 transporter efflux activity) with the ALDEFLUOR reagent (BAAA-DA; BODIPY-aminoacetaldehyde diethyl acetal) according to the manufacturer's instructions. BAAA-DA is dissolved in dimethylsulfoxide and exposed to hydrochloric acid to convert it to the ALDH substrate BAAA. As BAAA diffuses freely across the cell membrane, all the viable cells will be fluorescent. However, cells with high ALDH activity metabolize the substrate into BAA (BODIPY-aminoacetate) containing a charged carboxylate group and become intensely fluorescent. Including an inhibitor of ABCB1 transporter efflux activity throughout the assay ensures retention of the fluorescent BAA compound within the cell. Cells incubated in the presence of diethylaminobenzaldehyde (DEAB), a potent ALDH inhibitor, provide a control for background BAA fluorescence.

5. On completion of the assay (30–60 min), stain aliquots of the cells with combinations of anti-CD133/2–PE, anti-CD34–PerCP, and anti-CD38–APC at 2–8° as described previously. After 20 min, centrifuge all samples at 375g for 10 min at 4°. Decant the supernatant and drain. Resuspend samples in ALDEFLUOR assay buffer.

6. Analyze the samples with a flow cytometer equipped for excitation wavelengths of 488 and 633 nm. Detect scatter and fluorescence signals with 488/10 bandpass (BP) filters for SSC and forward scatter signals, 530/30 BP for BAA fluorescence, 576/26 BP for anti-CD133–PE, 675/20 BP for anti-CD34–PerCP, and 660/20 BP for anti-CD38–APC signals.

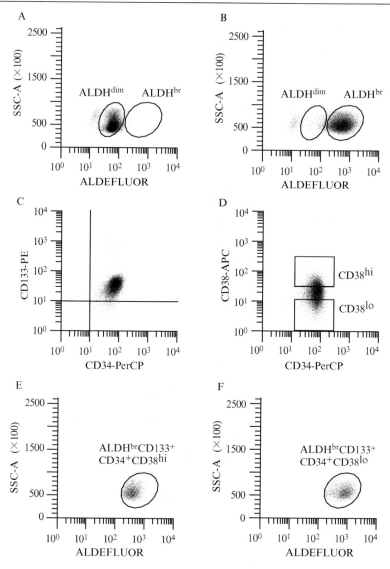

FIG. 1. Flow cytometric characterization of human CD133$^+$ cord blood cells expressing high levels of ALDH activity. (A) Negative control: Human cord blood cells highly enriched for CD133 expression (>95% CD133$^+$), exhibiting low side scatter (SSC^{lo}), and stained with BAAA (using the ALDEFLUOR reagent) in the presence of DEAB, a potent ALDH inhibitor, show background levels of BAA fluorescence (ALDH^{dim}). Events to the left of the ALDH^{dim} gate represent dead cells with no BAA fluorescence. (B) Human cord blood cells highly enriched for CD133 expression (>95% CD133$^+$), exhibiting low side scatter (SSC^{lo}) and stained with BAAA (using the ALDEFLUOR reagent) in the absence of DEAB, show

7. Fluorescence-activated cell sorting of CD133$^+$ cells expressing the highest levels of ALDH activity enriches for candidate human HSCs with a predominantly SSC^{lo}ALDH^{br}CD133$^+$CD34$^+$CD38^{lo} phenotype (Fig. 1). *Note:* Because dead and dying cells without intact cellular membranes cannot retain the fluorescent BAA derivative, only viable cells are identified by this method.

Functional Characterization of Candidate HSCs

Surrogate In Vivo Assays

Heterogeneity of the human HSC compartment and continued questions regarding cell surface phenotype necessitated the use of *in vivo* assays of HSC function (Baum *et al.*, 1992; Bhatia *et al.*, 1998; Dao *et al.*, 2003; Dorrell *et al.*, 2000; Gallacher *et al.*, 2000; Glimm *et al.*, 2001; Guenechea *et al.*, 2001; Mazurier *et al.*, 2003; Sieburg *et al.*, 2006; Wang *et al.*, 2003; Zanjani *et al.*, 1998).

Several xenogeneic transplant models have been developed as surrogate assays of human hematopoietic repopulating cells. The preimmune fetal sheep transplant assay has emerged as a useful large animal model (Civin *et al.*, 1996a; Zanjani *et al.*, 1996). However, the majority of functional assays of human HSC activity involve transplantation into immunodeficient mice with various degrees of residual natural immunity (Bhatia *et al.*, 1997b; Bock *et al.*, 1995; Cashman *et al.*, 1997; Cheng *et al.*, 1998; Gimeno *et al.*, 2004; Glimm *et al.*, 2001; Goldman *et al.*, 1998; Guenechea *et al.*, 2001; Hiramatsu *et al.*, 2003; Hogan *et al.*, 1997; Ishikawa *et al.*, 2002; Ito *et al.*, 2002; Kamel-Reid and Dick, 1988; Kollet *et al.*, 2000; Kyoizumi *et al.*, 1992; Lapidot *et al.*, 1992; Larochelle *et al.*, 1996; Lowry *et al.*, 1996; Mazurier *et al.*, 1999; McCune *et al.*, 1991; Meyerrose *et al.*, 2003; Nolta *et al.*, 1994; Pflumio *et al.*, 1996; Shultz *et al.*, 2005; Traggiai *et al.*, 2004; Vormoor *et al.*, 1994; Wang *et al.*, 1997). The most widely used of these small animal models is the

---

that almost all the cells expressed high levels of ALDH activity (ALDH^{br}). (C) The vast majority of ALDH^{br}CD133$^+$ cells coexpress the CD34 HSC surface antigen. (D–F) Flow cytometric analysis indicates that cells within the more primitive CD133$^+$CD34$^+$CD38^{lo} subpopulation express higher levels of ALDH activity than do cells within the CD133$^+$CD34$^+$CD38^{hi} subpopulation. (D) Gating strategy for CD133$^+$CD34$^+$CD38^{hi} and CD133$^+$CD34$^+$CD38^{lo} subpopulations. (E) CD133$^+$CD34$^+$CD38^{hi} cells are enriched for cells with the lowest levels of BAA fluorescence within the ALDH^{br} gate. (F) CD133$^+$CD34$^+$CD38^{lo} cells are enriched for cells with the highest levels of BAA fluorescence within the ALDH^{br} gate. Flow cytometry data were acquired with a FACSAria instrument (BD Biosciences Immunocytometry Systems, San Jose, CA) and analyzed with WinList 3D version 6.0 prerelease software (Verity Software House, Topsham, ME).

NOD.CB17-*Prkdc*^{scid} mouse—nonobese diabetic (NOD) mice crossed with severe combined immunodeficient (SCID) mice (Bhatia *et al.*, 1997b; Cashman *et al.*, 1997; Hogan *et al.*, 1997; Larochelle *et al.*, 1996; Lowry *et al.*, 1996; Pflumio *et al.*, 1996; Shultz *et al.*, 1995; Wang *et al.*, 1997). NOD/SCID mice support human cell engraftment because of defective rearrangement of T cell receptor and immunoglobulin (Ig) genes, resulting in defects of functional T and B cells; they also have low levels of natural killer cell cytotoxic activity, functionally immature macrophages, and an absence of hemolytic complement. Candidate HSCs collectively termed SCID-repopulating cells (SRCs) are scored positive for engraftment if ~1% CD45$^+$ human cells or >0.1% human DNA can be detected in the bone marrow of NOD/SCID recipients at or greater than 6 weeks posttransplantation. Under most conditions, the NOD/SCID xenograft assay does not require administration of exogenous human cytokines; however, a sublethal conditioning regimen of 250–400 cGy of irradiation is necessary, and cytokine administration or coadministration of accessory cells facilitates engraftment at limiting doses (Bonnet *et al.*, 1999). Under these conditions, the frequency of SRCs in human cord blood cells was determined to be 1 in 9.3×10^5 mononuclear cells (Wang *et al.*, 1997) and 1 in 617 CD34$^+$CD38$^-$Lin$^-$ cells (Bhatia *et al.*, 1997b). Although both lymphoid and myeloid cell populations are found, a shortcoming of the NOD/SCID xenograft assay is the general lack of T cell development, and differentiation of human hematopoietic precursors is limited mainly to immature cells belonging to the B cell and, to a lesser degree, myeloid lineages. Other disadvantages of the NOD/SCID mouse model include its high sensitivity to irradiation and relatively short life span (~80% of female and ~50% of male NOD/SCID mice develop lethal thymic lymphomas by 20 weeks of age).

Attempts to obtain an improved host for human HSC transplantation led to the development of a strain of immunodeficient mouse in which the residual low natural killer activity present in the NOD/SCID mouse was eliminated by backcrossing the β_2-microglobulin null allele onto the NOD/SCID background (NOD/SCID/B2m$^{-/-}$) (Kollet *et al.*, 2000). NOD/SCID/B2m$^{-/-}$ mice support a more than 11-fold higher level of SRC frequency than NOD/SCID mice, with transplantation of $\sim 8 \times 10^4$ human cord blood mononuclear cells resulting in multilineage differentiation in the mouse bone marrow (Kollet *et al.*, 2000). The enhanced SRC frequency in NOD/SCID/B2m$^{-/-}$ mice is due to short-term repopulation by myeloid-restricted CD34$^+$CD38$^+$ cells and a predominantly CD34$^+$CD38$^-$ population that has broader lymphomyeloid differentiation potential but that does not efficiently engraft NOD/SCID mice (Glimm *et al.*, 2001). A limitation of NOD/SCID/B2m$^{-/-}$ mice is a relatively short life span due to earlier onset and increased incidence of thymic lymphomas (the mean life span of

NOD/SCID/B2m$^{-/-}$ mice is ~11 weeks shorter than that of NOD/SCID mice) (Christianson *et al.*, 1997). New NOD/SCID models for human HSC engraftment have been reported that lack a functional X-linked common cytokine receptor γ-chain gene (NOD/SCID/γ_c^-) (Ito *et al.*, 2002; Shultz *et al.*, 2005; Yahata *et al.*, 2002). NOD/SCID/γ_c^- mice support ~6-fold higher percentages of human hematopoietic cells in the host bone marrow than do NOD/SCID mice, with precursors developing into mature human CD3$^+$CD4$^+$ and CD3$^+$CD8$^+$ T cells, Ig$^+$ B cells, natural killer cells, myeloid cells, and plasmacytoid dendritic cells. Notably, NOD/SCID/γ_c^- mice survive beyond 16 months of age and even after sublethal irradiation resist lymphoma development.

Other immunodeficient mouse models have been created by crossing mice with a deficient recombinase activating gene 2 (*Rag2*) with mice harboring the γ_c cytokine receptor gene deletion (Gimeno *et al.*, 2004; Goldman *et al.*, 1998; Mazurier *et al.*, 1999; Traggiai *et al.*, 2004; Weijer *et al.*, 2002). Rag2$^{-/-}\gamma_c^-$ mice are characterized by an absence of all T cell, B cell, and natural killer cell function and show no spontaneous lymphoma development. However, efficient human multilineage hematopoietic engraftment in Rag2$^{-/-}\gamma_c^-$ mice with a mixed H-2 major histocompatibility locus background requires exogenous human cytokines (Mazurier *et al.*, 1999).

As noted earlier, the TPO receptor c-Mpl is a selective marker of mouse and human HSCs (Hashiyama *et al.*, 1996; Ninos *et al.*, 2006; Solar *et al.*, 1998). Consistent with this observation, TPO has been demonstrated to be an important HSC supportive factor (Alexander *et al.*, 1996; Fox *et al.*, 2002; Kaushansky, 2003a; Petzer *et al.*, 1996; Solar *et al.*, 1998) in addition to being the physiologic regulator of megakaryocytopoiesis and thrombopoiesis (Kaushansky, 2003b). One report suggested that human TPO is a major limiting factor for multilineage outgrowth of human hematopoietic cells in NOD/SCID mice (Verstegen *et al.*, 2003). To assess the effects of human TPO on hematopoietic engraftment of candidate human HSCs in Rag2$^{-/-}\gamma_c^-$ mice, we generated human TPO-producing Rag2$^{-/-}\gamma_c^-$ mice by lentiviral vector-mediated transgenesis (Lois *et al.*, 2002; Ma *et al.*, 2003; Pfeifer *et al.*, 2002; Punzon *et al.*, 2004). A self-inactivating (SIN) HIV-1-based lentiviral vector, SINF-EF-hTPO-W, was developed that expresses the human TPO cDNA from an internal human elongation factor 1α (EF1α) promoter (Ramezani and Hawley, 2002a, 2003; Ramezani *et al.*, 2000, 2003). Concentrated vesicular stomatitis virus G glycoprotein (VSV-G)-pseudotyped lentiviral vector particles (10^8 transducing units/ml) were microinjected into the perivitelline space of single-cell H-2^b Rag2$^{-/-}\gamma_c^-$ embryos and implanted into pseudopregnant recipient H-2^b Rag2$^{-/-}\gamma_c^-$ mice (Lois *et al.*, 2002; Punzon *et al.*, 2004; Ramezani and Hawley, 2002b). Polymerase chain reaction analysis of genomic tail DNA, using a forward primer located within the EF1α promoter and a reverse primer located within the human TPO cDNA, was used to detect founder animals carrying the integrated transgene (H-2^b Rag2$^{-/-}\gamma_c^-$-hTPO mice). Serum levels of human TPO in the founder mice ranged between 100 and 500 pg/ml (R. Behnam, M. B. Chase, S. Soukharev, A. Ramezani, and R. G. Hawley, unpublished data). Human CD34$^+$ hematopoietic cells were isolated from cord blood as described later and intravenously injected into sublethally irradiated (350 cGy) H-2^b Rag2$^{-/-}\gamma_c^-$-hTPO and control H-2^b Rag2$^{-/-}\gamma_c^-$ mice. As a potential preclinical predictor of the rate of platelet recovery after transplantation and thus an indication of the quality of hematopoietic engraftment (Angelopoulou et al., 2004; Bruno et al., 2004; Perez et al., 2001; Yasui et al., 2003), human platelets were evaluated in peripheral blood from weeks 1 to 8 after transplantation. Human platelets were detected in the peripheral blood of H-2^b Rag2$^{-/-}\gamma_c^-$-hTPO but not in control H-2^b Rag2$^{-/-}\gamma_c^-$ mice by week 3 (1.2 $\pm$ 0.8%), reaching 8 $\pm$ 2% by week 8 (Fig. 2). Flow cytometric analysis of nucleated peripheral blood cells revealed that all the H-2^b Rag2$^{-/-}\gamma_c^-$-hTPO mice (15 of 15) but none of the control H-2^b Rag2$^{-/-}\gamma_c^-$ mice (0 of 6) engrafted with human hematopoietic cells (17 $\pm$ 7% CD45$^+$ human cells at 6 weeks posttransplantation Fig. 3A). Slightly higher engraftment levels were obtained in mice that received coadministration of CD34$^-$Lin$^+$ accessory cells (17 vs. 13%). Of the engrafted CD45$^+$ human hematopoietic cells, 13 $\pm$ 2% were CD19$^+$ cells belonging to the B cell lineage and 26 $\pm$ 4% were CD33$^+$ myeloid cells (Fig. 3B). In contrast to the negative results obtained with adult H-2^b Rag2$^{-/-}\gamma_c^-$ recipients, transplantation of CD34$^+$ human hematopoietic progenitor cells into sublethally irradiated H-2^d Rag2$^{-/-}\gamma_c^-$ newborns leads to de novo development of T cells, B cells, natural killer cells, myeloid cells, and plasmacytoid dendritic cells, formation of structured primary and secondary lymphoid organs, and production of functional immune responses (Gimeno et al., 2004; Traggiai et al., 2004).

Human Hematopoietic Repopulating Cell Assay Protocol

1. Isolate mononuclear cells from human cord blood as described in the previous section. Enrich for cells expressing CD34 with a CD34 MicroBead kit (cat. no. 130-046-703; Miltenyi Biotec) utilizing superparamagnetic beads conjugated to a monoclonal mouse anti-human CD34 antibody and a VarioMACS separator. Follow the manufacturer's recommendations and obtain ~1 $\times$ 10^6 CD34$^+$ cells/ml with >95% purity (if necessary, repeat the enrichment with a second MACS cell separation column). Retain the CD34$^-$Lin$^+$ flow-through cells for coadministration as accessory cells.

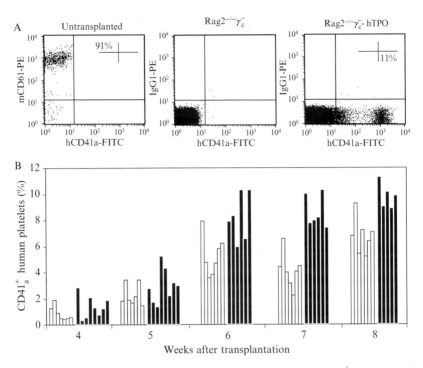

FIG. 2. Human platelet production in the peripheral blood of Rag2$^{-/-}\gamma_c^-$-hTPO mice transplanted with candidate human HSCs. (A) Sublethally irradiated (350 cGy) Rag2$^{-/-}\gamma_c^-$-hTPO and Rag2$^{-/-}\gamma_c^-$ mice were transplanted with 5×10^5 human CD34$^+$ cord blood cells. Human platelets were detected in the peripheral blood of all Rag2$^{-/-}\gamma_c^-$-hTPO mice but not Rag2$^{-/-}\gamma_c^-$ mice, determined by staining with an anti-human CD41a monoclonal antibody and gating on low forward and side scatter (platelet population gate). Shown are representative examples. Flow cytometry data were acquired with a FACSCalibur instrument and analyzed with CellQuest software (BD Biosciences Immunocytometry Systems). (B) Summary of the analysis of human CD41a$^+$ platelets within the platelet population in the peripheral blood of individual Rag2$^{-/-}\gamma_c^-$-hTPO mice 4 to 8 weeks after transplantation with 5×10^5 human CD34$^+$ cells plus (solid columns) or minus (open columns) 1×10^6 CD34$^-$Lin$^+$ accessory cells.

2. All animal procedures are carried out in accordance with Institutional Animal Care and Use Committee guidelines. H-2^b Rag2$^{-/-}\gamma_c^-$ ((C57BL/6J $\times$ C57BL/10SgSnAi)-[KO]γ_c-[KO]*Rag2*, cat. no. 004111; Taconic, Hudson, NY) and NOD/SCID (NOD.CB17-*Prkdc^{scid}*, cat. no. 001303; Jackson Laboratory, Bar Harbor, ME) immunodeficient mice are housed in sterile microisolator cages on laminar flow racks to minimize the chance of adventitious infections. Two to 6 h before transplantation, the mice are exposed to a single sublethal dose of total body γ irradiation from a ^{137}Cs source (350 cGy for H-2^b Rag2$^{-/-}\gamma_c^-$ mice; 250 cGy for NOD/SCID mice).

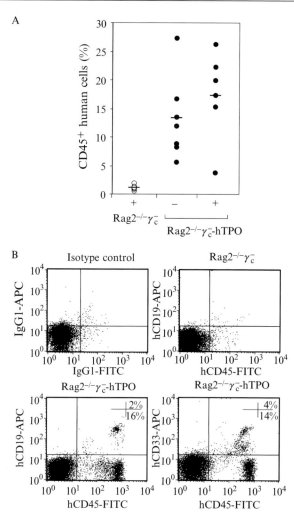

FIG. 3. Multilineage human hematopoietic engraftment in the peripheral blood of Rag2$^{-/-}\gamma_c^-$-hTPO mice transplanted with candidate human HSCs. (A) Sublethally irradiated (350 cGy) mice were transplanted with 5×10^5 human CD34$^+$ cord blood cells plus (+) or minus (–) 1×10^6 CD34$^-$Lin$^+$ accessory cells. Shown is a summary of the percentages of CD45$^+$ human cell engraftment in the peripheral blood of transplanted Rag2$^{-/-}\gamma_c^-$ (open circles) and Rag2$^{-/-}\gamma_c^-$-hTPO (solid circles) mice 6 weeks after transplantation. Each circle represents data for an individual mouse and the horizontal lines indicate the mean levels of human cells. (B) Flow cytometric analyses showing percentages of human CD45$^+$CD19$^+$ B cells and CD45$^+$CD33$^+$ myeloid cells in the peripheral blood of a representative Rag2$^{-/-}\gamma_c^-$-hTPO mouse 6 weeks posttransplantation. Flow cytometry data were acquired with a FACSCalibur instrument and analyzed with CellQuest software (BD Biosciences Immunocytometry Systems).

Baytril (active ingredient, enrofloxacin; Bayer HeathCare, Animal Health Division, Shawnee, KS) is added to the drinking water (2 ml/250 ml) immediately after irradiation and treatment is continued for 3 weeks as an additional prophylactic measure to prevent possible deaths due to adventitious infections.

3. Prepare aliquots of ~5 × 10^5 CD34$^+$ cells with or without 1 × 10^6 CD34$^-$Lin$^+$ cells (as accessory cells) in 200 μl of PBS and transplant into sublethally irradiated 8- to 10-week-old immunodeficient mice via intravenous tail vein injection, using a 27-gauge needle.

4. Detection of human platelets in mouse peripheral blood: Mouse bleeding (from the retroorbital venous sinus) is performed after inhalation anesthesia with isoflurane and administration to the eye of one drop of a local anesthetic (tetracaine ophthalmic solution, 0.5% solution; Phoenix Scientific, St. Joseph, MO). At weekly intervals posttransplantation, collect peripheral blood from the retroorbital venous sinus, using microhematocrit capillary tubes (cat. no. 22–362–566; Fisher Scientific, Pittsburgh, PA), and place ~100 μl of blood into microcollection tubes containing potassium ethylenediaminetetraacetic acid (EDTA, cat. no. 41.1395.105; Sarstedt, Nümbrecht, Germany). Centrifuge at 1500g for 5 min at 22° and resuspend the pellet in 500 μl of PBS containing 2% FBS. Stain 50-μl aliquots of the cell/platelet suspension for 30 min at 22° with the following monoclonal antibodies: fluorescein isothiocyanate (FITC)-conjugated anti-human CD41a (cat. no. 555466; BD Biosciences Pharmingen) or FITC-conjugated mouse IgG$_1$ isotype control (cat. no. 349041; BD Biosciences Pharmingen), and anti-mouse CD61–PE (cat. no. 553347; BD Biosciences Pharmingen) or PE-conjugated hamster IgG$_1$ isotype control (cat. no. 553972; BD Biosciences Pharmingen). Centrifuge at 750g for 5 min at 22°. Decant the supernatant and drain. Wash in 2 ml of PBS containing 2% FBS plus 0.1% NaN$_3$. Centrifuge at 750g for 5 min at 22°. Decant the supernatant and drain. Resuspend in 500 μl of PBS containing 2% FBS plus 0.1% NaN$_3$. Platelets are analyzed on a flow cytometer equipped for excitation wavelengths of 488 and 633 nm by gating for low SSC and forward scatter signals (Perez *et al.*, 2001).

5. Detection of human hematopoietic cells in mouse bone marrow: Mice are killed under inhalation anesthesia with isoflurane by cervical dislocation at or greater than 6 weeks posttransplantation. Single-cell bone marrow suspensions are prepared by flushing the femurs and tibias with PBS containing 2% FBS, using a 21-gauge needle. Erythrocytes are removed by hypotonic lysis in 0.15 M NH$_4$Cl, 1.0 mM KHCO$_3$, 0.1 mM EDTA, pH 7.2–7.4 (Eaker *et al.*, 2004; Ramezani *et al.*, 2000). Prepare aliquots of ~1 × 10^5 cells/100 μl in PBS containing 2% FBS and stain as described in the previous section with the following monoclonal antibodies

(all from BD Biosciences Pharmingen): anti-human CD45–FITC (cat. no. 555482), FITC-conjugated mouse IgG_1 isotype control (cat. no. 349041), anti-human CD19–APC (cat. no. 555415), anti-human CD33–APC (cat. no. 551378), and APC-conjugated mouse IgG_1 isotype control (cat. no. 555751). As an additional negative control, stain the bone marrow from an untransplanted mouse. Incubate at 4° for 20 min. Add 1 ml of PBS containing 2% FBS plus 0.1% NaN_3. Centrifuge at 375g for 10 min at 4°. Decant the supernatant and drain. Resuspend in 500 μl of PBS containing 2% FBS plus 0.1% NaN_3 and analyze by flow cytometry.

Long-Term Culture of Candidate HSCs and Progenitors

Overview

Although a variety of culture conditions support some self-renewal of human hematopoietic progenitors, long-term maintenance of HSCs *in vitro*

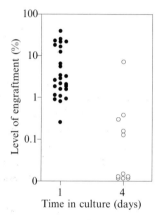

FIG. 4. Loss of human hematopoietic repopulating potential during short-term *in vitro* culture of CD34$^+$ cord blood cells. The potential of human CD34$^+$ cord blood cells to engraft in the bone marrow of NOD/SCID mice was compared for cells cultured *in vitro* for 1 or 4 days in X-VIVO-15 serum-free medium supplemented with 10% BIT 9500 serum substitute, 100 μM 2-mercaptoethanol, SCF (100 ng/ml), TPO (20 ng/ml), and Flt3 ligand (100 ng/ml). The cells (1.5 × 10^6) were harvested, mixed with 1 × 10^6 CD34$^-$Lin$^+$ accessory cells, and transplanted into sublethally irradiated (250 cGy) NOD/SCID mice. Twelve weeks after transplantation, the mice were killed and bone marrow cells were collected for flow cytometric analysis. Human cells in the mouse bone marrow were detected after staining with anti-human CD45–PE-Cy5 (cat. no. 555484; BD Biosciences Pharmingen) monoclonal antibody. Transplantation of mice with CD34$^+$ cord blood cells after 1 day of *in vitro* culture resulted in ~5% (0.5–50%) human hematopoietic cell engraftment (solid circles). The ability to repopulate NOD/SCID mouse bone marrow was significantly reduced (0–8%; mean, 0.2%) when the CD34$^+$ cord blood cells were cultured *in vitro* for 4 days (open circles).

remains a major challenge (Sauvageau *et al.*, 2004). As illustrated in Fig. 4, a significant decrease in human hematopoietic repopulating activity in NOD/SCID mice was observed after *in vitro* culture of CD34$^+$ cord blood cells for 4 days in serum-free medium supplemented with a combination of SCF, TPO, and Flt3 ligand (Larochelle *et al.*, 1996; Petzer *et al.*, 1996). However, modest expansion of SRCs (2- to 6-fold net increases) has been reported after short-term culture in serum-free medium in more complex cocktails of hematopoietic growth factors (Bhatia *et al.*, 1997a; Conneally *et al.*, 1997; Gammaitoni *et al.*, 2003). Evidence for a limited degree of *in vitro* expansion of candidate human HSCs has also been obtained under other culture conditions (Ando *et al.*, 2006; Chute *et al.*, 2005; Madlambayan *et al.*, 2005; Piacibello *et al.*, 1999; Ueda *et al.*, 2000).

We have explored another approach based on the observation that human embryonic stem cells circumvent cellular senescence by expressing the catalytic subunit of telomerase reverse transcriptase (hTERT), a specialized ribonucleoprotein complex that is responsible for adding telomeric DNA (repetitive TTAGGG sequences) to the ends of chromosomes to prevent shortening during replication (Smogorzewska and de Lange, 2004; Thomson *et al.*, 1998). Candidate human HSCs express relatively high levels of hTERT (Yui *et al.*, 1998), and telomere length analysis of human HSC subpopulations indicates that cells with the longest telomeres have the greatest proliferative potential (Bartolovic *et al.*, 2005; Van Ziffle *et al.*, 2003). Conversely, patients with aplastic anemia have short telomeres and mutations in telomerase have been identified as the cause of hematopoietic failure (Vulliamy *et al.*, 2001; Yamaguchi *et al.*, 2005). Besides progressive telomere shortening, human cells undergo senescence in response to various types of stress (Campisi, 2005). Regardless of the senescence-initiating stimuli, the signaling pathways triggered converge to varying extents on the p53 and retinoblastoma (Rb) tumor suppressors. Therefore, we employed HIV-1-based SIN lentiviral vectors to introduce the hTERT gene and the human papillomavirus type 16 (HPV16) E6 and E7 genes (Okamoto *et al.*, 2002), which accelerate the degradation of p53 and Rb, respectively (Munger *et al.*, 2004), into human CD34$^+$ cord blood cells. The transduced CD34$^+$ cells were then maintained under serum-free conditions in the presence of SCF, TPO, and Flt3 ligand, with or without IL-3 (Akimov *et al.*, 2005). Although this strategy did not result in the immortalization of human HSCs, several SCF-dependent cell lines resembling human myeloerythroid/mast cell progenitors were established in this manner, two of which express low levels of the HSC surface antigen CD133. It is important to point out, however, that the cell lines contain chromosomal aberrations (Table I). Abnormal karyotypes notwithstanding, the progenitor cell lines were not leukemogenic when injected into sublethally

TABLE I

CHARACTERISTICS OF hTERT- PLUS HPV16 E6/E7-IMMORTALIZED HUMAN CORD
BLOOD-DERIVED HEMATOPOIETIC PROGENITOR CELL LINES

Cell line	ET1a	ET2
Cell surface phenotype[a]	CD133^{lo}CD235a^{lo}CD71$^+$ CD203c$^+$ CD33$^+$CD13$^+$	CD133^{lo}CD235a^{lo}CD71$^+$ CD203c$^+$CD33$^+$CD13$^+$
Growth factor responsiveness[b]	SCF dependent	SCF dependent
Karyotype[c]	46,XY,der(22)t(17;22)[9]	45,XY,der(14)t(9;14), der(19)t(19;22),–22[8]

[a] In addition to the CD133 cell surface antigen, candidate human HSCs have been suggested to express CD33 and CD13 (Taussig et al., 2005).

[b] The ET1a and ET2 human hematopoietic progenitor cell lines require SCF for survival and proliferation but grow optimally in the presence of SCF, TPO, Flt3 ligand, and IL-3. On the basis of growth factor responsiveness, the cell lines are presumed to express CD117 (c-Kit receptor) and CD123 (the low-affinity binding subunit of the IL-3 receptor).

[c] See Akimov et al. (2005) for details.

irradiated NOD/SCID mice (Akimov et al., 2005). These findings establish the feasibility of bypassing senescence in human hematopoietic progenitors through genetic engineering, providing proof-of-principle for approaches that might eventually lead to the establishment of permanent human HSC lines. Accordingly, our future efforts will focus on extending these results by assessing the combinatorial effects of novel hematopoietic growth factors such as Notch ligands, Hedgehog proteins, Wnt molecules, bone morphogenic proteins, HOXB4 homeoprotein, and angiopoietin-like proteins (Sauvageau et al., 2004; Zhang et al., 2006a).

Protocol for Derivation of Human Hematopoietic Progenitor Cell Lines

1. The HIV-1-based SINF-MU3-hTERT-IRES-GFP-W-S and SINF-MU3-E6E7-IRES-YFP-W-S lentiviral vectors used to immortalize human hematopoietic progenitors have been described previously (Akimov et al., 2005). SINF-MU3-hTERT-IRES-GFP-W-S contains the hTERT cDNA upstream of an encephalomyocarditis virus internal ribosome entry site (IRES)–green fluorescent protein (GFP) gene cassette and SINF-MU3-E6E7-IRES-YFP-W-S contains the HPV16 E6/E7-coding region upstream of an IRES–yellow fluorescent protein (YFP) gene cassette. In both cases, transgene transcription is driven by an internal murine stem cell virus (MSCV) long terminal repeat (LTR) promoter (Hawley et al., 1994; Ramezani et al., 2000, 2003).

2. Culture human embryonic kidney 293T cells in Dulbecco's modified Eagle's medium (Invitrogen, Carlsbad, CA) supplemented with 10% heat-inactivated FBS, 2 mM L-glutamine, penicillin (50 IU/ml), and streptomycin (50 μg/ml) at 37° in a humidified atmosphere containing 5% CO_2. Plate the 293T cells (4 × 10^6) into 10-cm tissue culture dishes containing 7 ml of complete medium the day before transfection. Mix 15 μg of the transfer vector plasmid (SINF-MU3-hTERT-IRES-GFP-W-S or SINF-MU3-E6E7-IRES-YFP-W-S), 10 μg of the packaging plasmid pCMVΔR8.91 (Zufferey et al., 1997), and 5 μg of the VSV-G glycoprotein envelope plasmid pMD.G (Naldini et al., 1996). Bring the volume up to 450 μl with sterile water. Add 50 μl of 2.5 M $CaCl_2$ and mix. Add the DNA–$CaCl_2$ solution dropwise to 500 μl of 2× N-(2-hydroxyethyl)piperazine-N'-2-ethanesulfonic acid (HEPES)-buffered saline [0.283 M NaCl, 0.023 M HEPES (cat. no. H0887; Sigma-Aldrich), 1.5 mM Na_2HPO_4, pH 7.05] in a 15-ml conical tube. Use a 5-ml pipette to bubble the 2× HEPES-buffered saline while adding the DNA–$CaCl_2$ solution. Vortex immediately for 5 s and incubate for 20 min at 22°. Add the precipitate dropwise over the cells and mix gently. Incubate the cells overnight (16 h) at 37°. The next day, remove medium from the plate, rinse the cells with 5 ml of PBS, and add 7 ml of fresh medium. Collect the vector-containing medium after another 48 h, centrifuge at 2000g for 10 min to remove cellular debris, and filter through a 0.45-μm pore-size filter (Nalgene; Nalge Nunc International, Rochester, NY). Ultracentrifuge vector supernatants in 70-ml bottles (Beckman Coulter, Fullerton, CA) at 45,000g for 90 min at 4°. Resuspend pellets in 500 μl of medium by gentle vortexing for 2 h at 4°. Spin down the debris at 2000g for 5 min and store the concentrated vector particles at –80°. Titer vector stocks on human fibrosarcoma HT1080 cells and assay for the presence of replication-competent virus as previously described (Ramezani and Hawley, 2002b, 2003).

3. Coat 24-well non-tissue culture-treated plates (Lux suspension dish, cat. no. ICNLX171099; Fisher Scientific) with recombinant fibronectin fragment (RetroNectin, 2 μg/cm^2, cat. no. TAK_T100A; Takara Mirus Bio, Madison, WI). Culture CD34$^+$ cells isolated as described in the previous section at a density of 1 × 10^6 cells/ml for 24 h in X-VIVO-15 serum-free medium (cat. no. BW04-418Q; Fisher Scientific) supplemented with 10% BIT 9500 serum substitute (bovine serum albumin, insulin, and human transferrin) [cat. no. 09500; StemCell Technologies), 100 μM 2-mercaptoethanol, SCF (100 ng/ml), TPO (20 ng/ml), and Flt3 ligand (100 ng/ml), with or without IL-3 (20 ng/ml)] (all cytokines from PeproTech, Rocky Hill, NJ) at 37° in a humidified atmosphere containing 5% CO_2. Transduce the cells with lentiviral vector particles (2 × 10^6 transducing units/ml; multiplicity of infection, 2) in the presence of protamine sulfate (4 μg/ml; Sigma-Aldrich)

(Ramezani and Hawley, 2002b, 2003; Ramezani *et al.*, 2003). Change the medium after 24 h and continue culturing the cells.

4. Harvest the hematopoietic progenitor cells after an additional 48–72 h of culture (cell dissociation buffer, cat. no. 13151–014; Invitrogen), wash, and resuspend in PBS containing 2% FBS. Under the conditions employed, the majority of cells retain the CD34$^+$ phenotype (Ramezani *et al.*, 2000). Isolate GFP$^+$YFP$^+$ cells to >95% purity by fluorescence-activated cell sorting (Akimov *et al.*, 2005; Cheng *et al.*, 1997; Dorrell *et al.*, 2000; Hawley *et al.*, 2004). The YFP and GFP signals are separated with a 525-nm shortpass dichroic filter and collected with a 550/30 nm BP filter and a 510/20 nm BP filter, respectively (cat. no. XCY-500; Omega Optical, Brattleboro, VT).

5. Maintain the hematopoietic progenitor cells in continuous culture in X-VIVO-15 serum-free medium supplemented with 10% BIT 9500 serum substitute, 100 μM 2-mercaptoethanol, SCF (100 ng/ml), TPO (20 ng/ml), and Flt3 ligand (100 ng/ml), with or without IL-3 (20 ng/ml) at 37° in a humidified atmosphere containing 5% CO_2.

Acknowledgments

The authors gratefully acknowledge Reza Behnam, Michael Chase, and Serguei Soukharev for technical assistance. This work was supported in part by National Institutes of Health grants R01HL65519, R01HL66305, and R24RR16209, and by the King Fahd Endowment Fund (George Washington University School of Medicine and Health Sciences).

References

Abramson, S., Miller, R. G., and Phillips, R. A. (1977). The identification in adult bone marrow of pluripotent and restricted stem cells of the myeloid and lymphoid systems. *J. Exp. Med.* **145**, 1567–1579.

Akimov, S. S., Ramezani, A., Hawley, T. S., and Hawley, R. G. (2005). Bypass of senescence, immortalization, and transformation of human hematopoietic progenitor cells. *Stem Cells* **23**, 1423–1433.

Alexander, W. S., Roberts, A. W., Nicola, N. A., Li, R., and Metcalf, D. (1996). Deficiencies in progenitor cells of multiple hematopoietic lineages and defective megakaryocytopoiesis in mice lacking the thrombopoietic receptor c-Mpl. *Blood* **87**, 2162–2170.

Ando, K., Yahata, T., Sato, T., Miyatake, H., Matsuzawa, H., Oki, M., Miyoshi, H., Tsuji, T., Kato, S., and Hotta, T. (2006). Direct evidence for *ex vivo* expansion of human hematopoietic stem cells. *Blood* **107**, 3371–3377.

Angelopoulou, M. K., Rinder, H., Wang, C., Burtness, B., Cooper, D. L., and Krause, D. S. (2004). A preclinical xenotransplantation animal model to assess human hematopoietic stem cell engraftment. *Transfusion* **44**, 555–566.

Arai, F., Hirao, A., Ohmura, M., Sato, H., Matsuoka, S., Takubo, K., Ito, K., Koh, G. Y., and Suda, T. (2004). Tie2/angiopoietin-1 signaling regulates hematopoietic stem cell quiescence in the bone marrow niche. *Cell* **118**, 149–161.

Armstrong, L., Stojkovic, M., Dimmick, I., Ahmad, S., Stojkovic, P., Hole, N., and Lako, M. (2004). Phenotypic characterization of murine primitive hematopoietic progenitor cells isolated on basis of aldehyde dehydrogenase activity. *Stem Cells* **22,** 1142–1151.

Arndt-Jovin, D. J., and Jovin, T. M. (1977). Analysis and sorting of living cells according to deoxyribonucleic acid content. *J. Histochem. Cytochem.* **25,** 585–589.

Balazs, A. B., Fabian, A. J., Esmon, C. T., and Mulligan, R. C. (2006). Endothelial protein C receptor (CD201) explicitly identifies hematopoietic stem cells in murine bone marrow. *Blood* **107,** 2317–2321.

Bartolovic, K., Balabanov, S., Berner, B., Buhring, H. J., Komor, M., Becker, S., Hoelzer, D., Kanz, L., Hofmann, W. K., and Brummendorf, T. H. (2005). Clonal heterogeneity in growth kinetics of CD34$^+$CD38$^-$ human cord blood cells *in vitro* is correlated with gene expression pattern and telomere length. *Stem Cells* **23,** 946–957.

Baum, C. M., Weissman, I. L., Tsukamoto, A. S., Buckle, A.-M., and Peault, B. (1992). Isolation of a candidate human hematopoietic stem cell population. *Proc. Natl. Acad. Sci. USA* **89,** 2804–2808.

Bertoncello, I., and Williams, B. (2004). Hematopoietic stem cell characterization by Hoechst 33342 and rhodamine 123 staining. *Methods Mol. Biol.* **263,** 181–200.

Bertoncello, I., Hodgson, G. S., and Bradley, T. R. (1985). Multiparameter analysis of transplantable hemopoietic stem cells. I. The separation and enrichment of stem cells homing to marrow and spleen on the basis of rhodamine-123 fluorescence. *Exp. Hematol.* **13,** 999–1006.

Bertoncello, I., Hodgson, G. S., and Bradley, T. R. (1988). Multiparameter analysis of transplantable hemopoietic stem cells. II. Stem cells of long-term bone marrow-reconstituted recipients. *Exp. Hematol.* **16,** 245–249.

Bertoncello, I., Bradley, T. R., Hodgson, G. S., and Dunlop, J. M. (1991). The resolution, enrichment, and organization of normal bone marrow high proliferative potential colony-forming cell subsets on the basis of rhodamine-123 fluorescence. *Exp. Hematol.* **19,** 174–178.

Bhatia, M., Bonnet, D., Kapp, U., Wang, J. C. Y., Murdoch, B., and Dick, J. E. (1997a). Quantitative analysis reveals expansion of human hematopoietic repopulating cells after short-term *ex vivo* culture. *J. Exp. Med.* **186,** 619–624.

Bhatia, M., Wang, J. C., Kapp, U., Bonnet, D., and Dick, J. E. (1997b). Purification of primitive human hematopoietic cells capable of repopulating immune-deficient mice. *Proc. Natl. Acad. Sci. USA* **94,** 5320–5325.

Bhatia, M., Bonnet, D., Murdoch, B., Gan, O. I., and Dick, J. E. (1998). A newly discovered class of human hematopoietic cells with SCID-repopulating activity. *Nat. Med.* **4,** 1038–1045.

Bock, T. A., Orlic, D., Dunbar, C. E., Broxmeyer, H. E., and Bodine, D. M. (1995). Improved engraftment of human hematopoietic cells in severe combined immunodeficient (SCID) mice carrying human cytokine transgenes. *J. Exp. Med.* **182,** 2037–2043.

Bonnet, D. (2002). Haematopoietic stem cells. *J. Pathol.* **197,** 430–440.

Bonnet, D., Bhatia, M., Wang, J. C., Kapp, U., and Dick, J. E. (1999). Cytokine treatment or accessory cells are required to initiate engraftment of purified primitive human hematopoietic cells transplanted at limiting doses into NOD/SCID mice. *Bone Marrow Transplant.* **23,** 203–209.

Bruno, S., Gunetti, M., Gammaitoni, L., Perissinotto, E., Caione, L., Sanavio, F., Fagioli, F., Aglietta, M., and Piacibello, W. (2004). Fast but durable megakaryocyte repopulation and platelet production in NOD/SCID mice transplanted with *ex-vivo* expanded human cord blood CD34$^+$ cells. *Stem Cells* **22,** 135–143.

Campisi, J. (2005). Senescent cells, tumor suppression, and organismal aging: Good citizens, bad neighbors. *Cell* **120,** 513–522.

Capel, B., Hawley, R. G., and Mintz, B. (1990). Long- and short-lived murine hemato-poietic stem cell clones individually identified with retroviral integration markers. *Blood* **75,** 2267–2270.

Cashman, J. D., Lapidot, T., Wang, J. C., Doedens, M., Shultz, L. D., Lansdorp, P., Dick, J. E., and Eaves, C. J. (1997). Kinetic evidence of the regeneration of multilineage hematopoiesis from primitive cells in normal human bone marrow transplanted into immunodeficient mice. *Blood* **89,** 4307–4316.

Chaudhary, P. M., and Roninson, I. B. (1991). Expression and activity of P-glycoprotein, a multidrug efflux pump, in human hematopoietic stem cells. *Cell* **66,** 85–94.

Chen, C. Z., Li, M., de Graaf, D., Monti, S., Gottgens, B., Sanchez, M. J., Lander, E. S., Golub, T. R., Green, A. R., and Lodish, H. F. (2002). Identification of endoglin as a functional marker that defines long-term repopulating hematopoietic stem cells. *Proc. Natl. Acad. Sci. USA* **99,** 15468–15473.

Cheng, L., Du, C., Murray, D., Tong, X., Zhang, Y. A., Chen, B. P., and Hawley, R. G. (1997). A GFP reporter system to assess gene transfer and expression in viable human hematopoietic progenitors. *Gene Ther.* **4,** 1013–1022.

Cheng, L., Du, C., Lavau, C., Chen, S., Tong, J., Chen, B. P., Scollay, R., Hawley, R. G., and Hill, B. (1998). Sustained gene expression in retrovirally transduced, engrafting human hematopoietic stem cells and their lympho-myeloid progeny. *Blood* **92,** 83–92.

Christianson, S. W., Greiner, D. L., Hesselton, R. A., Leif, J. H., Wagar, E. J., Schweitzer, I. B., Rajan, T. V., Gott, B., Roopenian, D. C., and Shultz, L. D. (1997). Enhanced human CD4$^+$ T cell engraftment in β_2-microglobulin-deficient NOD-scid mice. *J. Immunol.* **158,** 3578–3586.

Chute, J. P., Muramoto, G. G., Fung, J., and Oxford, C. (2005). Soluble factors elaborated by human brain endothelial cells induce the concomitant expansion of purified human BM CD34$^+$ CD38$^-$ cells and SCID-repopulating cells. *Blood* **105,** 576–583.

Civin, C., Strauss, L. C., Brovall, C., Fackler, M. J., Schwartz, J. F., and Shaper, J. H. (1984). Antigenic analysis of haematopoiesis. III. A haematopoietic progenitor cell sur-face antigen defined by a monoclonal antibody raised against KG1a cells. *J. Immunol.* **133,** 157–165.

Civin, C. I., Almeida-Porada, G., Lee, M. J., Olweus, J., Terstappen, L. W., and Zanjani, E. D. (1996a). Sustained, retransplantable, multilineage engraftment of highly purified adult human bone marrow stem cells *in vivo*. *Blood* **88,** 4102–4109.

Civin, C. I., Trischmann, T., Kadan, N. S., Davis, J., Noga, S., Cohen, K., Duffy, B., Groenewegen, I., Wiley, J., Law, P., Hardwich, A., Oldham, F., and Gee, A. (1996b). Highly purified CD34-positive cells reconstitute hematopoiesis. *J. Clin. Oncol.* **14,** 2224–2233.

Conneally, E., Cashman, J., Petzer, A., and Eaves, C. (1997). Expansion *in vitro* of transplantable human cord blood stem cells demonstrated using a quantitative assay of their lympho-myeloid repopulating activity in nonobese diabetic-scid/scid mice. *Proc. Natl. Acad. Sci. USA* **94,** 9836–9841.

Dao, M. A., Arevalo, J., and Nolta, J. A. (2003). Reversibility of CD34 expression on human hematopoietic stem cells that retain the capacity for secondary reconstitution. *Blood* **101,** 112–118.

de Wynter, E. A., Buck, D., Hart, C., Heywood, R., Coutinho, L. H., Clayton, A., Rafferty, J. A., Burt, D., Guenechea, G., Bueren, J. A., Gagen, D., Fairbairn, L. J., Lord, B. I., and Testa, N. G. (1998). CD34$^+$AC133$^+$ cells isolated from cord blood are highly enriched in long-term culture-initiating cells, NOD/SCID-repopulating cells and dendritic cell progenitors. *Stem Cells* **16,** 387–396.

Dorrell, C., Gan, O. I., Pereira, D. S., Hawley, R. G., and Dick, J. E. (2000). Expansion of human cord blood CD34$^+$CD38$^-$ cells in *ex vivo* culture during retroviral transduction without a corresponding increase in SCID repopulating cell (SRC) frequency: Dissociation of SRC phenotype and function. *Blood* **95,** 102–110.

Eaker, S. S., Hawley, T. S., Ramezani, A., and Hawley, R. G. (2004). Detection and enrichment of hematopoietic stem cells by side population phenotype. *Methods Mol. Biol.* **263,** 161–180.

Fallon, P., Gentry, T., Balber, A., Boulware, D., Janssen, W., Smilee, R., Storms, R., and Smith, C. (2003). Mobilized peripheral blood SSCloALDHbr cells have the phenotypic and functional properties of primitive haematopoietic cells and their number correlates with engraftment following autologous transplantation. *Br. J. Haematol.* **122,** 99–108.

Fox, N., Priestley, G., Papayannopoulou, T., and Kaushansky, K. (2002). Thrombopoietin expands hematopoietic stem cells after transplantation. *J. Clin. Invest.* **110,** 389–394.

Gallacher, L., Murdoch, B., Wu, D. M., Karanu, F. N., Keeney, M., and Bhatia, M. (2000). Isolation and characterization of human CD34$^-$Lin$^-$ and CD34$^+$Lin$^-$ hematopoietic stem cells using cell surface markers AC133 and CD7. *Blood* **95,** 2813–2820.

Gammaitoni, L., Bruno, S., Sanavio, F., Gunetti, M., Kollet, O., Cavalloni, G., Falda, M., Fagioli, F., Lapidot, T., Aglietta, M., and Piacibello, W. (2003). *Ex vivo* expansion of human adult stem cells capable of primary and secondary hemopoietic reconstitution. *Exp. Hematol.* **31,** 261–270.

Gao, Z., Fackler, M. J., Leung, W., Lumkul, R., Ramirez, M., Theobald, N., Malech, H. L., and Civin, C. I. (2001). Human CD34$^+$ cell preparations contain over 100-fold greater NOD/SCID mouse engrafting capacity than do CD34$^-$ cell preparations. *Exp. Hematol.* **29,** 910–921.

Gimeno, R., Weijer, K., Voordouw, A., Uittenbogaart, C. H., Legrand, N., Alves, N. L., Wijnands, E., Blom, B., and Spits, H. (2004). Monitoring the effect of gene silencing by RNA interference in human CD34$^+$ cells injected into newborn RAG2$^{-/-}$ $\gamma_c^{-/-}$ mice: Functional inactivation of p53 in developing T cells. *Blood* **104,** 3886–3893.

Glimm, H., Eisterer, W., Lee, K., Cashman, J., Holyoake, T. L., Nicolini, F., Shultz, L. D., Von, K. C., and Eaves, C. J. (2001). Previously undetected human hematopoietic cell populations with short-term repopulating activity selectively engraft NOD/SCID-β_2 microglobulin-null mice. *J. Clin. Invest.* **107,** 199–206.

Goldman, J. P., Blundell, M. P., Lopes, L., Kinnon, C., Di Santo, J. P., and Thrasher, A. J. (1998). Enhanced human cell engraftment in mice deficient in RAG2 and the common cytokine receptor γ chain. *Br. J. Haematol.* **103,** 335–342.

Goodell, M. A., Brose, K., Paradis, G., Conner, A. S., and Mulligan, R. C. (1996). Isolation and functional properties of murine hematopoietic stem cells that are replicating *in vivo*. *J. Exp. Med.* **183,** 1797–1806.

Goodell, M. A., Rosenzweig, M., Kim, H., Marks, D. F., DeMaria, M., Paradis, G., Grupp, S. A., Sieff, C. A., Mulligan, R. C., and Johnson, R. P. (1997). Dye efflux studies suggest that hematopoietic stem cells expressing low or undetectable levels of CD34 antigen exist in multiple species. *Nat. Med.* **3,** 1337–1345.

Gordon, M. Y., Goldman, J. M., and Gordon-Smith, E. C. (1985). 4-Hydroperoxycyclophosphamide inhibits proliferation by human granulocyte-macrophage colony-forming cells (GM-CFC) but spares more primitive progenitor cells. *Leuk. Res.* **9,** 1017–1021.

Guenechea, G., Gan, O. I., Dorrell, C., and Dick, J. E. (2001). Distinct classes of human stem cells that differ in proliferative and self-renewal potential. *Nat. Immunol.* **2,** 75–82.

Gunji, Y., Nakamura, M., Osawa, H., Nagayoshi, K., Nakauchi, H., Miura, Y., Yanagisawa, M., and Suda, T. (1993). Human primitive hematopoietic progenitor cells are more enriched in KIT^{low} cells than in KIT^{high} cells. *Blood* **82,** 3283–3289.

Harrison, D. E. (1980). Competitive repopulation: A new assay for long-term stem cell functional capacity. *Blood* **55,** 77–81.

Harrison, D. E., Jordan, C. T., Zhong, R. K., and Astle, C. M. (1993). Primitive hemopoietic stem cells: Direct assay of most productive populations by competitive repopulation with simple binomial, correlation and covariance calculations. *Exp. Hematol.* **21,** 206–219.

Hashiyama, M., Iwama, A., Ohshiro, K., Kurozumi, K., Yasunaga, K., Shimizu, Y., Masuho, Y., Matsuda, I., Yamaguchi, N., and Suda, T. (1996). Predominant expression of a receptor tyrosine kinase, TIE, in hematopoietic stem cells and B cells. *Blood* **87,** 93–101.

Hawley, R. G. (2001). Progress toward vector design for hematopoietic stem cell gene therapy. *Curr. Gene Ther.* **1,** 1–17.

Hawley, R. G., Lieu, F. H. L., Fong, A. Z. C., and Hawley, T. S. (1994). Versatile retroviral vectors for potential use in gene therapy. *Gene Ther.* **1,** 136–138.

Hawley, T. S., Herbert, D. J., Eaker, S. S., and Hawley, R. G. (2004). Multiparameter flow cytometry of fluorescent protein reporters. *Methods Mol. Biol.* **263,** 219–238.

Hess, D. A., Meyerrose, T. E., Wirthlin, L., Craft, T. P., Herrbrich, P. E., Creer, M. H., and Nolta, J. A. (2004). Functional characterization of highly purified human hematopoietic repopulating cells isolated according to aldehyde dehydrogenase activity. *Blood* **104,** 1648–1655.

Hess, D. A., Wirthlin, L., Craft, T. P., Herrbrich, P. E., Hohm, S. A., Lahey, R., Eades, W. C., Creer, M. H., and Nolta, J. A. (2006). Selection based on CD133 and high aldehyde dehydrogenase activity isolates long-term reconstituting human hematopoietic stem cells. *Blood* **107,** 2162–2169.

Hill, B., Rozler, E., Travis, M., Chen, S., Zannetino, A., Simmons, P., Galy, A., Chen, B., and Hoffman, R. (1996). High-level expression of a novel epitope of CD59 identifies a subset of CD34$^+$ bone marrow cells highly enriched for pluripotent stem cells. *Exp. Hematol.* **24,** 936–943.

Hiramatsu, H., Nishikomori, R., Heike, T., Ito, M., Kobayashi, K., Katamura, K., and Nakahata, T. (2003). Complete reconstitution of human lymphocytes from cord blood CD34$^+$ cells using the NOD/SCID/γ_c^{null} mice model. *Blood* **102,** 873–880.

Hogan, C. J., Shpall, E. J., McNulty, O., McNiece, I., Dick, J. E., Shultz, L. D., and Keller, G. (1997). Engraftment and development of human CD34$^+$-enriched cells from umbilical cord blood in NOD/LtSz-*scid/scid* mice. *Blood* **90,** 85–96.

Ishikawa, F., Livingston, A. G., Wingard, J. R., Nishikawa, S., and Ogawa, M. (2002). An assay for long-term engrafting human hematopoietic cells based on newborn NOD/SCID/β_2-microglobulin^{null} mice. *Exp. Hematol.* **30,** 488–494.

Ito, M., Hiramatsu, H., Kobayashi, K., Suzue, K., Kawahata, M., Hioki, K., Ueyama, Y., Koyanagi, Y., Sugamura, K., Tsuji, K., Heike, T., and Nakahata, T. (2002). NOD/SCID/γ_c^{null} mouse: An excellent recipient mouse model for engraftment of human cells. *Blood* **100,** 3175–3182.

Ito, T., Tajima, F., and Ogawa, M. (2000). Developmental changes of CD34 expression by murine hematopoietic stem cells. *Exp. Hematol.* **28,** 1269–1273.

Iwama, A., Hamaguchi, I., Hashiyama, M., Murayama, Y., Yasunaga, K., and Suda, T. (1993). Molecular cloning and characterization of mouse TIE and TEK receptor tyrosine kinase genes and their expression in hematopoietic stem cells. *Biochem. Biophys. Res. Commun.* **195,** 301–309.

Johnson, L. V., Walsh, M. L., and Chen, L. B. (1980). Localization of mitochondria in living cells with rhodamine 123. *Proc. Natl. Acad. Sci. USA* **77,** 990–994.

Jones, R. J., Barber, J. P., Vala, M. S., Collector, M. I., Kaufmann, S. H., Ludeman, S. M., Colvin, O. M., and Hilton, J. (1995). Assessment of aldehyde dehydrogenase in viable cells. *Blood* **85,** 2742–2746.

Jones, R. J., Collector, M. I., Barber, J. P., Vala, M. S., Fackler, M. J., May, W. S., Griffin, C. A., Hawlekins, A. L., Zehnbauer, B. A., Hilton, J., Colvin, O. M., and Sharkis, S. J.

(1996). Characterization of mouse lymphohematopoietic stem cells lacking spleen colony-forming activity. *Blood* **88**, 487–491.

Jordan, C. T., and Lemischka, I. R. (1990). Clonal and systemic analysis of long-term hematopoiesis in the mouse. *Genes Dev.* **4**, 220–232.

Juliano, R. L., and Ling, V. (1976). A surface glycoprotein modulating drug permeability in Chinese hamster ovary cell mutants. *Biochim. Biophys. Acta* **455**, 152–162.

Kamel-Reid, S., and Dick, J. E. (1988). Engraftment of immune-deficient mice with human hematopoietic stem cells. *Science* **242**, 1706–1709.

Kastan, M. B., Schlaffer, E., Russo, J. E., Colvin, O. M., Civin, C. I., and Hilton, J. (1990). Direct demonstration of elevated aldehyde dehydrogenase in human hematopoietic progenitor cells. *Blood* **75**, 1947–1950.

Kaushansky, K. (2003a). Thrombopoietin: Accumulating evidence for an important biological effect on the hematopoietic stem cell. *Ann. N. Y. Acad. Sci.* **996**, 39–43.

Kaushansky, K. (2003b). Thrombopoietin: A tool for understanding thrombopoiesis. *J. Thromb. Haemost.* **1**, 1587–1592.

Kawashima, I., Zanjani, E. D., Almeida-Porada, G., Flake, A. W., Zeng, H., and Ogawa, M. (1996). CD34$^+$ human marrow cells that express low levels of Kit protein are enriched for long-term marrow-engrafting cells. *Blood* **87**, 4136–4142.

Keller, G., Paige, C., Gilboa, E., and Wagner, E. F. (1985). Expression of a foreign gene in myeloid and lymphoid cells derived from multipotent haematopoietic precursors. *Nature* **318**, 149–154.

Kiel, M. J., Yilmaz, O. H., Iwashita, T., Yilmaz, O. H., Terhorst, C., and Morrison, S. J. (2005). SLAM family receptors distinguish hematopoietic stem and progenitor cells and reveal endothelial niches for stem cells. *Cell* **121**, 1109–1121.

Kimura, T., Minamiguchi, H., Wang, J., Kaneko, H., Nakagawa, H., Fujii, H., and Sonoda, Y. (2004). Impaired stem cell function of CD34$^+$ cells selected by two different immuno-magnetic beads systems. *Leukemia* **18**, 566–574.

Kollet, O., Peled, A., Byk, T., Ben-Hur, H., Greiner, D., Shultz, L., and Lapidot, T. (2000). β_2 Microglobulin-deficient (B2m^{null}) NOD/SCID mice are excellent recipients for studying human stem cell function. *Blood* **95**, 3102–3105.

Krause, D. S., Fackler, M. J., Civin, C. I., and May, W. S. (1996). CD34: Structure, biology, and clinical utility. *Blood* **87**, 1–13.

Kyoizumi, S., Baum, C. M., Kaneshima, H., McCune, J. M., Yee, E. J., and Namikawa, R. (1992). Implantation and maintenance of functional human bone marrow in *SCID*-hu mice. *Blood* **79**, 1704–1711.

Lang, P., Bader, P., Schumm, M., Feuchtinger, T., Einsele, H., Fuhrer, M., Weinstock, C., Handgretinger, R., Kuci, S., Martin, D., Niethammer, D., and Greil, J. (2004). Transplantation of a combination of CD133$^+$ and CD34$^+$ selected progenitor cells from alternative donors. *Br. J. Haematol.* **124**, 72–79.

Lapidot, T., Pflumio, F., Doedens, M., Murdoch, B., Williams, D. E., and Dick, J. E. (1992). Cytokine stimulation of multilineage hematopoiesis from immature human cells engrafted in scid mice. *Science* **255**, 1137–1141.

Larochelle, A., Vormoor, J., Hannenberg, H., Wang, J. C. Y., Bhatia, M., Lapidot, T., Moritz, T., Murdoch, B., Xiao, X. L., Kato, I., Williams, D. A., and Dick, J. E. (1996). Identification of primitive human hematopoietic cells capable of repopulating NOD/SCID mouse bone marrow: Implications for gene therapy. *Nat. Med.* **2**, 1329–1337.

Leemhuis, T., Yoder, M. C., Grigsby, S., Aguero, B., Eder, P., and Srour, E. F. (1996). Isolation of primitive human bone marrow hematopoietic progenitor cells using Hoechst 33342 and rhodamine 123. *Exp. Hematol.* **24**, 1215–1224.

Lois, C., Hong, E. J., Pease, S., Brown, E. J., and Baltimore, D. (2002). Germline transmission and tissue-specific expression of transgenes delivered by lentiviral vectors. *Science* **295**, 868–872.

Lowry, P. A., Shultz, L. D., Greiner, D. L., Hesselton, R. M., Kittler, E. L., Tiarks, C. Y., Rao, S. S., Reilly, J., Leif, J. H., Ramshaw, H., Stewart, F. M., and Quesenberry, P. J. (1996). Improved engraftment of human cord blood stem cells in NOD/LtSz-*scid/scid* mice after irradiation or multiple-day injections into unirradiated recipients. *Biol. Blood Marrow Transplant* **2**, 15–23.

Ma, Y., Ramezani, A., Lewis, R., Hawley, R. G., and Thomson, J. A. (2003). High-level sustained transgene expression in human embryonic stem cells using lentiviral vectors. *Stem Cells* **21**, 111–117.

Madlambayan, G. J., Rogers, I., Kirouac, D. C., Yamanaka, N., Mazurier, F., Doedens, M., Casper, R. F., Dick, J. E., and Zandstra, P. W. (2005). Dynamic changes in cellular and microenvironmental composition can be controlled to elicit *in vitro* human hematopoietic stem cell expansion. *Exp. Hematol.* **33**, 1229–1239.

Matsubara, A., Iwama, A., Yamazaki, S., Furuta, C., Hirasawa, R., Morita, Y., Osawa, M., Motohashi, T., Eto, K., Ema, H., Kitamura, T., Vestweber, D., and Nakauchi, H. (2005). Endomucin, a CD34-like sialomucin, marks hematopoietic stem cells throughout development. *J. Exp. Med.* **202**, 1483–1492.

Matsuoka, S., Ebihara, Y., Xu, M., Ishii, T., Sugiyama, D., Yoshino, H., Ueda, T., Manabe, A., Tanaka, R., Ikeda, Y., Nakahata, T., and Tsuji, K. (2001). CD34 expression on long-term repopulating hematopoietic stem cells changes during developmental stages. *Blood* **97**, 419–425.

Matsuzaki, Y., Kinjo, K., Mulligan, R. C., and Okano, H. (2004). Unexpectedly efficient homing capacity of purified murine hematopoietic stem cells. *Immunity* **20**, 87–93.

Mazurier, F., Fontanellas, A., Salesse, S., Taine, L., Landriau, S., Moreau-Gaudry, F., Reiffers, J., Peault, B., Di Santo, J. P., and de, V. H. (1999). A novel immunodeficient mouse model—RAG2 × common cytokine receptor γ chain double mutants—requiring exogenous cytokine administration for human hematopoietic stem cell engraftment. *J. Interferon Cytokine Res.* **19**, 533–541.

Mazurier, F., Doedens, M., Gan, O. I., and Dick, J. E. (2003). Rapid myeloerythroid repopulation after intrafemoral transplantation of NOD-SCID mice reveals a new class of human stem cells. *Nat. Med.* **9**, 959–963.

McAlister, I., Wolf, N. S., Pietrzyk, M. E., Rabinovitch, P. S., Priestley, G., and Jaeger, B. (1990). Transplantation of hematopoietic stem cells obtained by a combined dye method fractionation of murine bone marrow. *Blood* **75**, 1240–1246.

McCormack, M. P., and Rabbitts, T. H. (2004). Activation of the T-cell oncogene LMO2 after gene therapy for X-linked severe combined immunodeficiency. *N. Engl. J. Med.* **350**, 913–922.

McCulloch, E. A., and Till, J. E. (2005). Perspectives on the properties of stem cells. *Nat. Med.* **11**, 1026–1028.

McCune, J. M., Peault, B., Streeter, P. R., and Rabin, L. (1991). Preclinical evaluation of human hematolymphoid function in the SCID-hu mouse. *Immunol. Rev.* **124**, 45–62.

Meyerrose, T. E., Herrbrich, P., Hess, D. A., and Nolta, J. A. (2003). Immune-deficient mouse models for analysis of human stem cells. *BioTechniques* **35**, 1262–1272.

Moayeri, M., Hawley, T. S., and Hawley, R. G. (2005). Correction of murine hemophilia A by hematopoietic stem cell gene therapy. *Mol. Ther.* **12**, 1034–1042.

Munger, K., Baldwin, A., Edwards, K. M., Hayakawa, H., Nguyen, C. L., Owens, M., Grace, M., and Huh, K. (2004). Mechanisms of human papillomavirus-induced oncogenesis. *J. Virol.* **78**, 11451–11460.

Naldini, L., Blomer, U., Gallay, P., Ory, D., Mulligan, R., Gage, F. H., Verma, I. M., and Trono, D. (1996). *In vivo* gene delivery and stable transduction of nondividing cells by a lentiviral vector. *Science* **272**, 263–267.

Naylor, C. S., Jaworska, E., Branson, K., Embleton, M. J., and Chopra, R. (2005). Side population/ABCG2-positive cells represent a heterogeneous group of haemopoietic cells: Implications for the use of adult stem cells in transplantation and plasticity protocols. *Bone Marrow Transplant.* **35**, 353–360.

Ninos, J. M., Jefferies, L. C., Cogle, C. R., and Kerr, W. G. (2006). The thrombopoietin receptor, c-Mpl, is a selective surface marker for human hematopoietic stem cells. *J. Transl. Med.* **4**, 9.

Nolta, J. A., Hanley, M. B., and Kohn, D. B. (1994). Sustained human hematopoiesis in immunodeficient mice by cotransplantation of marrow stroma expressing human interleukin-3: Analysis of gene transduction of long-lived progenitors. *Blood* **83**, 3041–3051.

Okamoto, T., Aoyama, T., Nakayama, T., Nakamata, T., Hosaka, T., Nishijo, K., Nakamura, T., Kiyono, T., and Toguchida, J. (2002). Clonal heterogeneity in differentiation potential of immortalized human mesenchymal stem cells. *Biochem. Biophys. Res. Commun.* **295**, 354–361.

Osawa, M., Hanada, K., Hamada, H., and Nakauchi, H. (1996). Long-term lymphohemato-poietic reconstitution by a single CD34-low/negative hematopoietic stem cells. *Science* **273**, 242–245.

Pearce, D. J., Ridler, C. M., Simpson, C., and Bonnet, D. (2004). Multiparameter analysis of murine bone marrow side population cells. *Blood* **103**, 2541–2546.

Perez, L. E., Rinder, H. M., Wang, C., Tracey, J. B., Maun, N., and Krause, D. S. (2001). Xenotransplantation of immunodeficient mice with mobilized human blood CD34$^+$ cells provides an *in vivo* model for human megakaryocytopoiesis and platelet production. *Blood* **97**, 1635–1643.

Petzer, A. L., Zandstra, P. W., Piret, J. M., and Eaves, C. J. (1996). Differential cytokine effects on primitive (CD34$^+$CD38$^-$) human hematopoietic cells: Novel responses to Flt3-ligand and thrombopoietin. *J. Exp. Med.* **183**, 2551–2558.

Pfeifer, A., Ikawa, M., Dayn, Y., and Verma, I. M. (2002). Transgenesis by lentiviral vectors: Lack of gene silencing in mammalian embryonic stem cells and preimplantation embryos. *Proc. Natl. Acad. Sci. USA* **99**, 2140–2145.

Pflumio, F., Izac, B., Katz, A., Shultz, L. D., Vainchenker, W., and Coulombel, L. (1996). Phenotype and function of human hematopoietic cells engrafting immune-deficient CB17-severe combined immunodeficiency mice and nonobese diabetic-severe combined immunodeficiency mice after transplantation of human cord blood mononuclear cells. *Blood* **88**, 3731–3740.

Piacibello, W., Sanavio, F., Severino, A., Dane, A., Gammaitoni, L., Fagioli, F., Perissinotto, E., Cavalloni, G., Kollet, O., Lapidot, T., and Aglietta, M. (1999). Engraftment in nonobese diabetic severe combined immunodeficient mice of human CD34$^+$ cord blood cells after *ex vivo* expansion: Evidence for the amplification and self-renewal of repopulating stem cells. *Blood* **93**, 3736–3749.

Preffer, F. I., Dombkowski, D., Sykes, M., Scadden, D., and Yang, Y. G. (2002). Lineage-negative side-population (SP) cells with restricted hematopoietic capacity circulate in normal human adult blood: Immunophenotypic and functional characterization. *Stem Cells* **20**, 417–427.

Punzon, I., Criado, L. M., Serrano, A., Serrano, F., and Bernad, A. (2004). Highly efficient lentiviral-mediated human cytokine transgenesis on the NOD/scid background. *Blood* **103**, 580–582.

Ramezani, A., and Hawley, R. G. (2003). Human immunodeficiency virus type 1-based vectors for gene delivery to human hematopoietic stem cells. *Methods Mol. Med.* **76,** 467–492.

Ramezani, A., and Hawley, R. G. (2002a). Overview of the HIV-1 lentiviral vector system. *In* "Current Protocols in Molecular Biology" (F. Ausubel, R. Brent, B. Kingston, D. Moore, J. Seidman, J. A. Smith, and K. Truhl, eds.), pp. 16.21.1–16.21.15. John Wiley & Sons, New York.

Ramezani, A., and Hawley, R. G. (2002b). Generation of HIV-1-based lentiviral vector particles. *In* "Current Protocols in Molecular Biology" (F. Ausubel, R. Brent, B. Kingston, D. Moore, J. Seidman, J. A. Smith, and K. Truhl, eds.), pp. 16.22.1–16.22.15. John Wiley & Sons, New York.

Ramezani, A., Hawley, T. S., and Hawley, R. G. (2000). Lentiviral vectors for enhanced gene expression in human hematopoietic cells. *Mol. Ther.* **2,** 458–469.

Ramezani, A., Hawley, T. S., and Hawley, R. G. (2003). Performance- and safety-enhanced lentiviral vectors containing the human interferon-β scaffold attachment region and the chicken β-globin insulator. *Blood* **101,** 4717–4724.

Sahovic, E. A., Colvin, M., Hilton, J., and Ogawa, M. (1988). Role for aldehyde dehydrogenase in survival of progenitors for murine blast cell colonies after treatment with 4-hydroperoxycyclophosphamide *in vitro. Cancer Res.* **48,** 1223–1226.

Sato, T., Laver, J. H., and Ogawa, M. (1999). Reversible expression of CD34 by murine hematopoietic stem cells. *Blood* **8,** 2548–2554.

Sauvageau, G., Iscove, N. N., and Humphries, R. K. (2004). *In vitro* and *in vivo* expansion of hematopoietic stem cells. *Oncogene* **23,** 7223–7232.

Scharenberg, C. W., Harkey, M. A., and Torok-Storb, B. (2002). The ABCG2 transporter is an efficient Hoechst 33342 efflux pump and is preferentially expressed by immature human hematopoietic progenitors. *Blood* **99,** 507–512.

Shizuru, J. A., Negrin, R. S., and Weissman, I. L. (2005). Hematopoietic stem and progenitor cells: Clinical and preclinical regeneration of the hematolymphoid system. *Annu. Rev. Med.* **56,** 509–538.

Shmelkov, S. V., St. Clair, R., Lyden, D., and Rafii, S. (2005). AC133/CD133/Prominin-1. *Int. J. Biochem. Cell Biol.* **37,** 715–719.

Shpall, E. J., Jones, R. B., Bearman, S. I., Franklin, W. A., Archer, P. G., Curiel, T., Bitter, M., Claman, H. N., Stemmer, S. M., Purdy, M., Myers, S. E., Hami, L., Taffs, S., Heimfeld, S., Hallagan, J., and Berenson, R. J. (1994). Transplantation of enriched CD34-positive autologous marrow into breast cancer patients following high-dose chemotherapy: Influence of CD34-positive peripheral blood progenitors and growth factors on engraftment. *J. Clin. Oncol.* **12,** 28–36.

Shultz, L. D., Schweitzer, P. A., Christianson, S. W., Gott, B., Schweitzer, I. B., Tennent, B., McKenna, S., Mobraaten, L., Rajan, T. V., Greiner, D. L., and Leiter, E. H. (1995). Multiple defects in innate and adaptive immunologic function in NOD/LtSz-scid mice. *J. Immunol.* **154,** 180–191.

Shultz, L. D., Lyons, B. L., Burzenski, L. M., Gott, B., Chen, X., Chaleff, S., Kotb, M., Gillies, S. D., King, M., Mangada, J., Greiner, D. L., and Handgretinger, R. (2005). Human lymphoid and myeloid cell development in NOD/LtSz-*scid* IL2Rγ^{null} mice engrafted with mobilized human hemopoietic stem cells. *J. Immunol.* **174,** 6477–6489.

Sieburg, H. B., Cho, R. H., Dykstra, B., Uchida, N., Eaves, C. J., and Muller-Sieburg, C. E. (2006). The hematopoietic stem compartment consists of a limited number of discrete stem cell subsets. *Blood* **107,** 2311–2316.

Smogorzewska, A., and de Lange, T. (2004). Regulation of telomerase by telomeric proteins. *Annu. Rev. Biochem.* **73,** 177–208.

Solar, G. P., Kerr, W. G., Zeigler, F. C., Hess, D., Donahue, C., de Sauvage, F. J., and Eaton, D. L. (1998). Role of c-mpl in early hematopoiesis. *Blood* **92**, 4–10.

Spangrude, G. J., and Johnson, G. R. (1990). Resting and activated subsets of mouse multipotent hematopoietic stem cells. *Proc. Natl. Acad. Sci. USA* **87**, 7433–7437.

Spangrude, G. J., Heimfeld, S., and Weissman, I. L. (1988). Purification and characterization of mouse hematopoietic stem cells. *Science* **241**, 58–62.

Steward, C. G., and Jarisch, A. (2005). Haemopoietic stem cell transplantation for genetic disorders. *Arch. Dis. Child* **90**, 1259–1263.

Stewart, A. K., Dube, I. D., and Hawley, R. G. (1999). Gene marking and the biology of hematopoietic cell transfer in human clinical trials. *In* "Blood Cell Biochemistry" (L. J. Fairbairn and N. Testa, eds.), Vol. 8, pp. 243–268. Kluwer Academic/Plenum, New York.

Storms, R. W., Trujillo, A. P., Springer, J. B., Shah, L., Colvin, O. M., Ludeman, S. M., and Smith, C. (1999). Isolation of primitive human hematopoietic progenitors on the basis of aldehyde dehydrogenase activity. *Proc. Natl. Acad. Sci. USA* **96**, 9118–9123.

Storms, R. W., Goodell, M. A., Fisher, A., Mulligan, R. C., and Smith, C. (2000). Hoechst dye efflux reveals a novel $CD7^+CD34^-$ lymphoid progenitor in human umbilical cord blood. *Blood* **96**, 2125–2133.

Sutherland, D. R., Anderson, L., Keeney, M., Nayar, R., and Chin-Yee, I. (1996). The ISHAGE guidelines for $CD34^+$ cell determination by flow cytometry. *J. Hematother.* **5**, 213–226.

Szilvassy, S. J., Humphries, R. K., Lansdorp, P. M., Eaves, A. C., and Eaves, C. J. (1990). Quantitative assay for totipotent reconstituting hematopoietic stem cells by a competitive repopulation strategy. *Proc. Natl. Acad. Sci. USA* **87**, 8736–8740.

Taussig, D. C., Pearce, D. J., Simpson, C., Rohatiner, A. Z., Lister, T. A., Kelly, G., Luongo, J. L., Net-Desnoyers, G. A., and Bonnet, D. (2005). Hematopoietic stem cells express multiple myeloid markers: Implications for the origin and targeted therapy of acute myeloid leukemia. *Blood* **106**, 4086–4092.

Thomson, J. A., Itskovitz-Eldor, J., Shapiro, S. S., Waknitz, M. A., Swiergiel, J. J., Marshall, V. S., and Jones, J. M. (1998). Embryonic stem cell lines derived from human blastocysts. *Science* **282**, 1145–1147.

Traggiai, E., Chicha, L., Mazzucchelli, L., Bronz, L., Piffaretti, J. C., Lanzavecchia, A., and Manz, M. G. (2004). Development of a human adaptive immune system in cord blood cell-transplanted mice. *Science* **304**, 104–107.

Uchida, N., Fujisaki, T., Eaves, A. C., and Eaves, C. J. (2001). Transplantable hematopoietic stem cells in human fetal liver have a $CD34^+$ side population (SP) phenotype. *J. Clin. Invest.* **108**, 1071–1077.

Ueda, T., Tsuji, K., Yoshino, H., Ebihara, Y., Yagasaki, H., Hisakawa, H., Mitsui, T., Manabe, A., Tanaka, R., Kobayashi, K., Ito, M., Yasukawa, K., and Nakahata, T. (2000). Expansion of human NOD/SCID-repopulating cells by stem cell factor, Flk2/Flt3 ligand, thrombopoietin, IL-6, and soluble IL-6 receptor. *J. Clin. Invest.* **105**, 1013–1021.

Van Ziffle, J. A., Baerlocher, G. M., and Lansdorp, P. M. (2003). Telomere length in subpopulations of human hematopoietic cells. *Stem Cells* **21**, 654–660.

Verstegen, M. M., Wognum, A. W., and Wagemaker, G. (2003). Thrombopoietin is a major limiting factor for selective outgrowth of human umbilical cord blood cells in non-obese diabetic/severe combined immunodeficient recipient mice. *Br. J. Haematol.* **122**, 837–846.

Visser, J. W., Bol, S. J., and van den Engh, G. (1981). Characterization and enrichment of murine hemopoietic stem cells by fluorescence activated cell sorting. *Exp. Hematol.* **9**, 644–655.

Vormoor, J., Lapidot, T., Pflumio, F., Risdon, G., Patterson, B., Broxmeyer, H. E., and Dick, J. E. (1994). Immature human cord blood progenitors engraft and proliferate to high levels in severe combined immunodeficient mice. *Blood* **83**, 2489–2497.

Vulliamy, T., Marrone, A., Goldman, F., Dearlove, A., Bessler, M., Mason, P. J., and Dokal, I. (2001). The RNA component of telomerase is mutated in autosomal dominant dyskeratosis congenita. *Nature* **413,** 432–435.

Wang, J., Kimura, T., Asada, R., Harada, S., Yokota, S., Kawamoto, Y., Fujimura, Y., Tsuji, T., Ikehara, S., and Sonoda, Y. (2003). SCID-repopulating cell activity of human cord blood-derived CD34⁻ cells assured by intra-bone marrow injection. *Blood* **101,** 2924–2931.

Wang, J. C., Doedens, M., and Dick, J. E. (1997). Primitive human hematopoietic cells are enriched in cord blood compared with adult bone marrow or mobilized peripheral blood as measured by the quantitative *in vivo* SCID-repopulating cell assay. *Blood* **89,** 3919–3924.

Weijer, K., Uittenbogaart, C. H., Voordouw, A., Couwenberg, F., Seppen, J., Blom, B., Vyth-Dreese, F. A., and Spits, H. (2002). Intrathymic and extrathymic development of human plasmacytoid dendritic cell precursors *in vivo. Blood* **99,** 2752–2759.

Wolf, N. S., Kone, A., Priestley, G. V., and Bartelmez, S. H. (1993). *In vivo* and *in vitro* characterization of long-term repopulating primitive hematopoietic cells isolated by sequential Hoechst 33342–rhodamine 123 FACS selection. *Exp. Hematol.* **21,** 614–622.

Yahata, T., Ando, K., Nakamura, Y., Ueyama, Y., Shimamura, K., Tamaoki, N., Kato, S., and Hotta, T. (2002). Functional human T lymphocyte development from cord blood CD34⁺ cells in nonobese diabetic/Shi-*scid*, IL-2 receptor γ null mice. *J. Immunol.* **169,** 204–209.

Yamaguchi, H., Calado, R. T., Ly, H., Kajigaya, S., Baerlocher, G. M., Chanock, S. J., Lansdorp, P. M., and Young, N. S. (2005). Mutations in TERT, the gene for telomerase reverse transcriptase, in aplastic anemia. *N. Engl. J. Med.* **352,** 1413–1424.

Yasui, K., Matsumoto, K., Hirayama, F., Tani, Y., and Nakano, T. (2003). Differences between peripheral blood and cord blood in the kinetics of lineage-restricted hematopoietic cells: Implications for delayed platelet recovery following cord blood transplantation. *Stem Cells* **21,** 143–151.

Yilmaz, O. H., Kiel, M. J., and Morrison, S. J. (2006). SLAM family markers are conserved among hematopoietic stem cells from old and reconstituted mice and markedly increase their purity. *Blood* **107,** 924–930.

Yin, A. H., Miraglia, S., Zanjani, E. D., Almeida-Porada, G., Ogawa, M., Leary, A. G., Olweus, J., Kearney, J., and Buck, D. W. (1997). AC133, a novel marker for human hematopoietic stem and progenitor cells. *Blood* **90,** 5002–5012.

Yui, J., Chiu, C. P., and Lansdorp, P. M. (1998). Telomerase activity in candidate stem cells from fetal liver and adult bone marrow. *Blood* **91,** 3255–3262.

Zanjani, E. D., Almeida-Porada, G., and Flake, A. W. (1996). The human/sheep xenograft model: A large animal model of human hematopoiesis. *Int. J. Hematol.* **63,** 179–192.

Zanjani, E. D., Almeida-Porada, G., Livingston, A. G., Flake, A. W., and Ogawa, M. (1998). Human bone marrow CD34⁻ cells engraft *in vivo* and undergo multilineage expression that includes giving rise to CD34⁺ cells. *Exp. Hematol.* **26,** 353–360.

Zhang, C. C., Kaba, M., Ge, G., Xie, K., Tong, W., Hug, C., and Lodish, H. F. (2006a). Angiopoietin-like proteins stimulate *ex vivo* expansion of hematopoietic stem cells. *Nat. Med.* **12,** 240–245.

Zhang, C. C., Steele, A. D., Lindquist, S., and Lodish, H. F. (2006b). Prion protein is expressed on long-term repopulating hematopoietic stem cells and is important for their self-renewal. *Proc. Natl. Acad. Sci. USA* **103,** 2184–2189.

Zhou, S., Schuetz, J. D., Bunting, K. D., Colapietro, A. M., Sampath, J., Morris, J. J., Lagutina, I., Grosveld, G. C., Osawa, M., Nakauchi, H., and Sorrentino, B. P. (2001). The ABC transporter Bcrp1/ABCG2 is expressed in a wide variety of stem cells and is a molecular determinant of the side-population phenotype. *Nat. Med.* **7,** 1028–1034.

Zhou, S., Morris, J. J., Barnes, Y., Lan, L., Schuetz, J. D., and Sorrentino, B. P. (2002). Bcrp1 gene expression is required for normal numbers of side population stem cells in mice, and confers relative protection to mitoxantrone in hematopoietic cells *in vivo. Proc. Natl. Acad. Sci. USA* **99**, 12339–12344.

Ziegler, B. L., Valtieri, M., Almeida Porada, G., De Maria, G., Muller, R., Masella, B., Gabbianelli, M., Casella, I., Pelosi, E., Bock, T., Zanjani, E. D., and Peschle, C. (1999). KDR receptor: A key marker defining hematopoietic stem cells. *Science* **285**, 1553–1558.

Zijlmans, J. M., Visser, J. W., Kleiverda, K., Kluin, P. M., Willemze, R., and Fibbe, W. E. (1995). Modification of rhodamine staining allows identification of hematopoietic stem cells with preferential short-term or long-term bone marrow-repopulating ability. *Proc. Natl. Acad. Sci. USA* **92**, 8901–8905.

Zufferey, R., Nagy, D., Mandel, R. J., Naldini, L., and Trono, D. (1997). Multiply attenuated lentiviral vector achieves efficient gene delivery *in vivo. Nat. Biotechnol.* **15**, 871–875.

[8] Hemangioblasts and Their Progeny

By Ursula M. Gehling

Abstract

In the developing embryo, the hemangioblast, a mesodermal precursor, gives rise to hematopoietic and endothelial cells. Recent work has shown that during postnatal life, a subset of hematopoietic progenitor cells also displays this dual differentiation capacity and can function as endothelial progenitor cells that contribute to neovascularization. Thus, this subset might be useful for therapy of various hematopoietic and vascular diseases. Here, we describe a two-step culture system that results in the generation of endothelial and hematopoietic cells from adult progenitor cells with hemangioblastic potential. We have developed growth conditions that allow retroviral gene marking of the adult hemangioblast. This culture system is amenable for single-cell analyses at distinct stages of endothelial and hematopoietic differentiation from mobilized CD133$^+$ progenitor cells.

Introduction

Identification of adult progenitor cells with endothelial differentiation potential has proved difficult because hematopoietic stem and progenitor cells and mature endothelial cells share expression of a number of different markers, such as CD34, Tie-2, and the vascular endothelial growth factor receptor-2 (VEGFR-2; reviewed by Rafii and Lyden, 2003). The human stem cell antigen CD133 (previously termed AC133) is expressed on a subset of hematopoietic CD34-positive (+) stem and progenitor cells but not on mature endothelial cells (Yin *et al.*, 1997). In addition, it has been

METHODS IN ENZYMOLOGY, VOL. 419
Copyright 2006, Elsevier Inc. All rights reserved.
0076-6879/06 $35.00
DOI: 10.1016/S0076-6879(06)19008-4

shown that the CD133$^+$ population contains precursors that can differentiate into functional endothelial cells (Gehling *et al.*, 2000; Peichev *et al.*, 2000). This chapter describes an *in vitro* system that has been used to investigate the bipotential hemangioblast activity of human adult CD133$^+$ cells.

Ex Vivo Expansion of CD133$^+$ Cells

In a previous study, we used stem cell growth factor (SCGF) and VEGF to induce endothelial differentiation of CD133$^+$ cells (Gehling *et al.*, 2000). Using these conditions, survival of CD133$^+$ progenitor cells was found to require high initial cell densities. To perform studies on the putative adult hemangioblast at the single-cell level, we have devised a culture system that allows for efficient retroviral gene marking of CD133$^+$ progenitor cells (Loges *et al.*, 2004). In this system, enriched CD133$^+$ cells were first expanded in the presence of VEGF, SCGF, and FLT-3 ligand. On day 4 of culture, one third of the amplified cells were subjected to transduction with a retroviral vector encoding the enhanced green fluorescent protein (EGFP). Single EGFP$^+$ cells were then transferred to cultures with the required numbers of corresponding expanded nontransduced cells from the same sample by limiting dilution. We also employed flow cytometry for isolation of the transduced cells for single-cell experiments. However, sorting severely affects their viability and is not recommended for this purpose (U. M. Gehling, unpublished observation). Further expansion of single EGFP$^+$ cells in coculture with corresponding nontransduced cells yields clones, comprising 50–100 genetically marked cells, in up to one quarter of these wells after 2 weeks.

Hematopoietic and Endothelial Differentiation of Single Cell-Derived EGFP$^+$ Clones

The hematopoietic potential of CD133$^+$ progenitor cells has been previously demonstrated (Gordon *et al.*, 2003; Miraglia *et al.*, 1997; Yin *et al.*, 1997). It is also well established that CD133$^+$ cells can differentiate into functional endothelial cells (Burt *et al.*, 2003; Fons *et al.*, 2004; Gehling *et al.*, 2000; Peichev *et al.*, 2000). Using our culture system, it is easy to study the hemangioblastic capacity of these cells (Loges *et al.*, 2004). Cells from individual wells containing EGFP$^+$ clones can be harvested, split, and replated. EGFP$^+$ cells can differentiate into endothelial cells when grown in fibronectin-coated wells in the presence of SCGF and VEGF. These cells form an adherent layer, display an endothelial cell morphology, and express endothelial markers, such as vascular endothelial-cadherin (VE-cadherin) and *Ulex europaeus* agglutinin-1. Hematopoietic differentiation can be induced when EGFP$^+$ cells are cultured in uncoated wells and stimulated with granulocyte colony-stimulating factor. Resulting cells do not become adherent during the culture period. Morphologically, they exhibit characteristics of mature granulocytes and express hematopoietic markers, for example, CD13. We found that approximately one-quarter of the EGFP$^+$ clones have the capacity to differentiate into both endothelial and hematopoietic cells. These findings clearly show that a subset of adult CD133$^+$ progenitor cells has hemangioblastic properties.

Phenotypical and Functional Studies of CD133-Derived Cells during *Ex Vivo* Expansion and Differentiation

In the described study, EGFP$^+$ cells were expanded and differentiated unpertubed. To analyze phenotypical changes and functional properties of CD133-derived cells during amplification and endothelial commitment, pure cultures with nontransduced cells were performed in parallel experiments (Loges *et al.*, 2004). As revealed by flow cytometry and immunocytochemistry, virtually all CD133$^+$ cells that were selected from granulocyte colony-stimulating factor (G-CSF)-mobilized peripheral blood coexpress the stem cell marker CD34 and the hematopoietic markers CD31 and CD45. The vast majority also stains positive for CD33. Thus, the adult hemangioblast has a hematopoietic phenotype. During the first week of expansion, a significant decrease in the number of CD133$^+$ and CD34$^+$ cells can be observed. After 2 weeks of culture, a novel population of CD133$^+$/CD34$^-$ cells develops and represents approximately one-quarter of the expanded cells at this time point. *Ex vivo* expansion does not change expression of CD31 and CD45, whereas the percentages of CD33$^+$ cells increase above 99% by day 24 of culture. In addition, the monocytic marker CD14, which is expressed only on a small subset of freshly isolated CD133$^+$ cells, is detectable on half of expanded cells after 1 week of culture. Transferring these findings to the development of expanding EGFP$^+$ cells, then half of the transduced cells are CD14$^+$ by the time the clones are split. Because CD14 has been shown to be expressed on human umbilical vein endothelial cells (Jersmann *et al.*, 2001), and because CD14$^+$ peripheral blood monocytes can differentiate into functional endothelial cells (Fernandez Pujol *et al.*, 2000; Harraz *et al.*, 2001; Moldovan *et al.*, 2000; Schmeisser *et al.*, 2001), endothelial cells derived from single EGFP$^+$ cells might differentiate via transdifferentiation of CD14$^+$ cells. However, expanded cells produce higher numbers of endothelial colonies in methylcellulose as compared with freshly isolated cells, indicating that the culture conditions used in our study support the expansion of endothelial progenitor cells. In line with this interpretation, immunocytochemical staining revealed a significant increase in the percentages of cells expressing VEGFR-2 during expansion culture of nontransduced cells, whereas staining for VE-cadherin and *Ulex europaeus* agglutinin-1 was negative. Immunocytochemistry was also performed to study the phenotype of endothelial cells generated from expanded nontransduced cells. After 2 weeks of *in vitro* differentiation, virtually all cells express the endothelial markers von Willebrand factor (vWF), VE-cadherin, and *Ulex europeaus* agglutinin-1. Approximately 90% of the cells stain positive for VEGFR-2. Expression of CD45, which is downregulated during endothelial differentiation, is still present on 50% of the cells at this time point and is lost after 4 weeks of differentiation. *In vitro* differentiated CD133-derived endothelial cells are functional and form capillaries when subcutaneously injected together with tumor cells into severe combined immunodeficient (SCID) mice. As mentioned previously, in our study we decided to use nontransduced cultures for further characterization of the expanded and differentiated CD133-derived cells. Nevertheless, the described two-step culture system is also amenable for the analysis of single cell-derived clones, for example, to study gene expression patterns at different times of endothelial and hematopoietic differentiation of CD133-derived cells with hemangioblastic properties.

Materials

CD133$^+$ Precursor Cells

Adult CD133$^+$ stem and progenitor cells can be selected from human bone marrow aspirates or G-CSF-mobilized blood. Because the described experiments require a high number of starting cells, we used whole leukapheresis products from patients undergoing high-dose chemotherapy with autologous hematopoietic progenitor cell support. For ethical reasons, only cryopreserved products that were not needed for patients' treatment were utilized. CD133$^+$ cells selected from bone marrow aspirates must be examined by the researcher for their ability to expand and differentiate into the endothelial lineage under the culture conditions chosen in our study.

Reagents

Isolation of CD133$^+$ Cells from Frozen Leukapheresis Products

Two sterile 50-ml syringes
Sterile 500-ml bottle
Buffer I: Phosphate-buffered saline lacking Ca^{2+} and Mg^{2+} (PBS$^-$) supplemented with 0.5% human serum albumin (HSA; Blood Donating Service of the German Red Cross, Hamburg, Germany),

0.6% anticoagulant citrate dextrose solution, formula A (ACD-A) (cat. no. 4B7891; Baxter, Munich, Germany), and deoxyribonuclease I (DNase I, 100 U/ml) (cat. no. D 4263; Sigma, Deisenhofen, Germany)

Buffer II: PBS$^-$ supplemented with 0.5% HSA and 0.6% ACD-A

Rotator shaker

Ficoll density gradient solution (density, 1.077) (Ficoll-Hypaque; Amersham Biosciences, Uppsala, Sweden)

CD133 cell isolation kit (cat. no. 130-050-801; Miltenyi Biotec, Bergisch Gladbach, Germany)

Nylon mesh filters (pore size, 30 μm) (Falcon; BD Biosciences Discovery Labware, Heidelberg, Germany)

autoMACS separator with Posselds program selected (cat. no. 201-01; Miltenyi Biotec)

autoMACS columns (cat. no. 130-021-101; Miltenyi Biotec)

autoMACS running buffer (cat. no. 130-091-221; Miltenyi Biotec)

autoMACS rinsing solution (cat. no. 130-091-222; Miltenyi Biotec)

autoMACS acrylic tube rack (cat. no. 130-090-348; Miltenyi Biotec).

Suspension Cultures

Fibronectin-coated 24-well plates (sterile)

Uncoated 24-well plates (sterile)

Opaque microtiter well plates (sterile)

Basal medium: Iscove's modified Dulbecco's medium (IMDM) plus GlutaMAX (Gibco/Invitrogen, Karlsruhe, Germany) supplemented with 10% fetal bovine serum (FBS; Sigma), 10% horse serum (Sigma), and 10^{-6} M hydrocortisone (Sigma)

Expansion medium: Basal medium supplemented with recombinant human (rh) SCGF-β (100 ng/ml) (cat. no. 100-22B; PeproTech via TEBU, Frankfurt, Germany), rhFLT-3 ligand (50 ng/ml) (cat. no. 300-19; TEBU), and rhVEGF (40 ng/ml) (cat. no. 100-20; TEBU)

Endothelial differentiation medium: Basal medium supplemented with recombinant rhSCGF-β (100 ng/ml; TEBU) and rhVEGF (40 ng/ml; TEBU)

Hematopoietic differentiation medium: Basal medium supplemented with rhG-CSF (10 ng/ml) (cat. no. 300-23; TEBU).

Retroviral Transduction of Expanding Cells

Replication-deficient retroviral vector SF11-EGFP (Kuhlcke *et al.*, 2002; can be purchased from EUFETS, Ida-Oberstein, Germany)

PG13 packing cell line [American Type Culture Collection (ATCC), Manassas, VA]

Dulbecco's modified Eagle's medium (DMEM) plus GlutaMAX (Gibco/
Invitrogen) supplemented with 10% FBS (Sigma) and 10^{-1} M sodium
pyruvate (Sigma)

Serum-free medium X-VIVO 10 (Cambrex Bio Science Walkersville,
Walkersville, MD)

Expansion medium (as described previously)

Fibronectin-coated 24-well plates (sterile) (fibronectin from TEBU)

Varifuge 3.0 RS (Heraeus Instruments, Hanau, Germany).

Clonogenic Assays for Hematopoietic and Endothelial Progenitor Cells

Complete methylcellulose medium: 1% methylcellulose in IMDM, 30%
FBS, 1% bovine serum albumin (BSA), 10^{-4} M mercaptoethanol,
2 mM L-glutamine, recombinant human stem cell factor (rhSCF,
50 ng/ml), recombinant human interleukin-3 (rhIL-3, 20 ng/ml), IL-6
(20 ng/ml), G-CSF (20 ng/ml), granulocyte/macrophage colony-
stimulating factor (GM-CSF, 20 ng/ml), and erythropoietin (3 U/ml)
(Methocult GF+ H4435; StemCell Technologies via Cell Systems,
St. Katharinen, Germany)

Methylcellulose medium without cytokines: 2.6% methylcellulose in
IMDM, 30% FBS, 1% BSA, 10^{-1} M mercaptoethanol, and 200 mM
L-glutamine (Methocult H4230; StemCell Technologies via Cell
Systems)

IMDM

rhSCGF (100 ng/ml; TEBU)

rhVEGF (50 ng/ml; TEBU)

Petri dishes (35 mm, sterile)

Petri dishes (90 mm, sterile)

Inverted microscope (Olympus IX 50; Olympus, Hamburg, Germany).

Immunocytochemical Staining of Freshly Isolated $CD133^+$ Cells and Cultured CD133-Derived Cells

Glass slides

Paper pads (Shandon filter cards; Thermo Electron, Pittsburgh, PA)

Cuvettes (Shandon)

Cytocentrifuge (Shandon)

Ice-cold methanol ($-20°$)

RPMI with 10% FBS

Anti-human VEGFR-2 monoclonal antibody (mAb) (clone 260.4, cat.
no. V3003; Sigma)

Anti-human CD31 mAb (BD Biosciences Pharmingen, Heidelberg,
Germany)

Anti-human von Willebrand factor (vWF) mAb (BD Biosciences Pharmingen)

Anti-human VE-cadherin mAb (BD Biosciences Pharmingen)

Ulex europaeus agglutinin-1 (Sigma)

Anti-human CD13 mAb (BD Biosciences Pharmingen)

Fluorescein isothiocyanate (FITC)-labeled goat anti-mouse immunoglobulin secondary antibody (Dako, Hamburg, Germany)

Texas red-conjugated goat anti-mouse immunoglobulin secondary antibody

Antibody diluent, background reducing (cat. no. S3022; Dako)

4',6-Diamidino-2-phenylindole dihydrochloride (DAPI; Roche Applied Sciences, Mannheim, Germany)

Antifading agent (VECTASHIELD; Vector Laboratories, Grünberg, Germany)

Fluorescence microscope (Zeiss Axioplan; Carl Zeiss, Jena, Germany)

Human umbilical vein endothelial cells [HUVECs; Cascade Biologics (Portland, OR) via TEBU)]

Medium 200 (Cascade Biologics via TEBU)

Low-serum growth supplement (Cascade Biologics via TEBU).

Flow Cytometric Analysis of Freshly Isolated and Expanded Cells

FcR blocking reagent, human (Miltenyi Biotec)

Phycoerythrin (PE)-labeled isotype control

FITC-labeled isotype control

Washing buffer (PBS⁻ containing 0.1% BSA)

PE-conjugated anti-CD133 mAb (Miltenyi Biotec)

PE-conjugated anti-CD14 mAb (BD Biosciences Pharmingen)

PE-conjugated anti-CD31 mAb (BD Biosciences Pharmingen)

PE-conjugated anti-CD117 mAb (Dako)

PE-conjugated anti-VEGFR-2 (R&D Systems, Wiesbaden, Germany)

FITC-conjugated anti-CD34 mAb (BD Biosciences Pharmingen)

FITC-conjugated anti-CD45 mAb (BD Biosciences Pharmingen)

FITC-conjugated anti-CD33 mAb (BD Biosciences Pharmingen)

FITC-conjugated anti-CD105 mAb (BD Biosciences Pharmingen)

Flow cytometer (FACSCalibur; BD Biosciences Immunocytometry Systems, Heidelberg, Germany)

Tubes (Falcon, 12 × 75 mm; BD Biosciences Discovery Labware).

Assessment of Functionality of CD133-Derived Endothelial Cells In Vivo

SCID mice

A549 lung cancer cell line (Sigma)

DMEM supplemented with 10% FBS.

Immunohistochemical Analysis of Mouse Tumor Tissues

> Bouin's fixative (Chemicon International, Planegg-Munich, Germany)
> Ethanol in ascending concentrations (50, 70, 95, and 100%)
> Paraffin wax (Sigma)
> Microtome (HM 335E; MICROM International, Walldorf, Germany)
> Chrome–gelatin-precoated glass slides
> Anti-human CEACAM-1 mAb (4D1/C2, CD66a) (cat. no. RDI-CD66Aabm-H2; RDI division of Fitzgerald Industries International, Concord, MA)
> 10 m*M* potassium phosphate buffer, pH 7.4.

Methods

Isolation of CD133⁺ Cells from Frozen Leukapheresis Products

1. Fill sterile 500-ml bottle with 200 ml of buffer I.
2. Thaw leukapheresis product at room temperature.
3. Immediately after thawing, aspirate the leukapheresis product with 50-ml syringes.
4. Transfer the product to the 500-ml bottle.
5. Fill up with buffer I so that the final leukapheresis product: buffer ratio is 1:3.
6. Shake the specimen gently for 30 min at room temperature.
7. Fill 50-ml tubes with 20 ml of Ficoll solution.
8. Add 28 ml of the specimen to each 50-ml tube.
9. Centrifuge the specimens for 25 min at 500*g*.
10. Add DNase at 100 U/ml to buffer I. It is important to add fresh DNase to the buffer every 30 min.
11. Collect the mononuclear cell (MNC) layers in one 50-ml tube and wash the cells twice in buffer I for 10 min at 300*g* to remove platelets.
12. Place a 30-μm nylon mesh filter on top of a fresh 50-ml tube and filter the specimen.
13. Centrifuge the specimen for 10 min at 500*g* and resuspend in 1 ml of buffer II.
14. Add 100 μl of FcR blocking reagent per 1×10^8 cells and shake gently for 1 min.
15. Add 100 μl of CD133-conjugated paramagnetic microbeads per 1×10^8 cells and shake gently for 1 min.
16. Incubate the specimen for 30 min at room temperature. Shake the specimen gently every 30 min.
17. Wash and resuspend the specimen in buffer II.

18. Perform immunomagnetic selection of CD133$^+$ cells. Use an automatic magnetic separation device (autoMACS) with the Posselds program selected.
19. Keep an aliquot of the selected cells for flow cytometric assessment of purity.

Suspension Cultures of Human CD133$^+$ Cells

Expansion Cultures

1. Incubate 24-well plates with fibronectin/PBS$^-$ (50 μg/ml) for 2 h at room temperature.
2. Remove the PBS.
3. Transfer freshly isolated cells to fibronectin-coated well plates at a density of 2×10^6 cells/ml in expansion medium.
4. Expand the cells for 4 days. Replace half the medium every second day. Incubate the cultures at 37° in 5% CO_2.
5. Split the cells into three portions.
6. Use a one-third portion for retroviral transduction as described later.
7. Culture the other two-thirds in expansion medium for an additional 2 days.
8. On day 6 of culture, use one-third for cocultures with transduced cells. Harvest transduced and nontransduced cells by gently pipetting. Replate nontransduced cells at 2×10^6 cells/ml in opaque microtiter 96-well plates. Add single transduced cells to these cultures by limiting dilution. Expand the cocultured single transduced cells for at least six cell doublings at 37° in 5% CO_2. Replace half the medium every second day.
9. Use the remaining one-third for pure cultures of nontransduced cells. Culture these cells for an additional 8 days in expansion medium at 37° in 5% CO_2. Cells will maintain a round morphology and will grow loosely adherent (Fig. 1). Change the medium depending on cell proliferation, but at least every second day. Count cells in the supernatant. Centrifuge cells in the supernatant for 5 min at 500g and retransfer to the cultures.

Differentiation Cultures

1. Harvest all cells from those wells showing single cell-derived clones with 50–100 cells by gentle pipetting.
2. Harvest the cells of each well separately.
3. Split the cells obtained from each well into two portions.

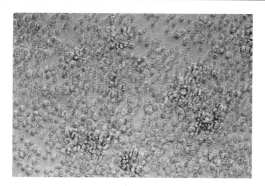

FIG. 1. Morphology of CD133-derived cells after 14 days of culture in expansion medium.

4. Transfer one part into fibronectin-coated microtiter wells and culture the cells in endothelial differentiation medium for 14 days at 37° in 5% CO_2. Replace half the medium every second day.

5. Culture the remaining half in microtiter wells in hematopoietic differentiation medium for 7 days at 37° in 5% CO_2 and again, replace half the medium every second day.

6. Harvest expanded cells from pure cultures of cells nontransduced by gentle pipetting and wash the cells in PBS⁻ for 5 min at 500g. Resuspend the cells in 1 ml of basal medium and divide the cells into two portions.

7. Culture one part in endothelial differentiation medium for 28 days as described previously. Under these conditions, expanded cells will begin to grow adherently within 1 to 2 days of culture. After 14 days of culture, many cells will have developed a morphology that is typical of mature endothelial cells (Fig. 2).

8. Culture the other half in hematopoietic differentiation medium for 14 days as described previously. Hematopoietic differentiation medium may be supplemented with additional hematopoietic growth factors, such as IL-3, erythropoietin, and thrombopoietin, to induce differentiation into the erythroid and megakaryocytic lineages.

Retroviral Transduction of Expanded CD133⁺ Cells

1. Keep the PG13 packaging cell line at $1–2 \times 10^4/cm^2$ in DMEM with GlutaMAX supplemented with 10% heat-inactivated FBS and 1 mM sodium pyruvate.

2. For introductions of the retroviral vector S11-EGFP, culture PG13 cells in X-VIVO 10 at the same density.

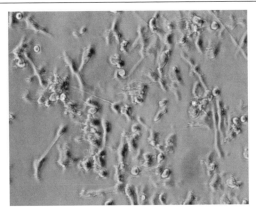

FIG. 2. Morphology of expanded CD133-derived cells after 14 days of differentiation with SCGF and VEGF.

3. Collect the vector-containing supernatant and preload fibronectin-coated 24-well plates with the supernatant by four to six centrifugation steps in a Varifuge 3.0 RS at 1000g for 30 min at 4°.
4. For transduction, incubate CD133$^+$ cells in the preloaded wells at a density of 2×10^6 cells/ml in expansion medium overnight without further centrifugation.
5. For *in vitro* studies with transduced CD133$^+$ cells, perform two transduction cycles on consecutive days. This will result in a mean transduction rate of approximately 21%.
6. For *in vivo* studies with transduced CD133$^+$ cells, perform three to five consecutive transduction cycles. This will increase the transduction rate up to 73%.

Clonogenic Assays for Hematopoietic and Endothelial Progenitor Cells

Clonogenic Assay for Hematopoietic Progenitor Cells

1. Use complete methylcellulose medium supplemented with several hematopoietic cytokines.
2. Make aliquots of 5 ml. Use a 50-ml tube for each aliquot.
3. Store aliquots at −20° until use.
4. Thaw aliquots at room temperature.
5. Keep six 35-mm Petri dishes and two 90-mm Petri dishes ready.
6. To study the clonogenic potential of freshly isolated CD133$^+$ cells, use 1×10^3 cells/ml methylcellulose medium, that is, add 5×10^3 CD133$^+$ cells to a 5-ml aliquot.

7. To study the clonogenic potential of expanded CD133-derived cells, use 5×10^4 cells/ml methylcellulose medium, that is, add 2.5×10^5 CD133$^+$ cells to a 5-ml aliquot.
8. Resuspend the cells in 300 μl of PBS$^-$.
9. Use a 1000-μl pipette to add cells to the medium.
10. Gently mix the cells in the medium with the pipette tip.
11. Use a 5-ml pipette to transfer the cell-containing medium into the 35-mm Petri dishes.
12. Add 1 ml of cell-containing medium per Petri dish. This will yield four dishes per experiment.
13. Place the cell-containing medium in the center of the Petri dish.
14. Swirl the dish to homogeneously distribute the medium on the bottom of the disk.
15. Place two dishes into one 90-mm Petri dish.
16. Fill two 35-mm dishes with distilled water.
17. Place one dish with water into one 90-mm dish.
18. Incubate the cells for 14 days at 37° in 5% CO_2 and 95% humidity.
19. Use an inverted microscope to score the colonies.

Clonogenic Assay for Endothelial Progenitor Cells

1. Use methylcellulose medium without cytokines.
2. Add 20 ml of IMDM to the medium.
3. Gently mix the medium with a 25-ml pipette.
4. Add SCGF (100 ng/ml) and VEGF (50 ng/ml) to the medium.
5. Gently mix the medium with a 25-ml pipette.
6. Make aliquots of 5 ml. Use a 50-ml tube for each aliquot.
7. Store aliquots at −20° until use.
8. Thaw aliquots at room temperature.
9. Perform the assay in quadruplicate as described in Clonogenic Assay for Hematopoietic Progenitor Cells.
10. After 14 days of culture, visualize and count the colonies, using an inverted microscope. Colony-forming units-endothelial cell (CFU-ECs) can be identified as loosely packed colonies consisting of 20–200 very small round cells growing three-dimensionally in the methylcellulose (Gehling *et al.*, 2000).

Immunocytochemical Staining of Freshly Isolated CD133$^+$
Cells and Cultured CD133-Derived Cells

1. Wash freshly isolated CD133$^+$ cells in buffer II.
2. Wash cultured CD133-derived cells twice in PBS$^-$.

3. Resuspend the cells in 200 μl of PBS$^-$ per 1×10^5 cells.
4. Deposit cells onto slides (1×10^5 cells per slide) for 5 min at 500 rpm, using a Shandon cytocentrifuge.
5. Let the specimens air dry for at least 10 min.
6. Fix the specimens in methanol for 15 min at $-20°$.
7. Block the cells with 10% FBS–RPMI. Incubate the cells with the primary antibody for 30 min at room temperature.
8. Wash the specimens three times in PBS$^-$.
9. Incubate the cells with the secondary antibody for 30 min at room temperature in the dark.
10. Wash the specimens three times in PBS$^-$.
11. Counterstain the cells with DAPI embedded in VECTASHIELD for 5 min in the dark.
12. Use a fluorescence microscope to evaluate the number of positive cells.
13. Use HUVECs as positive control.
14. Use peripheral blood smears as negative controls.
15. Also perform negative controls by using mouse anti-human immunoglobulin (IgG) instead of the primary antibody.

Flow Cytometric Analysis of Freshly Isolated and Expanded Cells

1. Transfer cells to 12×75 mm Falcon tubes (if possible, 1×10^6 cells per tube).
2. Wash the cells twice by centrifuging in washing buffer at 500g for 5 min at $4°$.
3. Decant the supernatant and resuspend the cells in the remaining volume.
4. Add 10 μl of FcR blocking reagent to each tube and mix by gentle pipetting.
5. Add 20 μl of fluorochrome-conjugated antibody to the tubes. Add two antibodies per tube to perform two-color flow cytometry. The chosen antibodies may also be used with other labeled fluorochromes, such as allophycocyanin (APC) and peridinin chlorophyll protein (PerCP)-Cy5.5, to perform multicolor flow cytometry.
6. Incubate the cells with the antibodies for at least 15 min at $4°$ in the dark.
7. Wash the cells in washing buffer at 500g for 5 min at $4°$.
8. Resuspend the cells in 0.5 ml of PBS$^-$.
9. Perform flow cytometric analysis of the stained cells.
10. Use isotype controls to set gates for analysis.

Functionality of CD133-Derived Endothelial Cells In Vivo

1. Maintain the A549 cells in DMEM supplemented with 10% FBS.
2. Harvest A549 cells by trysinization.
3. Wash tumor cells with PBS⁻ for 10 min at 500g.
4. Harvest by trypsinization adherent CD133-derived cells that were expanded for 14 days and cultured in endothelial differentiation medium for an additional 14 days.
5. Also harvest nonadherent CD133-derived cells from the same culture by gentle pipetting.
6. Mix adherent and nonadherent cells and wash them twice in PBS⁻.
7. In parallel experiments, inject SCID mice subcutaneously with either 1×10^6 tumor cells, 1×10^6 CD133-derived endothelial cells, or a mixture of 1×10^6 tumor cells and 1×10^6 CD133-derived endothelial cells.
8. Measure tumor growth in all groups at weekly intervals.
9. Kill all mice at week 5 after transplantation.
10. Remove subcutaneous tumors.
11. Fix tumor tissue blocks with Bouin's fixative for 24 h at room temperature.
12. Dehydrate the tissue blocks in ascending ethanol concentrations.
13. Embed the tissue blocks in paraffin.
14. Mount 6-μm sections on chrome–gelatin-precoated slides.
15. Deparaffinate sections in descending concentrations of ethanol.
16. Rehydrate the sections in ascending ethanol concentrations.
17. Incubate the tissue sections with human-specific primary monoclonal antibody 4D1/C2 overnight at 4° in a humid chamber.
18. Wash the tissue sections three times in PBS.
19. Incubate the tissue sections with biotinylated secondary for 1 h at room temperature.
20. Wash the tissue sections three times in PBS.
21. Incubate the tissue sections with avidin–peroxidase anti-peroxidase complex for 30 min at room temperature.
22. Wash the tissue sections in PBS for 10 min.
23. Wash the tissue sections in phosphate buffer, pH 7.4, for 10 min.
24. Incubate the tissue sections with glucose oxidase and nickel for up to 30 min. Control positive reactions every 5 min to avoid overstaining.
25. Wash the tissue sections in phosphate buffer, pH 7.4, for 10 min.
26. Counterstain the tissue sections with calcium red.
27. Use HUVECs as positive controls.
28. Also perform negative controls by using mouse anti-human immunoglobulin (IgG) instead of the primary antibody.

References

Burt, R., Pearce, W., Luo, K., Oyama, Y., Davidson, C., Beohar, N., and Gheorghiade, M. (2003). Hematopoietic stem cell transplantation for cardiac and peripheral vascular disease. *Bone Marrow Transplant.* **32**, S29–S31.

Fernandez Pujol, B., Lucibello, F. C., Gehling, U. M., Lindemann, K., Weidner, N., Zuzarte, M. L., Adamkiewicz, J., Elsasser, H. P., Muller, R., and Havemann, K. (2000). Endothelial-like cells derived from human CD14 positive monocytes. *Differentiation* **65**, 287–300.

Fons, P., Herault, J. P., Delesque, N., Tuyaret, J., and Herbert, J. M. (2004). VEGF-R2 and neuropilin-1 are involved in VEGF-A-induced differentiation of human bone marrow progenitor cells. *J. Cell. Physiol.* **200**, 351–359.

Gehling, U. M., Ergün, S., Schumacher, U., Wagener, C., Pantel, K., Otte, M., Schuch, G., Schafhausen, P., Mende, T., Kilic, N., Kluge, K., Schafer, B., Hossfeld, D. K., and Fiedler, W. (2000). *In vitro* differentiation of endothelial cells from AC133-positive progenitors. *Blood* **95**, 3106–3112.

Gordon, P. R., Leimig, T., Babarin-Dorner, A., Houston, J., Holladay, M., Mueller, I., Geiger, T., and Handgretinger, R. (2003). Large-scale isolation of CD133+ progenitor cells from G-CSF mobilized peripheral blood stem cells. *Bone Marrow Transplant.* **31**, 17–22.

Harraz, M., Jiao, C., Hanlon, H. D., Hartley, R. S., and Schatteman, G. C. (2001). CD34− blood-derived human endothelial progenitors. *Stem Cells* **19**, 304–312.

Jersmann, H. P., Hii, C. S., Hodge, G. L., and Ferrante, A. (2001). Synthesis and surface expression of CD14 by human endothelial cells. *Infect. Immun.* **69**, 479–485.

Kuhlcke, K., Fehse, B., Schilz, A., Loges, S., Lindemann, C., Ayuk, F., Lehmann, F., Stute, N., Fauser, A. A., Zander, A. R., and Eckert, H. G. (2002). Highly efficient retroviral gene transfer based on centrifugation-mediated vector preloading of tissue culture vessels. *Mol. Ther.* **5**, 473–478.

Loges, S., Heil, G., Bruweleit, M., Schoder, V., Butzal, M., Fischer, U., Gehling, U. M., Schuch, G., Hossfeld, D. K., and Fiedler, W. (2004). Analysis of concerted expression of angiogenic growth factors in acute myeloid leukemia: Expression of angiopoietin-2 represents an independent prognostic factor for overall survival. *J. Clin. Oncol.* **23**, 1109–1117.

Miraglia, S., Godfrey, W., Yin, A. H., Atkins, K., Warnke, R., Holden, J. T., Bray, R. A., Waller, E. K., and Buck, D. W. (1997). A novel five-transmembrane hematopoietic stem cell antigen: Isolation, characterization, and molecular cloning. *Blood* **90**, 5013–5021.

Moldovan, N. I., Goldschmidt-Clermont, P. J., Parker-Thornburg, J., Shapira, S. D., and Kolattukudy, P. E. (2000). Contribution of monocytes/macrophages to compensatory neovascularization: The drilling of metalloelastase-positive tunnels in ischemic myocardium. *Circ. Res.* **87**, 378–384.

Peichev, M., Naiyer, A. J., Pereira, D., Zhu, Z., Lane, W. J., Williams, M., Oz, M. C., Hicklin, D. J., Witte, L., Moore, M. A., and Rafii, S. (2000). Expression of VEGFR-2 and AC133 by circulating human CD34+ cells identifies a population of functional endothelial precursors. *Blood* **95**, 952–958.

Rafii, S., and Lyden, D. (2003). Therapeutic stem and progenitor cell transplantation for organ vascularization and regeneration. *Nat. Med.* **9**, 702–712.

Schmeisser, A., Garlichs, C. D., Zhang, H., Eskafi, S., Graffy, C., Ludwig, J., Strasse, R. H., and Daniel, W. G. (2001). Monocytes coexpress endothelial and macrophagocytic lineage markers and form cord-like structures in Matrigel under angiogenic conditions. *Cardiovasc. Res.* **49**, 671–680.

Yin, A., Miraglia, S., Zanjani, E. D., Almeida-Porada, G., Ogawa, M., Leary, A. G., Olweus, J., Kearney, J., and Buck, D. W. (1997). AC133, a novel marker for human hematopoietic stem and progenitor cells. *Blood* **95**, 5002–5012.

[9] Kidney Epithelial Cells

By Peter L. Smith, Deborah A. Buffington, and H. David Humes

Abstract

Kidney tubules are an essential component of an organism's blood clearance mechanism, recovering essential metabolites from glomerular filtration by active transport. Tubules are subject to injury, usually as the result of ischemia–reperfusion events that damage the polarized tubular cell layer that coats the tubule basement membrane, causing dysfunction and necrosis that is often associated with acute renal failure. However, tubules are capable of self-repair, forming new proximal tubular cells to replace failing or necrotic cells. The origin of the progenitor cells that give rise to new tubular cells is unknown. At one extreme, it is possible that all or a fraction of tubular cells can undergo a form of dedifferentiation and subsequent mitosis to form new tubular cells, or alternatively, it is possible that tubular regeneration follows the stem cell/transit-amplifying cell paradigm described for more rapidly regenerating organ systems. Regardless of the mechanism employed to generate new tubular cells, human tubular cells are readily grown in primary cultures and can recapitulate many of the metabolic, endocrine, and immunological properties attributable to endogenous renal proximal tubules when engrafted into bioartificial devices.

Introduction

The adult kidney is composed of between 250,000 and 1 million nephrons made of specialized epithelial cells that form the Bowman's capsule, proximal and distal tubules, and collecting ducts (Tisher and Madsen, 1986). Nephrons are intricately associated with vascular endothelium that directs approximately 20% of cardiac output for filtration as well as provides the energy to support the large metabolic consumption of the renal tubules (Brenner et al., 1986). Although the kidney is not an organ that undergoes rapid turnover, like the hematopoietic, intestinal, and skin systems, its tubular epithelium is subject to injury, usually coming in the form of ischemia. Deprivation of the required metabolites can result in tubular dysfunction and necrosis, leading to acute renal failure and reduced glomerular filtration rates (Schrier et al., 2004). However, the tubule epithelium is capable of regeneration and can repair itself to a functional state, indicating that progenitor cells residing either within or outside the

METHODS IN ENZYMOLOGY, VOL. 419
Copyright 2006, Elsevier Inc. All rights reserved.
0076-6879/06 $35.00
DOI: 10.1016/S0076-6879(06)19009-6

boundaries of the tubule basement membrane can provide newly formed and fully differentiated tubules.

In this chapter, we describe a method to culture human primary kidney tubules with properties that recapitulate many of the functional attributes of native renal proximal tubules (Humes *et al.*, 2002). In addition, we consider some of the more recent literature in an effort to understand the origin of these tubule progenitor cells and how they may contribute to renal repair.

Analysis of Steady State Tubular Epithelium Maintenance

The steady state proliferation of tubular cells has been examined by light and electron microscopy. By randomly scanning perfusion fixed rat tubules in the cortex and outer medulla of the kidney, Vogetseder and colleagues (2005) documented cells undergoing mitosis within the boundaries of the tubular basement membrane. Cells undergoing mitosis resembled the neighboring quiescent tubules, including a brush border on cells in the proximal tubule and tight junctions. Immunostaining of mitotic cells revealed by 4',6-diamidino-2-phenylindole (DAPI) staining showed abundant basolateral staining of the Na^+,K^+-ATPase and all proximal tubule mitotic cells reacted positively with antibodies directed against sodium-dependent phosphate transporter type IIa (NaPi-IIa), megalin, or a peroxisome-specific protein (PMP70). No mitotic or quiescent tubules expressed the mesenchymal marker vimentin. Double staining with proliferating cell nuclear antigen (PCNA) to detect cells in the late G_1 and early S_1 phases of the cell cycle confirmed that all cells undergoing mitosis expressed cell surface markers characteristic of mature, functional tubules.

It is logical that a newly minted tubular cell would be equipped to handle the microenvironment of the tubule lumen. Without the proper establishment of transporters to regulate its internal homeostasis, a new tubule, and for that matter its parental cell, would find it difficult to survive. This leads to the question of whether all mature tubular cells within the tubular basement membrane possess the ability to give rise to progeny, or are only a subset of tubular cells capable of giving rise to progeny and are thus progenitor cells although phenotypically indistinguishable from terminally differentiated tubular cells. One approach to distinguish between these possibilities is to look for cells that are slowly cycling. By analogy to the definition of a stem cell, slowly cycling cells are long lived and give rise to progeny that are rapidly cycling transit-amplifying cells (Potten and Loeffler, 1990). Thus researchers have effectively used detection reagents that incorporate into the genomic DNA of cycling cells, such as the thymidine analog 5-bromo-2-deoxyuridine (BrdU), or ^{3}H labeling of thymidine, looking for so-called label-retaining cells (LRCs). These cells have by definition undergone mitosis, thus incorporating the DNA label, but remain in a quiescent state for long periods of time, thus retaining the label (and presumably can reenter the cell cycle when required). For example, this approach has been used to define the stem cells and transit-amplifying cells in epithelial repair of the cornea (Lehrer *et al.*, 1998).

In addition to histologically examining normal rat tubular cells as described above, Vogetseder and colleagues (2005) also injected BrdU into newborn rats and looked for LRCs 8, 14, or 35 weeks later. Although LRCs could be found only in the cortex tubules, because the tubules in the medulla had not been established at the time of BrdU injection, they did identify a total of 431 label-retaining tubular cells. All but three of these cells were costained with antibody to the Na^+,K^+-ATPase or NaPi-IIa. This suggests that the vast majority of LRCs within cortex kidney tubules are phenotypically similar to, if not mature, functional renal tubular cells. Of course, the three LRCs that did not label with the antibodies used in this study could constitute a progenitor cell niche that is responsible for the generation of all newly generated tubular cells. A second study using BrdU labeling to look at the steady state LRCs of the rat kidney found LRCs to be scattered in both the proximal and distal tubules (Maeshima *et al.*, 2003). In this experiment, adult rats were labeled daily for 7 days and sacrificed 14 days later. No staining of cells in the glomeruli or capillary vessels was noted, but some LRCs were found in the *Dolichos biflorus* agglutinin (DBA)-staining collecting ducts. Most of the LRCs were colocalized with *Lotus tetragonolobus* agglutinin (LTA), a lectin with specificity for the proximal tubule (Laitinen *et al.*, 1987), and were located adjacent to capillary endothelial cells. In addition, limited staining was observed in Tamm–Horsfall protein-expressing distal tubules. These experiments confirm that the majority of LRCs are localized to the renal tubular epithelium.

Regeneration of Experimentally Injured Kidney Tubules

Witzgall and colleagues (1994) examined the identity of cells expressing PCNA and vimentin after creating a unilateral ischemic injury to the rat kidney. After 48 h of reperfusion, as many as 80% of tubules in the outer stripe of the inner medulla expressed detectable levels of PCNA staining. This portion of the kidney contains the S3 segment of the proximal kidney tubules, which also corresponds to the portion of tubules most damaged by the ischemic insult. Sections of the earlier proximal tubule residing in the cortex (S1 and S2) undergo more muted damage in this model and likewise contain fewer cells undergoing mitosis (approximately 20% at 48 h post-reperfusion) as assessed by PCNA staining. The majority of viable cells in the S3 segment 48 h after reperfusion transiently express vimentin. Vimentin is a developmentally regulated intermediate filament protein expressed in the mesenchymal precursor cells before their conversion to the epithelial elements that make up the mature kidney (Holthofer et al., 1984). It is not normally expressed in mature tubule epithelial cells and is not expressed in the contralateral kidney (undamaged) in these studies. Likewise, Pax2, a transcription factor expressed during the early establishment of the metanephric mesenchyme, but not in mature tubular cells (Davies and Fisher, 2002), is reexpressed in regenerating tubules (Imgrund et al., 1999) after folic acid-induced tubular necrosis and is expressed in the tubular cells and interstitial cells of the S3 segment of the postischemic tubule as early as 12 h post-reperfusion (Maeshima et al., 2002). Acute BrdU labeling to analyze proliferating cells showed a significant overlap of proliferating cells with those expressing Pax2. Furthermore, it was shown that Pax2 expression preceded BrdU labeling.

Induction of focal depletion of tubular cells in the proximal three-quarters of the S3 segment was generated by intravenously injecting low amounts of uranyl acetate (UA) into rats (Fujigaki et al., 2006). By injecting BrdU into rats 1 h before sacrifice, mitotic cells were identified and cataloged into four separate zones of the S3 proximal tubule segment. Not surprisingly, the greatest number of mitotic tubular cells was found in the proximal-most zones, where damage was the greatest. Regeneration began as early as 2 days after injury and by day 5 mitotic cells appeared to lift brush border-containing proximal tubules, forming a hyperplastic appearance. When treating rats with higher doses of UA, abundant necrosis of the first three zones of the proximal tubules was observed, with less damage to the distal-most zone. BrdU injections showed that initial tubule regeneration began exclusively in the less damaged, distal-most zone 2 days postinjury and began to progress to the proximal-most zones on days 3–5. Brush-borderless cells with large oval nuclei were observed in the transient zone between the distal-most part of the proximal tubule and the thin descending loop of Henle. By day 7, under either UA regimen, tubule regeneration was nearly complete. Consistent with the observations of other researchers, initially regenerating tubular cells expressed vimentin and this continued in BrdU-positive and nonpositive cells on day 7.

To define LRCs after UA injection, [^{3}H]thymidine was injected 2–3 days after UA administration and chased for 7, 21, or 42 days. In the low-dose UA focal damage regimen, LRCs were found to populate primarily the proximal-most zones on day 42. These cells were often found in pairs or clusters. Increasing tubular damage resulted in a shift of the LRCs to the distal-most zone of the proximal tubule, where they persisted for at least 42 days post-injury.

Examination of BrdU-labeled cells (adult rats; 7 days of label, 14-day chase) that were subjected to bilateral ischemia–reperfusion showed that the number of LRCs increased approximately 24 h after reperfusion and that many LRCs were found in clusters of two cells, suggesting that they had recently divided (Maeshima et al., 2003). Like the previously described study, numerous PCNA-positive cells were found in tubular cells of the outer medulla (the S3 segment of the proximal tubule). Interestingly, 75% of the LRCs were positive for PCNA after 24 h of reperfusion, with all but 14% of PCNA-positive cells containing BrdU (Maeshima et al., 2006). In this experiment, chasing for 10 days post-reperfusion revealed the elimination of PCNA-expressing cells and a significant diminution of LRCs, although some clearly remain. Eighteen hours after reperfusion one of each pair of BrdU-labeled cells expressed vimentin and after 24 h of reperfusion only weakly labeled BrdU cells (label diluted) expressed vimentin, whereas strongly BrdU-expressing cells did not. After 10 days, vimentin was widely expressed by tubular cells of the S3 segment of the proximal tubule and E-cadherin was also detectable on a subset of cells. Although LRCs lay in close proximity to vimentin- and E-cadherin-expressing cells, LRCs only weakly expressed vimentin and no E-cadherin.

Following the identical protocol of 7 days of BrdU labeling followed by a 2-week chase, but instead using a unilateral ureteral obstruction (UUO) model to induce tubular cell damage, Yamashita and colleagues observed a significant increase in the number of BrdU-labeled cells 3–7 days postinjury (Yamashita et al., 2005). In addition, they noted a significant increase in LRCs in the interstitium. This appeared to be the result of destruction of the tubule basement membrane, as assessed by a loss of laminin staining near the damaged tubule epithelium. There was a diminution of E-cadherin staining on regenerating tubules, including LRCs, and an appearance of the matrix metalloproteinase-2 on tubular cells, but not on the LRCs. There was also detectable matrix metalloproteinase-9 in interstitial cells, but again, not on interstitial LRCs. Neither metalloproteinase is expressed on normal kidney tubular or interstitial cells. Tubular and interstitial cells of injured kidneys, including LRCs, expressed vimentin and heat shock protein 47 (HSP-47), a molecular chaperone that binds collagen I. Some interstitial cells, including a number of the interstitial LRCs, expressed α-smooth muscle actin. When isolated tubules from UUO, BrdU-treated animals were placed in three-dimensional collagen gels some of the BrdU cells would migrate out of the tubules and into the collagen gel, suggesting that LRCs have migratory ability. The authors suggest that this indicates that LRCs contribute to fibrosis by migrating through the ruptured tubule basement membrane and undergoing an epithelial-to-mesenchymal transition (EMT; Iwano et al., 2002), although it is not clear from the experiments presented if LRCs from the tubules are migrating out of the tubule or if LRCs from the interstitium are migrating into the tubule, undergoing a mesenchymal-to-epithelial transition (MET; Zeisberg *et al.*, 2005), or neither. This does leave open the possibility that a kidney progenitor or a noncommitted multipotent cell population residing outside of the tubule may contribute progeny to tubule epithelial repair, including kidney interstitial cells and bone marrow.

Maeshima and colleagues (2006) have isolated the LRCs from normal adult rat kidney cortex labeled for 1 week continuously with BrdU and chased for 2 weeks by flow cytometry (Mozdziak *et al.*, 2000). These cells, representing approximately 6% of the viable tubular cells isolated, would proliferate 2- to 4-fold over 10 days in the presence of epidermal growth factor (EGF), transforming growth factor (TGF)-α, insulin-like growth factor I (IGF-I), or hepatocyte growth factor (HGF), but not in the presence of TGF-β, activin A, bone morphogenetic protein (BMP)-2, BMP-4, BMP-7, fibroblast growth factor (FGF)-1, FGF-2, FGF-7, glial cell line-derived neurotrophic factor (GDNF), or leukemia inhibitory factor (LIF). However, non-LRCs behaved in a similar manner. LRCs cultured *in vitro* in the presence of HGF formed epithelial sheets and expressed epithelial cell surface markers E-cadherin, cytokeratin, F-actin, and zonula occludens 1 (ZO-1). In addition, these cells also expressed the mesenchymal-associated marker vimentin. Conversely, the isolated LRCs would also expand in conditioned medium collected from a metanephric mesenchymal-derived cell line (BSN cells), whereas non-LRCs would not. The LRCs cultured in this manner remained as spherical dissociated cells and remained viable for greater than 2 months when cultured on plastic but formed spindle-like fibroblasts when cultured on collagen I or laminin. When cultured on fibronectin the cells remained spherical but formed clusters. When LRCs were introduced into collagen I gel in the presence of either HGF or the conditioned medium from UB cells (a cell line derived from the uteric bud), cells formed tubule-like structures with lumens. When injected into cultured rat metanephros and cultured to allow the epithelialization process to proceed, the donor cells appeared to engraft in the tubules (as identified by costaining with LTA), collecting duct (costaining with DBA), and interstitium. Non-LRCs remained as single cells in the interstitium. These latter data should be interpreted with caution as fluorescence colocalization of donor injected cells is easy to misinterpret without the use of confocal microscopy (Brazelton and Blau, 2005).

There are at least two models available to explain the vast majority of the results presented here, although none of the evidence presented in these studies provides compelling evidence in favor of either. The simplest model is that every somatic tubular cell is capable of undergoing mitosis and replacing damaged neighboring tubular cells when necessary. As shown in several of these studies, the regenerating tubular cells express a more primitive intermediate filament protein, as has been reported in other regenerating tissue (Gilles *et al.*, 1999), and a transcription factor associated with kidney development, but not normally expressed in mature tubules, demonstrating that regenerating tubules are able to express proteins associated with a more primitive state of differentiation than mature kidney tubular cells. This can be considered a form of dedifferentiation, but may not be a strict reversion to the more primitive precursor cell that gave rise to the epithelia during embryonic development; rather, it may be a preprogrammed blueprint for accessing the machinery necessary to replicate tubular cells, migrate, and form polar epithelial tubule cells with tight intercellular junctions.

The second model is derived from that established for more rapidly regenerating tissues such as the epidermis (Blanpain *et al.*, 2004; Jensen *et al.*, 1999) and the intestine (Brittan and Wright, 2004). In this model, there is a relatively rare, long-lived group of progenitor cells that undergo mitosis only occasionally, giving rise to a transit-amplifying population of cells that is relatively short lived but highly proliferative, residing in an advantageous niche. These transit-amplifying cells then give rise to the somatic cells. The data presented here would be consistent with this idea, although not proof of it. In kidney tubules, either scattered throughout the tubule, or perhaps localized to the distal-most portion of the proximal tubule, reside long-lived cells with kidney progenitor potential. These cells would cycle only occasionally and thus could be labeled with BrdU, but would most certainly be only a subset of the replicating cells labeled. When large numbers of new tubular cells are required to replace those injured, these tubular progenitors give rise to a transit-amplifying population of cells that rapidly proliferate and migrate, finally producing postmitotic mature tubular cells. The transit-amplifying population of cells would label readily with BrdU, and would be the cells observed in the studies summarized here that express Pax2 and vimentin. These cells would then essentially disappear after the tubule is regenerated, appearing in normal tubules only at a low level to provide progeny to replace the occasional necrotic cell. In the case of the kidney tubule, the rare kidney tubule progenitor cell, and transit-amplifying cells derived from it, by necessity carry the functional attributes of a somatic tubular cell to regulate its internal stasis while residing along the tubule lumen, thus making the LRCs appear phenotypically similar to postmitotic tubules, at least at the initial level of detection used in these studies. Contrary to this, it remains possible that the brush-borderless cells with large oval nuclei residing at the junction of the distal-most proximal tubule and the thin loop of Henle

(Fujigaki *et al.*, 2006) could be the tubular progenitor cells and this particular location may provide its niche. If this progenitor exists, the differentiation potential of this cell remains unknown.

Extratubular Progenitor Cells

AC133 is an antibody that has gained popularity as a unique cell surface epitope expressed on hematopoietic stem cells as well as somatic tissue progenitor cells derived from human tissues (Yin *et al.*, 1997). AC133 recognizes a subset of cells expressing an otherwise ubiquitously expressed cell surface protein, CD133 (prominin-1), perhaps through alternative splicing of this molecule or a specific posttranslational modification (Shmelkov *et al.*, 2005). Immunostaining of adult human kidney section with AC133 labeled rare cells residing in the interstitium (Boussolati *et al.*, 2005). This antibody was then used to purify these cells for characterization and *in vitro* culture. AC133-positive cells represented approximately 1% of the viable cells recovered from dissected human kidney cortex after mechanical sieving to generate a single-cell suspension. Isolated cells were grown on fibronectin in serum-free medium including EGF and platelet-derived growth factor-BB (PDGF-BB), medium and extracellular matrix used for culturing human multipotent adult progenitor cells (MAPCs) isolated from bone marrow (Reyes *et al.*, 2001). Cells isolated and cultured under these conditions could be clonally derived and continued to express CD133 for seven to nine passages (estimated at approximately 20 to 25 cell doublings) before becoming senescent. Freshly isolated cells expressed cell surface markers characteristic of bone marrow mesenchymal stem cells (MSCs), including CD44, integrin β_1 (CD29), and SH3 (CD73) (Pittenger *et al.*, 1999) and also expressed HLA class I along with the kidney developmental transcription factor Pax2.

Expanded clonal isolates of CD133-expressing cells would upregulate several proteins characteristic of kidney tubular epithelium when cultured for 10 days in the presence of HGF and FGF-4, including E-cadherin, ZO-1, alkaline phosphatase, aminopeptidase A, and thiazide-sensitive NaCl cotransporter, and a distinct minority expressed calbindin-D. Cells cultured in this manner also expressed cytokeratin and vimentin, while losing expression of CD133. When cultured on collagen-coated Transwell plates and examined by transmission electron microscopy, cells displayed morphological aspects expected for kidney tubules including apical microvilli and junctional complexes. Differentiated cells also exhibited transepithelial resistance comparable to that of isolated kidney tubules. This same population of clonally derived cells would develop characteristics of endothelium when plated on endothelial cell attachment factor and grown in the presence of vascular endothelial growth factor (VEGF), including mucin (Muc)-18, KDR (kinase insert domain receptor), CD105, VE-cadherin, and von Willebrand factor (vWF). When cultured on Matrigel, cells differentiated under endothelial conditions rapidly aligned to form capillary-like ring structures expressing vWF. Of note, Pax2 continued to be expressed in cells differentiated under either the epithelial or endothelial differentiation program.

When undifferentiated CD133-expressing cells were implanted subcutaneously, in Matrigel plugs, into severe combined immunodeficient (SCID) mice, HLA class I-expressing cells assembled into tubule-like structures and expressed tubular epithelial markers, microvilli, and tight junctions. When cultured under conditions that induce endothelial differentiation before subcutaneous injection, cells formed functional vessels connecting to the mouse vasculature, as the presence of intralumenal red blood cells could be observed. CD133-expressing cells also incorporated into regenerating glycerol-induced injured tubules, both proximal and distal. Some of these HLA class I-staining cells also expressed PCNA and cytokeratin.

BrdU labeling of neonatal rat pups was used to identify LRCs in the adult rat (Oliver et al., 2004). The majority of LRCs after 2 months of chase were found in the papilla region of the kidney, primarily in the interstitium, but also incorporated into the tubules of the papilla, costaining with ZO-1. Cells were isolated and cultured from collagenase-digested isolated papilla, of which 40% of the cells in culture were LRCs. Initially cells expressed epithelial features, but unless cultured under low-oxygen conditions began expressing α-smooth muscle actin. Isolated cells tended to form spheres when cultured on plastic or in serum-free media, whereas culture on fibronectin or in the presence of LIF caused the cells to form primarily monolayers. As spheres and after clonal isolation, some cells expressed nestin and the neural marker class III β-tubulin, suggesting possible multipotency.

When transient unilateral ischemia was imposed in this model of BrdU labeling, there was a drastic diminution of LRCs in the papilla of the ischemic kidney, suggesting the LRCs were proliferating and thus diluting the BrdU label beyond detection. Proliferation of LRCs was confirmed by a significant increase in cells staining with Ki-67, a marker of proliferating cells (Scholzen and Gerdes, 2000), in the papilla. Rats were also acutely labeled with BrdU 36 h after ischemic injury to identify proliferating cells. Although proliferating cells were identified in the cortex and medulla, the papilla had abundant BrdU-labeled cells almost exclusively adjacent to the urinary space of the outer papilla (closest to the medullary portion of the kidney), suggesting migration toward the medulla and potential contribution to its repair. To examine the potential contribution of these cells to tubule repair, isolated cells were labeled with a fluorescent dye and injected into the subcapsular region of the kidney of normal rats. After 7 days labeled cells could be found in clumps, scattered throughout the interstitium, and incorporated into tubules, suggesting that isolated, cultured papillary cells could readily migrate and contribute to tubule maintenance.

In addition, cells can be isolated from adult mouse kidney on the basis of the differential exclusion of Hoechst dyes, called side population cells. Side population cells from diverse tissues appear to have *in vitro* hematopoietic differentiation potential (Asakura and Rudnicki, 2002a), but can contribute to the formation of myocytes and satellite cells when injected intramuscularly (Asakura *et al.*, 2002b). Side population cells isolated from the kidney appear to contribute to tubular repair after cisplatin-induced acute renal failure, but not by providing progeny to regenerating tubules; rather, they contribute regenerative growth factors to the process (Hishikawa *et al.*, 2005b). When these cells are isolated and cultured in collagen gels in the presence of LIF, side population cells upregulate cadherin 16 (Hishikawa *et al.*, 2005a), a kidney-specific extracellular matrix protein (Thompson *et al.*, 1995). However, when isolated side population cells were injected intravenously into normal rats, they did not appear to home to the kidney but rather migrated to the skeletal muscle, liver, and bone marrow (Iwatani *et al.*, 2004).

Finally, it has been noted that tissue committed progenitor cells may be able to circulate and take up residence in the bone marrow (Ratajczak *et al.*, 2004), based on the observation that transcripts normally associated with precursors for muscle, liver, and neural tissue can be detected in bone marrow by reverse transcription-polymerase chain reaction (RT-PCR). Several attempts have been made to test whether bone marrow cells are capable of contributing to renal tubule repair after various tubular injuries, resulting in controversial conclusions (e.g., see Duffield *et al.*, 2005; Lin *et al.*, 2005). It should be noted, however, that our group has been unable to detect any Pax2 expression by RT-PCR in low-density bone marrow (data not shown), so it remains unknown whether kidney committed progenitors capable of contributing to renal tubule repair reside outside the confines of the kidney.

Culture of Human Tubular Progenitor Cells

Tubular cells can be isolated from a number of species (Humes and Cieslinski, 1992; Humes *et al.*, 1996) including human cadaveric sources (Humes *et al.*, 2002). Isolated human tubular cells have been incorporated into a bioartificial device that has been employed to treat experimentally induced acute renal failure and has been tested in phase I/II clinical trials to treat clinical acute renal failure (Humes *et al.*, 2004). The cells incorporated into this unit were shown to help regulate the plasma levels of HCO_3^-, P_i, and K^+ and the active transport of glucose, HCO_3^-, and K^+ as well as ammonia excretion, glutathione processing, and 25-hydroxyvitamin D_3 conversion in animals undergoing experimentally induced acute renal failure. When used in clinical phase I/II trials on patients with multiple organ failure and acute renal failure, glutathione degradation and 25-hydroxyvitamin D_3 conversion were maintained, and for patients with excessive proinflammatory levels, significant declines in granulocyte colony-stimulating factor, interleukin (IL)-10, and IL-6:IL-10 ratios were noted. This suggests that the transport and metabolic and immunological regulatory properties of renal proximal tubules are recapitulated in the primary cells cultured from human kidney.

Despite our lack of understanding of the origin of these cells, human renal tubular cells are readily isolated by the following procedure.

1. Human kidneys rejected for organ transplant on the basis of anatomic or fibrotic defects are dissected to remove the cortex from the medulla. The cortex is excised and minced and up to 25 g is placed in a 250-ml vented Erlenmeyer flask containing 20 ml of a 37° prewarmed sterile incubation solution (1.14 mM $CaCl_2$, 0.73 mM $MgSO_4$ in Dulbecco's modified Eagle's medium [DMEM]). Before tissue addition, 3.5 ml of a 4° collagenase, class IV (Worthington, Freehold, NJ) solution (10 mg/ml in 1.14 mM $CaCl_2$, 0.73 mM $MgSO_4$ in DMEM) and 1.5 ml of a 4° DNase (Sigma, St. Louis, MO) solution (5.5 mg/ml in 1.14 mM $CaCl_2$, 0.73 mM $MgSO_4$ in DMEM) are added to the 20-ml incubation solution. A sterile stir bar is added to the flask for suspension mixing. Multiple flasks are used as needed, dependent on kidney mass, for each kidney digest.

2. Prepared flasks are placed into a 2-liter beaker containing 1 liter of prewarmed 37° water, which is placed into a 37° CO_2 incubator. Digestions are allowed to proceed for 15 min.

3. To terminate the digestion, 25 ml of 4° DMEM is added and the tissue is strained through an 850-μm pore size sieve.

4. The unfilterable tissue remaining on top of the strainer is redigested with DNase–collagenase as described in steps 1–3 for between four and six cycles, until minimal undigested tissue remains. Undigested tissue from multiple flasks is combined for subsequent digestions, as the undigested tissue mass is reduced.

5. Digested cortex is poured into 50-ml conical tubes and collected by centrifugation at 53 × g for 5 min. The pellets are resuspended in 4° DMEM and strained sequentially through 710- and 600-μm pore size sieves. The collected filtrate is again collected by centrifugation.

6. For every 1 ml of pellet obtained after the 600-μm sieve straining, 24 ml of growth medium (GM) containing UltraMDCK medium (Cambrex

Bio Science Walkersville, Walkersville, MD) supplemented with insulin, transferrin, ethanolamine, selenium (ITES, 1 ml/liter; Cambrex Bio Science Walkersville), epidermal growth factor (60 μg/liter; R&D Systems, Minneapolis, MN), and triiodothyronine (0.010 ml/liter; Cambrex Bio Science Walkersville) is used for resuspension. Two milliliters of resuspended pellet is transferred per 100-mm^2 tissue culture dish containing 10 ml of GM to allow for attachment of the renal cortical fragments and cellular growth.

7. Cells are cultured at 37° in 5% CO_2, with medium exchanges every 1 to 3 days. The initial medium exchange is performed the day after the isolation in order to minimize glomerular cell growth. Cells are lifted from the culture dish with 0.25% trypsin–EDTA and replated on two culture dishes after reaching confluence. Twenty-four hours before initial passage, the medium is exchanged for GM containing retinoic acid (0.033 mg/liter; Sigma). GM with retinoic acid is used for all subsequent medium changes. Cells are passed 1:2 every 3–7 days, depending on the rate of growth. Cells are typically employed in bioartificial devices within five passages.

References

Asakura, A., and Rudnicki, M. A. (2002a). Side population cells from diverse adult tissues are capable of *in vitro* hematopoietic differentiation. *Exp. Hematol.* **30,** 1339–1345.

Asakura, A., Seale, P., Girgis-Gabardo, A., and Rudnicki, M. A. (2002b). Myogenic specification of side population cells in skeletal muscle. *J. Cell Biol.* **159,** 123–134.

Blanpain, C., Lowry, W. E., Geoghegan, A., Polak, L., and Fuchs, E. (2004). Self-renewal, multipotency, and the existence of two cell populations within an epithelial stem cell niche. *Cell* **118,** 635–648.

Boussolati, B., Bruno, S., Grange, C., Buttiglieri, S., Deregibus, M. C., Cantino, D., and Camussi, G. (2005). Isolation of renal progenitor cells from adult human kidney. *Am. J. Pathol.* **166,** 545–555.

Brazelton, T. R., and Blau, H. M. (2005). Optimizing techniques for tracking transplanted stem cells *in vivo*. *Stem Cells* **23,** 1251–1265.

Brenner, B. M., Zatz, R., and Ichikawa, I. (1986). The renal circulations. *In* "The Kidney" (B. M. Brenner and F. C. Rector, eds.), 3rd ed., pp. 93–123. W.B. Saunders, New York.

Brittan, M., and Wright, N. A. (2004). Stem cell in gastrointestinal structure and neoplastic development. *Gut* **53,** 899–910.

Davies, J. A., and Fisher, C. E. (2002). Genes and proteins in renal development. *Exp. Nephrol.* **10,** 102–113.

Duffield, J. S., Park, K. M., Hsiao, L.-L., Kelley, V. R., Scadden, D. T., Ichimura, T., and Bonventre, J. V. (2005). Restoration of tubular epithelial cells during repair of the postischemic kidney occurs independently of bone marrow-derived stem cells. *J. Clin. Invest.* **115,** 1743–1755.

Fujigaki, Y., Goto, T., Sakakima, M., Fukasawa, H., Miyaji, T., Yamamoto, T., and Hishida, A. (2006). Kinetics and characterization of initially regenerating proximal tubules in S3 segment in response to various degrees of acute tubular injury. *Nephrol. Dial. Transplant.* **21,** 41–50.

Gilles, C., Polette, M., Zahm, J. M., Tournier, J. M., Volders, L., Foidart, J. M., and Birembaut, P. (1999). Vimentin contributes to human mammary epithelial cell migration. *J. Cell Sci.* **112**, 4615–4625.

Hishikawa, K., Marumo, T., Miura, S., Nakanishi, A., Masuzaki, Y., Shibata, K., Kohike, H., Komori, T., Hayashi, M., Nakaki, T., Nakauchi, H., Okano, J., and Fujita, T. (2005a). Leukemia inhibitory factor induces multi-lineage differentiation of adult stem-like cells in kidney via kidney-specific cadherin 16. *Biochem. Biophys. Res. Commun.* **328**, 288–291.

Hishikawa, K., Marumo, T., Miura, S., Nakanishi, A., Masuzaki, Y., Shibata, K., Ichiyanagi, T., Kohike, H., Komori, T., Takahashi, I., Takase, O., Imai, N., Yoshikawa, M., Inowa, T., Hayashi, M., Nakaki, T., Nakauchi, H., Okano, J., and Fujita, T. (2005b). Musculin/MyoR is expressed in kidney side population cells and can regulate their function. *J. Cell Biol.* **169**, 921–928.

Holthofer, H., Miettinen, A., Lehto, V. P., Lehtonen, E., and Virtanen, I. (1984). Expression of vimentin and cytokeratin types of intermediate filament proteins in developing and adult human kidneys. *Lab. Invest.* **50**, 552–559.

Humes, H. D., and Cieslinski, D. A. (1992). Interaction between growth factors and retinoic acid in the induction of kidney tubulogenesis in tissue culture. *Exp. Cell Res.* **201**, 8–15.

Humes, H. D., Krauss, J. C., Cieslinski, D. A., and Funke, A. J. (1996). Tubulogenesis from isolated single cells of adult mammalian kidney: Clonal analysis with a recombinant retrovirus. *Am. J. Physiol.* **271**, F42–F49.

Humes, H. D., Fissell, W. H., Weitzel, W. F., Buffington, D. A., Westover, A. J., MacKay, S. M., and Gutierrez, J. M. (2002). Metabolic replacement of kidney function in uremic animals with a bioartificial kidney containing human cells. *Am. J. Kidney Dis.* **39**, 1078–1087.

Humes, H. D., Weitzel, W. F., Bartlett, R. H., Swaniker, F. C., Paganni, E. P., Luderer, J. R., and Sobota, J. (2004). Initial clinical results of the bioartificial kidney containing human cells in ICU patients with acute renal failure. *Kidney Int.* **66**, 1578–1588.

Imgrund, M., Grone, E., Grone, H.-J., Kretzler, M., Holzman, L., Schlondorff, D., and Rothenpieler, U. W. (1999). Re-expression of the developmental gene Pax-2 during experimental acute tubular necrosis in mice. *Kidney Int.* **56**, 1423–1431.

Iwano, M., Plieth, D., Danoff, T. M., Xue, C., Okada, H., and Neilson, E. G. (2002). Evidence that fibroblasts derive from epithelium during tissue fibrosis. *J. Clin. Invest.* **110**, 341–350.

Iwatani, H., Ito, T., Imai, E., Matsuzaki, Y., Suzuki, A., Yamato, M., Okabe, M., and Hori, M. (2004). Hematopoietic and nonhematopoietic potentials of Hoechst[low]/side population cells isolated from adult rat kidney. *Kidney Int.* **65**, 1604–1614.

Jensen, U. B., Lowell, S., and Watt, F. M. (1999). The spatial relationship between stem cells and their progeny in the basal layer of the human epidermis: A new view based on whole-mount labelling and lineage analysis. *Development* **126**, 2409–2418.

Laitinen, L., Virtanen, I., and Saxen, L. (1987). Changes in the glycosylation pattern during embryonic development of mouse kidney as revealed with lectin conjugates. *J. Histochem. Cytochem.* **35**, 55–65.

Lehrer, M. S., Sun, T.-T., and Lavker, R. M. (1998). Strategies of epithelial repair: Modulation of stem cell and transit amplifying cell proliferation. *J. Cell Sci.* **111**, 2867–2875.

Lin, F., Moran, A., and Igarashi, P. (2005). Intrarenal cells, not bone marrow-derived cells are the major source for regeneration in postischemic kidney. *J. Clin. Invest.* **115**, 1756–1764.

Maeshima, A., Maeshima, K., Nojima, Y., and Kojima, I. (2002). Involvement of Pax-2 in the action of Activin A on tubular cell regeneration. *J. Am. Soc. Nephrol.* **13**, 2850–2859.

Maeshima, A., Yamashita, S., and Nojima, Y. (2003). Identification of renal progenitor-like tubular cells that participate in the regeneration processes of the kidney. *J. Am. Soc. Nephrol.* **14**, 3138–3146.

Maeshima, A., Sakurai, H., and Nigam, S. K. (2006). Adult kidney tubular cell population showing phenotypic plasticity, tubulogenic capacity, and integration capability into developing kidney. *J. Am. Soc. Nephrol.* **17,** 188–198.

Mozdziak, P. E., Pulvermacher, P. M., Schultz, E., and Schell, K. (2000). Hoechst fluorescence intensity can be used to separate viable bromodeoxyuridine-labeled cells from viable non-bromodeoxyuridine-labeled cells. *Cytometry* **41,** 89–95.

Oliver, J. A., Maarouf, O., Cheema, F. H., Martens, T. P., and Al-Awqati, Q. (2004). The renal papilla is a niche for adult kidney stem cells. *J. Clin. Invest.* **114,** 795–804.

Pittenger, M. F., Mackay, A. M., Beck, S. C., Jaiswal, R. K., Douglass, R., Mosca, J. D., Moorman, M. A., Simonetti, D. W., Craig, S., and Marshak, D. R. (1999). Multilineage protential of adult human mesenchymal stem cells. *Science* **284,** 143–147.

Potten, C. S., and Loeffler, M. (1990). Stem cells: Attributes, cycles, spirals, pitfalls and uncertainties: Lessons for and from the crypt. *Development* **110,** 1001–1020.

Ratajczak, M. Z., Kucia, M., Reca, R., Majka, M., Janowska-Wieczorek, A., and Rataczak, J. (2004). Stem cell plasticity revisited: CXCR4-positive cells expressing mRNA for early muscle, live and neural cells 'hide out' in the bone marrow. *Leukemia* **18,** 29–40.

Reyes, M., Lund, T., Lenvik, T., Aguiar, D., Koodie, L., and Verfaillie, C. M. (2001). Purification and *ex vivo* expansion of postnatal human marrow mesodermal progenitor cells. *Blood* **98,** 2615–2625.

Schrier, R. W., Wang, W., Poole, B., and Mitra, A. (2004). Acute renal failure: Definitions, diagnosis, pathogenesis, and therapy. *J. Clin. Invest.* **114,** 5–14.

Shmelkov, S. V., St Clair, R., Lyden, D., and Rafii, S. (2005). AC133/CD133/Prominin-1. *Int. J. Biochem. Cell. Biol.* **37,** 715–719.

Thompson, R. B., Igarashi, P., Biemesderfer, D., Kim, R., Abu-Alfa, A., Soleimani, M., and Aronson, P. S. (1995). Isolation and cDNA cloning of Ksp-cadherin, a novel kidney-specific member of the cadherin multigene family. *J. Biol. Chem.* **270,** 17594–17601.

Tisher, C. C., and Madsen, K. M. (1986). Anatomy of the kidney. *In* "The Kidney" (B. M. Brenner and F. C. Rector, eds.), 3rd ed., pp. 3–60. W.B. Saunders, New York.

Vogetseder, A., Karadeniz, A., Kaissling, B., and Le Hir, M. (2005). Tubular cell proliferation in the healthy rat kidney. *Histochem. Cell. Biol.* **124,** 97–104.

Witzgall, R., Brown, D., Schwarz, C., and Bonventre, J. V. (1994). Localization of proliferating cell nuclear antigen, vimentin, c-Fos, and clusterin in the postischemic kidney. *J. Clin. Invest.* **93,** 2175–2188.

Yamashita, S., Maeshima, A., and Nojima, Y. (2005). Involvement of renal progenitor tubular cells in epithelial-to-mesenchymal transition in fibrotic rat kidneys. *J. Am. Soc. Nephrol.* **16,** 2044–2051.

Yin, A. H., Miraglia, S., Zanjani, E. D., Almeida-Porada, G., Ogawa, M., Leary, A. G., Olweus, J., Kearney, J., and Buck, D. W. (1997). AC133, a novel marker for human hematopoietic stem and progenitor cells. *Blood* **90,** 5002–5012.

Zeisberg, M., Shah, A., and Kalluri, R. (2005). Bone morphogenic protein-7 induces mesenchymal to epithelial transition in adult renal fibroblasts and facilitates regeneration of injured kidney. *J. Biol. Chem.* **280,** 8094–8100.

[10] Ovarian Germ Cells

By ANTONIN BUKOVSKY, IRMA VIRANT-KLUN, MARTA SVETLIKOVA, and
ISABELLE WILLSON

Abstract

Surface cells in adult ovaries represent germ line-competent embryonic
stem cells. They are a novel type of totipotent progenitors for distinct cell
types including female germ cells/oocytes, with the potential for use in the
autologous treatment of ovarian infertility and stem cell therapy. Ovarian
infertility and stem cell therapy are complex scientific, therapeutic, and
socioeconomic issues, which are accompanied by legal restrictions in many
developed countries. We have described the differentiation of distinct cell
types and the production of new eggs in cultures derived from adult human
ovaries. The possibility of producing new eggs from ovarian surface epithelium
representing totipotent stem cells supports new opportunities for the treatment of premature ovarian failure, whether idiopathic or after cytostatic
chemotherapy treatment, as well as infertility associated with aged primary
follicles, and infertility after natural menopause. The stem cells derived from
adult human ovaries can also be used for stem cell research and to direct
autologous stem cell therapy. This chapter describes general considerations
regarding the egg origin from somatic progenitor cells, oogenesis and follicle
formation in fetal and adult human ovaries (follicular renewal), including
the promotional role of the immune system-related cells *in vivo*, and possible causes of ovarian infertility. It then provides detailed protocols for the
separation and cultivation of adult ovarian stem cells.

Introduction

Adult stem cells have the advantage of being produced and used in an
autologous manner in regenerative medicine, and some of them show pluripotent capacity. A wide variety of multipotent and pluripotent adult stem
cells, which may be capable of differentiating into distinct cell types, have
been reported (reviewed in Prentice, 2004). However, only adult ovarian
surface epithelium (OSE) cells appear to have the capacity of totipotent
germ line-competent embryonic stem cells (Bukovsky *et al.*, 2005b). They
were shown to be capable of differentiating into oocytes, fibroblasts, and
epithelial and neural cell types. Differentiation of oocytes from adult human
OSE cells is an additional proof of the somatic origin of germ cells and eggs.
In vivo, human female germ cells and oocytes are formed during the second

METHODS IN ENZYMOLOGY, VOL. 419   0076-6879/06 $35.00
Copyright 2006, Elsevier Inc. All rights reserved.   DOI: 10.1016/S0076-6879(06)19010-2

trimester of intrauterine life and then throughout the prime reproductive period from OSE cells influenced by immune system-related cells (tissue macrophages and T lymphocytes) (Bukovsky *et al.*, 1995c, 2004, 2005a). The oocytes, surrounded by a layer of follicular (granulosa) cells, form primary follicles, which are able to survive within the ovaries for an extended period of time, possibly up to the 15 years. During that period, however, the oocytes may accumulate endogenously and environmentally induced genetic alterations, which could make them unsuitable for fertilization and/or production of healthy progeny (reviewed in Bukovsky *et al.*, 2004). Such and other types of ovarian infertility could be overcome by production of fresh eggs in ovarian cultures (Bukovsky *et al.*, 2005b). Here we review the available information and current views on germ cell origin from somatic precursors, and ovarian follicle origin in fetal and adult ovaries, as well as the pathophysiology of follicular renewal and possible causes of ovarian infertility. We then describe the use of ovarian samples for *in vitro* production of autologous adult ovarian stem cells and new eggs, and provide detailed protocols for the separation and cultivation of adult ovarian stem cells.

Origin of Primordial Germ Cells

For successful differentiation *in vitro* of adult stem cells into a certain cell type, it is necessary to understand the natural progenitors and conditions of cell differentiation *in vivo*. This is particularly important for the development of germ cell precursors and their *in vitro* differentiation into the stage suitable for fertilization.

It is now well documented that mammalian primordial germ cells originate from uncommitted (totipotent) somatic stem cells, and that their sex commitment is determined by the local gonadal environment, that is, signals produced by neighboring somatic cells (Alberts *et al.*, 2002). Studies of mouse embryos, in which genetically marked cells were introduced at the four- and eight-cell stage blastomere, have shown that such cells can become either germ cells or somatic cells (Kelly, 1977). This suggests that no specific germ cell commitment exists before implantation. During the postimplantation period, mouse germ cells are not identifiable before ~7 days after fertilization (Ginsburg *et al.*, 1990), suggesting that germ cells differentiate from the somatic lineage (Lawson and Hage, 1994). It also has been shown that cellular differentiation of grafted embryonic cells does not depend on where the grafts were taken, but rather on where they were placed (Tam and Zhou, 1996).

After the primordial germ cells enter the developing embryonic gonad, they commit to a developmental pathway that will lead them to become

either eggs or sperm, depending not on their own sex chromosome constitu-
tion, but on whether the gonad has begun to develop into an ovary or a testis,
respectively. The sex chromosomes in the gonadal somatic cells determine
which type of gonad will develop, as a single SRY (sex- determining region
Y) gene on the Y chromosome can redirect a female embryo to become a
male (reviewed in Alberts *et al.*, 2002).

Formation of Human Fetal Oocytes and
  Primordial and Primary Follicles

Unlike other cells and tissue structures, human oocytes and primary
follicles do not differentiate/regenerate throughout the entire life, but only
during two relatively limited periods. These are the second trimester of
intrauterine life and the prime reproductive period—from menarche until
about the end of the third decade.

Ovarian differentiation begins before follicles form, and it is characterized by the development of oocytes, organization of the rete ovarii, and
evolution of the OSE. In human embryos, primordial germ cells arise
outside the urogenital ridge, in the dorsal endoderm of the yolk sac at
24 days of age. They migrate by ameboid movements to indifferent gonadal
primordia at 28–35 days (Peters and McNatty, 1980). Differentiation of the
indifferent gonad into an ovary or a testis takes place during the second
fetal month (Simkins, 1932).

The human fetal OSE contains numerous germ cells (10 μm in diameter)
from 7 weeks of intrauterine life until the neonatal period, and it has been
suggested that these cells are extruded into the peritoneal cavity (Motta and
Makabe, 1982, 1986). This could happen after the cessation of oogenesis [6–7
months of fetal life (Peters and McNatty, 1980; Simkins, 1932)], when germ
cells emerging in the OSE may be prevented from entering the cortex by the
developing ovarian tunica albuginea (TA). However, former views indicate
that the OSE is a source of germ cells differentiating into oocytes in human
fetal ovaries (Simkins, 1928, 1932), and we did not observe germ cells leaving the ovary in midpregnancy human fetuses (Bukovsky *et al.*, 2005a).
Our observations indicate the presence of small (10-μm) germ cells within
the OSE, but such cells are smaller when compared with those positioned
under the OSE. In deeper ovarian cortex, germ cells with well-defined
cytoplasm and plasma membrane show a further increase in size (Bukovsky
et al., 2005a). They lie among smaller neighboring cells with round or
elongated nuclei. Beneath the layer of well-defined germ cells lie nuclear
clusters or syncytia of germ cells, and the entire area is surrounded by
mesenchymal cell cords, that is, extensions of centrally located rete cords
into the ovarian cortex (Peters and McNatty, 1980).

Formation of the follicle requires the attachment of granulosa cells to the oocyte surface and closure of the basement membrane around this unit. At 5 months of fetal age, the ovary contains its peak population of germ cells. Between the sixth and seventh months of intrauterine life, the last oogonia enter meiosis and formation of new follicles is terminated (Peters and McNatty, 1980; Simkins, 1932).

Primary versus Primordial Follicles

Cleveland Sylvester Simkins was the first to distinguish two types of follicles in human fetal ovaries: the smaller ($\sim$20 μm in diameter), which he called "primordial" follicles, and the larger ($\sim$50 μm), which he designated as "primary" follicles (Simkins, 1932) (see also Fig. 1F). The primordial follicles consist of oocytes with often altered nuclei, which are surrounded by an inconstant and incomplete layer of small ellipsoid granulosa cells (see also Fig. 1E). These follicles are found only in the cortex of the fetal and infant ovaries, where they constitute the most conspicuous part of the gland. The primary follicles are found only in the peripheral margin of the ovarian medulla. Their oocytes take deeper stain and are always surrounded by at least one row of large, round, and regular granulosa cells. Primary follicles are the follicles that grow and develop into definitive structures, whereas the primordial follicles gradually disappear during childhood and are not detected in adult human ovaries (Simkins, 1932).

Role of Rete Ovarii

At the embryonic age of 9 weeks, female gonads show a marked development of rete cords with lumen formation, and the rete reaches the center of the ovary at 12 weeks. The first follicles are formed after the fourth fetal month, and follicle formation always begins in the inner-most part of the cortex, close to the rete ovarii. This structure is essential for follicular development, because if it is removed before formation of follicles has started, follicles will not form (Byskov et al., 1977).

Immune System-Related Cells and Oogenesis in Fetal Ovaries

The reasons for the limited time window for the differentiation of germ cells and primary follicles appear to come from two participating events. One is that not each OSE cell differentiates into a germ cell in a given time, but only those that are associated with immune system-related cells (tissue macrophages and T cells). The other is that a certain hormonal milieu, such as high levels of estradiol (E_2) and luteinizing hormone (LH) [or human chorionic gonadotropin (hCG)], might be required.

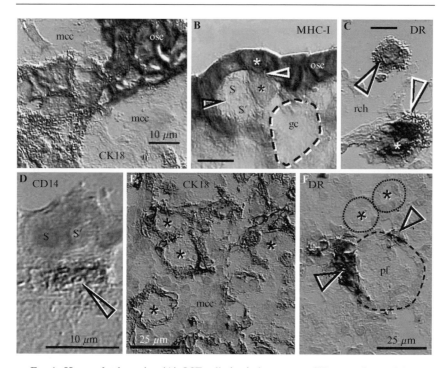

FIG. 1. Human fetal ovaries. (A) OSE cells (ose) show strong CK expression and descent among mesenchymal cell cords (mcc) to give rise to moderately CK$^+$ primitive granulosa (pg) cells. (B) MHC class I$^+$ OSE cell (white asterisk) undergoes asymmetric division (white arrowhead) to give rise to the MHC class I$^-$ germ cell (black asterisk). After symmetric division (black arrowhead and s & s' cells) the tadpole-like germ cells (gc, dashed line) enter the ovarian cortex. No hematoxylin counterstain. (C) Rete channels (rch) show HLA-DR monocytes (black arrowhead) interacting (white arrowhead) with resident cells (white asterisk). (D) Germ cells undergoing symmetric division (s and s') in the fetal OSE and accompanying CD14 (primitive) tissue macrophage (arrowhead). (E) Primordial follicles (asterisks) adjacent to mesenchymal cell cord show CK$^+$ granulosa cells but no Balbiani bodies (compare with Fig. 4C). (F) Development of primary follicle (pf) is accompanied by activated (HLA-DR$^+$) tissue macrophages (arrowheads); asterisks indicate small primordial follicles. Reproduced from Bukovsky *et al.* (2005a), with permission. (See color insert.)

The immune system is traditionally viewed as a system with an exclusive role in the protection of the body from foreign (nonself) substances, including bacterial and viral infections and allogeneic tissue transplants. It sometimes affects an individual's own tissues (autoimmune diseases) and possibly prevents cancer. From this simplified point of view, this system exercises immune defense or it does nothing. However, cells belonging to the immune system, such as monocyte-derived tissue macrophages and

T lymphocytes, are present and differentiate among epithelial cells and also in most parenchymal tissues, and autoantibodies bind to normal self tissues. In addition, immune cells are a source of growth factors and cytokines, not only involved in the development of immune surveillance, but also required for the differentiation and regeneration of nonimmune (epithelial and parenchymal) cells. With this in mind, we suggested that the immune system (or the more widely defined tissue control system—see later) plays a dual role within the body, consisting of (1) stimulation and regulation of cell differentiation and (2) elimination of aged, infected, or nonself cells and substances (reviewed in Bukovsky et al., 2001).

Our observations (Bukovsky et al., 2005a) indicate that in the ovaries of midpregnancy human fetuses, the primitive granulosa cells originate from sprouts of OSE cells extending into the ovary between mesenchymal cell cords (Fig. 1A). The fetal germ cells originate from the OSE by asymmetric division (asterisks in Fig. 1B) and undergo subsequent symmetric division (s and s', Fig. 1B) followed by migration of tadpole-like germ cells (dashed line, Fig. 1B) into the cortex. These processes are similar to those described in adult human ovaries (Bukovsky et al., 1995c, 2004).

The mesenchymal cell cords are rich in expression of Thy-1 differentiation protein (Thy-1 dp), an ancestral member of the immunoglobulin gene superfamily of molecules (Williams and Barclay, 1988). Thy-1 dp plays an important role in the stimulation of early steps in cellular differentiation (Bukovsky et al., 2001; Williams, 1985; Williams et al., 1989). Activated tissue macrophages (HLA-DR$^+$) reside in rete cords, and lymphocyte and monocyte type cells are found in rete channels (black arrowhead, Fig. 1C) and show interactions with resident macrophages. (white arrowhead, Fig. 1C). The rete cords also show prominent Thy-1 expression. The accumulation of activated tissue macrophages is apparent under the OSE. Rete channels also contain CD8$^+$ T cells, found in the vicinity of the OSE. Primitive macrophages expressing CD14 show association with OSE cells (arrowhead, Fig. 1D) and have been found to accompany the appearance of intraepithelial germ cells (s and s') (Bukovsky et al., 2005a). It is possible that primitive macrophages contribute to the emergence of germ cells while activated macrophages induce the formation of primitive granulosa cells from the OSE.

The granulosa cells associate with oocytes in the deeper ovarian cortex to form fetal follicles (asterisks, Fig. 1E). However, in contrast to adult ovaries (Bukovsky et al., 2004), no cytokeratin (CK)-positive Balbiani bodies were detected in follicular oocytes of midpregnancy fetal ovaries. Activated tissue macrophages are also associated with larger (primary) follicles (arrowheads, Fig. 1F), but not with small (primordial) ones (asterisks, Fig. 1F). In contrast to OSE cells, the intraepithelial germ cells do not show a binding of immunoglobulins (Bukovsky et al., 2005a).

These observations indicate that the OSE is a source of germ and primitive granulosa cells. Hence, as in adult ovaries (Bukovsky et al., 2004), the midpregnancy human OSE stem cells are bipotent progenitors with a commitment to both cell types. It is possible that tissue macrophages residing in rete cords carry a memory of the characteristics of germ cells populating the ovary during embryonal period of life. Such memory could be transferred to monocytes and T cells migrating through the rete channels in midpregnancy ovaries, and the migrating cells reaching the OSE may stimulate transformation of some OSE stem cells into germ cells. In addition, the mesenchymal cell cords rich in Thy-1 dp may participate in the transformation of OSE cells into primitive granulosa cells. In this way, different potentials of OSE cells may be realized, depending on the local influence of migrating and resident mesenchymal cells. Pluripotency of progenitor cells is not unusual. It persists in bone marrow throughout life, and the "one cell, two fates" phenomenon has also been described for vascular progenitor cells (Yamashita et al., 2000).

Fetal Programming of Follicular Renewal during Adulthood

During the last two centuries, public health and genetics shared common ground through similar approaches to health promotion in the population. By the middle of the last century, there was a division between public health and genetics, with eugenicists estranged and clinical genetics focused on single gene disorders, usually relevant only to small numbers of people. Advances in molecular biology hold great promise for complex conditions such as cardiovascular disease. However, there has been little tangible success in defining specific mutations that can explain the more common forms of cardiovascular disease. Of the numerous genes tested, inconsistent results are a recurring theme (Halliday et al., 2004; Harrap, 2000).

It is to be expected that genetic mutations are manifested throughout life and not with a delay of several decades. However, it is now taken for granted that early-life environmental factors influence prenatal development and may cause structural and functional changes that may manifest later in the life. This organizational phenomenon is termed "early-life programming" (Seckl, 2004).

Accumulating evidence does support the view that the function of tissues and organs in adult individuals is programmed during the fetal period of life (Baker, 1994; Barker et al., 1989a,b; Lucas, 1991; Ozanne, 2001; Seckl, 1998). In contrast to the more conventional theories that cardiovascular and metabolic disorders of middle age are caused by specific lifestyle-derived risk factors acting in adulthood on an individual's genetic background (Harrap, 1994), a series of provocative epidemiologic findings suggests that environmental factors in early life are of substantial importance to disease risk in later years. The number of adult diseases that may have their major origin in the course of fetal growth and development grows steadily. It includes cardiovascular disease, hypertension, stroke, non-insulin-dependent diabetes mellitus, chronic renal failure, chronic obstructive lung disease, osteoporosis, schizophrenia, depression, dyslipidemia, cancer, obesity, and polycystic ovary syndrome (Ozanne et al., 2004).

To explain these findings, the idea of early life physiologic programming or imprinting has been advanced. Fetal programming has been documented in a variety of systems, and it is triggered when a stimulus of insult occurs at a gestational age critical for target organ differentiation, growth, or development, and induces permanent changes in organ size, structure, or function. Programming agents might include growth factors, hormones, and nutrients. These factors may produce adaptations that permanently alter adult metabolism and responses in a direction optimizing survival under continued conditions of malnutrition, stress, or other deprivations (Barker and Bagby, 2005; Seckl, 2004).

An excess of sex steroids during fetal programming causes alteration of ovarian development, and subsequent or delayed alteration of ovarian function during adulthood (Abbott et al., 2005; Birch et al., 2003; Dumesic et al., 2002; Zachos et al., 2002). Yet, the level at which the alteration of organ function in adulthood is programmed during fetal life remains unknown. It is unlikely that the ovary or hypothalamo–pituitary system is permanently altered per se, because ovarian dysfunction may be manifested with a delay, after a period of normal ovarian ovulatory function (Arai, 1972; Handa et al., 1985; Mobbs et al., 1985; Swanson and van der Werff ten Bosch, 1964).

Administration of androgens to rat females during the critical postnatal window (days 5 or 7) caused temporary acceleration of ovarian development toward the aged phenotype (overexpression of androgen receptor). However, from day 14 until puberty (day 35) these ovaries showed no differences as compared with controls, but premature aging of rat ovaries was apparent from the beginning of sexual maturity and accompanied by distinct behavior of immune system-related cells in the ovaries (Bukovsky et al., 2000a, 2002).

Hence, it is possible that ovarian development and function during adulthood are influenced by mesenchymal–epithelial interactions, which accompany differentiation of epithelial cells in adult tissues (Bukovsky et al., 2001). This may depend on the ability of ovary-specific mesenchymal cells (o-SMCs), such as monocyte-derived cells (MDCs) and T cells, first to recognize and memorize the character of primordial germ cells that populate the gonadal anlage from the extragonadal source during embryonal development, and then to induce replication of this process within fetal ovaries. Ovarian surface epithelium, because of its rapid proliferation and pleomorphism, as well as its capacity to differentiate into various cell types (ovarian cancers), might be a target of o-SMCs in this process.

In normal individuals, the first organ affected by aging is the thymus (Kay, 1979), and next is the ovary (Kirkwood, 1998; Talbert, 1968). There is a striking correlation between the period at which an organ is present during early ontogeny and the functional longevity of that organ. For instance, the heart, which differentiates very early, can in human beings function over 100 years. However, the ovary, which differentiates later, does not function for more than half of that period. We have suggested that the later the differentiation of a certain type of tissue occurs during early ontogeny, the earlier its function expires during adulthood (Bukovsky and Caudle, 2002).

During immune adaptation (through the end of the second trimester of intrauterine life in humans; Klein, 1982), the differentiating tissues are recognized by the developing lymphoid (immune) system as self. However, depending on the time at which a certain tissue arises during immune adaptation, a memory could be built for how long such tissue would be supported by tissue-specific mesenchymal cells to function thereafter. Immune system-related cells (MDCs and T cells) are present in peripheral tissues, and may influence the differentiation of tissue cells (Bukovsky et al., 2001), including formation of germ and granulosa cells and differentiation of primary follicles (Bukovsky et al., 1995c, 2005a).

Monocyte-derived cells play an important role in regulation of the immune system. These cells control the function of tissue lymphocytes associated with differentiation of tissue-specific cells (Bukovsky et al., 2001). Lymphoid tissues not only produce cells promoting differentiation of tissue-specific epithelial and parenchymal cells, but also receive information from peripheral tissues via the afferent lymph. This information is transmitted by veiled MDCs, a subpopulation of tissue macrophages. They are low phagocytic, HLA-DR positive, and highly immunogenic, because they present antigens to T cells in the draining lymph nodes (Balfour et al., 1981; Hoefsmit et al., 1982; Howard and Hope, 2000; Knight et al., 1986).

In the fetal ovary, presumptive memory cells reside in the rete ovarii, and immature MDCs and T cells migrate through rete channels toward the ovarian surface and participate in the development of germ cells from the OSE (Bukovsky et al., 2005a). A similar interaction of immune cells with the OSE was described in the ovaries of adult women (Bukovsky et al., 1995c). During adulthood, however, no rete is present in ovaries, so the memory cells may reside in the lymphoid tissues, sources of circulating immune cells. The immune system shows a significant functional decrease between 35 and 40 years of age (Mathe, 1997), and concomitantly ovarian follicular renewal appears to cease (Bukovsky *et al.*, 2004).

The thymus plays an important role in the immune system, and it has been suggested that thymic peptides play a role in determining the reproductive life span of females (Rebar, 1982; Suh *et al.*, 1985). A relationship of age-associated thymic involution with diminution of ovarian function is supported by the alteration of ovarian function in neonatally thymectomized mice (Nishizuka and Sakakura, 1969). In addition, in congenitally athymic (nude) mice, follicular loss is first evident at 2 months of age, and this is specifically due to a reduction in the numbers of primary follicles. The first ovulation is delayed until 2.5 months of age, compared with the first ovulation in 1.5-month-old normal mouse females. By 4 months, an overall reduction in all fractions of the follicle population occurs in nude mice, and ovulation ceases (Lintern Moore and Pantelouris, 1975). Interestingly, the absence of the thymus might also be responsible for the lack of hair in nude mice, because of the lack of thymus-derived T cells, which might be required for hair development.

During the third trimester of pregnancy, the formation of new oocytes in human fetal ovaries ceases (Peters and McNatty, 1980). The reason for this discontinuation of oogenesis at a certain period of fetal development, while other tissues continue to grow, remains unclear. The source of germ cells in midpregnancy fetal ovaries appears to be OSE cells, and the transition of OSE cells into germ cells occurs in the presence of immune system-related cells (Bukovsky *et al.*, 2005a), resembling events during oogenesis in ovaries of adult human females (Bukovsky *et al.*, 1995c).

OSE cells represent bipotent progenitor cells with a capacity to differentiate into both primitive granulosa and germ cells (Bukovsky *et al.*, 2004, 2005a,b; Simkins, 1932). These transitions into one or another cell type may reflect the plasticity of progenitor cells in a particular microenvironment. They may occur under the influence of the extracellular matrix, cytokines (many of which are secreted by immune system cells), adhesion molecules, membrane receptors, intercellular junctions, signaling pathways, or transcription factors, commonly produced in the embryo and less frequently in adult organisms. Such transitions are examples of manifestations of cell plasticity and subsequent dramatic changes resulting in lineage commitment into certain cell types (reviewed in Bukovsky *et al.*, 2004; Prindull and Zipori, 2004).

We speculated that the particular microenvironment required for the transformation of OSE cells into germ cells (association of immune system-related cells with OSE) also requires certain hormonal conditions. The OSE expresses LH/hCG receptor (Bukovsky *et al.*, 2003), and the high levels of hCG seen in early pregnancy decline as pregnancy progresses (Hershkovitz *et al.*, 2003). In addition, development of the placental hCG barrier causes

TABLE I

WORKING MODEL FOR AGE-ASSOCIATED CHANGES IN OVARY-SPECIFIC MESENCHYMAL CELLS AND
HORMONAL SIGNALS[a] REQUIRED FOR THE INITIATION AND RESUMPTION OF OOGENESIS IN
HUMAN OVARIES[b]

Period of life	o-SMCs[c]	LH/hCG[d]	E$_2$[e]	Oogenesis
First trimester–midpregnancy	Yes	Yes	Yes	Yes[f]
Last trimester–newborn	Yes	No	Yes	No[f]
Postnatal–menarche	Yes	No	No	No[g]
Reproductive period[h]	Yes	Yes	Yes	Yes[f]
Premenopause[i]	No	Yes	Yes	No[g]
Postmenopause	No	Yes	No	No[g]

[a] LH/hCG and estradiol.

[b] From Bukovsky et al. (2005a), with permission.

[c] Specialized mesenchymal cells (tissue macrophages and T cells) with commitment for stimulation of OSE cells to germ cell transformation.

[d] Levels corresponding to the midcycle LH peak, or more; hCG levels should be 10 times more, because it has a 10% affinity for the LH receptor compared with that of LH (Bousfield et al., 1996).

[e] Levels corresponding to the preovulatory E$_2$ peak, or more.

[f] Confirmed.

[g] Predicted.

[h] From menarche until 38 ± 2 years of age.

[i] From 38 ± 2 years until menopause.

Abbreviations: E$_2$, estradiol; hCG, human chorionic gonadotropin; LH, luteinizing hormone; o-SMCs, ovary-specific mesenchymal cells.

virtually zero hCG values in the blood of human fetuses at term (Penny et al., 1976). If the association of immune system-related cells with the OSE requires LH/hCG binding to its receptor, it may resume during ovulatory LH peak, which does not occur until menarche. In other words, we proposed that there is a lack of immune system-induced oogenesis between 7 months of intrauterine life and menarche, because of the lack of saturation of LH receptors in fetal and postnatal OSE. The preparation for oogenesis may also require high levels of estrogens, corresponding to the preovulatory peak of 17β-estradiol, as we demonstrated in our studies of human oogenesis in vitro (Bukovsky et al., 2005b). A working model on age-associated changes of immune system-related cells and hormonal signals (LH/hCG and E$_2$) required for the initiation and resumption of oogenesis in human ovaries (Bukovsky et al., 2005a) is given in Table I.

Role of Ovarian Tunica Albuginea–OSE Transitions

In adult human ovaries, mesenchymal cells in the TA are progenitors of OSE cells, which are committed to differentiate into either primitive granulosa or germ cells (Bukovsky *et al.*, 1995c, 2004). In culture, OSE cells undergo an epithelio–mesenchymal transition (Auersperg *et al.*, 2001). On introduction of E-cadherin, the OSE-derived mesenchymal cells differentiate back into the epithelial phenotype (Auersperg *et al.*, 1999; Dyck *et al.*, 1996). The human TA is a thick fibrous subepithelial layer of loose connective tissue cells. It does not begin to form until the end of intrauterine life (Motta and Makabe, 1986), and even then it is not a true membrane, but merely a collection of loose connective tissue cells (Simkins, 1932).

The ovarian cortex is usually covered by a layer of irregularly shaped special epithelial-like mesothelial cells (Van Blerkom and Motta, 1979), commonly referred to as the ovarian "germinal" or surface epithelium. This layer is attached to the basal lamina, which is continuous with the subjacent TA by means of collagenous fibrils. Except in the ovaries of newborn animals, mitoses are essentially absent in OSE (Motta *et al.*, 1980). In functional human ovaries the OSE is found in certain areas only, but in women with anovulatory cycles, or patients with polycystic or sclerotic ovaries, the ovarian surface is completely covered with OSE (Makabe *et al.*, 1980). The TA shows a variable thickness, ranging from almost undetectable to more than 50 μm. Cytokeratin expression of various density is detected in mesenchymal cells of some TA segments, particularly in segments showing an appearance of OSE cells. In contrast, no CK expression is detected in mesenchymal cells of the ovarian cortex (Bukovsky *et al.*, 2004). The development of OSE-derived cortical nests of primitive granulosa cells (Motta *et al.*, 1980; Van Blerkom and Motta, 1979) coincides with the appearance of OSE cells directly connected to the ovarian cortex (without interference from the TA), and with the appearance of TA flaps extending over the OSE (Bukovsky *et al.*, 2004).

The TA mesenchymal cells, which show CK immunoexpression, differentiate into OSE cells by a process of mesenchymal–epithelial transition (see Fig. 4A). However, depending on additional, unknown factors, this process may result in either the formation of primitive granulosa cell nests or the differentiation of OSE cells covering the TA surface (precursors of germ cells). The nests of primitive granulosa cells descend into the deeper ovarian cortex, where they eventually associate with oocytes to form primary follicles (Bukovsky *et al.*, 2004). Figure 2, a panoramic view of the ovarian surface and adjacent cortex (ovc), shows a thick segment of TA (ta) and CK expression in TA fibroblasts associated with differentiation of the OSE (white arrow on left). No OSE is seen if the TA lacks CK[+] fibroblasts (white arrow on right). The upper ovarian cortex contains a row of longitudinally (black arrows) or perpendicularly viewed solid epithelial cords (arrowheads; as evidenced from serial sections), precursor structures of epithelial nests.

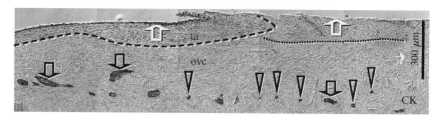

Fig. 2. Ovarian surface and adjacent tunica albuginea (ta) and cortex in adult human ovary. Parallel view (black arrows) and perpendicular view (arrowheads) of epithelial cords in the upper ovarian cortex (ovc) are shown. Dashed line indicates a segment of CK^+ tunica albuginea with flap and differentiating OSE cover (left white arrow). Dotted line indicates CK^- segment lacking OSE cover (right white arrow). Reprinted from Bukovsky et al. (2004), © Antonin Bukovsky.

Figure 3 shows the formation of epithelial channels (black arrowhead), longitudinally viewed (dashed arrowhead) and perpendicularly viewed solid epithelial cords (white arrowheads and right inset), and follicle-like structures containing stromal elements (solid box, and see left top inset for detail) in the upper cortex (uc). These structures were arranged in a surprisingly straight row (see also Fig. 2). In some ovaries a similar orientation of primary follicles was observed in the lower ovarian cortex (lc).

Hence, the formation of new primary follicles in adult human ovaries is more complex as compared with the more primitive association of oocytes with clusters of granulosa cells available in human fetal or adult rat ovaries (Bukovsky et al., 2005a).

Follicular Renewal

Studies by Motta and collaborators (Motta et al., 1980; Van Blerkom and Motta, 1979) have shown that in some areas of the ovary, epithelial cords fragment and appear as small nests of epithelial cells that lie in proximity to primary follicles. We observed that in the lower ovarian cortex, some epithelial nests associated with the lumen of ovarian venules, and these nests exhibited a stretch of intravascular oocytes (Fig. 4B). During oocyte–nest assembly the Balbiani body was formed by granulosa cell extensions penetrating deep into the ooplasm (Fig. 4C).

Oocytes in primary (resting) follicles show a single Balbiani body (named for the nineteenth-century Dutch microscopist) in the cytoplasmic region near the nucleus, where the majority of oocyte organelles are concentrated (reviewed in Zamboni, 1970). The Balbiani body contains aggregated mitochondria and can be observed in primary follicles in both fetal and adult ovaries (Carlson et al., 1996; Cox and Spradling, 2003;

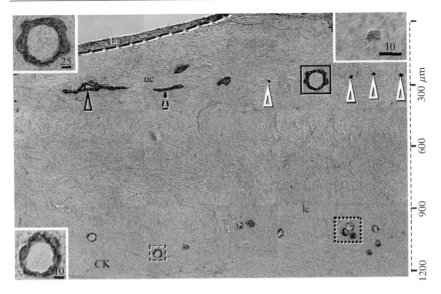

FIG. 3. Distribution of epithelial cords and primary follicles in the ovarian cortex. Panoramic view composed of nine images shows ovarian tunica albuginea (ta) filled with CK18$^+$ mesenchymal cells, upper cortex (uc) with epithelial channels (black arrowhead), cords (dashed and white arrowheads; see right inset), and follicle-like structure lacking oocyte (solid box; see upper left inset for detail). Lower cortex (lc) shows isolated (dashed box; see lower left inset for detail) and grouped primary follicles (dotted box). Scale bars in insets indicate micrometers. Cytokeratin 18 immunostaining, hematoxylin counterstain. Reprinted from Bukovsky *et al.* (2004), © Antonin Bukovsky.

Motta *et al.*, 2000). In a study of turkey hens, no Balbiani body was detected in stage I oocytes, appeared in stage II oocytes, and diminished in the oocytes of growing follicles, coinciding with the dispersion of mitochondria throughout the ooplasm (Carlson *et al.*, 1996). Similar observations were reported in human oocytes (Motta *et al.*, 2000).

Balbiani bodies show immunostaining for CK8, −18, and −19 (Bukovsky *et al.*, 1995c; Santini *et al.*, 1993). In primary follicles of fetal and adult human ovaries, follicular (granulosa) cell extensions penetrate deep into the ooplasm, much like a sword in its sheath. There may be as many as three to five "intraooplasmic processes," even in one scanning micros-copy plane. These intraoocytic invaginations are closely associated with a variety of organelles. They are close to the nuclear zone, and may help activate growth of the oocyte (Motta *et al.*, 1994).

In mouse fetal ovaries, germ cells are arranged in special clusters (germline cysts) and dividing germ cells remain connected by intercellular bridges called "ring canals." The cysts may allow certain germ cells to

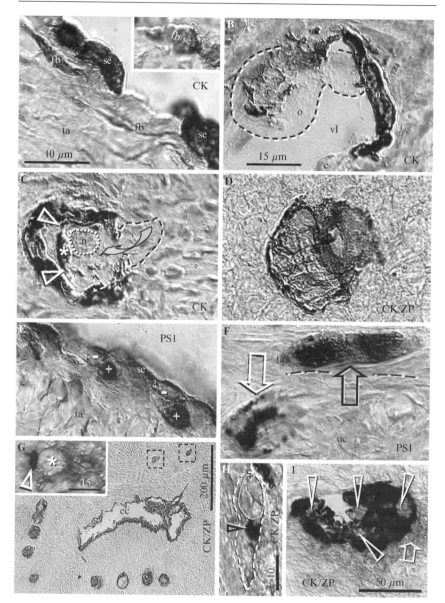

Fig. 4. Follicular renewal in adult human ovaries. (A) Some fibroblasts (fb) in the tunica albuginea (ta) show CK18 expression and transition into OSE (se) cells. Inset: A cell in mesenchymal–epithelial transition (fb/se). (B) The CK$^+$ epithelial nest (n) wall inside the vascular lumen (vl) pocket, which is lined with endothelial cells (e), extends an arm (a) to catch the oocyte (o, dashed line). (C) The nest body (n) with closing "gate" and a portion of the oocyte (dashed line) still outside the nest–oocyte complex (arched arrow). The oocyte contains

specialize as nurse cells (Pepling and Spradling, 2001). One possibility is that such nurse cells in germline cysts help provide oocytes with mitochondria (Pepling and Spradling, 1998). It has been proposed that mitochondria with functional and defective genomes would be actively transported into different germ cells, and the quality of the mitochondria of each cell might then determine whether it survived or entered apoptosis (Pepling and Spradling, 2001). A study by Cox and Spradling indicates that during *Drosophila* oogenesis, follicular cells are a source of mitochondria, which enter the oocyte cytoplasm via the ring canal to form the Balbiani body, thereby supplying virtually all the mitochondria of the oocyte (Cox and Spradling, 2003).

It appears that the Balbiani body contributes to the resting state of the oocyte, because oocyte mitochondria are not released until the initiation of the follicular growth. In addition, the contribution of granulosa cells to the formation of the Balbiani body in the oocyte cytoplasm is an indicator of the use of these cells in ongoing oocyte assembly.

We also visualized the assembly of oocytes with epithelial nests by double-color immunohistochemistry for CK in epithelial nests and zona pellucida (ZP) glycoproteins expressed in oocytes, and showed an association of the oocytes with epithelial nests resembling occupied bird's nests (Bukovsky *et al.*, 2004) (Fig. 4D).

Most of the OSE cells showed CK immunostaining, but immunoexpression of ZP proteins was restricted to certain OSE segments. Although most ZP proteins were also detected in the zona pellucida of oocytes during and after assembly within epithelial nests (including primary, secondary, preantral, and antral follicles), the meiotically expressed porcine ZP oocyte carbohydrate antigen named PS1 (Skinner *et al.*, 1997) was detected in OSE segments but not in the zona pellucida of oocytes in human ovarian follicles (contrary to porcine ovaries).

---

intraooplasmic CK^+ (brown color) extensions from the nest wall (arrowheads), which contribute to the formation of CK^+ paranuclear (Balbiani) body (asterisk); the nucleus is indicated by a dotted line. (D) Occupied "bird's nest" indicates a half-formed oocyte–nest assembly. CK (brown)/ZP (blue). (E) Segments of OSE show cytoplasmic PS1 (brown) expression (nuclei) and give rise to cells exhibiting nuclear PS1 (+ nuclei, asymmetric division), which descend into the tunica albuginea. (F) In the tunica albuginea, the putative germ cells show symmetric division (black arrow) and also exhibit development of cytoplasmic PS1 when entering (white arrow) the upper cortex (uc). (G) Primary follicles develop around the OSE cortical epithelial crypt (ec); dashed boxes indicate unassembled epithelial nests. Inset: The emergence of CK^- germ cell (asterisk) with ZP^+ intermediate segment (arrowhead) among crypt CK^+ epithelial cells. (H) Migrating tadpole-like germ cell shows no CK staining but ZP^+ intermediate segment (arrowhead). (I) Accumulation of multiple ZP^+ oocytes with unstained nuclei (arrowheads) in a medullary vessel. (A–C and E) Hematoxylin counterstain. Reprinted from Bukovsky *et al.* (2004), © Antonin Bukovsky. (See color insert.)

The immunoexpression of PS1 in human OSE cells was cytoplasmic. However, cells descending from the OSE into the TA also showed nuclear PS1 staining (Fig. 4E). The dividing OSE cells showed an asymmetric distribution of meiotically expressed nuclear PS1, suggesting meiotic activity. Double-color immunostaining for PS1 and CK revealed an asymmetric distribution of PS1 in putative germ cells descending from the OSE, resulting in two distinct (CK^+ or $PS1^+$) daughter cells. Larger putative germ cells with nuclear PS1 staining were detected in the TA. Such cells divided symmetrically (both daughter cells expressed PS1; Fig. 4F) and entered the adjacent ovarian cortex. In the cortex, the putative germ cells showed a translocation of nuclear PS1 immunoexpression to cytoplasmic staining, and an association with cortical vasculature showing minute amounts of PS1 immunoexpression in adjacent endothelial cells. In addition, PS1 immunoexpression was detected in germ cells migrating from CK^+ epithelial crypts into the adjacent cortex—an alternative pathway for germ cell origin (Bukovsky et al., 2004).

These observations indicate that in adult human females, there are no persisting oogonia or germline stem cells, but germ cells nonetheless originate from asymmetrically dividing OSE cells. Such germ cells symmetrically divide in the TA (crossing over), enter the ovarian vasculature, or migrate from OSE cortical crypts to reach nests of primitive granulosa cells and form new primary follicles. Such follicular renewal replaces earlier primary follicles undergoing atresia.

Follicular Atresia

Follicular atresia and luteal regression are essential mechanisms required for the elimination of unnecessary and aged structures, as well as for normal ovarian function. Elimination of antral follicles undergoing atresia and of degenerating corpora lutea during reproductive years in human females is a fast process, associated with infiltration by activated macrophages (Bukovsky et al., 1995a,b), and there is no reason to expect that the similar process accompanying regression of primary and secondary follicles (Bukovsky et al., 2004) will last longer than several days. If at least 60% of oocytes in adult ovaries are in various stages of degeneration (Ingram, 1962), one may conclude that without follicular renewal the ovarian function will cease in human females within a few months. However, in aging ovaries, the elimination of degenerating ovarian structures appears to be affected (Bukovsky et al., 1996), possibly due to the age-induced alterations of immune system function (Bukovsky et al., 1996). Hence, atresia may not affect primary follicles in aging ovaries (Gougeon et al., 1994), and such follicles may persist in spite of an accumulation of spontaneously arising or environmentally induced genetic alterations of oocytes.

During the prime reproductive period, the degeneration may affect groups of primary and secondary follicles. Immunohistochemically, the follicles undergoing atresia release ZP proteins into the neighboring stroma. This is associated with an altered oocyte morphology and disorganization of the follicular CK^+ granulosa layer. In addition, there is a considerable influx of large macrophages into the area from accompanying vessels (Bukovsky et al., 2004). Some investigators have claimed that characteristic morphological features of primary follicle atresia are often difficult to determine (reviewed in Baker, 1972), whereas others are more confident (Gougeon et al., 1994).

In our immunohistochemical study, the assembly of oocytes with epithelial nests was also associated with some release of ZP proteins. Formation of new follicles was characterized by a well-defined oocyte nucleus, intraooplasmic CK^+ extensions from the nest cell wall, and formation of the Balbiani body, that is, structures and processes not apparent during follicular regression. Resting primary follicles and growing secondary/preantral follicles show a regular morphology, with no leakage of ZP proteins, and only occasional small tissue macrophages associated with the developing theca (Bukovsky et al., 2004).

Enhanced follicular atresia was accompanied by the appearance of epithelial nests (fragmented epithelial cords) in adjacent segments of the ovarian cortex. We have shown that these nests are small CK^+ spheroidal cell clusters 20–30 μm in diameter. There were also epithelial crypts, likely originating from deep OSE invaginations. These do not communicate with the ovarian surface, as evidenced from serial sections. The movements of epithelial nests and crypts appear to be caused by a rearrangement of stromal bundles, and their migration is probably guided by HLA-DR$^+$ (activated) tissue macrophages. We have also described an alternative germ cell origin from epithelial crypts in the lower ovarian cortex (inset, Fig. 4G), accompanied by migration of tadpole-like germ cells with a ZP$^+$ intermediate segment (arrowhead, Fig. 4H), the presence of epithelial cell nests without oocytes (dashed boxes, Fig. 4G), and accumulation of primary follicles in the neighborhood of epithelial crypts (Bukovsky et al., 2004) (Fig. 4G).

These observations indicate that adult human ovaries exhibiting atresia of primary and secondary follicles initiate formation of new epithelial nests with granulosa cell features (Motta et al., 1980). This is one of the prerequisites for formation of new primary follicles. Cortical crypts, consisting of epithelial cells retaining a relatively embryonic structure of OSE cells (Mossman and Duke, 1973; Motta et al., 1980), appear to be an alternative source of germ cells. Germ cells entering the vasculature may reach epithelial nests at distant destinations, although vascular proximity is not a requirement for follicular development from nests approaching the cortical epithelial crypts producing germ cells (Bukovsky et al., 2004).

Adult Oogenesis Exceeds Follicular Renewal

An important question is whether the number of newly formed follicles in adult human ovaries is determined by the number of available epithelial nests or the number of generated germ cells. In the first instance, the isolated nests will either persist or degenerate. In the second, degenerating oocytes not used in new follicle formation might be detected. To answer this question we used double-color immunohistochemistry to search for ZP^+ oocytes not assembled with CK^+ structures.

We observed (Bukovsky *et al.*, 2004) an accumulation of degenerating ZP^+ oocytes in medullary vessels (Fig. 4I). Accumulation of oocytes in occasional ovarian medullary vessels was observed in 4 of 12 cases studied (33%). This suggests that the differentiation of oocytes during the reproductive period is a relatively frequent event. These ovarian samples showed either preparation for, or the ongoing formation of, new primary follicles. The accumulation of oocytes in some medullary vessels was present in the samples from both ovaries. Yet, there were eight cases showing no such activity. These ovaries rarely showed primary follicles in the ovarian cortex (Bukovsky *et al.*, 2004). This agrees with the observations of Block (1952), who observed a similarity between oocyte numbers in the right and left ovaries but wide variation between cases during the prime reproductive period.

Our observations support the idea that the formation of new primary follicles in adult human ovaries is a periodical process, which may occur during a certain phase of the ovarian cycle. This idea was first proposed by Evans and Swezy (1931). Although we did not study a large number of patients, our observations indicate that new primary follicles are likely to be formed during the late luteal phase, as evidenced from the patient's history, ovarian corpus luteum immunohistochemistry, and endometrium morphology (Bukovsky *et al.*, 2004). However, the preparation for follicular renewal consists of two sequential steps: formation of primitive granulosa nests, followed later by differentiation of new germ cells, which may be initiated earlier during the ovarian cycle. For example, follicular atresia is most prominent during the late follicular phase (follicular selection), when the formation of epithelial nests may be initiated, followed by an appearance of germ cells during midcycle (estradiol and LH peaks), and formation of primary follicles thereafter.

"Ovary within the Ovary" Pattern and Thy-1 Differentiation Protein

Why do primary follicles form in the lower ovarian cortex, not just near the origin of their components, primitive granulosa and germ cells? We have reported previously that groups of follicles lie in isolated areas of the cortex, exhibiting an oval arrangement of stromal elements (Bukovsky

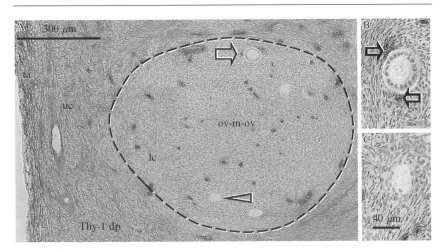

FIG. 5. Distribution of Thy-1 differentiation protein (dp) in the ovarian cortex—an "ovary within the ovary" pattern. (A) Thy-1 dp is strongly expressed by tunica albuginea fibroblasts (ta), and moderately in the upper ovarian cortex (uc) and lower ovarian cortex (lc) except for areas showing an "ovary within the ovary" pattern (ov-in-ov) with virtually no Thy-1 dp immunoexpression except for vascular pericytes. These areas characteristically contained primary follicles [arrowhead; detail in (C)], some of which showed an increase in size accompanied by Thy-1 dp$^+$ pericytes [arrows in (A) and (B)]. Hematoxylin counterstain; details in text. Reprinted from Bukovsky *et al.* (2004), © Antonin Bukovsky.

et al., 1995c). In one study, staining for Thy-1 dp revealed that groups of primary follicles reside in the center of rounded areas, extending ~400–1200 μm from the ovarian surface, exhibiting low stromal Thy-1 dp immunoexpression, and showing an ovary-within-the ovary pattern (ov-in-ov, Fig. 5A). In addition, growth of some follicles in a given cohort is associated with Thy-1 dp$^+$ vasculature [arrow in Fig. 5A (detail in Fig. 5B) vs. arrowhead in Fig. 5A (detail in Fig. 5C)]. Hence, a lack of Thy-1 dp may be required to maintain primary follicles in the resting state, and the presence of Thy-1 dp may stimulate follicular growth. Note the strong Thy-1 dp immunostaining of TA fibroblasts (ta, Fig. 5A).

Cessation of Adult Oogenesis

Our observations (Bukovsky *et al.*, 2004) indicate that epithelial nests of primitive granulosa cells derived from the OSE cells contribute to the follicular renewal in adult ovaries. We also studied ovaries of premeno-pausal and postmenopausal women for CK expression in order to determine whether formation of these nests is associated exclusively with follicular renewal, or whether it persists after cessation of this process (Bukovsky, 2005b). During the optimal reproductive period (<40 years of age), both epithelial nests and primary follicles have been detected in deeper ovarian cortex. In older (premenopausal) females, only primary follicles were detected. During the premenopausal period, most ovaries (92%) did not show the formation or presence of epithelial nests. Only a single case, from 12 patients investigated (8%), showed the formation of epithelial nests in the proximity of the OSE, as well as the occurrence of epithelial nests in the deeper cortex, but these nests showed degenerative changes and no adjacent primary follicles were detected.

Therefore, it appears that the formation of epithelial nests and their migration into the deeper cortex during the premenopausal period usually does not occur. Premenopausal women show OSE segments with more or less pronounced hyperplasia, and such OSE behavior persists in postmenopausal women. One may speculate that hyperplasia of the OSE in aging women reflects the attempt of a homeostatic mechanism [tissue control system (TCS); see later] to initiate formation of new primary follicles in the ovaries. In other words, a failure of the TCS to stimulate formation of new primordial follicles in adult ovaries may result in the stimulation of OSE proliferation and the development of ovarian cancer.

Another important question is whether the premenopausal and postmenopausal ovaries are able to produce new germ cells and oocytes from persisting OSE stem cells *in vivo*. They are certainly able to do so under *in vitro* conditions, even when derived from the ovaries of over 60-year-old females (our unpublished observations), and we have occasionally observed degenerating oocytes in the medullary vessels of perimenopausal ovaries lacking epithelial cell nests (our unpublished observations).

Altogether, it appears that pre- and postmenopausal ovaries may have a capacity to produce new oocytes, and rarely epithelial nests, but events required for their association and follicle formation are missing, either because the oocytes and germ cell do not have a chance to assemble with epithelial nests, or vice versa.

Oogenesis and Follicular Renewal in Adult Ovaries

The currently prevailing view that all oocytes and primary follicles in adult mammalian ovaries originate from fetal oogenesis (storage theory) is apparently contradictory to Darwinian evolutionary theory. Why should adult mammalian females carry their oocytes from the fetal period of life, as compared with the lower vertebrates (fish and frogs) and males of all species, including mammals, with persisting gametogenesis during adulthood (continued formation theory)?

By the end of the nineteenth century, and during the first half of the twentieth century, two opinions concerning oogenesis were discussed.

The first view was introduced by Waldeyer (1870) and elaborated by Kingery (1917), who proposed that new oocytes were formed from the germinal (surface) epithelium of adult ovaries. The second view was based on the studies of Beard (1900) and on an article by Pearl and Schoppe (1921), who suggested that all oocytes present in adult mammalian individuals originate from the period when the ovaries were formed and are stored and used until menopause.

During the 1950s and 1960s, the belief that all primary follicles in adult mammalian females were formed during the fetal period (Baker, 1972; Franchi et al., 1962) prevailed, primarily because of the overall diminution of primary follicle numbers with age reported by Block in the early 1950s (Block, 1952). However, Erik Block wrote: "In the age range eighteen to thirty-eight, the relation between the patient's age and the number of primary follicles cannot be established statistically" (Block, 1952). This suggests that during the 20 years of the human female prime reproductive period, there is no significant change in the number of primary follicles. However, this important part of Block's conclusions has not been appreciated by interpreters (Baker, 1972; Franchi et al., 1962).

We performed a statistical analysis (one-way analysis of variance and post hoc test) of data presented by Block (including neonatal ovaries; Block, 1951, 1952, 1953) and Gougeon et al. (1994) on the variation of the number of human ovarian primary follicles at different ages, between birth and 50 years of age (Bukovsky et al., 2004). The analysis showed that no significant change was apparent during the 20-year prime reproductive period, between 18 and 38 $\pm$ 4.1 (SD) years. However, age groups 40–44 and 45–50 years showed significantly lower numbers compared with other age groups.

Nevertheless, it remained unclear why the follicle numbers were not constant from the beginning of sexual maturity ($\sim$14 years of age) and continue to fall for several more years. One possible explanation comes from the different fate of primordial and primary follicles described in a laborious work and observations by Simkins (1932). The primordial follicles represent a majority of follicular structures in fetal ovaries. They never differentiate but gradually degenerate during childhood and the early stages of sexual maturity, and are absent in adult ovaries. In contrast, primary follicles are capable of differentiation into mature follicles and, although less numerous in fetal ovaries, their number remains unchanged throughout childhood (Simkins, 1932).

A compilation of the data from observations by Simkins (1932), Block (1951, 1952, 1953), and Gougeon et al. (1994) presented in a schematic manner is provided in Fig. 6A, where the dashed line indicates persisting primary follicles, the dotted line represents the diminution of primordial follicles, the dash-dotted line denotes the total follicle numbers—the sum of primary and

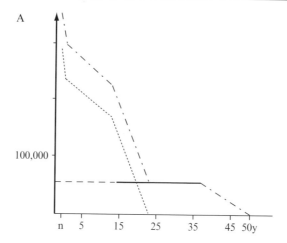

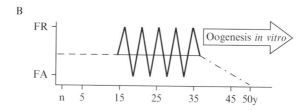

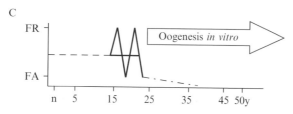

FIG. 6. Schematic comparison of primordial and primary follicle numbers. Compiled from Simkins (1932), Block (1952), and Gougeon *et al.* (1994). (A) The follicular pool (dashed/dotted line) is depleted during childhood and early adulthood by diminution of primordial follicles (dotted line), but not primary follicles [dashed line and (C)]. (B) The number of primary follicles remains constant during childhood (lack of atresia). During adulthood (prime reproductive period) periodical atresia of older follicles (FA) is compensated by follicular renewal (FR). During premenopause, the follicular pool is depleted by consumption of aging primary follicles (A and B). (C) Premature ovarian failure could be caused by an earlier discontinuation of follicular renewal and accelerated diminution of the number and functionality of persisting and aging primary follicles. Age-independent oogenesis *in vitro* could be considered for autologous IVF purposes. Reprinted from Bukovsky *et al.* (2006), with permission.

primordial follicles during childhood and the early years of sexual maturity and diminution of primary follicles by use after cessation of follicular renewal, and the thick solid line indicates the period of primary follicle renewal.

One may assume that the number of primary follicles formed during fetal period persists unaffected during childhood. In other words, either atresia of primary follicles does not occur before menarche (no follicular renewal is required), or follicular renewal should already exist during childhood. During the prime reproductive period, the primary follicle numbers markedly fluctuate (Block, 1952), possibly due to the waves of atresia and renewal (Bukovsky et al., 2004) (FA and FR, Fig. 6B). By the end of the third decade, follicular renewal ceases. Primary follicles persisting in aging ovaries accumulate genetic alterations within the oocytes (Bukovsky et al., 2004), and are supposed to diminish as a result of their periodical use (Gougeon et al., 1994), that is, growth of follicular cohorts, atresia during follicular selection, and dominant follicle ovulation. Developing ovarian infertility due to oocyte aging might be compensated by age-independent production of new oocytes from OSE stem cells in vitro, and their use in in vitro fertilization (IVF) treatment.

If follicular renewal diminishes earlier (Fig. 6C), two consequences may follow: (1) waves of persisting atresia without follicular renewal will cause fast depletion of existing follicular pool, or (2) ovarian aging, as described previously, will begin earlier.

Altogether, mammalian oogenesis is not a permanent event, and the origination of germ cells from somatic stem cells and their differentiation into oocytes varies with age, with the particularly complex situation of periodic oocyte and follicular atresia and renewal being apparent during the prime reproductive period in human females.

Working Hypothesis on the Role of the Gonadal Environment

Our working hypothesis on the role of the gonadal environment in the regulation of human oogenesis is presented in Fig. 7. After the indifferent gonad is populated by primordial germ cells (Fig. 7A), the rete ovarii develops and stimulates the differentiation of oocytes from primordial germ cells (Fig. 7B). During immune adaptation, the rete is populated by uncommitted mesenchymal cells (u-MCs), from which the MDCs may differentiate into the veiled cells. The veiled cells transmit information on oocytes from the rete into the developing lymphoid tissues (curved arrow, Fig. 7B). The MDCs in rete ovarii then become ovarian memory cells able to convert u-MCs passing through the rete channels into committed o-SMCs. These o-SMCs, along with appropriate hormonal

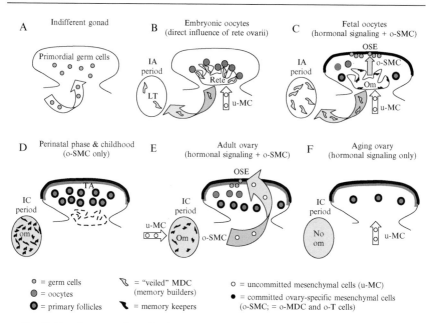

FIG. 7. Development of ovaries during immune adaptation (IA) and immune competence (IC). LT, lymphoid tissues; om, ovarian memory. Details in text. Adapted from Bukovsky (2006), with permission.

signaling, induce the development of germ cells from the OSE (Fig. 7C). The number of veiled cells populating lymphoid tissues increases further.

When immune adaptation is terminated, the rete ovarii degenerates and oogenesis ceases because of the diminution of hormonal signaling [fetal hCG barrier (Bukovsky et al., 2005a); Table I]. The ovarian TA develops from OSE cells (epithelial–mesenchymal transformation), and the number of ovarian memory cells (om; the transformed veiled cells) in lymphoid tissues is set (Fig. 7D). Around menarche, and during the prime reproductive period, hormonal signaling and o-SMCs stimulate cyclic oogenesis to replace aging primary follicles undergoing atresia (Fig. 7E). The cyclic follicular renewal during adulthood requires a cyclic supply of o-SMCs, and their generation in lymphoid tissues causes depletion of the pool of memory cells. Hence, from this point of view, the pool of ovarian memory cells in lymphoid tissues, but not the pool of primary fetal follicles, is what is set during mammalian fetal development. Once the available pool of ovarian memory cells is consumed, oogenesis and follicular renewal cease, in spite of the presence of hormonal signaling (Fig. 7F and Table I). Remaining primary follicles persist and are used until gone. However, the aging oocytes accumulate genetic alterations and may become unsuitable for ovulation and fertilization. The postmenopausal ovaries were reported to carry occasional follicles with degenerated oocytes (reviewed in Talbert, 1968).

Differentiation of Stem Cells and Tissue Control System Theory

The tissue control system (TCS) theory (Bukovsky *et al.*, 1991, 1995b, 1997, 2000b, 2001, 2002) deals with the role of vascular pericytes releasing Thy-1 dp, MDCs, and T and B cells in the regulation of tissue function. It proposes that MDCs participate in the stimulation of early differentiation of tissue-specific (epithelial, parenchymal, neural, and muscle) cells. The MDCs also regulate expression of epitopes of specific tissue cells, and in that way control their recognition by circulating tissue-committed T cells and antibodies. Such T cells and antibodies are suggested to participate in the stimulation of advanced differentiation of tissue cells, which may ultimately result in the aging and degeneration of these cells (Bukovsky *et al.*, 2001).

The function of the TCS is expected to be programmed during immune adaptation in early ontogeny, during fetal (larger mammals) or perinatal (small rodents) programming of tissue physiology or pathology. The term fetal programming of tissue pathology has been used to describe the process whereby a stimulus or insult, when applied at a critical or sensitive period of development, results in long-term or permanent changes in organ size, structure, or function (Lucas, 1991).

By the end of immune adaptation, the MDCs are supposed to encounter the most differentiated tissue cells in a tissue-specific manner, and then prevent them from differentiating beyond the encoded state by the so-called stop effect. The nature of the stop effect may reside in the inability of monocyte-derived cells to stimulate differentiation of tissue cells beyond the encoded stage. Retardation or acceleration of the differentiation of a certain tissue during immune adaptation is suggested to cause a rigid and persisting alteration of the function of this tissue. The ability of monocytes to preserve tissue cells in the functional state declines with age, and this is accompanied by a functional decline of various tissues within the body, including the ovary, resulting in menopause and increased incidence of degenerative diseases in aging human beings.

Animal Models

In large mammals, including primates, immune adaptation is terminated during intrauterine life, whereas in small laboratory rodents (rats and mice) immune adaptation continues for several postnatal days, ending about 1 week after birth (Klein, 1982). Estrogens given to neonatal rats or mice inhibit ovarian development, and during adulthood these female rats exhibit persisting ovarian immaturity characterized by retardation of follicular development (Bukovsky et al., 1997, 2002) in spite of normal serum levels of gonadotropins (Matsumoto et al., 1975; Nagasawa et al., 1973). This indicates that suppression of early ovarian development results in persisting ovarian immaturity, which resembles premature ovarian failure (POF) associated with gonadotropin resistance of ovarian follicles.

Injection of estrogens into neonatal mice (days 0–3) caused permanent anovulation, but mice injected later (days 3–6; closer to the end of immune adaptation) showed resumption of ovulatory cycles after initial anovulation (Deshpande et al., 1997). Hence, persisting ovarian immaturity can result in a delay of normal ovarian function. Because the incidence of degenerative diseases increases with age, one may expect that there is a tendency of the stop effect to shift with age (arrowheads, Fig. 8). This may explain how persisting ovarian immaturity may switch to the functioning ovary.

On the other hand, injection of androgens causes premature ovarian aging persisting during adulthood (Bukovsky et al., 2000a). However, androgen-induced anovulation can be prevented by neonatal injection of a thymic cell suspension from immunocompetent prepubertal normal female donors, but not if obtained from animals before completion of immune adaptation (Kincl, 1965; Kincl et al., 1965). This suggests that certain thymic cells (thymocytes, or thymic MDCs) of normal immunocompetent females carry information on appropriate differentiation of ovarian structures, and this information can be transferred to immunologically immature neonatal rats.

Nevertheless, when a low dose of androgens is injected during immune adaptation, the rats exhibit so-called delayed anovulatory syndrome. Ovaries exhibit onset of normal function after puberty ($\sim$40 days of age), but premature aging of the ovary occurs between 60 and 100 days (Swanson and van der Werff ten Bosch, 1964). Similar premature reproductive senescence was reported in sheep. Females androgenized during the intrauterine period showed normal onset of puberty, with ovarian cycles beginning at the same time and of the same duration during the first breeding season as control females. Differences appeared during the second breeding season, when the reproductive cycles were either absent or disrupted in androgenized ewes (Birch et al., 2003). This delayed manifestation of ovarian dysfunction indicates that androgenization during the critical period of tissue development did not affect target tissue permanently (from the beginning), but that manifestation of the programmed pathology may occur with a delay. These observations resemble human POF with secondary amenorrhea, as well as some human degenerative diseases with autoimmune character, which also occur after a shorter (juvenile diabetes mellitus) or longer (Alzheimer's disease) period of normal tissue function.

Fetal programming of tissue physiology and pathology

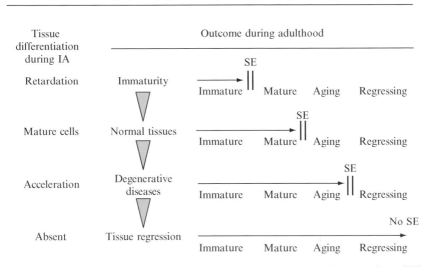

Fig. 8. Stage of cell differentiation during immune adaptation sets TCS "stop effect" (SE) for tissue physiology and pathology during adulthood. Arrowheads indicate a tendency of the SE to "shift" with age. Adapted from Bukovsky *et al.* (1997, 2000a,b) and Bukovsky (2000).

In addition, fetal androgenization of rhesus monkeys caused premature aging of oocytes during adulthood, because only 13% of fertilized eggs from adult females progressed to blastocysts as compared with controls (Dumesic *et al.*, 2002). This suggested that premature aging of oocytes resulting in ovarian infertility, possibly due to the earlier diminution of follicular renewal (Fig. 7C), can be programmed during the fetal period in nonhuman primates.

Critical periods for the alteration of certain organs or tissue functions may vary, depending on development. For instance, in the rat, ovarian follicles are not formed before birth. Accordingly, the experiments with androgens in rat females also indicated that the effective window for the induction of altered ovarian function is not only closed after postnatal day 7, but also not open before birth. The effectiveness of androgen adminis-tration before birth was studied by giving a single injection of 2500 μg of testosterone propionate (TP) to pregnant rats on days 19 to 22 after conception (Swanson and van der Werff ten Bosch, 1964). Prenatal admin-istration had no effect on the ovarian function of female offspring, although the dose was sufficient to cause masculinization of the external genitalia. The effects of postnatal TP administration on ovarian function varied with the dose (single dose of 5, 10, 50, or 500 μg of TP), and with the time of administration [day 1 (day of birth) and postnatal days 2, 4, or 5]. Threshold doses (5 and 10 μg) were more effective the earlier they were given after birth. With these small doses, most of the rats had normal luteinized ovaries at 10 weeks and were able to bear and suckle normal litters. Some time later, ovulation ceased, so that at 21 weeks they were no longer fertile (Swanson and van der Werff ten Bosch, 1964).

A simplified application of the TCS theory to the regulation of tissue function via the stop effect is depicted in Fig. 8 (see also Bukovsky et al., 2000a, 2001, 2002). In normal tissues, the mature cells are present during immune adaptation and the tissue-specific cells are "parked" in the mature state during adulthood. Retardation of cell differentiation during adaptation results in persisting immaturity (POF with primary amenorrhea), and acceleration results in premature aging (POF with secondary amenorrhea, degenerative diseases). If the self tissues were absent during adaptation (corpus luteum), no stop effect is established and they are handled as a "graft."

These observations indicate that in some instances, such as systemic degenerative diseases, injection of stem cells may not necessarily have the expected beneficial effect, unless the systemic TCS memory is properly set. To do so, one may need to erase the existing TCS memory and reopen an adaptation window by use of a high single dose of cytostatic chemotherapy, alone or in combination with subsequent (several days later) injection of stem cells (optimally autologous) adequately differentiated for a tissue in need. When immune competence is reestablished (several weeks later), the bulk of autologous uncommitted stem cells could be used for local or systemic regenerative therapy. Such a complex procedure could also be required in attempts to rejuvenate ovaries in patients with premature ovarian failure, or aged premenopausal or postmenopausal ovaries. On the other hand, for the repair of local tissue injury, such as damage after a brain stroke or heart attack, uncommitted autologous stem cell therapy alone is more likely to be efficient.

Potential Treatment of Ovarian Infertility by Producing New Primary Follicles In Vivo

Oocytes encased in primary follicles can survive for about 12–15 years, because follicular renewal ceases between 35 and 40 years of age and ovaries can still exhibit cyclic function until menopause at about 50 years. Hence, one can imagine that a single wave of follicular renewal, if it occurred before menopause, would extend ovarian function for another 12–15 years.

Yet, follicular renewal in adult ovaries is a complex process, requiring not only the formation of new germ cells but also nests of primitive granulosa cells, and assembly of these basic components of primary follicles (see also previously). From this point of view, unless we understand all aspects of follicle formation during adulthood, including involvement of ovarian macrophages and T cells, the idea that transplantation of germ cells developed in the ovarian culture from OSE stem cells would induce a new wave of follicular renewal in aging ovaries remains theoretically plausible but practically unpredictable, especially when considering the possible need for resetting TCS memory, as discussed previously.

However, there might be a better chance to induce wave(s) of follicular renewal with germ cells derived from ovarian stem cells in younger females with POF, who may still have the capacity to develop nests of primitive granulosa cells required for the formation of new primary follicles.

On the other hand, because ovarian surface stem cells persist in aging ovaries, we speculate that autologous white blood cells preserved from the prime reproductive period may induce a single wave of follicular renewal even after menopause.

Mystery of the Origin of Oocytes in Adult Mammalian Ovaries

The article in *Nature* on the persistence of germline stem cells (GSCs) in adult mouse ovaries (Johnson *et al.*, 2004) raised doubts (Albertini, 2004; Greenfeld and Flaws, 2004; Gosden, 2004; Telfer, 2004) on whether there is a need to revise the generally accepted 50-year-old dogma that all oocytes in adult mammalian females are formed during the fetal period (Franchi *et al.*, 1962; Zuckerman, 1951). In addition, the claim by the same group of scientists in *Cell* on the origin of GSCs in adult mammalian females from bone marrow (Johnson *et al.*, 2005a) has been found even more controversial (Ainsworth, 2005; Bukovsky, 2005a; Powel, 2005; Telfer *et al.*, 2005; Vogel, 2005).

In Powel's commentary (Powel, 2005), David Albertini of the University of Kansas (Kansas City, KS) indicated: "There are so many inconsistencies between the first paper (Johnson *et al.*, 2004) and this one (Johnson *et al.*, 2005a) that it makes it very difficult to believe in these findings," referring to Tilly's earlier work, which suggested that the stem cells reside in the outer covering of ovaries. When the researchers could not isolate the stem cells from ovaries, Tilly says, their data led them to the ovarian blood supply, and then to the bone marrow (Powel, 2005). Evidence that germ cell formation in bone marrow disappears in ovariectomized mice (Johnson *et al.*, 2005a) raises by itself solid doubts on the bone marrow origin of oocytes (Bukovsky, 2005a).

In our opinion, the 2004 article by Tilly's group (Johnson *et al.*, 2004) confirmed earlier (Allen, 1923; Bukovsky *et al.*, 1995c; Butler and Juma, 1970; Evans and Swezy, 1931; Ioannou, 1968; Kingery, 1917; Simkins, 1932) and concurrent (Bukovsky *et al.*, 2004) reports that follicular renewal exists during adulthood in mammalian ovaries. However, the idea that large germline stem cells persisting in the OSE (i.e., not originating from OSE cells) are a source of new oocytes in adult mammalian ovaries (Johnson *et al.*, 2004) has been found inappropriate by others (Bukovsky *et al.*, 2004; Gosden, 2004), and by the authors themselves (Johnson *et al.*, 2005a).

The article by Johnson *et al.* (2005a) indicates that blood in the peripheral circulation carries cells capable of initiating the formation of new primary follicles in adult ovaries. However, the claim that germ cells and oocytes have an extragonadal origin from bone marrow also appears to be incorrect (Bukovsky, 2005a; Powel, 2005; Telfer *et al.*, 2005). In fetal (Bukovsky *et al.*, 2005a) and adult human ovaries (Bukovsky *et al.*, 1995c, 2004), the germ cells originate by asymmetric division from OSE cells, with the assistance of immune system-related cells (o-SMC signaling), and possibly under certain hormonal conditions (Bukovsky *et al.*, 2005a). The germ cells in adult ovaries enter the blood circulation to saturate distant epithelial nests (Bukovsky *et al.*, 2004). Before that, the o-SMCs are delivered from lymphoid tissues to the ovary by the peripheral circulation to initiate the formation of new germ cells from the OSE (see later).

Our observations and views on oogenesis in fetal and adult human ovaries are summarized in Fig. 9. Fetal human OSE is a source of primitive granulosa cells (pg, Fig. 9A; see also Fig. 1A) and germ cells (dashed box in Fig. 9A; see also Fig. 1B and D). The germ cells differentiate into oocytes that associate with available granulosa cells to form fetal follicles. Primitive germ cells originate by asymmetric division (ad, Fig. 9B) from OSE cells influenced by hormonal signaling (HS; elevated hCG and E_2 levels) and cellular signaling (CS; o-SMCs). The primitive germ cells undergo symmetric division (sd), which may be associated with chromosomal crossing over (co), and tadpole-like germ cells (gc) migrate into the cortex, where they differentiate into oocytes and form ovarian follicles (see Fig. 9A). In perinatal ovaries, the OSE cells form a loose subepithelial layer of TA by epithelial–mesenchymal transition (arched arrows; Fig. 9C). Around menarche and during the prime reproductive period, mesenchymal cells of the TA are influenced by blood-delivered CS (o-SMCs) and differentiate back into segments of OSE cells (arched arrow, Fig. 9D), forming cords of primitive granulosa cells, which descend into the cortex and fragment in the primitive granulosa nests (pgn, Fig. 9D). When hormonal signaling (HS, Fig. 9E) is also present, the TA mesenchymal cells form segments above the TA (arched arrow in Fig. 9E; see also Fig. 4A) and produce germ cells (see Fig. 4E and F), as in fetal ovaries (dashed box, Fig. 9E). Next, the tadpole-like germ cells migrate to enter adjacent blood vessels (bv). The OSE-derived epithelial crypts (ec, Fig. 9F and Fig. 4G) in the deep cortex are an alternative source of germ cells (gc) (see Fig. 9H and inset in Fig. 9G),

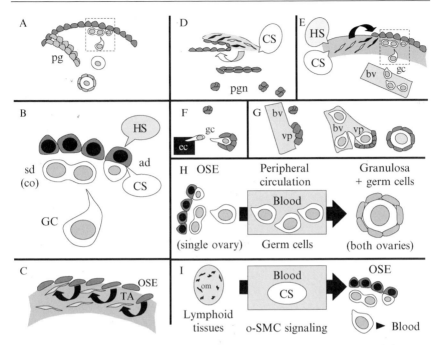

Fig. 9. Survey of oogenesis and formation of ovarian follicles in human ovaries. (A) OSE in midpregnancy fetal ovaries is a source of primitive granulosa (pg) and germ cells (dashed box) differentiating into oocytes and forming follicles. (B) Detail from (A) shows asymmetric division (ad) of OSE cells and symmetric division (sd), possibly with crossing over (co), of germ cells entering the cortex (gc). HS, hormonal signals (E_2, hCG); cs, cellular signals (o-SMCs). (C) Perinatal period is associated with formation of subepithelial TA layer by transition of OSE into mesenchymal-type cells (arched arrows). Around menarche and during the prime reproductive period, TA cells differentiate back into OSE cells, forming (D) primitive granulosa cell nests (pgn), or germ cells entering blood vessels [bv; (E)]. (F) The OSE-derived deep cortical epithelial crypts (ec) are an alternative source of germ cells, which either associate with adjacent nests or also enter blood vessels. (G) Other nests associate with deep cortical vessels and form vascular pockets (vp) to catch the germ cells from the blood stream and form primary follicles. (H) The OSE–blood–follicle pathway. (I) The lymphoid–blood–OSE pathway, continuing as in (H). Reprinted from Bukovsky *et al.* (2006), with permission.

which may assemble with adjacent nests to form primary follicles (see Fig. 4G). If epithelial crypts are absent, the nests associate with vessels and form vascular pockets (vp, Fig. 9G) to catch circulating germ cells (see Fig. 4B) and form primary follicles (see Fig. 4C and D). Hence, once OSE-derived germ cells enter the peripheral circulation during midcycle, they are ready to form primary follicles without delay [no need for the "black box" (Johnson *et al.*, 2005b) speculation], assuming the nests of primitive granulosa cells are available within the ovaries (Fig. 9H). Even if oogenesis is initiated in only one of the ovaries (single ovary), the circulating germ cells are capable of saturation the epithelial nests in both of them (both ovaries). Because of the ability of germ cells to settle in ovarian vessels, supernumerary germ cells often differentiate into oocytes, accumulating and degenerating in ovarian medullary vessels (see Fig. 4I). Before the production of germ cells from OSE, the committed ovary-specific mesenchymal cells (o-SMCs), generated in lymphoid tissues and carrying ovarian memory (om; see also Fig. 6), are available in the circulation to reach the OSE and provide cellular signaling (CS) for initiation of oogenesis (Fig. 9I). Resulting germ cells enter the peripheral circulation and the process continues as indicated in Fig. 9H.

The bone marrow, which was proposed to be an extraovarian source of germ cells (Johnson et al., 2005a), represents a tissue highly supplied by blood. Therefore, it may show the presence of many OSE-derived germ cells contaminating the blood at the time when they enter peripheral circulation in the ovaries. That is why the bone marrow shows no such "oogenetic" properties during other periods of the ovarian cycle, as reported by Tilly's group (Johnson et al., 2005a).

In a rebuttal (Johnson et al., 2005b) to the tremendous amount of comments on their work, Tilly's group divided precursors of germ cells into germ stem cells (GSCs), allegedly residing in the bone marrow, and germ progenitor cells (GPCs) in the peripheral blood, and created a magic "black box" for the following questions remaining to be addressed: (1) what are the factors that control GPC trafficking out of the bone marrow into the circulation, and how long do these cells take to home to the ovaries; (2) how many GPCs are present in the peripheral circulation at any given time, and how is this number controlled; (3) what is the life span once in the peripheral blood; and (4) how long does it take for GPCs to differentiate into primordial oocytes, and is the differentiation process initiated during transport in the blood?

Answers for most of these questions are available from our studies of human ovaries *in vivo* (Bukovsky et al., 1995c, 2004): (1) there is no need to control trafficking of GPCs from bone marrow because germ cells originate from the OSE or cortical OSE crypts around midcycle (possibly for several days after the priming role of E_2 and LH peaks in humans), enter the blood circulation, and hence also contaminate the bone marrow; (2) the number of OSE-derived germ cells in the peripheral circulation is maximal around midcycle, higher than the number of newly formed primary follicles, and determined by the number of o-SMCs inducing development of germ cells in ovarian segments expressing OSE, and there is no need to expect circulating germ cells in other phases of the ovarian cycle; (3) once in the peripheral blood, the germ cells should quickly (within ~1 day) assemble with granulosa cells. If this does not happen, they differentiate into oocytes and degenerate in ovarian medullary vessels; and (4) the germ cells are converted into oocytes exhibiting Balbiani bodies (see Fig. 4C) (which is characteristic for oocytes in primary follicles; Motta *et al.*, 2000) just during their association with nests of primitive granulosa cells.

Altogether, we always believed that the current dogma on the fetal origin of oocytes in adult mammalian ovaries does not fit with Darwinian evolutionary theory (Bukovsky and Presl, 1977), and provided immunohistochemical evidence on the oogenesis from OSE cells and follicular renewal in adult human ovaries (Bukovsky *et al.*, 1995c, 2004). The effort to raise doubts regarding the current dogma was joined by Tilly's group in 2004 (Johnson *et al.*, 2004), but, in our opinion, there is no solid evidence for the origin of oocytes from persisting germline stem cells in mouse ovaries or extragonadal origin of germ cells from bone marrow in adult mouse and human females (Johnson *et al.*, 2005a,b). In contrast to the use of autologous blood transfusion in infertile women for follicular renewal (Johnson *et al.*, 2005a), with its questionable effectiveness (Telfer *et al.*, 2005), the age-independent *in vitro* production of new autologous eggs from the OSE cells of human ovaries (Bukovsky, 2005c; Bukovsky *et al.*, 2005a,b), and their *in vitro* fertilization and use of embryos for intrauterine implantation, may represent a more suitable variant for providing genetically related children to women with ovarian infertility, worthy of consideration and further exploration (Bukovsky, 2005a).

Development of Distinct Cell Types from Totipotent Ovarian Stem Cells *In Vitro*

Cells of OSE in adult human ovaries were reported to retain an embryonic structure (Mossman and Duke, 1973), and therefore they are excellent candidates for totipotent adult stem cells. We were first to demonstrate that cultured OSE cells can differentiate into oocytes and other distinct cell types (Bukovsky *et al.*, 2005b).

Oogenesis In Vitro

Primary ovarian cultures exhibited occasional fibroblasts and rounded stem cells, but no presence of oocyte type cells during the first 3 days. However, the undifferentiated cells varied in the density of organelles and size, the latter ranging from 15 to 50 μm in diameter. Between days 4 and 5 cells of the oocyte phenotype were found in OSE cultures. They showed moderate (100 μm) size without zona pellucida, and were accompanied by fibroblasts. Larger oocytes (120 μm) accompanied by fibroblasts showed a developing zona layer. On the other hand, an association of small round cells resembling granulosa cells was characteristic for large oocytes (200 μm). The oocytes (accompanied by fibroblasts) with thick zona pellucida observed on day 4 showed no change in size on day 5. Staining for ZP proteins showed strong nuclear localization but no surface ZP expression. A nuclear envelope in the oocyte (separation between the nucleus and cytoplasm) was well defined (Bukovsky *et al.*, 2005b). Isolated oocyte-type cells exhibited germinal vesicle breakdown (arrowhead, Fig. 10A), and some showed characteristics of secondary oocytes with surface ZP expression, expulsion of the polar body, and poor nuclear–cytoplasmic separation (Bukovsky *et al.*, 2005b).

At present, we are unsure whether the o-SMCs are required for the differentiation of new oocytes in ovarian cultures. The ability of OSE

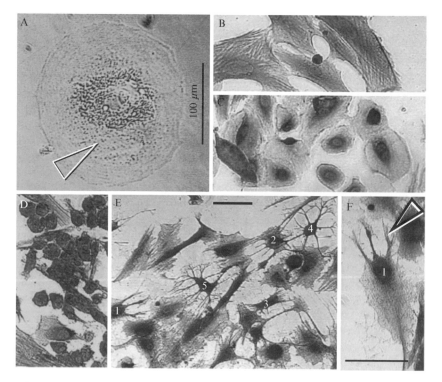

FIG. 10. Differentiation of OSE cells *in vitro*. Development of oocyte (A), fibroblast (B), epithelial (C), granulosa (D), and neural-type cells (E and F) in primary OSE cultures. Arrowhead in (A) indicates initial breakdown of the germinal vesicle. Numbers in (E) show stages of neural cell differentiation. (F) Detail from (E) of transition of an epithelial cell type into a neural cell type (arrowhead). (A) live culture in phase contrast; (B–I) immunohistochemical staining for zona pellucida glycoproteins. Scale bars: 100 μm. Reproduced from Bukovsky *et al.* (2005b), © Antonin Bukovsky. (See color insert.)

cultures to produce new oocytes from ovaries lacking follicular renewal raises the possibility that OSE cells retain this differentiation potential, but are inhibited from doing so *in vivo*. Because the putative stimulatory potential of o-SMCs is absent, uncommitted mesenchymal cells may exhibit a suppressive effect. For instance, $CD8^+$ T lymphocytes, which are a predominant type of T cell present in epithelial and other nonlymphoid tissues (Beagley and Husband, 1998), can function as effector or suppressor cells (Doherty *et al.*, 1997). Interestingly, marked changes in the circulating subsets of immune cells were described in postmenopausal women and women with premature ovarian failure (Giglio *et al.*, 1994).

Pluripotency of Ovarian Stem Cells

Depending on culture conditions (type of medium used), processing of the collected cells, age of the ovaries, commitment of neighboring cells, and other local and hormonal factors, the progenitor mesenchymal cells in ovarian cultures may differentiate into additional cell types. Separated clusters of fibroblasts, epithelial cells, granulosa cells, and neural type cells (Fig. 10B–F) were all observed to differentiate from the uncommitted stem cells (Bukovsky *et al.*, 2005b). Hence, the ovarian stem cells may serve as progenitor cells for several cell types for stem cell research and autologous stem cell therapeutic applications.

Totipotent ovarian stem cells are a new type of adult stem cell, which are capable of producing, beside oocytes (Fig. 10A), also fibroblasts (Fig. 10B), epithelial-type cells (Fig. 10C), granulosa cells (Fig. 10D), and neural-type cells in various stages of differentiation (1–5, Fig. 10E and F). These groups of distinct cell types (Fig. 10B–F) were found in a single well (24-well plate) of ovarian OSE culture, and we suggested that, if a single cell becomes committed toward a certain specific differentiation, other cells in the group will follow (Bukovsky *et al.*, 2005b). Early transition of an epithelial into a neural cell type is apparent in Fig. 10F, a detail from Fig. 10E (turned 90°).

These observations strongly indicate that OSE stem cells are capable of fast differentiation into distinct cell types under the influence of already-committed cells. They support the idea that uncommitted totipotent stem cells injected into the blood stream will differentiate into cell types required in different tissues, or into a single cell type required most in the given individual.

Altogether, in contrast to males, human females with preserved ovaries have the advantage of totipotent adult ovarian stem cells persisting until advanced age, which can be used for autologous stem cell therapy without involvement of allogeneic embryonic stem cells and somatic cell nuclear transfer with questionable therapeutic outcome (Ainsworth, 2005).

Ovarian Stem Cell Culture Protocol

Here we describe ovarian stem cell culture protocol used for basic research purposes. These studies are intended to study differentiation of oocytes and occasionally other cell types from adult ovarian stem cells, that is, TA precursors and OSE stem cells, which may be collected from adult ovaries. Our experience is that the oocytes will differentiate more quickly in phenol red-containing (PhR$^+$) media (exogenous estrogenic effect), although the same may occur in PhR$^-$ media, possibly due to the endogenous production of estrogens, particularly in mixed (see later) ovarian cultures. Use of OSE cultures for therapeutic purposes will require certain modifications. All chemicals, except where specified otherwise, were purchased from Sigma (St. Louis, MO).

Tissue Culture Media Used

> Dulbecco's modified Eagle's medium/nutrient mixture F-12 Ham (cat. no. D2906; Sigma), with L-glutamine and 15 mM HEPES, without phenol red, without sodium bicarbonate; referred to later as DMEM/F-12, PhR$^-$. Adjustment: sodium bicarbonate was added (3.7 g/liter instead of recommended 1.2 g/liter), adjusted to pH 7.4
> Dulbecco's modified Eagle's medium–high glucose (cat. no. D1152; Sigma), with L-glutamine, glucose at 4500 mg/liter, and 35 mM HEPES, without sodium bicarbonate; referred to later as DMEM HG, PhR$^+$. Adjustment: sodium bicarbonate was added (3.7 g/liter) as recommended, and adjusted to pH 7.4

Collection and Culture of OSE Stem Cells

1. After surgery, the surface epithelium of intact left and right ovary is gently scraped in an aseptic laminar flow hood with a sterile stainless steel surgery knife blade no. 22 (Becton Dickinson, AcuteCare, Franklin Lakes, NJ) into separate sterile Petri dishes (pure OSE cultures; L1 and R1, Fig. 11) with medium DMEM/F-12, PhR$^-$ supplemented with antibiotics [gentamicin (50 μg/ml), penicillin (100 U/ml), and streptomycin (100 μg/ml)] and sterilized by filtration through the membrane with 0.22-μm pores (Millipore, Billerica, MA).

2. Ovaries are cut lengthwise in two halves by the Department of Pathology; one-half is kept for diagnostic evaluation and the second is processed separately from each ovarian sample for a second type of culture, in which the cells are collected from the surface and stromal compartment (mixed cultures; L2 and R2, Fig. 11).

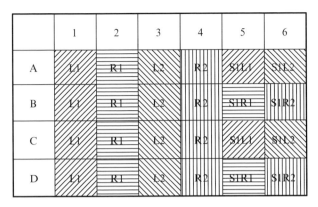

Fig. 11. Seeding ovarian cultures in 24-well plates. Media are supplemented with 20% FBS and antibiotics, 350 ml/well. Rows A and B, phenol red (PhR) negative; rows C and D, medium PhR positive. L, left ovary; R, right ovary. First day seeding cells: L1 and R1, scrapings from ovarian surfaces; L2 and R2, scrapings from ovarian surface plus cortex (mixed cultures). Second day 24 h after seeding: Supernatants (medium plus unattached cells) from wells described previously. Well 5A, supernatants from wells 1A plus 1B; well 5B, wells 1C plus 1D; well 6A, wells 3A plus 3B; well 6C, wells 3C plus 3D; and so on. New media (PHR$^-$ or PhR$^+$) put to the first four columns (1–4).

3. Each cell sample with contaminating erythrocytes is collected into DMEM/F-12, PhR$^-$ medium, transferred to a 15-ml tube, centrifuged (1000g, 5 min, 25°), and resuspended in 1.5 ml of medium (DMEM/F-12, PhR$^-$, rows A and B, Fig. 11 or DMEM HG, PhR$^+$, rows C and D) supplemented with 20% heat-inactivated fetal bovine serum (FBS; GIBCO Invitrogen, Grand Island, NY). The cells are seeded in columns 1–4 of the 24-well plate (350 μl/well; Fisher Scientific, Pittsburgh, PA) as indicated in Fig. 11, and cultured in a humidified atmosphere with 5% CO$_2$ at 37°. The number of seeded cells cannot be evaluated because of the strong contamination with erythrocytes, but varies between 100 and 10,000 cells per well after medium replacement (see later).

4. The medium with nonadherent cells and most erythrocytes is collected 24 h later and transferred into new wells (columns 5 and 6, as indicated in Fig. 11). Removal of erythrocytes is better if the plate is slightly shaken before collection of medium with floating cells. New medium is added to columns 1–4. Medium in columns 5 and 6 is also changed 24 h later.

5. Ovarian samples remaining after scraping are mounted in cryomolds with optimum cutting temperature (O.C.T.) compound (Sakura Finetek USA, Torrance, CA) as soon as possible, frozen by floating on liquid nitrogen, and stored at −80° until used for immunohistochemical evaluation of cryostat sections.

Monitoring of Ovarian Cultures

Cellular content and differentiation is monitored daily by visualization with an inverted microscope equipped with a heated stage. Images are captured with a high-resolution camera with detail enhancement (DEI-470 CCD video camera system; Optronics Engineering, Goleta, CA) and CG-7 color frame grabber (Scion, Frederick, MD) supported by Scion Image public software (Wayne Rasband, National Institutes of Health, Bethesda, MD).

Cultures are maintained for 5 to 12 days, until differentiation of some cells into oocytes is apparent (positive cultures). If no oocytes appear until day 12, the cultures are considered negative. Next, the cultures are fixed and processed by peroxidase immunohistochemistry for final evaluation.

Fixation of Cells and Peroxidase Immunohistochemistry

After experimenting with several fixation techniques, we found that the best immunohistochemical staining is obtained when the plates are first subjected to Tris-buffered saline containing 0.05% Tween 20 (TBST, pH 7.4) for 10 min at room temperature, then fixed in cold methanol for 10 min, and finally stored wet in phosphate-buffered saline (PBS), pH 7.22, at +4°. The Tween acts as a detergent and helps remove some cellular by-products that otherwise form a film and often prevent efficient staining of the cell clusters in culture. Also, it seems that keeping the cells wet after fixation helps keep them better preserved for immunostaining. Longer preservation requires PBS supplemented with 0.01% sodium azide as a preservative.

For immunohistochemistry, the PBS is removed from the wells and replaced directly with primary antibody. Depending on the antibody, the incubation time varies from 20 to 60 min at 37°. Some of the most commonly used primary antibodies are cytokeratin (mouse anti-human CK18, clone CY-90; Sigma), diluted to 5 μg/ml in PBS; noncommercial rabbit antibody to heat-solubilized porcine zona (HSPZ), diluted 1:20 in PBS; sheep antibody to heat-solubilized rabbit zona (HSRZ), diluted 1:20 as well; and other zona pellucida (ZP) antibodies (mouse monoclonal supernatant antibodies are usually applied undiluted).

After incubation, the wells are repeatedly washed in PBS at room temperature. The appropriate peroxidase-labeled secondary antibodies are applied and incubated for 20 min at room temperature. Mouse primary antibodies are followed by swine anti-mouse IgG–peroxidase conjugate (SwAM/Px; SEVA-PHARMA, Prague, Czech Republic) diluted 1:50 and preabsorbed with rat kidney homogenate in order to remove nonspecific background. Rabbit primary antibodies are followed by goat Fab' anti-rabbit (GAR) IgG (H+L) preabsorbed with human serum (Protos Immunoresearch, Burlingame, CA) and diluted 1:50, and sheep primary antibodies are followed by AffiniPure rabbit anti-sheep (RAS) IgG, Fc Fragment Specific for minimal cross-reaction to Human Serum Proteins (Jackson Immunoresearch Laboratories, West Grove, PA), also diluted 1:50.

The wells are then again extensively washed in PBS and the antigen–antibody complexes are detected with either diaminobenzidine (DAB), Vector SG, or Vector VIP detection kits (Vector Laboratories, Burlingame, CA) according to the manufacturer's instructions, giving the substrate brown, blue, or violet color. No hematoxylin counterstain was used in cell culture immunostaining. Finally, the washed cells were covered with PBS containing 0.01% sodium azide.

For double-color immunostaining, the procedure is the same through DAB staining. The cells are then incubated again with a different primary, and its corresponding secondary antibody, and detection is performed with Vector SG or VIP in place of DAB for the second round. The diaminobenzidine reaction product masks the antigen and catalytic sites of the first sequence of immunoreagents, therefore preventing interactions with the reagents of the second sequence (Sternberger, 1979), which allows for selective double staining of the cells.

Lessons from Ovarian Tissue Cultures

There is a natural tendency in OSE cultures to switch between the fibroblast-type cells with CK expression, which are characteristic for the ovarian TA, and epithelial cell types. Some of the latter may differentiate into oocytes or other cell types. In "mixed" cultures, the CK$^-$ fibroblasts are often present, which appear to support early oocyte differentiation. However, persisting interaction may cause a lack of terminal maturation into secondary oocytes.

The oocytes need for development and maturation a double amount of organelles (mitochondria). Mammalian oocytes are supplied with additional organelles by granulosa cells during the homing of oocytes in the nests (formation of primary follicles), resulting in the accumulation of mitochondria in the paranuclear Balbiani body (see Fig. 4C). The organelles are released during the initiation of the oocyte growth. A more primitive mechanism is used by the *Drosophila* oocyte, which undergoes a nuclear endoreplication and produces nurse cells connected to the oocyte by intercellular bridges (Cox and Spradling, 2003). These bridges then serve to supply additional organelles to the oocyte. Such a mechanism can also be observed in OSE cultures, where at least two cells develop in parallel (Bukovsky *et al.*, 2005b), one of which may serve as a source of additional organelles.

OSE cultures also show that advanced differentiation of oocytes requires that the cell lacks contact with neighboring cells. In epithelial sheets, the larger oocytes are rarely found inside but usually differentiate at the sheet periphery, or in the free space.

Contamination with one or two small oocytes from primary follicles per well was rarely observed in mixed, but not in pure, OSE cultures derived from the ovaries of younger women. However, oocyte-type cells differentiated in both culture types. The follicular oocytes are easy to recognize as small (30- to 40-μm) oocytes with thick ZP, and they always degenerated. Otherwise, it will be an excellent procedure for the multiplication of primary follicle oocytes (not observed) and their maturation, which was not described to be achievable (Gardner et al., 2000).

Distinct cell types, like neural cells, do not necessarily differentiate in vitro, because the stem cells may need a single cell decision as a template to be followed. This may more easily occur in vivo, when the stem cells home into tissue in need of the initiation of regeneration.

Ovarian Stem Cell Cultures and Autologous Treatment of POF

Differentiation of fresh oocytes from OSE stem cells in vitro may be worth further investigation and consideration for the autologous IVF treatment of ovarian infertility associated with POF and diminished functionality of oocytes in aging ovaries.

Suitability of Patients for Clinical Trial

Patients with a diagnosis of premature ovarian failure could be included in a clinical trial. Optimally, these patients would have failed to conceive because of they lacked their own functional oocytes during previous standard IVF therapy, or because such therapy was impossible because of the lack of oocytes within the ovaries, and they would be considering new options to have a genetically related child before undergoing conventional IVF with donated oocytes. Patients should provide detailed medical history and available laboratory results for consideration for the trial. Ultrasound or magnetic resonance imaging (MRI) of ovaries may be requested, and patients may be advised to use certain hormonal therapy, including replacement of the existing one, several weeks before admission.

Prospective patients and their partners should not carry any genetic alterations that can be transmitted to the child. Of particular importance is the exclusion of POF with fragile X premutation ($>$200 CGG trinucleotide repeats in the FMR1 gene), because the birth of a child in such women may result in mental retardation of the progeny (Corrigan et al., 2005). Genetic alterations are detected in a proportion of patients with POF, particularly those with primary amenorrhea (Rebar, 2000), and fragile X premutation was detected in 4.8% of patients with POF (Gersak *et al.*, 2003). Therefore, evidence on the lack of genetic abnormality should be provided, or the patients will be tested.

If needed, additional laboratory investigation of blood and urine, as well as imaging procedures, would be done after admission. All considered women should have a male partner with normal semen quality. Women with infertile partners (i.e., with azoospermia) should be excluded.

Therapy of ovarian infertility with cultured ovarian stem cells should be explained to the patient by a specialist in gynecology and obstetrics, who is familiar with the new technique. The medical documentation of each patient and her male partner should be evaluated by an interdisciplinary committee for *in vitro* fertilization, which would decide inclusion in the trial.

Collection of Ovarian Stem Cells and Use of Oocytes

Ovarian surface epithelium cells and small ovarian biopsy are collected during laparoscopy. The OSE cells and cells collected by scraping of tissue biopsy are cultured for 5 to 10 days to determine whether they can produce new oocytes. During culture, any nonhuman proteins (e.g., FBS) should be avoided. In general, the tissue culture medium should conform with the IVF protocol or heat-inactivated human serum, optimally collected from the patient or husband, should be used, or heat-inactivated umbilical cord serum reflecting fetal bovine serum.

Oocyte quality should be evaluated (morphology, maturity, and genetic status). If oocytes are normal, they can be fertilized by classical IVF, or by intracytoplasmic sperm injection (ICSI) with the partner's semen. Embryos can be cultured to the blastocyst stage and, before transfer into the uterus, evaluated by preimplantation genetic diagnosis. At most two normal blastocysts are transferred into the uterus and supernumerary blastocysts are cryopreserved for a potential later need of the patient. In case of a pregnancy, amniocentesis should be performed (genetic evaluation of embryo/fetus).

Potential Pitfalls

During the clinical trial, the following complications of cultured cells could occur: oocytes will not develop, oocytes will not be appropriate for fertilization (abnormal genetic status), oocytes will not be fertilized, fertilized oocytes will not develop into embryos, or embryos will not be transferred into the uterus because they will be genetically abnormal. If the oocyte culture is not successful, the infertility treatment could be continued by usual treatment with donated oocytes.

Autologous Ovarian Stem Cells and Treatment
  of Degenerative Diseases

The capacity of totipotent adult OSE stem cells to differentiate into distinct cell types can open new pathways for stem cell research and for autologous stem cell therapy of degenerative diseases. It is expected that intravenous injections of strained uncommitted stem cells from short-term (2–3 days) primary OSE cultures can be used. They may be able to home where needed and differentiate along the cellular pathway required for the given tissue/organ. Ovarian cultures for this purpose should also avoid any contamination with nonhuman proteins (see previously). Ovarian surface epithelium stem cells are present even in advanced age, so it appears that women with preserved ovary(ies) may have an advantage over men by carrying a substrate for autologous stem cell therapies.

Conclusion

A better understanding of the conditions required for oocyte and follicular renewal may form a basis for the extension of women's prime reproductive period, treatment of premature ovarian failure, rejuvenation of ovaries after chemotherapy, and delaying menopause and the manifestation of diseases characteristic of postmenopausal years. Oogenesis persists during adulthood in lower vertebrates, and for more than 100 years debate has continued on whether the oogenesis and formation of new primary follicles occurs in mammals. Most of the former and more recent studies were performed in laboratory rodents, whose ovaries have a more simple structure compared with humans.

About 10 years ago, the first evidence was presented that OSE cells in adult human ovaries produce new germ and granulosa cells and primary follicles with the assistance of immune system-related cells. Similar observations were reported from the ovaries of midpregnancy fetuses, and it is possible that the time at which fetal primordial and primary follicles are formed during immune adaptation determines how long follicular renewal will persist in the adult ovaries.

Additional, more recent observations indicate that during the prime reproductive period in human females, new eggs and granulosa cells differentiate from OSE cells to provide periodic replacement of aging follicles. The role of periodic follicular renewal is not just to maintain the number of primary follicles depleted by periodic atresia, but also to make sure that fresh eggs are always available for the creation of healthy babies.

Age-related changes in the immune system by the end of the third decade of life may cause a diminution of follicular renewal in the ovaries. Existing primary follicles are able to persist in aging ovaries for another

10 to 12 years, but their oocytes accumulate endogenous and environmentally induced genetic alterations. This results in the increased incidence of altered pregnancy outcome not detected in normal human females before 35 years of age, or in the inability of aging oocytes to respond to hormonal stimulation and develop into mature gametes. The lack of follicles with functional oocytes results in natural menopause. Similarly, cytostatic chemotherapy, which besides germ cells also affects the immune system, may cause ovarian failure due to the lack of follicular renewal.

When follicular renewal ceases earlier, POF develops, either because of the lack of primary follicles or because of their inability to respond to hormonal stimuli. Idiopathic POF may affect women of any age, and it represents a serious socioeconomic burden to the family and society. Although treatment with donated oocytes and IVF is an available option, most families prefer to have genetically related children. Therefore, attempts should be made either to preserve autologous frozen ovarian samples (e.g., before cytostatic chemotherapy), to use autologous stem cells for the production of fresh eggs *in vitro* for IVF and implantation of embryo(s) into the uterus, or to rejuvenate (prematurely) aged ovaries with autologous OSE stem cells.

Freezing of ovarian samples and their later subcutaneous implantation is reserved for when POF can be predicted, such as in patients scheduled for cancer chemotherapy, and possibly those with a genetic disposition to develop POF in certain families.

In vitro oogenesis, although relatively simple, has not been attempted yet. Therefore, many questions remain to be answered. For instance, is the quality of *in vitro*-developed eggs similar to that of a younger women? Do they develop into the ready-to-be-fertilized stage, and if not how can their maturation be stimulated? If they are fertilized, do the embryos exhibit normal genetic features? Will the embryos develop in normal fetuses? However, many of these questions are common to the standard IVF techniques and, therefore, might be investigated and answered.

Rejuvenation of aged ovaries with autologous OSE stem cells, if successful, would provide the most wanted yield, including POF treatment and delay of premature or natural menopause. One may imagine that if primary follicles are able to survive in ovaries for more than 10 years during the postnatal or premenopausal period, a single wave of follicular renewal before or after menopause may extend ovarian function for another decade. This will not only provide extension of ovarian function, but also delay the manifestation of diseases characteristic of postmenopausal years, such as osteoporosis and pelvic floor diseases (urinary and fecal incontinence, with or without pelvic organ prolapse). However, many serious questions remain to be answered when compared with the more feasible testing of oogenesis *in vitro*. For instance, follicular renewal requires not only availability of germ cells but also nests of granulosa cells, the development of which in premenopausal ovaries is usually missing. Also, even if formed, will new primary follicles persist during postmenopausal years as they do before menopause occurs?

The capacity of totipotent adult OSE stem cells to differentiate into distinct cell types can open new pathways for stem cell research and for autologous stem cell therapy of degenerative diseases. It is expected that intravenous injections of strained uncommitted stem cells from short-term (2–3 days) primary OSE cultures can be used. They may be able to home where needed and to differentiate along the cellular pathway required for the given tissue/organ. Ovarian surface epithelium stem cells are present even in advanced age, so it appears that women with preserved ovary(ies) may have an advantage over men by carrying a substrate for autologous stem cell therapies.

Acknowledgment

This research is supported by the Physicians' Medical Education and Research Foundation (Knoxville, TN).

References

Abbott, D. H., Barnett, D. K., Bruns, C. M., and Dumesic, D. A. (2005). Androgen excess fetal programming of female reproduction: A developmental aetiology for polycystic ovary syndrome? *Hum. Reprod. Update* **11,** 357–374.

Ainsworth, C. (2005). Bone cells linked to creation of fresh eggs in mammals. *Nature* **436,** 609.

Albertini, D. F. (2004). Micromanagement of the ovarian follicle reserve: Do stem cells play into the ledger? *Reproduction* **127,** 513–514.

Alberts, B., Johnson, A., Lewis, J., Raff, M., Roberts, K., and Walter, P. (2002). "Molecular Biology of the Cell." Garland Science, New York.

Allen, E. (1923). Ovogenesis during sexual maturity. *Am. J. Anat.* **31,** 439–481.

Arai, Y. (1972). Some aspects of the mechanisms involved in steroid-induced sterility. *UCLA Forum Med. Sci.* **15,** 185–191.

Auersperg, N., Pan, J., Grove, B. D., Peterson, T., Fisher, J., Maines-Bandiera, S., Somasiri, A., and Roskelley, C. D. (1999). E-cadherin induces mesenchymal-to-epithelial transition in human ovarian surface epithelium. *Proc. Natl. Acad. Sci. USA* **96,** 6249–6254.

Auersperg, N., Wong, A. S., Choi, K. C., Kang, S. K., and Leung, P. C. (2001). Ovarian surface epithelium: Biology, endocrinology, and pathology. *Endocr. Rev.* **22,** 255–288.

Baker, D. J. P. (1994). "Mothers, Babies and Health in Later Life." Churchill Livingstone, Edinburgh.

Baker, T. G. (1972). Oogenesis and ovarian development. *In* "Reproductive Biology" (H. Balin and S. Glasser, eds.), pp. 398–437. Excerpta Medica, Amsterdam.

Balfour, B. M., Drexhage, H. A., Kamperdijk, E. W., and Hoefsmit, E. C. (1981). Antigen-presenting cells, including Langerhans cells, veiled cells and interdigitating cells. *Ciba Found. Symp.* **84,** 281–301.

Barker, D. J., and Bagby, S. P. (2005). Developmental antecedents of cardiovascular disease: A historical perspective. *J. Am. Soc. Nephrol.* **16,** 2537–2544.

Barker, D. J., Osmond, C., Golding, J., Kuh, D., and Wadsworth, M. E. (1989a). Growth *in utero*, blood pressure in childhood and adult life, and mortality from cardiovascular disease. *BMJ* **298,** 564–567.

Barker, D. J., Osmond, C., and Law, C. M. (1989b). The intrauterine and early postnatal origins of cardiovascular disease and chronic bronchitis. *J. Epidemiol. Community Health* **43,** 237–240.

Beagley, K. W., and Husband, A. J. (1998). Intraepithelial lymphocytes: Origins, distribution, and function. *Crit. Rev. Immunol.* **18,** 237–254.

Beard, J. (1900). The morphological continuity of the germ cells in *Raja batis. Anat. Anz.* **18,** 465–485.

Birch, R. A., Padmanabhan, V., Foster, D. L., Unsworth, W. P., and Robinson, J. E. (2003). Prenatal programming of reproductive neuroendocrine function: Fetal androgen exposure produces progressive disruption of reproductive cycles in sheep. *Endocrinology* **144,** 1426–1434.

Block, E. (1951). Quantitative morphological investigations of the follicular system in women: Methods of quantitative determinations. *Acta Anat. (Basel)* **12,** 267–285.

Block, E. (1952). Quantitative morphological investigations of the follicular system in women: Variations at different ages. *Acta Anat. (Basel)* **14,** 108–123.

Block, E. (1953). A quantitative morphological investigation of the follicular system in newborn female infants. *Acta Anat. (Basel)* **17,** 201–206.

Bousfield, G. R., Butnev, V. Y., Gotschall, R. R., Baker, V. L., and Moore, W. T. (1996). Structural features of mammalian gonadotropins. *Mol. Cell. Endocrinol.* **125,** 3–19.

Bukovsky, A. (2000). Mesenchymal cells in tissue homeostasis and cancer. *Mod. Aspects Immunobiol.* **1,** 43–47.

Bukovsky, A. (2005a). Can ovarian infertility be treated with bone marrow- or ovary-derived germ cells? *Reprod. Biol. Endocrinol.* **3,** 36.

Bukovsky, A. (2005b). Immune system involvement in the regulation of ovarian function and augmentation of cancer. *Microsc. Res. Tech.* **69,** 482–500.

Bukovsky, A. (2005c). Origin of germ cells and follicular renewal in adult human ovaries [invited presentation]. Presented at the Microscopy and Microanalysis Conference, July 31–August 4, 2005, Honolulu, HI.

Bukovsky, A. (2006). Oogenesis from somatic stem cells and a role of immune adaptation in premature ovarian failure. *Curr. Stem Cell Res. Ther.* **1,** 289–303.

Bukovsky, A., and Caudle, M. R. (2002). Immunology: Animal models. *In* "Encyclopedia of Aging" (D. J. Ekerdt, ed.), pp. 691–695. Macmillan Reference USA, New York.

Bukovsky, A., and Presl, J. (1977). Origin of "definitive" oocytes in the mammal ovary. *Cesk. Gynekol.* **42,** 285–294.

Bukovsky, A., Copas, P., and Virant-Klun, I. (2006). Potential new strategies for the treatment of ovarian infertility and degenerative diseases with autologous ovarian stem cells. *Expert Opin. Biol. Ther.* **6,** 341–365.

Bukovsky, A., Michael, S. D., and Presl, J. (1991). Cell-mediated and neural control of morphostasis. *Med. Hypotheses* **36,** 261–268.

Bukovsky, A., Caudle, M. R., Keenan, J. A., Wimalasena, J., Foster, J. S., and Van Meter, S. E. (1995a). Quantitative evaluation of the cell cycle-related retinoblastoma protein and localization of Thy-1 differentiation protein and macrophages during follicular development and atresia, and in human corpora lutea. *Biol. Reprod.* **52,** 776–792.

Bukovsky, A., Caudle, M. R., Keenan, J. A., Wimalasena, J., Upadhyaya, N. B., and Van Meter, S. E. (1995b). Is corpus luteum regression an immune-mediated event? Localization of immune system components, and luteinizing hormone receptor in human corpora lutea. *Biol. Reprod.* **53,** 1373–1384.

Bukovsky, A., Keenan, J. A., Caudle, M. R., Wimalasena, J., Upadhyaya, N. B., and Van Meter, S. E. (1995c). Immunohistochemical studies of the adult human ovary: Possible contribution of immune and epithelial factors to folliculogenesis. *Am. J. Reprod. Immunol.* **33,** 323–340.

Bukovsky, A., Caudle, M. R., Keenan, J. A., Wimalasena, J., Upadhyaya, N. B., and Van Meter, S. E. (1996). Is irregular regression of corpora lutea in climacteric women caused by age-induced alterations in the "tissue control system"? *Am. J. Reprod. Immunol.* **36,** 327–341.

Bukovsky, A., Caudle, M. R., and Keenan, J. A. (1997). Regulation of ovarian function by immune system components: The tissue control system (TCS). *In* "Microscopy of Reproduction and Development: A Dynamic Approach" (P. M. Motta, ed.), pp. 79–89. Antonio Delfino Editore, Rome.

Bukovsky, A., Ayala, M. E., Dominguez, R., Keenan, J. A., Wimalasena, J., McKenzie, P. P., and Caudle, M. R. (2000a). Postnatal androgenization induces premature aging of rat ovaries. *Steroids* **65,** 190–205.

Bukovsky, A., Caudle, M. R., and Keenan, J. A. (2000b). Dominant role of monocytes in control of tissue function and aging. *Med. Hypotheses* **55,** 337–347.

Bukovsky, A., Caudle, M. R., Keenan, J. A., Upadhyaya, N. B., Van Meter, S., Wimalasena, J., and Elder, R. F. (2001). Association of mesenchymal cells and immunoglobulins with differentiating epithelial cells. *BMC Dev. Biol.* **1,** 11.

Bukovsky, A., Ayala, M. E., Dominguez, R., Keenan, J. A., Wimalasena, J., Elder, R. F., and Caudle, M. R. (2002). Changes of ovarian interstitial cell hormone receptors and behavior of resident mesenchymal cells in developing and adult rats with steroid-induced sterility. *Steroids* **67,** 277–289.

Bukovsky, A., Indrapichate, K., Fujiwara, H., Cekanova, M., Ayala, M. E., Dominguez, R., Caudle, M. R., Wimalasena, J., Foster, J. S., Henley, D. C., and Elder, R. F. (2003). Multiple luteinizing hormone receptor (LHR) protein variants, interspecies reactivity of anti-LHR mAb clone 3B5, subcellular localization of LHR in human placenta, pelvic floor and brain, and possible role for LHR in the development of abnormal pregnancy, pelvic floor disorders and Alzheimer's disease. *Reprod. Biol. Endocrinol.* **1,** 46.

Bukovsky, A., Caudle, M. R., Svetlikova, M., and Upadhyaya, N. B. (2004). Origin of germ cells and formation of new primary follicles in adult human ovaries. *Reprod. Biol. Endocrinol.* **2,** 20.

Bukovsky, A., Caudle, M. R., Svetlikova, M., Wimalasena, J., Ayala, M. E., and Dominguez, R. (2005a). Oogenesis in adult mammals, including humans: A review. *Endocrine* **26,** 301–316.

Bukovsky, A., Svetlikova, M., and Caudle, M. R. (2005b). Oogenesis in cultures derived from adult human ovaries. *Reprod. Biol. Endocrinol.* **3,** 17.

Butler, H., and Juma, M. B. (1970). Oogenesis in an adult prosimian. *Nature* **226,** 552–553.

Byskov, A. G., Skakkebaek, N. E., Stafanger, G., and Peters, H. (1977). Influence of ovarian surface epithelium and rete ovarii on follicle formation. *J. Anat.* **123,** 77–86.

Carlson, J. L., Bakst, M. R., and Ottinger, M. A. (1996). Developmental stages of primary oocytes in turkeys. *Poult. Sci.* **75,** 1569–1578.

Corrigan, E. C., Raygada, M. J., Vanderhoof, V. H., and Nelson, L. M. (2005). A woman with spontaneous premature ovarian failure gives birth to a child with fragile X syndrome. *Fertil. Steril.* **84,** 1508.

Cox, R. T., and Spradling, A. C. (2003). A Balbiani body and the fusome mediate mitochondrial inheritance during *Drosophila* oogenesis. *Development* **130,** 1579–1590.

Deshpande, R. R., Chapman, J. C., and Michael, S. D. (1997). The anovulation in female mice resulting from postnatal injections of estrogen is correlated with altered levels of CD8$^+$ lymphocytes. *Am. J. Reprod. Immunol.* **38,** 114–120.

Doherty, P. C., Topham, D. J., Tripp, R. A., Cardin, R. D., Brooks, J. W., and Stevenson, P. G. (1997). Effector CD4$^+$ and CD8$^+$ T-cell mechanisms in the control of respiratory virus infections. *Immunol. Rev.* **159,** 105–117.

Dumesic, D. A., Schramm, R. D., Peterson, E., Paprocki, A. M., Zhou, R., and Abbott, D. H. (2002). Impaired developmental competence of oocytes in adult prenatally androgenized female rhesus monkeys undergoing gonadotropin stimulation for *in vitro* fertilization. *J. Clin. Endocrinol. Metab.* **87,** 1111–1119.

Dyck, H. G., Hamilton, T. C., Godwin, A. K., Lynch, H. T., Maines-Bandiera, S., and Auersperg, N. (1996). Autonomy of the epithelial phenotype in human ovarian surface epithelium: Changes with neoplastic progression and with a family history of ovarian cancer. *Int. J. Cancer* **69,** 429–436.

Evans, H. M., and Swezy, O. (1931). Ovogenesis and the normal follicular cycle in adult mammalia. *Mem. Univ. Calif.* **9,** 119–224.

Franchi, L. L., Mandl, A. M., and Zuckerman, S. (1962). The development of the ovary and the process of oogenesis. *In* "The Ovary" (S. Zuckerman, ed.), pp. 1–88. Academic Press, London.

Gardner, D. K., Weissman, A., Howles, C. M., and Shoham, Z. (2000). "Textbook of Assisted Reproductive Techniques: Laboratory and Clinical Perpectives." Taylor & Francis, New York.

Gersak, K., Meden-Vrtovec, H., and Peterlin, B. (2003). Fragile X premutation in women with sporadic premature ovarian failure in Slovenia. *Hum. Reprod.* **18,** 1637–1640.

Giglio, T., Imro, M. A., Filaci, G., Scudeletti, M., Puppo, F., De Cecco, L., Indiveri, F., and Costantini, S. (1994). Immune cell circulating subsets are affected by gonadal function. *Life Sci.* **54,** 1305–1312.

Ginsburg, M., Snow, M. H., and McLaren, A. (1990). Primordial germ cells in the mouse embryo during gastrulation. *Development* **110,** 521–528.

Gosden, R. G. (2004). Germline stem cells in the postnatal ovary: Is the ovary more like a testis? *Hum. Reprod. Update* **10,** 193–195.

Gougeon, A., Echochard, R., and Thalabard, J. C. (1994). Age-related changes of the population of human ovarian follicles: Increase in the disappearance rate of non-growing and early-growing follicles in aging women. *Biol. Reprod.* **50,** 653–663.

Greenfeld, C., and Flaws, J. A. (2004). Renewed debate over postnatal oogenesis in the mammalian ovary. *Bioessays* **26,** 829–832.

Halliday, J. L., Collins, V. R., Aitken, M. A., Richards, M. P., and Olsson, C. A. (2004). Genetics and public health: Evolution, or revolution? *J. Epidemiol. Commun. Health* **58,** 894–899.

Handa, R. J., Nass, T. E., and Gorski, R. A. (1985). Proestrous hormonal changes preceding the onset of ovulatory failure in lightly androgenized female rats. *Biol. Reprod.* **32,** 232–240.

Harrap, S. B. (1994). Hypertension: Genes versus environment. *Lancet* **344,** 169–171.

Harrap, S. B. (2000). Cardiovascular disease: Genes and public health. *Ann. Acad. Med. Singapore* **29,** 279–283.

Hershkovitz, R., Erez, O., Sheiner, E., Landau, D., Mankuta, D., and Mazor, M. (2003). Elevated maternal mid-trimester chorionic gonadotropin ≥4 MoM is associated with fetal cerebral blood flow redistribution. *Acta Obstet. Gynecol. Scand.* **82,** 22–27.

Hoefsmit, E. C., Duijvestijn, A. M., and Kamperdijk, E. W. (1982). Relation between Langerhans cells, veiled cells, and interdigitating cells. *Immunobiology* **161,** 255–265.

Howard, C. J., and Hope, J. C. (2000). Dendritic cells, implications on function from studies of the afferent lymph veiled cell. *Vet. Immunol. Immunopathol.* **77,** 1–13.

Ingram, D. L. (1962). Atresia. *In* "The Ovary" (S. Zuckerman, ed.), pp. 247–273. Academic Press, London.

Ioannou, J. M. (1968). Oogenesis in adult prosimians. *J. Embryol. Exp. Morphol.* **17,** 139–145.

Johnson, J., Canning, J., Kaneko, T., Pru, J. K., and Tilly, J. L. (2004). Germline stem cells and follicular renewal in the postnatal mammalian ovary. *Nature* **428,** 145–150.

Johnson, J., Bagley, J., Skaznik-Wikiel, M., Lee, H. J., Adams, G. B., Niikura, Y., Tschudy, K. S., Tilly, J. C., Cortes, M. L., Forkert, R., Spitzer, T., Iacomini, J., Scadden, D. T., and Tilly, J. L. (2005a). Oocyte generation in adult mammalian ovaries by putative germ cells in bone marrow and peripheral blood. *Cell* **122,** 303–315.

Johnson, J., Skaznik-Wikiel, M., Lee, H. J., Niikura, Y., Tilly, J. C., and Tilly, J. L. (2005b). Setting the record straight on data supporting postnatal oogenesis in female mammals. *Cell Cycle* **4,** e36–e42.

Kay, M. M. (1979). An overview of immune aging. *Mech. Ageing Dev.* **9,** 39–59.

Kelly, S. J. (1977). Studies of the developmental potential of 4- and 8-cell stage mouse blastomeres. *J. Exp. Zool.* **200,** 365–376.

Kincl, F. A. (1965). Permanent Atrophy of Gonadal Glands Induced by Steroid Hormones. Ph.D. thesis, Charles University, Prague.

Kincl, F. A., Oriol, A., Folch Pi, A., and Maqueo, M. (1965). Prevention of steroid-induced sterility in neonatal rats with thymic cell suspension. *Proc. Soc. Exp. Biol. Med.* **120,** 252–255.

Kingery, H. M. (1917). Oogenesis in the white mouse. *J. Morphol.* **30,** 261–315.

Kirkwood, T. B. (1998). Ovarian ageing and the general biology of senescence. *Maturitas* **30,** 105–111.

Klein, J. (1982). "Immunology: The Science of Self–Nonself Discrimination." John Wiley & Sons, New York.

Knight, S. C., Farrant, J., Bryant, A., Edwards, A. J., Burman, S., Lever, A., Clarke, J., and Webster, A. D. (1986). Non-adherent, low-density cells from human peripheral blood contain dendritic cells and monocytes, both with veiled morphology. *Immunology* **57,** 595–603.

Lawson, K. A., and Hage, W. J. (1994). Clonal analysis of the origin of primordial germ cells in the mouse. *Ciba Found. Symp.* **182,** 68–84.

Lintern Moore, S., and Pantelouris, E. M. (1975). Ovarian development in athymic nude mice: The size and composition of the follicle population. *Mech. Ageing Dev.* **4,** 385–390.

Lucas, A. (1991). Programming by early nutrition in man. *In* "The Childhood Environment and Adult Disease" (G. R. Bock and J. Whelan, eds.), pp. 38–55. Wiley, Chichester, 1991.

Makabe, S., Iwaki, A., Hafez, E. S. E., and Motta, P. M. (1980). Physiomorphology of fertile and infertile human ovaries. *In* "Biology of the Ovary" (P. M. Motta and E. S. E. Hafez, eds.), pp. 279–290. Martinus Nijhoff, The Hague.

Mathe, G. (1997). Immunity aging. I. The chronic perduration of the thymus acute involution at puberty? Or the participation of the lymphoid organs and cells in fatal physiologic decline? *Biomed. Pharmacother.* **51,** 49–57.

Matsumoto, A., Asai, T., and Wakabayashi, K. (1975). Effects of X-ray irradiation on the subsequent gonadotropin secretion in normal and neonatally estrogenized female rats. *Endocrinol. Jpn.* **22,** 233–241.

Mobbs, C. V., Kannegieter, L. S., and Finch, C. E. (1985). Delayed anovulatory syndrome induced by estradiol in female C57BL/6J mice: Age-like neuroendocrine, but not ovarian, impairments. *Biol. Reprod.* **32,** 1010–1017.

Mossman, H. W., and Duke, K. L. (1973). Some comparative aspects of the mammalian ovary. *In* "Handbook of Physiology," Sect. 7: "Endocrinology" (R. O. Greep, ed.), pp. 389–402. American Physiological Society, Washington, D.C.

Motta, P. M., and Makabe, S. (1982). Development of the ovarian surface and associated germ cells in the human fetus. *Cell Tissue Res.* **226,** 493–510.

Motta, P. M., and Makabe, S. (1986). Germ cells in the ovarian surface during fetal development in humans: A three-dimensional microanatomical study by scanning and transmission electron microscopy. *J. Submicrosc. Cytol.* **18,** 271–290.

Motta, P. M., Van Blerkom, J., and Makabe, S. (1980). Changes in the surface morphology of ovarian 'germinal' epithelium during the reproductive cycle and in some pathological conditions. *J. Submicrosc. Cytol.* **12,** 407–425.

Motta, P. M., Makabe, S., Naguro, T., and Correr, S. (1994). Oocyte follicle cells association during development of human ovarian follicle: A study by high resolution scanning and transmission electron microscopy. *Arch. Histol. Cytol.* **57,** 369–394.

Motta, P. M., Nottola, S. A., Makabe, S., and Heyn, R. (2000). Mitochondrial morphology in human fetal and adult female germ cells. *Hum. Reprod.* **15**(Suppl. 2), 129–147.

Nagasawa, H., Yanai, R., Kikuyama, S., and Mori, J. (1973). Pituitary secretion of prolactin, luteinizing hormone and follicle-stimulating hormone in adult female rats treated neonatally with oestrogen. *J. Endocrinol.* **59,** 599–604.

Nishizuka, Y., and Sakakura, T. (1969). Thymus and reproduction: Sex-linked dysgenesis of the gonad after neonatal thymectomy in mice. *Science* **166,** 753–755.

Ozanne, S. E. (2001). Metabolic programming in animals. *Br. Med. Bull.* **60,** 143–152.

Ozanne, S. E., Fernandez-Twinn, D., and Hales, C. N. (2004). Fetal growth and adult diseases. *Semin. Perinatol.* **28,** 81–87.

Pearl, R., and Schoppe, W. F. (1921). Studies on the physiology of reproduction in the domestic fowl. XVIII. Further observations on the anatomical basis of fecundity. *J. Exp. Zool.* **34,** 101–118.

Penny, R., Olambiwonnu, O., and Frasier, S. D. (1976). Measurement of human chorionic gonadotropin (HCG) concentrations in paired maternal and cord sera using an assay specific for the β subunit of HCG. *Pediatrics* **58,** 110–114.

Pepling, M. E., and Spradling, A. C. (1998). Female mouse germ cells form synchronously dividing cysts. *Development* **125,** 3323–3328.

Pepling, M. E., and Spradling, A. C. (2001). Mouse ovarian germ cell cysts undergo programmed breakdown to form primordial follicles. *Dev. Biol.* **234,** 339–351.

Peters, H., and McNatty, K. P. (1980). "The Ovary: A Correlation of Structure and Function in Mammals." University of California Press, Berkeley, CA.

Powel, K. (2005). Sceptics demand duplication of controversial fertility claim. *Nat. Med.* **11,** 911.

Prentice, D. A. (2004). Adult stem cells. *Issues Law Med.* **19,** 265–294.

Prindull, G., and Zipori, D. (2004). Environmental guidance of normal and tumor cell plasticity: Epithelial mesenchymal transitions as a paradigm. *Blood* **103,** 2892–2899.

Rebar, R. W. (1982). The thymus gland and reproduction: Do thymic peptides influence the reproductive lifespan in females? *J. Am. Geriatr. Soc.* **30,** 603–606.

Rebar, R. W. (2000). Premature ovarian failure. *In* "Menopause Biology and Pathobiology" (R. A. Lobo, J. Kesley, and R. Marcus, eds.), pp. 135–146. Academic Press, San Diego, CA.

Santini, D., Ceccarelli, C., Mazzoleni, G., Pasquinelli, G., Jasonni, V. M., and Martinelli, G. N. (1993). Demonstration of cytokeratin intermediate filaments in oocytes of the developing and adult human ovary. *Histochemistry* **99**, 311–319.

Seckl, J. R. (1998). Physiologic programming of the fetus. *Clin. Perinatol.* **25**, 939–962.

Seckl, J. R. (2004). Prenatal glucocorticoids and long-term programming. *Eur. J. Endocrinol.* **151**(Suppl. 3), U49–U62.

Simkins, C. S. (1928). Origin of the sex cells in man. *Am. J. Anat.* **41**, 249–253.

Simkins, C. S. (1932). Development of the human ovary from birth to sexual maturity. *J. Anat.* **51**, 465–505.

Skinner, S. M., Lee, V. H., Kieback, D. G., Jones, L. A., Kaplan, A. L., and Dunbar, B. S. (1997). Identification of a meiotically expressed carbohydrate antigen in ovarian carcinoma. I. Immunohistochemical localization. *Anticancer Res.* **17**, 901–906.

Sternberger, L. A. (1979). The unlabeled antibody method: Hormone receptor, Golgi-like and dual color immunocytochemistry. *J. Histochem. Cytochem.* **27**, 1658–1659.

Suh, B. Y., Naylor, P. H., Goldstein, A. L., and Rebar, R. W. (1985). Modulation of thymosin β_4 by estrogen. *Am. J. Obstet. Gynecol.* **151**, 544–549.

Swanson, H. E., and van der Werff ten Bosch, J. J. (1964). The "early-androgen" syndrome: Differences in response to prenatal and postnatal administration of various doses of testosterone propionate in female and male rats. *Acta Endocrinol. (Copenh)* **47**, 37–50.

Talbert, G. B. (1968). Effect of maternal age on reproductive capacity. *Am. J. Obstet. Gynecol.* **102**, 451–477.

Tam, P. P., and Zhou, S. X. (1996). The allocation of epiblast cells to ectodermal and germ-line lineages is influenced by the position of the cells in the gastrulating mouse embryo. *Dev. Biol.* **178**, 124–132.

Telfer, E. E. (2004). Germline stem cells in the postnatal mammalian ovary: A phenomenon of prosimian primates and mice? *Reprod. Biol. Endocrinol.* **2**, 24.

Telfer, E. E., Gosden, R. G., Byskov, A. G., Spears, N., Albertini, D., Andersen, C. Y., Anderson, R., Braw-Tal, R., Clarke, H., Gougeon, A., McLaughlin, E., McLaren, A., McNatty, K., Schatten, G., Silber, S., and Tsafriri, A. (2005). On regenerating the ovary and generating controversy. *Cell* **122**, 821–822.

Van Blerkom, J., and Motta, P. M. (1979). "The Cellular Basis of Mammalian Reproduction." Urban & Schwarzenberg, Baltimore, MD.

Vogel, G. (2005). Reproductive biology: Controversial study finds an unexpected source of oocytes. *Science* **309**, 678–679.

Waldeyer, W. (1870). "Eierstock und Ei." Engelmann, Leipzig, Germany.

Williams, A. F. (1985). Immunoglobulin-related domains for cell surface recognition. *Nature* **314**, 579–580.

Williams, A. F., and Barclay, A. N. (1988). The immunoglobulin superfamily: Domains for cell surface recognition. *Annu. Rev. Immunol.* **6**, 381–405.

Williams, A. F., Davis, S. J., He, Q., and Barclay, A. N. (1989). Structural diversity in domains of the immunoglobulin superfamily. *Cold Spring Harb. Symp. Quant. Biol.* **54**, 637–647.

Yamashita, J., Itoh, H., Hirashima, M., Ogawa, M., Nishikawa, S., Yurugi, T., Naito, M., Nakao, K., and Nishikawa, S. (2000). Flk1-positive cells derived from embryonic stem cells serve as vascular progenitors. *Nature* **408**, 92–96.

Zachos, N. C., Billiar, R. B., Albrecht, E. D., and Pepe, G. J. (2002). Developmental regulation of baboon fetal ovarian maturation by estrogen. *Biol. Reprod.* **67**, 1148–1156.

Zamboni, L. (1970). Ultrastructure of mammalian oocytes and ova. *Biol. Reprod. Suppl.* **2**, 44–63.

Zuckerman, S. (1951). The number of oocytes in the mature ovary. *Recent Prog. Horm. Res.* **6**, 63–109.

[11]   Spermatogonial Stem Cells

By Jon M. Oatley and Ralph L. Brinster

Abstract

The biological activities of spermatogonial stem cells (SSCs) are the foundation for spermatogenesis and thus sustained male fertility. Therefore, understanding the mechanisms governing their ability to both self-renew and differentiate is essential. Moreover, because SSCs are the only adult stem cell to contribute genetic information to the next generation, they are an excellent target for genetic modification. In this chapter, we discuss two important approaches to investigate SSCs and their cognate niche microenvironment in the mouse, the SSC transplantation assay and the long-term serum-free SSC culture method. These techniques can be used to enhance our understanding of SSC biology as well as to produce genetically modified animals.

Introduction

In mammals, spermatozoa are the vehicle by which a male's genetic information is passed to the next generation. Thus, the process of sperm production, termed spermatogenesis, is essential for species preservation and genetic diversity. Spermatogenesis is one of the most productive cell-producing systems in adult animals, generating approximately 100 million sperm each day in adult humans (Sharpe, 1994). The entire process is composed of three main phases. The first phase, termed the proliferative phase, consists of mitotic amplifying divisions of spermatogonia. These cells ultimately become primary spermatocytes that enter the second phase of spermatogenesis, termed the meiotic phase, in which meiosis and genetic recombination occur. After two meiotic divisions, haploid spermatids are produced that undergo the third phase, termed spermiogenesis, consisting of a dramatic transformation from round germ cells to specialized spermatozoa. This complex process occurs within seminiferous tubules of the testis and is supported by the somatic Sertoli cells, which are evenly spaced throughout the seminiferous tubules and form a complex architecture (Russell, 1993). The Sertoli cells are anchored to a basement membrane, which is generated by contributions from both Sertoli and myoid cells. Together, Sertoli cells and their intricate association with germ cells, form the seminiferous epithelium. The tight junctions formed between adjacent Sertoli cells divide the seminiferous epithelium into a basal compartment,

METHODS IN ENZYMOLOGY, VOL. 419   0076-6879/06 $35.00
Copyright 2006, Elsevier Inc. All rights reserved.   DOI: 10.1016/S0076-6879(06)19011-4

exposed to many lymph- and blood-borne substances, and an adlumenal compartment, to which blood-borne substances have limited direct access. The tight junction separation of these two compartments is often referred to as the blood–testis barrier. The adlumenal compartment contains late meiotic stage germ cells, and postmeiotic spermatids and spermatozoa, and is often considered to be an immune-privileged site. Similar to other adult self-renewing tissues that rely on differentiated cells to be replenished at a constant rate or rapidly after toxic injury, the testis contains an adult tissue-specific stem cell population.

Spermatogonial Stem Cells

The spermatogonial stem cells (SSCs) are the supporting foundation for continual sperm production throughout the majority of a male's life span. SSCs first arise in the testis from development of more undifferentiated germ cells called gonocytes, which are themselves derived from primordial germ cells or PGCs that migrate from the urogenital ridge to the embryonic gonad during prenatal development (McLaren, 2003). The developmental transition of gonocytes to SSCs occurs between postnatal days 0 and 5 in the mouse (McLean et $al.$, 2003), and single spermatogonia (termed A_{single} or A_s) appear at about 6 days of age in the mouse (Bellve et $al.$, 1977; Huckins and Clermont, 1968). This A_s spermatogonia population is generally agreed to contain the SSCs. Like other adult stem cells, SSCs have the ability to undergo both self-renewal and differentiation. Self-renewal is the key function needed for SSCs to maintain tissue continuity throughout the lifetime of a male. The first step in spermatogenesis is the fate decision of an SSC to produce differentiated daughter progeny that are destined to eventually become spermatozoa.

Theoretically, 4096 spermatozoa are capable of being produced from a single SSC in the adult rat testis (Russell et $al.$, 1990). However, the overall efficiency of spermatogenesis has been estimated to be only 10 to 25% (Barratt, 1995; Tegelenbosch and de Rooij, 1993) because of germ cell apoptosis, thus considerably fewer than the potential number are actually generated. Spermatogenesis is initiated when SSC division results in the production of daughter progeny that are committed to differentiation rather than self-renewal. These newly formed differentiating daughters are termed A_{paired} (A_{pr}) spermatogonia because of their connecting intercellular bridge. On the other hand, self-renewing divisions of SSCs rather than production of A_{pr} daughters generate more SSCs (A_s) and provide the basis for continued spermatogenesis. The A_{pr} spermatogonia then undergo a series of mitotic cell divisions, becoming $A_{aligned}$ (A_{al}) spermatogonia; the number of these divisions varies among mammalian species. The A_{al}

spermatogonia give rise to differentiating spermatogonia, termed A_1 through A_4 spermatogonia in the mouse, indicating the number of mitotic amplifying divisions they undergo. The differentiating A_4 spermatogonia are capable of further maturation into intermediate and type B spermatogonia that enter meiosis, becoming primary and secondary spermatocytes, reduce their DNA content, and become haploid spermatids that undergo a transformation into fertilization-competent spermatozoa (Russell *et al.*, 1990).

Collectively, the A_s (including SSCs), A_{pr}, and A_{al} germ cells are referred to as undifferentiated spermatogonia and all share similar phenotypic and likely molecular characteristics. At present, there are no phenotypic, biochemical, or molecular characteristics that distinguish these undifferentiated spermatogonia populations from one another. The differentiating spermatogonia (A_1–A_4) are distinguishable from undifferentiated spermatogonia on the basis of several phenotypic characteristics including being c-kit positive (Ohta *et al.*, 2003). Research has clearly demonstrated that undifferentiated spermatogonia, including SSCs, are c-kit negative (Kubota *et al.*, 2003). Moreover, reciprocal expression of c-kit and Thy1 (CD90) occurs in the spermatogonia subpopulation (Kubota *et al.*, 2003). Essentially all SSC activity is found in the $Thy1^+c$-kit^- cells, indicating that Thy1 is a surface marker specifically expressed on undifferentiated spermatogonia. Thus, c-kit and Thy1 expression on the cell surface can be used as distinguishing markers between undifferentiated and differentiating spermatogonia. There are likely many other phenotypical, biochemical, and molecular differences between the different types of spermatogonia in the mammalian testis that await discovery.

After prepubertal development of the seminiferous epithelium and the initial wave of spermatogenesis, SSC biological function is supported by a niche microenvironment. SSC and niche development during the prepubertal growth period is an essential process for shaping the framework of a normal functioning SSC-niche unit in the adult. The beginning of SSC niche formation occurs during early postnatal development, coinciding with Sertoli cell maturation. The number of available SSC niches changes according to age, increasing during the developmental period from birth to sexual maturity, at which point the number becomes stable in the mouse (Shinohara *et al.*, 2001), but may increase in species in which testes enlarge during adult life, as in the rat (Ryu *et al.*, 2003). A normal functioning SSC-niche unit in the adult animal, in which SSC self-renewal and differentiation are supported, is essential for continual fertility and thus preservation of a species. Also, declining fertility that is often associated with aging in males may be due to reduced support of the SSC niche by somatic cell contributions (Ryu *et al.*, 2006), which may be secondary to impaired endocrine support of Sertoli cell function.

A key contribution to the SSC niche is growth factors produced and secreted by Sertoli cells. One such factor is glial cell line-derived neurotrophic factor (GDNF), a related member of the transforming growth factor (TGF)-β superfamily of growth factors produced by Sertoli cells in the mammalian testis (Meng *et al.*, 2000; Tadokoro *et al.*, 2002). When GDNF was overexpressed by both germ cells and somatic cells of the testis *in vivo*, a dramatic accumulation of undifferentiated spermatogonia occurred, resulting in germ cell seminoma in the mouse (Meng *et al.*, 2000). Also, mice with one GDNF-null allele showed an impairment of spermatogenesis (Meng *et al.*, 2000). These studies demonstrated that GDNF has a dramatic effect on the undifferentiated spermatogonia population and the process of spermatogenesis. In subsequent *in vivo* experiments, forced expression of human GDNF by Sertoli cells was shown to result in amplification of SSCs in the mouse testis (Yomogida *et al.*, 2003), strongly suggesting that GDNF affects the proliferation of SSCs. Subsequently, critical experiments were done *in vitro* to establish the essential role of GDNF. A serum-free culture medium for SSCs was developed (Kubota *et al.*, 2004a), which was able to maintain SSCs *in vitro* without loss of stem cell activity for a short period of time. Subsequent improvements in the culture system allowed expansion in number of the SSC population and culture for extended periods. These culture studies demonstrated that GDNF is essential for mouse SSC self-renewal *in vitro* (Kubota *et al.*, 2004b). Previous *in vitro* experiments aimed at culturing SSCs had also included GDNF as part of growth factor cocktails in less defined serum-containing culture conditions (Kanatsu-Shinohara *et al.*, 2003; Nagano *et al.*, 2003; Oatley *et al.*, 2004), suggesting its importance in SSC proliferation and survival. All these studies, culminating with Kubota *et al.* (2004b), have now clearly established GDNF as a central factor regulating the self-renewal fate decision of SSCs in the mouse and rat (Ryu *et al.*, 2005). In studying GDNF-regulated changes in gene expression within SSC-enriched populations of germ cells, the molecular mechanisms governing SSC self-renewal have begun to be identified (Oatley *et al.*, 2006). There are undoubtedly multiple other niche factors that influence SSC fate decisions of self-renewal, differentiation, and survival to be discovered.

Even though SSCs are of critical importance for male fertility, little is known about mechanisms regulating their function. They are a rare cell type in the testis, constituting only approximately 0.03% of the total testicular cell population in the adult mouse (Tegelenbosch and de Rooij, 1993). At present, there are no known specific markers that can be used to distinguish SSCs from other types of undifferentiated spermatogonia. Also, the self-renewing proliferation rate of SSCs is slow, calculated to be approximately 6 days *in vitro* (Kubota *et al.*, 2004b). For these reasons it has been difficult to study SSC biology accurately. At present, the only means to directly study SSCs is by functional transplantation (Brinster and Avarbock, 1994; Brinster and Zimmermann, 1994). Also, the ability to maintain self-renewing populations of mouse and rat SSCs *in vitro* has been achieved (Hamra *et al.*, 2005; Kanatsu-Shinohara *et al.*, 2003, 2005a; Kubota *et al.*, 2004a,b; Ryu *et al.*, 2005). These two techniques, *in vitro* maintenance of SSCs and transplantation, provide essential tools that can be used to make new discoveries and define mechanisms regulating SSC biological functions at the cell and molecular levels. This chapter contains detailed protocols on how to perform the SSC transplantation assay technique and to establish long-term cultures of self-renewing mouse SSCs.

SSC Transplantation

For many years knowledge of SSC biology was mainly theory, based on morphological observations because direct assays to identify and study them were lacking. In 1994, a transplantation technique was described (Fig. 1) that not only validated the presence of SSCs in the mammalian testis but also provided a means to investigate directly their functional characteristics (Brinster and Avarbock, 1994; Brinster and Zimmermann, 1994). At present, it is the only means for accurately identifying SSCs, and studies reporting biological characteristics of putative SSCs should require stem cell identity to be confirmed by transplantation. The technique involves injection of a donor testicular cell suspension containing SSCs into the seminiferous tubules of a germ cell-depleted or -deficient recipient male. After injection, donor SSCs migrate/relocate from the center of the seminiferous tubules and colonize stem cell niches on the basement membrane. The donor SSCs are then capable of expansion and forming colonies of donor-derived spermatogenesis. On breeding of recipient males, offspring with donor haplotype can be generated and, thus, must be derived from donor stem cell-derived sperm produced within the recipient testis. Each colony of donor-derived spermatogenesis within a recipient testis, in general, is clonally derived from a single transplanted SSC (Fig. 2A) (Dobrinski *et al.*, 1999; Kanatsu-Shinohara *et al.*, 2006a; Nagano *et al.*, 1999; Zhang *et al.*, 2003). For this reason the SSC transplantation technique can be used as an accurate and reproducible assay to study directly the characteristics of SSC biology.

Using the transplantation assay to study SSC biology involves (1) choice of donor animals, (2) preparation of recipient mice, (3) microinjection of the donor cell suspension, and (4) analysis of the recipient testis for donor-derived spermatogenesis. All four of these points are described to provide researchers with the information necessary to use the transplantation assay in mice.

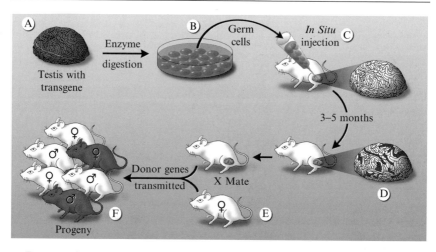

FIG. 1. The spermatogonial stem cell transplantation technique in mice. (A) The testis from a male in which a reporter transgene is expressed in germ cells is digested to generate a single-cell suspension. (B and C) The isolated cells are subsequently cultured or microinjected as a fresh cell suspension into the seminiferous tubules of an infertile germ cell-depleted recipient. (D) Colonies of donor-derived spermatogenesis can be detected in the recipient testis on the basis of reporter transgene expression by donor germ cells. Only spermatogonial stem cells are capable of establishing colonies of spermatogenesis after transplantation. Each donor-derived colony of spermatogenesis in a recipient testis is generated from a single transplanted spermatogonial stem cell. (E and F) On mating of the recipient male to a wild-type female, offspring can be produced containing the donor haplotype and thus must have been produced by sperm generated from a donor spermatogonial stem cell-derived colony of spermatogenesis. Reproduced with permission from Brinster (2002) © American Association for the Advancement of Science. (See color insert.)

Choice of Donor Animals

The choice of donor and recipient animal is essential for effective use of the transplantation assay. The most accurate and direct assessment of SSCs in a cell population is made by counting the number of donor cell-derived colonies of spermatogenesis and measuring the length of each colony. The following points must be taken into consideration for the transplantation technique to be used accurately as an assay for studying SSC biology.

• Colony number is directly correlated to the number of SSCs in a population and colony length provides a measure of the ability of an SSC to proliferate and reestablish spermatogenesis (Dobrinski *et al.*, 1999; Nagano *et al.*, 1999). Thus, donor-derived spermatogenesis after transplantation must be capable of being unequivocally identified. The best means to achieve positive identification is use of a donor animal that expresses a marker transgene in germ cells, for example, LacZ-expressing mice

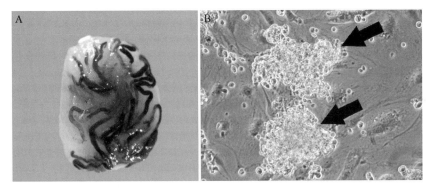

Fig. 2. Transplantation and culture of mouse spermatogonial stem cells. (A) Busulfan-treated recipient testis transplanted with cultured Thy1$^+$ germ cells from ROSA donors. Donor-derived spermatogenesis is easily detectable by staining for β-galactosidase expression. Each blue colony of spermatogenesis represents the colonization of a single donor spermatogonial stem cell. Assessing the number and length of blue colonies provides unequivocal quantitative determination of spermatogonial stem cell presence, number, and biological function in the injected cell population. (B) An established spermatogonial stem cell-enriched germ cell culture from 6-day-old ROSA donor mice. The germ cells grow as clumps of stem cells with tightly adhering membranes (arrows) and are loosely attached to the underlying STO feeder cell monolayer. In healthy, robust cultures of self-renewing spermatogonial stem cells the germ cells form tight clumps where individual cells are difficult to distinguish. (See color insert.)

(B6.129S7-Gtrosa26, designated ROSA; Jackson Laboratory, Bar Harbor, ME) or green fluorescent protein (GFP)-expressing mice (C57BL/6-TgN ACTbEGFP 1Osb; Jackson Laboratory).

• Methods for depleting endogenous germ cells in recipient testes generally are not completely effective; the possibility that some endogenous spermatogenesis will return is always present. Thus, simply identifying spermatogenesis in the testis of a recipient animal after transplantation is not an accurate means for assessing donor SSC biological function because donor-derived spermatogenesis cannot be unequivocally distinguished from reestablished endogenous spermatogenesis of the recipient, unless the donor cell expresses a transgene.

• Use of a naturally sterile recipient such as homozygous W mice (W^v/W^v or W/W^v) (Brinster and Zimmermann, 1994; Silvers, 1979) can overcome the problem of reestablished recipient spermatogenesis, and thus can be used to confirm SSC presence in a cell population without a reporter gene. However, colony number cannot be determined; therefore, transplantation cannot be used for determining the number of SSC in the injected cell population when unmarked donor cells are used.

Consequently, it is essential to use a donor strain in which donor-derived spermatogenesis can be identified visually by transgene expression.

Preparation of Recipient Mice

In the normal adult mouse testis active spermatogenesis limits, but does not completely block, the access of donor SSCs to the basal compartment of the recipient testis (Shinohara *et al.*, 2002). In order for the injected SSCs to migrate efficiently from the lumen to the basement membrane, endogenous germ cells must be depleted. Several techniques have been used to achieve this, including irradiation (Meistrich *et al.*, 1978; van Beek *et al.*, 1990; Withers *et al.*, 1974) and chemotherapeutic drugs (Brinster and Avarbock, 1994; Bucci and Meistrich, 1987). In our experience the most effective means for depleting endogenous germ cells to prepare an adult recipient is use of the chemotherapeutic drug busulfan (Brinster and Avarbock, 1994). These recipients consistently produce reliable and reproducible results. Neonatal *W* pup recipients may also be used to enhance colonization efficiency and generation of offspring; however, busulfan-treated adults are more ideal when using transplantation as an assay to study SSCs.

Busulfan-Treated Adult Recipients

• Choice of recipient strain is important. Even though the adlumenal compartment of the seminiferous tubules is immune privileged, the basal compartment to which SSCs migrate and then reside is not protected by the blood–testis barrier. Also, Sertoli cells have immunological protective characteristics (Bellgrau *et al.*, 1995; Tung, 1993). For these reasons it is essential that the recipient strain used be immunologically compatible with the transplanted donor cells. Use of nude mice as recipients is a means for avoiding immunological rejection when suitable recipients may not be readily available for the donor cells used (e.g., transplantation of rat SSCs into mouse testes). In our experiments an F_1 hybrid cross of 129 SvCP × C57BL/6 (129 × C57) males is used as recipients. These animals are suitable as recipients for both ROSA (129 × C57 ROSA and C57 homozygous ROSA) and GFP (C57 GFP) donor cells.

• Age of the recipient mouse influences the effectiveness of busulfan treatment. In our experience the optimal age is 6–8 weeks.

• Busulfan is prepared by first dissolving it in dimethyl sulfoxide (DMSO). An equal volume of warm ($\sim$40°) distilled water is then added slowly, to avoid mixing with the DMSO, to a final volume in which the concentration of busulfan will be 4 mg/ml. The DMSO and water phases are mixed just before injection to avoid busulfan precipitation. The solution is then maintained at 35 to 40°. Mice are administered an intraperitoneal injection at a dose of 60 mg/kg of body weight. This dose will eliminate nearly all endogenous spermatogenesis. Nude mice or other less hardy strains should be administered a lower dose of 45 mg/kg of body weight to avoid excessive stress to the animal.

• After busulfan treatment, males are maintained for at least 5 weeks before transplantation, thus providing sufficient time for endogenous germ cells to be depleted from the seminiferous tubules.

Pup Recipients

• The normal pup testis is more receptive to colonization by donor SSCs than adult busulfan-treated males (Shinohara *et al.*, 2001). This is likely due to immature Sertoli cells lacking tight junctions, coupled with receptive niches at the time of transplantation. Donor SSCs can be injected into the seminiferous tubules of pup recipients without any prior treatment for eliminating endogenous germ cells and the stem cells will form colonies of spermatogenesis. However, the consistency and reproducibility of the transplantation procedure are variable (Brinster *et al.*, 2003). Thus, the busulfan-treated adult recipient is more reliable for using transplantation as an assay system to study SSCs. Both endogenous and donor-derived spermatogenesis will occur in mice that were transplanted with donor SSCs as pups.

• W mice lack endogenous spermatogenesis because of a mutation of the c-kit receptor gene. These animals can therefore be used as recipients for transplantation of wild-type donor SSCs that lack a reporter gene. W mouse pups are receptive to colonization and unlike wild-type pups endogenous spermatogenesis will not occur when the mice reach puberty. For this reason, W pups are excellent recipients for generating offspring containing the donor haplotype after transplantation. Homozygous W pup recipients are difficult to generate, microinjection of donor cells is challenging, and colonization can be variable; therefore, they are less than ideal for use in the quantitative transplantation assay.

Microinjection of Donor Cell Suspension

There are multiple routes for introducing donor germ cells into the tubules of recipient mouse testes, including injection into the efferent ducts, the rete testis, or seminiferous tubules. The most effective injection method in mice and rats is injection into the rete testis via the efferent duct bundle. The rete testis is a central collecting network of tubuli that connects both ends of all seminiferous tubules. In the mouse, the rete lies cranial to the vascular pedicle and extends out from the vessels, becoming the efferent ducts that lead into the caput epididymis. For injection, a micropipette is inserted into an efferent duct at a point preceding the rete and gently threaded a short way into the duct. The testis is relatively small in size, and the efferent ducts or rete testis are not clearly visible. Thus, microinjection of donor cell suspensions is facilitated by using a dissecting microscope. An effective system has been described by Ogawa *et al.* (1997). For the purpose of this protocol the basic needs are presented.

Injection Pipette

The microinjection pipette is created from a 3-in. length of borosilicate glass (internal diameter, 0.75 mm; external diameter, 1 mm). A pipette puller is used to draw out the glass tube into two pipettes, and a rotating grinding stone is then used to create a sharpened beveled point. Suitable pipettes typically have a tip with an external diameter of 40–60 μm and a beveled angel of approximately 30° (Ogawa *et al.*, 1997).

Preparation of Donor Cell Suspension

A number of different techniques can be used to prepare donor cell suspensions for transplantation. We do not describe a specific method for this protocol; however, collection of an SSC-enriched cell suspension by Thy1$^+$ magnetic activated cell-sorting selection is described in the SSC culture protocol of this chapter. Donor cells are suspended at a concentration of $1–10 \times 10^6$ cells/ml. The choice of concentration is dependent on SSC concentration in the cell suspension. Use of a concentrated SSC testis cell suspension may result in excessive colonization, where individual colonies cannot be counted or analyzed accurately. Typically a mouse Thy1$^+$ cell population, which is highly enriched with SSCs, is injected at a concentration of 10^6 cells/ml. Testicular cell suspensions that have not been enriched for SSCs can be injected at higher concentrations. Several different injection media have been used on the basis of experiments performed. In most experiments we use a minimal essential medium α (MEM-α)-based serum-free medium (Kubota *et al.*, 2004b) to suspend cells for injection. Before injection, cell suspensions should be maintained at 4° and can be stored up to 4 h without significant cell death.

Microinjection

An essential piece of equipment is a dissecting microscope to visualize the testis and efferent ducts. The recipient mouse is anesthetized and placed in dorsal recumbence on the microscope stage. A midline incision is made through the abdomen; the testis is then brought through the incision, immobilized outside the body, and visualized at ×6 to ×20 magnification. The micropipette is then filled with 10 μl of donor cell suspension, inserted into the efferent duct, and gently threaded almost to the rete. This step is critical as the efferent ducts are translucent and difficult to immobilize, but can be visualized running from the caput of the epididymis to the rete testis. Often, a fat pad covers all or a portion of the ducts and must be gently dissected away. Care must be taken not to destroy the ducts as this will make injection nearly impossible. Positioning the needle can be done with a micromanipulator as described by Ogawa *et al.* (1997). Alternatively, the micropipette can be placed into the efferent ducts manually; however, this requires a steady hand. Piercing of the efferent ducts with a 30-gauge needle to create an entry point may facilitate insertion and positioning of the micropipette. Once in place, pressure is applied to infuse the donor cell suspension into the seminiferous tubules. Addition of trypan blue will facilitate visualizing advancement of the donor cell suspension. Pressure for injection can be generated by attaching Silastic or polyethylene tubing between the micropipette and a mouth pipette, syringe, or pressure injector. In general, the flow rate should be relatively slow to avoid harmful effects to the tubules. Injecting 10 μl of donor cell suspension typically fills 75–85% of the surface tubules of the mouse testis. It is important that the cell suspension be visually seen coursing through the tubules, thus signifying a successful injection into the tubules. Injection into the interstitial space will result in a cloudy appearance, with tubules outlined by the trypan blue. This scenario can be deceptive and lead the investigator to believe a successful injection into the seminiferous tubules has been achieved. However, an injection of this type will result in poor colonization even if some tubules were filled, because fibrosis of the testis often occurs.

Analysis of Recipient Testes for Donor-Derived Spermatogenesis

Using the transplantation technique as an assay to investigate SSC biology requires that colony number and length be determined after transplantation. In general, each colony of donor-derived spermatogenesis is clonally derived from a single SSC (Dobrinski *et al.*, 1999; Kanatsu-Shinohara *et al.*, 2006a; Nagano *et al.*, 1999; Zhang *et al.*, 2003), thus colony number is directly related to the number of SSCs in the cell suspension injected and reflects the efficiency of colonization. It is estimated that approximately 5% of the stem cells injected will produce colonies of spermatogenesis (Ogawa *et al.*, 2003; Shinohara *et al.*, 2001). Because many factors can influence this efficiency, it is critical to have adequate experimental controls and particularly to standardize recipient males for accurate results. Measuring colony length provides an assessment of the ability of a donor SSC to proliferate and reestablish spermatogenesis. To obtain both these measurements, colonies of donor-derived spermatogenesis must be identified after transplantation. Useful donor cells are obtained from LacZ-expressing ROSA mice; however, GFP donor cells may also be used but analysis is technically more difficult. Colonies of spermatogenesis from donor ROSA mice will stain a strong blue, making identification and quantification simple with a dissecting microscope. The following protocol is for recipient testes transplanted with germ cells from donor ROSA mice.

1. Recipient males are generally maintained for 8 weeks after transplantation to allow time for robust donor-derived spermatogenesis to occur.
2. Testes are recovered and the overlying tunica albuginea is gently removed, followed by fixation in 4% paraformaldehyde at 4° for 2 h. A small piece of epididymis is included as a positive control for staining because it has endogenous β-galactosidase activity.
3. The testes are then washed in three changes of LacZ rinse buffer at 4° for 30- to 60-min intervals and stained with 5-bromo-4-chloro-3-indolyl-β-D-galactosidase (X-Gal) overnight at 32° on a mechanical shaker as previously described (Brinster and Avarbock, 1994). The next day, stained testes are postfixed in 10% neutral buffered formalin for 24 h.
4. Blue-stained donor-derived colonies of spermatogenesis are counted, using a dissecting microscope at ×6–12 magnification. The tubules are gently separated to visually inspect each colony and digital pictures are captured.
5. The length of each colony can be determined with a software program (e.g., NIH Image 1.62).
6. Histological cross-sections of blue-stained colonies may be made to assess the degree of spermatogenesis. Complete spermatogenesis should be observed 8 weeks after transplantation.
7. Statistical comparisons of colony numbers and lengths generated from experimental and control cell populations provide accurate assessments of treatment effects on SSC biological function.

SSC Culture

A key tool for studying SSC biology at the cell and molecular level is the ability to maintain an enriched population of stem cells *in vitro* for extended periods of time. Also, genetic modification and preservation of a male's genotype are facilitated by long-term maintenance *in vitro*. Early reports indicated that SSCs could be maintained *in vitro* with STO feeder cells and a testis cell suspension (Nagano *et al.*, 1998). However, SSC expansion in number by self-renewal *in vitro* required a more specific system (Kanatsu-Shinohara *et al.*, 2003, 2005a; Kubota *et al.*, 2004a,b), as described later.

An important prerequisite for establishing a long-term culture of SSCs is the purity of the starting cell population. Early experiments demonstrated that the first 2 weeks of a culture period is critical, during which SSC survival is sensitive to culture conditions (Kubota *et al.*, 2004a; Nagano *et al.*, 2003). If the concentration of the contaminating somatic cell population is too high they will outgrow the SSCs. In addition, some somatic cells have a deleterious affect on SSC self-renewal and expansion. Thus, it is imperative to begin with an enriched SSC population. There are several different methods for collecting an SSC-enriched population from the testis, including differential plating (Shinohara *et al.*, 2000a), use of experimental cryptorchid donors (Shinohara *et al.*, 2000b), fluorescence-activated cell sorting (FACS) (Kubota *et al.*, 2003; Shinohara *et al.*, 1999), and magnetic activated cell sorting (MACS) (Kubota *et al.*, 2004a,b). The purest population of SSCs can be obtained by FACS or MACS selection based on cell surface phenotype. In fact, FACS has been used to define the cell surface phenotype of SSCs from mice (Kubota *et al.*, 2003; Shinohara *et al.*, 1999) and rats (Ryu *et al.*, 2004). The Thy1 (CD90) cell population in the mouse testis has been shown to contain nearly all the SSCs at all developmental ages from neonate to adult (Kubota *et al.*, 2004b). Selection of cells from mouse testes, based on Thy1 expression and using FACS or MACS, results in significant SSC enrichment in which approximately 1 in 15 cells is an SSC (Kubota *et al.*, 2003, 2004a). Even though several surface markers have been identified on mouse SSCs, this is the highest level of enrichment that has been achieved. It is likely that all undifferentiated spermatogonia (A_s, A_{pr}, and A_{al}) have a similar surface phenotype. Thus, no currently known markers can be used to specifically identify or isolate SSCs. The markers basically enrich the testis cell population for undifferentiated spermatogonia.

A second critical component for establishing a long-term expanding culture of SSCs is ideal conditions that support SSC self-renewal and survival without promoting contaminating somatic cell growth. A serum-free condition was shown to be important for providing a proper *in vitro* environment (Kubota *et al.*, 2004a,b). Classically, serum has been used to promote proliferation of many different types of cells. However, the use of serum is less than ideal because its components are undefined and vary among lots. In addition, serum can be toxic to some cell types and promotes the growth of contaminating somatic fibroblast cells that overgrow germ cells *in vitro*. Moreover, fetal bovine serum (FBS) has been shown to have a deleterious effect on both mouse and rat SSC proliferation *in vitro* (Kubota *et al.*, 2004b; Ryu *et al.*, 2005). However, the absence of

FBS in the culture medium increases the reliance of the SSCs on added growth factors.

In past attempts many growth factors have been tested for their ability to promote SSC expansion *in vitro*, including GDNF, basic fibroblast growth factor (bFGF), epidermal growth factor (EGF), transforming growth factor (TGF)-β, insulin-like growth factor I (IGF-I), leukemia inhibitory factor (LIF), and stem cell factor (SCF, kit ligand). One report has now demonstrated that GDNF is the critical growth factor for maintaining and stimulating SSC self-renewal *in vitro* (Kubota *et al.*, 2004b). Earlier *in vitro* experiments aimed at culturing SSCs had also included GDNF as part of growth factor cocktails or in less defined serum-containing medium (Kanatsu-Shinohara *et al.*, 2003; Nagano *et al.*, 2003; Oatley *et al.*, 2004), providing the initial suggestion of its importance in SSC proliferation and survival. However, the definitive proof of the central importance of GDNF came from Kubota *et al.* (2004a,b). In addition, self-renewal of rat SSCs *in vitro* has also been shown to be stimulated by GDNF (Hamra *et al.*, 2005; Ryu *et al.*, 2005). Because SSCs from many species, including humans, proliferate on the basement membrane of mouse seminiferous tubules (Brinster, 2002), it is likely that GDNF is a key regulator of SSC self-renewal in most mammalian species. Therefore, culture conditions described for mouse SSCs should be applicable as a starting point to develop similar systems in other species.

Using a serum-free chemically defined medium containing GDNF, soluble GDNF family receptor $\alpha 1$ (GFR$\alpha 1$), and bFGF, Kubota *et al.* (2004a) demonstrated a dramatic expansion of SSCs *in vitro* from several different strains of mice. This study clearly established that GDNF is an essential factor that promotes self-renewal of mouse SSCs. A similar expansion of SSCs *in vitro* under serum-free conditions was reported by Kanatsu-Shinohara *et al.* (2005a), using a more complex medium containing proprietary supplements and a cocktail of growth factors including GDNF; however, this system appears to be restricted to mice with a DBA background and does not support expansion of SSCs from other strains.

In the chemically defined system reported by Kubota *et al.* (2004b) an enriched self-renewing population of SSCs was maintained for up to 6 months *in vitro* on STO feeder layers. In this culture system the SSCs are present in clumps of densely packed germ cells with tightly adhering membranes (Fig. 2B). The germ cell clumps are thought to contain SSCs at a high concentration ($\sim$1 in 15 cells), similar to Thy1$^+$c-kit$^-$MHC$^-$ cells recovered by FACS (Kubota *et al.*, 2003, 2004b). The remaining non-SSCs are likely A_{pr} and A_{al} spermatogonia, which are the differentiating progeny of SSCs described previously in this chapter that are incapable of reestablishing spermatogenesis after transplantation. On the basis of transplantation studies, more than a 5000-fold increase in SSC number over a 10-week period was demonstrated, and a stem cell doubling time of approximately 6 days was determined. In this section we describe the methods necessary for establishing and maintaining enriched self-renewing populations of mouse SSCs for extended periods of time (Kubota *et al.*, 2004b). There are three main aspects: (1) isolation of an SSC-enriched cell population from the testis, (2) serum-free culture conditions using mitotically arrested feeder cells, and (3) subculturing practices that optimize expansion of SSC numbers. Once a culture of self-renewing SSCs is established, it can be used to investigate the mechanisms regulating SSC biological functions. However, it is essential to frequently confirm the stem cell potential of the *in vitro*-manipulated cells, using the transplantation assay to authenticate the biological relevance of the results from experimental manipulations.

Isolation of SSC-Enriched Population from Mouse Testis

In this protocol we describe isolation of an SSC-enriched $Thy1^+$ cell population from the mouse testis, using MACS separation (Kubota *et al.*, 2004a).

Testis Digestion

1. The donor age at which cells are collected has an effect on the overall purity of the collected cell population. In our experience 6- to 8-day-old mouse pups produce the most enriched population of stem cells. In a typical preparation, 8–10 pups are used; however, a greater number can be used if the volumes are adjusted appropriately.
2. Testes are collected in sterile Hanks' balanced salt solution (HBSS), using aseptic technique, and the tunica albuginea is manually removed.
3. The tissue is then placed in 4.5 ml of trypsin–EDTA (0.25% trypsin and 1 mM EDTA) solution and 0.5 ml of DNase I solution (7 mg/ml in HBSS) is added.
4. The tubules are gently dispersed and incubated at 37° for 5 min followed by gentle pipetting and addition of another 0.5 ml of DNase I solution.
5. Digestion is continued at 37° for another 5 min followed by addition of 1 ml of fetal bovine serum (FBS) and 0.5 ml of DNase I solution.
6. Pipetting is then used to ensure a single-cell suspension is achieved. More DNase solution can be added if needed.
7. The cell suspension is passed through a 40-μm pore size cell strainer, which is subsequently washed with HBSS, and the cells are pelleted by centrifugation at 600g for 7 min at 4°.

Percoll Selection

1. After centrifugation, the supernatant is removed, the pellet is resuspended in 10 ml of HBSS, and the cell concentration is determined.
2. The cell suspension is then slowly layered on top of a 30% Percoll solution [phosphate-buffered saline (PBS) with 1% FBS, penicillin (50 U/ml), streptomycin (50 μg/ml), and 30% Percoll] in 15-ml conical centrifuge tubes. Five milliliters of cell suspension is layered onto 2 ml of Percoll solution per tube. No more than 20×10^6 total cells should be layered on top of 2 ml of Percoll. It is beneficial to tilt the tube at an angle while pipetting to minimize mixing of the cell suspension with the Percoll solution.
3. Tubes are then centrifuged at 600g for 8 min at 4°.
4. The top solution phases (HBSS and Percoll) are removed, and the pellets are each resuspended in 2 ml of PBS-S [PBS with 1% FBS, 10 mM HEPES, 1 mM pyruvate, glucose (1 mg/ml), penicillin (50 U/ml), and streptomycin (50 μg/ml)]. At this point, resuspended cell suspensions from all tubes are combined and cell concentration is determined. A typical recovery after Percoll selection is 60–80% of the cells.
5. The cells are again pelleted by centrifugation at 600g for 7 min at 4°.
6. The supernatant is removed and the pellet is resuspended in 180 μl of PBS-S.

MACS Thy1 Selection

1. Thy1 antibody-conjugated microbeads [anti-mouse CD90 (Thy1.2); Miltenyi Biotech, Auburn, CA] are added to the pelleted cells resuspended in PBS-S at a dilution of 1:10 and mixed gently followed by incubation at 4° for 20 min, with mixing after 10 min (see the manufacturer's instructions).
2. An additional volume of 2 ml of PBS-S is then added, and the cells are pelleted by centrifugation at 600g for 7 min at 4°. The pellet is then resuspended in 1 ml of PBS-S.
3. The MACS columns are set up and prewashed with 0.5 ml of PBS-S followed by addition of the Thy1-labeled cell suspension. The number of columns needed is determined on the basis of total cell number, as described by the manufacturer. Typically only one column is needed for a preparation involving 8–10 mouse pups that are 6 days of age.
4. After the suspension has passed through the column, it is washed three times with 0.5 ml of PBS-S and the Thy1$^+$ cells are eluted from

the column in 1 ml of serum-free medium (SFM; for formulation see Kubota *et al.*, 2004a,b).

5. The cells are pelleted by centrifugation (600*g* for 7 min at 4°), resuspended in 1 ml of SFM, and the cell concentration is determined. A typical yield should be approximately 20–25 × 10^4 Thy1$^+$ cells per testis. Microscopically, the cell population should be relatively homogeneous, and 80–90% of the cells are large spherical germ cells and 70% are Thy1$^+$ spermatogonia (Kubota *et al.*, 2004a,b).

Serum-Free Culture on Feeder Cell Monolayers

Serum-free conditions provide the most suitable environment for expansion of SSCs *in vitro* (Kubota *et al.*, 2004b). Also, use of a chemically defined medium simplifies investigation of molecular and biochemical regulatory mechanisms of SSC biology. Development of this system clearly demonstrated that GDNF is an essential regulator of SSC self-renewal. The following protocol describes an ideal culture method for supporting mouse SSC self-renewal and expansion of their numbers.

STO Feeder Cell Monolayer Preparation

1. STO (SIM mouse embryo-derived, thioguanine and ouabain resistant) cells are grown in 10-cm dishes in Dulbecco's modified Eagle's medium (DMEM) growth medium [DMEM with 7% FBS, 100 m*M* 2-mercaptoethanol, penicillin (50 U/ml), and streptomycin (50 µg/ml)].

2. When confluent, the cells are treated with mitomycin-C (10 µg/ml in DMEM growth medium) for 3 h at 37°.

3. The cells are then washed three times with 10 ml of HBSS to remove traces of mitomycin-C and collected by trypsin–EDTA (0.25% trypsin and 1 m*M* EDTA) digestion at 37° for 5 min.

4. Growth medium containing FBS is added and the cells are pipetted to generate a single-cell suspension. Cells are pelleted by centrifugation (600*g* for 7 min at 4°), resuspended in freeze medium (DMEM growth medium containing 10% dimethyl sulfoxide) at a concentration of 4 × 10^6 cells/ml, and 1 ml is aliquoted to each cryovial and stored at –70°.

5. Frozen stocks can be stored for several months without significant loss of cell viability.

6. STO monolayers for SSC culture are created with either fresh or thawed mitomycin-C-treated cells plated at a concentration of 0.5 × 10^5 cells/cm^2 in 12-well tissue culture plates precoated with 0.1% gelatin. The feeders are maintained in DMEM growth medium before adding SSCs.

7. The optimal time for plating STO monolayers is 3–4 days before addition of SSCs.

SSC Culture

1. After MACS selection, $5–8 \times 10^4$ Thy1$^+$ cells are placed in individual wells of 12-well tissue culture plates (BD Biosciences Discovery Labware, Bedford, MA) containing STO feeder monolayers.
2. Before adding germ cell, STO monolayers are washed with HBSS to reduce residual FBS from the STO medium.
3. It is important to avoid placing more than 10^5 cells per well in the initial culture. A high number of cells per well often results in contaminating somatic cells overgrowing the SSCs, and the culture will fail.
4. The cells are cultured in SFM with the addition of recombinant human GDNF (20 ng/ml; R&D Systems; Minneapolis, MN), rat GFRα1–Fc fusion protein (150 ng/ml; R&D Systems), and recombinant human bFGF (1 ng/ml; BD Biosciences Discovery Labware).

Note: The SFM components are described by Kubota *et al.* (2004a,b). A major component of the medium is bovine serum albumin (BSA), constituting 0.2%. The source, type, and lot of BSA used are important. We have found that BSA from MP Biomedicals (Solon, OH; cat. no. 810661, lot no. 2943C) and Sigma (St. Louis, MO; cat. no. A3803, lot nos. 064K0720, 025K1497, and 124K0729) support SSC survival and expansion. Other BSA types and lots will likely also be effective; however, each should be independently tested before use in experiments.

5. Cultures are maintained at 37° in a humidified incubator with an atmosphere of 21% O_2, 5% CO_2, and the balance N_2.
6. Medium is changed with the addition of the three growth factors every 2–3 days.

Subculturing Practices That Optimize Expansion of SSC Numbers

The initial 2-week period is essential in establishing a long-term culture of enriched self-renewing SSCs. During this time SSC survival is greatly influenced by the culture conditions and contaminating somatic cells. If conditions are not ideal, contaminating somatic cells often overgrow SSCs, and a long-term culture cannot be established. Therefore, considerable attention must be paid to assessing the proliferation of the germ cells during this time. If successful, an established culture will contain a highly enriched population of self-renewing SSCs after 4 or 5 weeks. Before this time, especially during the first 2 weeks, cultures are not as enriched for SSCs, because contaminating cells often grow vigorously during this time. After several passages during a 3- or 4-week period, the majority of contaminating somatic cells disappear, leaving an enriched SSC population. The following points should be emphasized for establishing a long-term SSC-enriched culture.

• Purity of the starting cell preparation is essential to reduce the chances that somatic fibroblasts will outgrow SSCs. Thus, Thy1 selection or a similar SSC enrichment strategy is imperative for isolating a cell population intended for culture.

• SSCs grow as cell clumps that are loosely attached to the underlying feeders. Assessment of clump morphology provides a good indication of SSC support. Healthy germ cell clumps have tightly adhering membranes with a smooth and robust appearance in which individual cells are not easily discernible. Unhealthy clumps are loose, and often have a dark and grainy appearance.

• In our experience a 12-well plate format is ideal for supporting SSC expansion. When larger wells are used cell proliferation declines and clumps break apart, resulting in cell death.

• Germ cell clumps should become evident 3–4 days after initial plating. The first subculturing should be performed after 6–8 days, and the cells placed in the same size well as the original culture on a new mitomycin-C STO feeder layer.

• Subculturing typically involves trypsin–EDTA digestion (0.5 ml/well of a 12-well plate) for 5 min at 37° followed by addition of 50–100 μl of FBS. A single-cell suspension is then created by gentle pipetting. Cells are pelleted by centrifugation at 600g for 7 min at 4°, the supernatant is removed, and cells are resuspended in 2 ml of SFM. The cells are again pelleted by centrifugation, resuspended in SFM with GDNF, GFRα1, and bFGF, and plated onto a fresh STO cell feeder monolayer.

• In healthy cultures clumps will begin to reform and grow 3–4 days after subculturing.

• The second and third passages are usually conducted 6–8 days after the first subculturing and involve a split ratio of 1:1.5 or 1:2. After the first three passages, cultures can be split 1:2 to 1:4 every 7 days. The split ratio is determined subjectively on the basis of visual assessment of the size and number of clumps.

• When using established SSC cultures for experimentation, cell concentration should be determined, and an equal number of cells should be plated per well.

- In some instances, such as contamination with a high concentration of somatic cells, it may be beneficial to use gentle pipetting to remove the germ cell clumps from the feeder as described by Ryu *et al.* (2005). This technique will leave most of the contaminating cells and STO feeders in the well and allow collection of the majority of clump cells. The clumps can then be pelleted by centrifugation, digested with trypsin–EDTA, and subcultured as previously described. In our experience this subculturing technique is less effective than digesting the entire well when a high concentration of contaminating cells is not present.

- The gentle pipetting technique to remove germ cell clumps from feeders results in the collection of a cell population in which greater than 90% of the cells have the SSC cell surface phenotype (Oatley *et al.*, 2006). Thus, it is a good method for obtaining a pure cell population for molecular and biochemical experimentation.

Implications

Male fertility is dependent on SSCs, and they are the only stem cell type in the adult that contributes genetic information to the next generation. Thus, SSCs are essential for preservation of a species and genetic diversity. Therefore, understanding the mechanisms that regulate their biological functions is of critical importance. In addition, SSCs have enormous potential for use in generating genetically modified animals and gene therapy. The transplantation assay and *in vitro* culture system for long-term expansion described in this chapter are two techniques that greatly aid in the ability to study SSCs and take advantage of their applications. Using established cultures enriched for self-renewing SSCs in a chemically defined environment provides a means to conduct biochemical and molecular studies that are impossible to achieve *in vivo* because the SSC is extremely rare in the testis. By combining *in vitro* experiments with the transplantation assay, biological experiments can now be performed on these important stem cells.

The initial demonstration that SSCs could survive *in vitro* for long periods of time and then generate spermatogenesis after transplantation (Nagano *et al.*, 1998) laid the foundation for development of subsequent systems that support continual self-renewal of SSCs (Kanatsu-Shinohara *et al.*, 2003, 2005a; Kubota *et al.*, 2004a). These culture advances have paved the way for genetic modification of SSCs (Kanatsu-Shinohara *et al.*, 2005b; Nagano *et al.*, 2000, 2001, 2002) including gene-targeting strategies (Kanatsu-Shinohara *et al.*, 2006b). Similar techniques have been used to introduce genes into rat SSCs (Hamra *et al.*, 2002; Orwig *et al.*, 2002). Moreover, sophisticated experiments regarding gene activity involved in self-renewal and fate determination of

SSC can now be undertaken (Oatley *et al.*, 2006). Knowledge gained from investigating SSCs, using the culture and transplantation strategies described in this chapter, will enhance our understanding of male fertility and general stem cell biology in mammals. Because spermatogenesis is a highly conserved process in mammals, discoveries made with mice will likely be applicable to other species including humans.

Acknowledgments

We thank Drs. H. Kubota and J. A. Schmidt for manuscript review and helpful comments. Research support has been from the National Institutes of Health and Robert J. Kleberg, Jr. and Helen C. Kleberg Foundation.

References

Barratt, C. L. R. (1995). Spermatogenesis. *In* "Gametes: The Spermatozoon" (J. G. Grudzinskas and J. L. Yovich, eds.), pp. 250–267. Cambridge University Press, New York.

Bellgrau, D., Gold, D., Selawry, H., Moore, J., Franzusoff, A., and Duke, R. C. (1995). A role of CD95 ligand in preventing graft rejection. *Nature* **377,** 630–632.

Bellve, A. R., Cavicchia, J. C., Millette, C. F., O'Brien, D. A., Bhatnagar, Y. M., and Dym, M. (1977). Spermatogenic cells of the prepubertal mouse: Isolation and morphological characterization. *J. Cell Biol.* **74,** 68–85.

Brinster, C. J., Ryu, B.-Y., Avarbock, M. R., Karagenc, L., Brinster, R. L., and Orwig, K. E. (2003). Restoration of fertility by germ cell transplantation requires effective recipient preparation. *Biol. Reprod.* **69,** 412–420.

Brinster, R. L. (2002). Germline stem cell transplantation and transgenesis. *Science* **296,** 2174–2176.

Brinster, R. L., and Avarbock, M. R. (1994). Germline transmission of donor haplotype following spermatogonial transplantation. *Proc. Natl. Acad. Sci. USA* **91,** 11303–11307.

Brinster, R. L., and Zimmermann, J. W. (1994). Spermatogenesis following male germ cell transplantation. *Proc. Natl. Acad. Sci. USA* **91,** 11298–11302.

Bucci, L. R., and Meistrich, M. L. (1987). Effects of busulfan on murine spermatogenesis: Cytotoxicity, sterility, sperm abnormalities and dominant lethal mutations. *Mutat. Res.* **176,** 159–268.

Dobrinski, I., Ogawa, T., Avarbock, M. R., and Brinster, R. L. (1999). Computer assisted image analysis to assess colonization of recipient seminiferous tubules by spermatogonial stem cells from transgenic donor mice. *Mol. Reprod. Dev.* **53,** 142–148.

Hamra, F. K., Gatlin, J., Chapman, K. M., Grellhesl, D. M., Garcia, J. V., Hammer, R. E., and Garbers, D. L. (2002). Production of transgenic rats by lentiviral transduction of male germ-line stem cells. *Proc. Natl. Acad. Sci. USA* **99,** 14931–14936.

Hamra, F. K., Chapman, K. M., Nguyen, D. M., Williams-Stephens, A. A., Hammer, R. E., and Garbers, D. L. (2005). Self renewal, expansion, and transfection of rat spermatogonial stem cells in culture. *Proc. Natl. Acad. Sci. USA* **102,** 17430–17435.

Huckins, C., and Clermont, Y. (1968). Evolution of gonocytes in the rat testis during late embryonic and early post-natal life. *Arch. Anat. Histol. Embryol.* **51,** 341–354.

Kanatsu-Shinohara, M., Ogonuki, N., Inoue, K., Miki, H., Ogura, A., Toyokuni, S., and Shinohara, T. (2003). Long-term proliferation in culture and germline transmission of mouse male germline stem cells. *Biol. Reprod.* **69,** 612–616.

Kanatsu-Shinohara, M., Miki, H., Inoue, K., Ogonuki, N., Toyokuni, S., Ogura, A., and Shinohara, T. (2005a). Long-term culture of mouse male germline stem cells under serum- or feeder-free conditions. *Biol. Reprod.* **72,** 985–991.

Kanatsu-Shinohara, M., Toyokuni, S., and Shinohara, T. (2005b). Genetic selection of mouse male germline stem cells *in vitro*: Offspring from single stem cells. *Biol. Reprod.* **72,** 236–240.

Kanatsu-Shinohara, M., Inoue, K., Miki, H., Ogonuki, N., Takehashi, M., Morimoto, T., Ogura, A., and Shinohara, T. (2006a). Clonal origin of germ cell colonies after spermatogonial transplantation in mice. *Biol. Reprod.* **75,** 68–74.

Kanatsu-Shinohara, M., Ikawa, M., Takehashi, M., Ogonuki, N., Miki, H., Inoue, K., Kazuki, Y., Lee, J., Toyokuni, S., Oshimura, M., Ogura, A., and Shinohara, T. (2006b). Production of knockout mice by random or targeted mutagenesis in spermatogonial stem cells. *Proc. Natl. Acad. Sci. USA* **103,** 8018–8023.

Kubota, H., Avarbock, M. R., and Brinster, R. L. (2003). Spermatogonial stem cell share some, but not all, phenotypic and functional characteristics with other stem cells. *Proc. Natl. Acad. Sci. USA* **100,** 6487–6492.

Kubota, H., Avarbock, M. R., and Brinster, R. L. (2004a). Culture conditions and single growth factors affect fate determination of mouse spermatogonial stem cells. *Biol. Reprod.* **71,** 722–731.

Kubota, H., Avarbock, M. R., and Brinster, R. L. (2004b). Growth factors essential for self-renewal and expansion of mouse spermatogonial stem cells. *Proc. Natl. Acad. Sci. USA* **101,** 16489–16494.

McLaren, A. (2003). Primordial germ cells in the mouse. *Dev. Biol.* **262,** 1–15.

McLean, D. J., Friel, P. J., Johnston, D. S., and Griswold, M. D. (2003). Characterization of spermatogonial stem cell maturation and differentiation in neonatal mice. *Biol. Reprod.* **69,** 2085–2091.

Meistrich, M. L., Hunter, N. R., Suzuki, N., Trostle, P. K., and Withers, H. R. (1978). Gradual regeneration of mouse testicular stem cell after exposure to ionizing radiation. *Radiat. Res.* **74,** 349–362.

Meng, X., Lindahl, M., Hyvonen, M. E., Parvinen, M., de Rooij, D. G., Hess, M. W., Raatikainen-Ahokas, A., Sainio, K., Rauvala, H., Lakso, M., Pichel, J. G., Westphal, H., Saarma, M., and Sariola, H. (2000). Regulation of fate decision of undifferentiated spermatogonia by GDNF. *Science* **287,** 1489–1493.

Nagano, M., Avarbock, M. R., Leonida, E. B., Brinster, C. J., and Brinster, R. L. (1998). Culture of mouse spermatogonial stem cell. *Tissue Cell* **30,** 389–397.

Nagano, M., Avarbock, M. R., and Brinster, R. L. (1999). Pattern and kinetics of mouse donor spermatogonial stem cell colonization in recipient testes. *Biol. Reprod.* **60,** 1429–1436.

Nagano, M., Shinohara, T., Avarbock, M. R., and Brinster, R. L. (2000). Retrovirus-mediated gene delivery into male germ line stem cells. *FEBS Lett.* **475,** 7–10.

Nagano, M., Brinster, C. J., Orwig, K. E., Ryu, B.-Y., Avarbock, M. R., and Brinster, R. L. (2001). Transgenic mice produced by retroviral transduction of male germ-line stem cells. *Proc. Natl. Acad. Sci. USA* **98,** 13090–13095.

Nagano, M., Watson, D. J., Ryu, B.-Y., Wolfe, J. H., and Brinster, R. L. (2002). Lentiviral vector transduction of male germ line stem cells in mice. *FEBS Lett.* **524,** 111–115.

Nagano, M., Ryu, B.-Y., Brinster, C. J., Avarbock, M. R., and Brinster, R. L. (2003). Maintenance of mouse male germ line stem cell *in vitro*. *Biol. Reprod.* **68,** 2207–2214.

Oatley, J. M., Reeves, J. J., and McLean, D. J. (2004). Biological activity of cryporeserved bovine spermatogonial stem cell during *in vitro* culture. *Biol. Reprod.* **71,** 942–947.

Oatley, J. M., Avarbock, M. R., Teleranta, A. I., Fearon, D. T., and Brinster, R. L. (2006). Identifying genes important for spermatogonial stem cell self-renewal and survival. *Proc. Natl. Acad. Sci. USA* **103**, 9524–9529.

Ogawa, T., Arechaga, J. M., Avarbock, M. R., and Brinster, R. L. (1997). Transplantation of testis germinal cells into mouse seminiferous tubules. *Int. J. Dev. Biol.* **41**, 111–122.

Ogawa, T., Ohmura, M., Yumura, Y., Sawada, H., and Kubota, Y. (2003). Expansion of murine spermatogonial stem cells through serial transplantation. *Biol. Reprod.* **68**, 316–322.

Ohta, H., Tohda, A., and Nishimune, Y. (2003). Proliferation and differentiation of spermatogonial stem cells in the *W/W^v* mutant mouse testis. *Biol. Reprod.* **69**, 1815–1821.

Orwig, K. E., Avarbock, M. R., and Brinster, R. L. (2002). Retrovirus-mediated modification of male germline stem cells in rats. *Biol. Reprod.* **67**, 874–879.

Russell, L. D. (1993). Form, dimension, and cytology of mammalian Sertoli cells. *In* "The Sertoli Cell" (L. D. Russell and M. D. Griswold, eds.), pp. 1–37. Cache River Press, Clearwater, FL.

Russell, L. D., Ettlin, R. A., Hikim, A. P., and Clegg, E. D. (1990). Mammalian spermatogenesis. *In* "Histological and Histopathological Evaluation of the Testis," pp. 1–40. Cache River Press, Clearwater, FL.

Ryu, B.-Y., Orwig, K. E., Avarbock, M. R., and Brinster, R. L. (2003). Stem cell and niche development in the postnatal rat testis. *Dev. Biol.* **263**, 253–263.

Ryu, B.-Y., Orwig, K. E., Kubota, H., Avarbock, M. R., and Brinster, R. L. (2004). Phenotypic and functional characteristics of spermatogonial stem cells in rats. *Dev. Biol.* **274**, 158–170.

Ryu, B.-Y., Kubota, H., Avarbock, M. R., and Brinster, R. L. (2005). Conservation of spermatogonial stem cell renewal signaling between mouse and rat. *Proc. Natl. Acad. Sci. USA* **102**, 14302–14307.

Ryu, B.-Y., Orwig, K. E., Oatley, J. M., Avarbock, M. R., and Brinster, R. L. (2006). Effects of aging and niche microenvironment on spermatogonial stem cell self-renewal. *Stem Cells* **24**, 1505–1511.

Sharpe, R. (1994). Regulation of spermatogenesis. *In* "The Physiology of Reproduction" (E. Knobil and J. D. Neill, eds.), pp. 1363–1434. Raven, New York.

Shinohara, T., Avarbock, M. R., and Brinster, R. L. (1999). β_1- and α_6-integrin are surface markers on mouse spermatogonial stem cells. *Proc. Natl. Acad. Sci. USA* **96**, 5504–5509.

Shinohara, T., Orwig, K. E., Avarbock, M. R., and Brinster, R. L. (2000a). Spermatogonial stem cell enrichment by multiparameter selection of mouse testis cells. *Proc. Natl. Acad. Sci. USA* **97**, 8346–8351.

Shinohara, T., Avarbock, M. R., and Brinster, R. L. (2000b). Functional analysis of spermatogonial stem cells in Steel and cryptorchid infertile mouse models. *Dev. Biol.* **220**, 401–411.

Shinohara, T., Orwig, K. E., Avarbock, M. R., and Brinster, R. L. (2001). Remodeling of the postnatal mouse testis is accompanied by dramatic changes in stem cell number and niche accessibility. *Proc. Natl. Acad. Sci. USA* **98**, 6186–6191.

Shinohara, T., Orwig, K. E., Avarbock, M. R., and Brinster, R. L. (2002). Germ line stem cell competition in postnatal mouse testes. *Biol. Reprod.* **66**, 1491–1497.

Silvers, W. K. (1979). "The Coat Colors of Mice." Springer-Verlag, New York.

Tadokoro, Y., Yomogida, K., Ohta, H., Tohda, A., and Nishimune, Y. (2002). Homeostatic regulation of germinal stem cell proliferation by the GDNF/FSH pathway. *Mech. Dev.* **113**, 29–39.

Tegelenbosch, R. A. J., and de Rooij, D. G. (1993). A quantitative study of spermatogonial multiplication and stem cell renewal in the C3H/101 F_1 hybrid mouse. *Mutat. Res.* **290**, 193–200.

Tung, K. S. K. (1993). Regulation of testicular autoimmune disease. *In* "Cell and Molecular Biology of the Testis" (C. Desjardins and L. L. Ewing, eds.), pp. 474–490. Oxford University Press, New York.

van Beek, M. E. A. B., Meistrich, M. L., and de Rooij, D. G. (1990). Probability of self-renewing divisions of spermatogonial stem cells in colonies formed after fission neuron irradiation. *Cell Tissue Kinet.* **23,** 1–16.

Withers, H. R., Hunter, N., Barkley, H. T., Jr., and Reid, B. O. (1974). Radiation survival and regeneration characteristics of spermatogenic stem cells of mouse testis. *Radiat. Res.* **57,** 88–103.

Yomogida, K., Yagura, Y., Tadokoro, Y., and Nishimune, Y. (2003). Dramatic expansion of germinal stem cell by ectopically expressed human glial cell line-derived neurotrophic factor in mouse Sertoli cells. *Biol. Reprod.* **69,** 1303–1307.

Zhang, X., Ebata, K. T., and Nagano, M. C. (2003). Genetic analysis of the clonal origin of regenerating mouse spermatogenesis following transplantation. *Biol. Reprod.* **69,** 1872–1878.

Section III

Endoderm

[12]   Stem Cells in the Lung

By XIAOMING LIU, RYAN R. DRISKELL, and JOHN F. ENGELHARDT

Abstract

The lung is composed of two major anatomically distinct regions—the conducting airways and gas-exchanging airspaces. From a cell biology standpoint, the conducting airways can be further divided into two major compartments, the tracheobronchial and bronchiolar airways, while the alveolar regions of the lung make up the gas-exchanging airspaces. Each of these regions consists of distinct epithelial cell types with unique cellular physiologies and stem cell compartments. This chapter focuses on model systems with which to study stem cells in the adult tracheobronchial airways, also referred to as the proximal airway of the lung. Important in such models is an appreciation for the diversity of stem cell niches in the conducting airways that provide localized environmental signals to both maintain and mobilize stem cells in the setting of airway injury and normal cellular turnover. Because cellular turnover in airways is relatively slow, methods for analysis of stem cells *in vivo* have required prior injury to the lung. In contrast, *ex vivo* and *in vitro* models for analysis of airway stem cells have used genetic markers to track lineage relationships together with reconstitution systems that mimic airway biology. Over the past decades, several widely acceptable methods have been developed and used in the characterization of adult airway stem/progenitor cells. These include localization of label-retaining cells (LRCs), retroviral tagging of epithelial cells seeded into xenografts, air–liquid interface cultures to track clonal proliferative potential, and multiple transgenic mouse models. This chapter reviews the biologic context and use of these models while providing detailed methods for several of the more broadly useful models for studying adult airway stem/progenitor cell types.

Introduction

The adult lung is lined with numerous distinct types of epithelial cells in various anatomical regions, progressing proximally in the order trachea $\rightarrow$ bronchi $\rightarrow$ bronchioles $\rightarrow$ alveoli. The conducting airways (including the trachea, bronchi, and bronchioles) constitute a minority of the surface area in the human lung, amounting to approximately 0.25 m^2, whereas the alveolar regions of the lung make up approximately 100 m^2. In a normal steady state airway, epithelial cell proliferation in the adult lung is much lower compared with highly proliferative compartments in the gut, skin,

METHODS IN ENZYMOLOGY, VOL. 419   0076-6879/06 $35.00
Copyright 2006, Elsevier Inc. All rights reserved.   DOI: 10.1016/S0076-6879(06)19012-6

and hematopoietic system. As in other adult tissues and organs, stem cells in the adult lung are a subset of undifferentiated cells with the capacity to maintain multipotency in the context of the physiologic domain in which they reside. In this context, adult lung stem cells give rise to transient amplifying (TA) progenitor cells capable of abundant self-renewal and regeneration of specific cell lineages in the lung. According to their position within the pulmonary tree, several cell types in the lung have been suggested to act as stem/progenitor cells in response to injury, and to serve to effect local repair (Emura, 2002; Engelhardt, 2001; Otto, 2002). Basal cells, Clara cells, and cells that reside in submucosal glands have been shown to function as progenitors or stem cells in the conducting airway of mice (Borges *et al.*, 1997; Borthwick *et al.*, 2001; Engelhardt, 2001; Engelhardt *et al.*, 1995; Hong *et al.*, 2004a). In addition, a subset of variant Clara cells residing within neuroepithelial bodies (NEBs) (Hong *et al.*, 2001; Peake *et al.*, 2000; Reynolds *et al.*, 2000a,b) or bronchoalveolar duct junctions (Giangreco *et al.*, 2002; Kim *et al.*, 2005) has been considered stem cells in bronchioles. In contrast, alveolar epithelial type II cells (AEC II) are thought to function as the progenitor of the alveolar epithelium, based on their capacity to replicate and to give rise to terminally differentiated alveolar type I cells (AEC I) (Mason *et al.*, 1997). More recently, "side population" (SP) cells, exhibiting Hoechst dye efflux properties similar to those of hematopoietic stem cells, have been isolated and identified as a putative adult stem cell population for alveolar regions of the adult lung (Giangreco *et al.*, 2004; Summer *et al.*, 2003). These SP cells appear to have both mesenchymal and epithelial potential.

Despite these advances in the study of lung stem cells, a single lung stem cell that can give rise to multiple epithelial lineages in both the proximal and distal airways of the lung has not been identified. To date, the concept of a pluripotent stem cell for all regions of the adult lung has not been widely accepted, based on a developing concept that local regional niches are required to control both the phenotype and expansion of stem cells across vast distances of biologically distinct airways. Thus, stem cell biology in the adult lung is most widely investigated in distinct regions of tracheobronchial, bronchiolar, and alveolar (AEC II) epithelium (Berns, 2005; Bishop, 2004; Emura, 1997, 2002; Engelhardt, 2001; Liu *et al.*, 2004; Mason *et al.*, 1997; Neuringer and Randell, 2004; Otto, 2002).

Most of our current understanding of stem cells in the adult lung originates from research using classic lung injury models and/or epithelial reconstitution models. These models will likely continue to be the most useful approaches to identify stem cell populations and to understand the regulation of stem cell fates in the adult lung. However, as the molecular understanding of genes that control lung stem cell populations unravels, improved genetic approaches to study stem cell phenotypes in their niches will likely emerge in combination with more traditional approaches of analysis. This chapter first reviews the diversity of cell types in the adult lung. We then summarize the current existing knowledge and describe classic experiments used to characterize potential stem cell populations and stem cell niches in the adult lung. Last, we describe several methods for characterizing stem cells in the adult airway and provide detailed protocols for stem cell labeling, the use of lung injury models to mobilize airway stem cells, and the reconstitution of airway epithelia and submucosal glands, using *in vitro* and *ex vivo* models.

Anatomical and Cellular Diversity of Adult Lung

The adult lung can be functionally and structurally divided into three epithelial domains: the proximal cartilaginous airway (trachea and bronchi), distal bronchioles (bronchioles, terminal bronchioles, and respiratory bronchioles) and gas-exchanging airspaces (alveoli). Each of these domains has historically been classified on the basis of morphologic criteria that define the unique type of airway epithelium within each domain of the lung (Fig. 1). Consequently, defining functional criteria for the diversity of epithelial cell types in the airway has been a major focus.

In the murine proximal conducting airway and major bronchi, basal, Clara, and ciliated cells are predominant cell types, although less frequent pulmonary neuroendocrine cells (PNECs) also reside in this domain. The nonciliated, columnar Clara cells comprise the majority of the distal bronchiolar epithelium in mice, whereas the alveolar epithelium is lined by squamous alveolar type I pneumocytes (AEC I) and cuboidal type II pneumocytes (AEC II). Unique to the proximal cartilaginous airways are submucosal glands (SMGs) that are contiguous with surface airway epithelium (Fig. 1). Although less abundant in mice and confined to the trachea, SMGs are a major secretory structure located in the interstitium beneath the cartilaginous airway of other nonrodent species (Choi *et al.*, 2000; Liu *et al.*, 2004; Plopper *et al.*, 1986; Widdicombe *et al.*, 2001). SMGs are composed of an interconnecting network of serous and mucous tubules that secrete antibacterial factors, mucus, and fluid into the airway lumen. In contrast, the alveolar epithelium of adult lung is highly similar among mammalian species (Ten Have-Opbroek and Plopper, 1992).

Most of the current studies on stem cell biology in the adult airway have employed rodent animal models. To this end, it is important to note the anatomical and cellular differences between rodent and human airways. In mice, Clara cells reside throughout the tracheobronchial and bronchiolar epithelium and are the predominant secretory cell type in airways of this species. In contrast, Clara cells are limited to the bronchioles of human

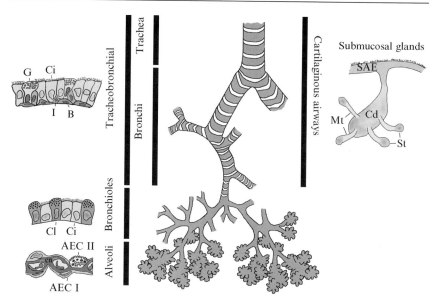

Fig. 1. Cellular diversity of epithelial cell types in the human adult lung. When discussing progenitor/stem cells in adult airways, one must always consider the diversity of cell phenotypes that exist in spatially distinct epithelia of the lung. Three main levels of conducting airways exist in the lung, including the trachea, bronchi, and bronchioles. Predominant cell types in the human pseudostratified, columnar tracheal, and bronchial epithelia include basal (B), intermediate (I), goblet (G), and ciliated (Ci) cells; less abundant nonciliated and neuroendocrine cells are not shown. Submucosal glands are also present only in the cartilaginous airways of the trachea and bronchi. Predominant cell types in the human columnar bronchiolar epithelia include Clara cells (Cl) and ciliated cells; less abundant neuroendocrine cells are not shown. Predominant cell types in the gas-exchanging alveolar airspaces include alveolar epithelial type II cells (AEC II), alveolar epithelial type I cells (AEC I), and capillary endothelial cells of the capillary networks (cn). Mt, mucous tubule; St, serous tubule; SAE, surface airway epithelium; Cd, collecting duct. (See color insert.)

airways, and goblet cells are the predominant secretory cell type in the human tracheobronchial airway. In contrast to humans, the predominant secretory cell type in the rat conducting airway is the serous cell. However, goblet cells in mice and rats can be induced by specific cytokine stimuli and/ or injury (Liu *et al.*, 2004). A summary of cellular differences in the conducting airway epithelium and SMGs between rodents (mouse and rat) and primates (monkey and human) is listed in Table I. Despite the notable differences between rodents and primates in the cell biology of the airway epithelium and the abundance of SMGs in the conducting airway, studies in mice have played significant roles in investigating stem cell biology in proximal and distal airways because of their versatile genetics.

TABLE I
SUMMARY OF PREDOMINANT EPITHELIAL CELL TYPES AND SMGs IN THE CONDUCTING AIRWAYS BETWEEN RODENTS AND PRIMATES

Species	Basal	Intermediate	Goblet	Serous	Clara	NCC	Ciliated	SMG	References
Mouse	T,B	–	–[a]	–	T,B,Br	–	T,B,Br	T	Hansell and Moretti, 1969; Pack et al., 1980; Widdicombe et al., 2001
Rat	T,B	T,B	–[a]	T,B	Br	T,B	T,B,Br	T	Plopper et al., 1983; Souma, 1987; Widdicombe et al., 2001
Monkey	T,B	T,B	T,B	NA	NA	T,B	T,B,Br	T,B	Castleman et al., 1975; Plopper et al., 1983; Plopper et al., 1986
Human	T,B	T,B	T,B	–	Br	T,B	T,B,Br	T,B	Jeffery, 1983; Plopper et al., 1980

[a] Summary for rodents applies to pathogen-free animals; the abundance of goblet cells may increase in the setting of infection or cytokine simulation.

NA: Unknown or under investigation; T: Trachea; B: Bronchi; Br: Bronchioles; NCC: Nonciliated, nonsecretory columnar cells; SMG: Submucosal glands.

Stem Cell Phenotypes and Niches in Adult Lung

Studies in stem cell biology have demonstrated that cell fate and the maintenance of stem cell populations are regulated by their local anatomical and chemical microenvironment, or niche. Niches are discrete microenvironments of specialized cell types, matrix, and diffusible factors such as cytokines and growth factors, which are critical for maintaining stem cells and promoting appropriate cell fate and migration decisions (Watt and Hogan, 2000). In the adult lung, candidate stem cell populations have been identified that are restricted to the tracheal submucosal gland ducts, neuroepithelial bodies (NEBs) of the bronchi and bronchioles, and bronchoalveolar duct junction (BADJ) of the terminal bronchioles, suggesting distinct regional stem cell niches in the adult lung (Borthwick et al., 2001; Engelhardt, 2001; Engelhardt et al., 1995; Giangreco et al., 2002; Hong et al., 2001, 2004a; Kim et al., 2005).

Because stem cells divide infrequently, injury to the lung has been necessary to study lung stem cell phenotypes and their niches in the airway. This slow-cycling characteristic has allowed for the use of DNA-labeling techniques with detectable nucleotide analogs to track stem cells in situ. Bromodeoxyuridine (BrdU) is one such common nucleotide analog that is classically used to track label-retaining cells (LRCs) after a prolonged "washout" period that dilutes the label within the more rapidly cycling transient-amplifying (TA) cells.

Two methods of injury used to amplify mobilization of the stem cell compartment in the proximal airway and to identify LRCs have included intratracheal instillation of polidocanol or inhalation of SO_2 (Borthwick et al., 2001). Intratracheal instillation of polidocanol or inhalation of SO_2, followed by in vivo BrdU administration, labels airway epithelial cells throughout the entire trachea at early time points (Borthwick et al., 2001). However, longer time points postlabeling revealed LRC accumulation in submucosal gland (SMG) ducts of mouse trachea, suggesting that this region provides a protective niche for stem cells (Borthwick et al., 2001; Engelhardt, 2001). An example of this LRC patterning after naphthalene injury in mouse trachea is shown in Fig. 2. The ability of glandular cell types to repopulate the surface airway epithelium of the trachea was also studied by subcutaneous transplantation of epithelium-denuded tracheas into recipient mice (Borthwick et al., 2001). By 28 days after transplantation, outgrowth of glandular cells led to repopulation of a ciliated surface airway epithelium, suggesting that gland or gland duct cells could regenerate the tracheal epithelium.

It is commonly thought that basal cells have the capacity to produce all the major cell phenotypes found in mouse trachea, including basal, ciliated,

Fig. 2. Naphthalene-induced mouse tracheal epithelial injury and the subacute proliferative response. Hematoxylin and eosin (H&E)-stained sections of proximal trachea 1 day (A) and 7 days (B) after intraperitoneal injection of naphthalene (275 mg/kg). Naphthalene treatment induced complete ablation of tracheal epithelium (arrows in A). By 7 days postinjury, a regenerated tracheal epithelium is observed (arrows in B). (C) BrdU labeling of mice during the first week of naphthalene injury and harvested at 28 days postinjury. Immunofluorescence detection of BrdU reveals high levels of incorporation in epithelia of submucosal glands and infrequent labeling of cells in proximal surface tracheal epithelium.

goblet, and granular secretory cells (Hong et al., 2004b; Schoch et al., 2004). In humans, where the proximal airway is more pseudostratified, parabasal cells (which are located just above the basal cells, and may also be considered intermediate cells) are thought to contribute more to the proliferative cell population than basal cells, based on higher-level expression of the proliferation marker MIB-1 (Ki-67) (Boers et al., 1998). Hence, parabasal cells (or intermediate cells) are thought to act as a TA cell population in the proximal airway that is derived from a subset of basal stem cells. Studies using human bronchial xenografts and clonal analysis with retroviral vectors also support this concept (Engelhardt et al., 1995).

Clara cells have long been thought to act as progenitor cells of the airway. This type of cell resides throughout all levels of the mouse conducting airways, but is confined to bronchiolar airways in humans. A number of studies have suggested that a subset of Clara cells are stem cells in the bronchiolar epithelium of mice. Lung injury studies using naphthalene depletion of Clara cells revealed that a subset of variant Clara cells shows multipotent differentiation and the ability to regenerate the bronchiolar epithelium. This population of Clara cells is cytochrome P-450 2F2 (CYP2F2) negative, resides in discrete pools associated with NEBs and at BADJs. Further characterization of these cells has revealed that they express stem cell antigen (Sca-1), and have the ability to efflux Hoechst dye (Reynolds et al., 2000a,b).

Pulmonary neuroendocrine cells (PNECs) reside within NEBs at cartilage–intercartilage junctions in the proximal airway and airway junctions in the bronchioles. These cells or group of cells are thought to play a role in lung development and postnatal lung maintenance through the release of neuropeptides. After naphthalene injury, proliferative cells that express either Clara cell secretory protein (CCSP) or the PNEC marker calcitonin gene-related peptide (CGRP) accumulate within NEBs (Hong *et al.*, 2001). To exclude the possibility that CGRP-expressing PNECs are the potential stem cell compartment in bronchioles, studies were performed with transgenic mice expressing the thymidine kinase gene under the direction of the CCSP promoter. Expression of thymidine kinase specifically in Clara cells allowed for the specific ablation of naphthalene-induced replicating CCSP-expressing cells. Cell-specific ablation was facilitated by administration of the prodrug ganciclovir, which is converted by thymidine kinase to a toxic nucleotide analog that prevents DNA replication and hence kills cells. Under these conditions that ablated all CCSP-expressing Clara cells after naphthalene injury, hyperplasia of CGRP-expressing PNECs occurred; however, a ciliated epithelium could not be regenerated. Consequently, these studies demonstrated that CGRP-expressing PNECs are not a stem/progenitor cell population in the distal airway. Rather, PNECs are thought to provide a niche that regulates expansion of a CCSP-expressing stem cell population in mouse distal airways (Hong *et al.*, 2001).

More recent studies have also found, in another region in the terminal bronchioles, that is, the bronchioalveolar duct junctions, bronchioalveolar stem cells (BASCs) that accumulate LRCs. BASCs were previously termed DPCs (double-positive cells that express both SP-C and CCSP) (Giangreco *et al.*, 2002; Kim *et al.*, 2005). These cells appear to be resistant to bronchiolar and alveolar damage, proliferate during epithelial cell renewal *in vivo*, and are thought to function to maintain the bronchiolar Clara cell and alveolar cell populations of the distal lung. BASCs also possess characteristics of regional stem cells, such as self-renewal and multipotency in clonal assays (Kim *et al.*, 2005).

In the gas-exchanging regions of the lung, alveolar type II cells (AEC II) have been considered the stem/progenitor cell of the alveolar epithelium, based on their ability to repopulate both AEC II and AEC I after injury (Giangreco *et al.*, 2002; Kim *et al.*, 2005; Reynolds *et al.*, 2004). Studies using a rat lung injury model have suggested that there may be four AEC II groups, based on their expression of markers. The subpopulation of AEC II that is E-cadherin negative, proliferative, and expresses high levels of telomerase activity was considered a stem cell candidate for alveolar epithelium (Reddy *et al.*, 2004).

In contrast to insights regarding candidate stem cells in the respiratory epithelium of adult lung, much less information is available regarding stem cells in the vascular compartment of the lung. One study demonstrated that an extremely small "spore-like" cell population in the lungs of adult sheep and rats with low oxygen demand can generate lung-like alveolar tissue *in vitro* (Vacanti *et al.*, 2001). Side population (SP) cells isolated from the lung have also been shown to express stem cell markers indicative of epithelial and mesenchymal lineages. SP cells isolated from whole lung, using Hoechst 33342 efflux and other stem cell markers, comprise 0.03–0.07% of mouse lung cells and are Sca-1 antigen positive, lin negative, heterogeneous for CD45, and express the vascular marker CD31 (Giangreco *et al.*, 2004; Summer *et al.*, 2003). On the basis of the expression of hematopoietic marker CD45, lung SP cells were further subdivided into hematopoietic (CD45-positive) and non-hematopoietic (CD45-negative) subpopulations. Nonhematopoietic SP cells express markers of epithelial and mesenchymal cells, share some character-istics with airway stem cells, and are currently under further investigation (Giangreco *et al.*, 2004). Consequently, there is not a single unique approach to isolate stem cell populations from the intact adult lung; however, isolation of BASCs and SP cells from the lung has been described (Giangreco *et al.*, 2004; Kim *et al.*, 2005). We have summarized in Table II the current knowl-edge of epithelial cell types in the conducting and respiratory airways, their specific cellular markers, their potential as stem/progenitor cells, and their potential lineage relationship.

In Vivo Injury Models of the Lung

Overview

Stem cells are slow cycling in the setting of normal tissue turnover, giving rise to transient amplifying (TA) progenitor cells that retain the capacity to replicate and impart most of the tissue renewal in the presence of injury. However, unlike stem cells that have unlimited proliferative capacity, TA cells are eventually incapable of proliferation and become terminally differentiated (TD) cells. Injury models have played key roles in mobilizing stem cell compartments in the lung. Different epithelial cell types in distinct areas of the lung are sensitive to various agents and have been used in lung injury models to investigate stem cells in the adult lung (Table III). These include hyperoxic agents (Smith, 1985), radiation (Theise *et al.*, 2002), SO_2 or polidocanol (Borthwick *et al.*, 2001; Randell, 1992), bleomycin (Bigby *et al.*, 1985; Izbicki *et al.*, 2002), elastase (Dubaybo *et al.*, 1991), and naphthalene (Kim *et al.*, 2005; Van Winkle *et al.*, 1999). Detailed methods for polidocanol and naphthalene injury are described later, and the others are briefly introduced here.

Exposing the lung to high concentrations of oxygen (O_2) or oxidant gases such as nitrogen dioxide (NO_2) or ozone (O_3) has been used to generate hyperoxic lung injury models. Cell death from hyperoxic injury may occur

TABLE II

Epithelial Cell Types and Identified Potential Progenitor Cell Types in Mouse Adult Lung

Cell type	Markers	Progenitor potential	Daughter cells	Pollutant sensitivity	References
Basal	Cytokeratin 5, Cytokeratin 14	Yes	Basal, PNEC? Secretory, Mucous, Ciliated	NA	Hong et al., 2004a; Schoch et al., 2004
Clara	CCSP, CyP450 2F2	Yes	Clara, PNEC? Mucous, Ciliated, AEC I, II	Naphthalene	Hong et al., 2001; Kim et al., 2005; Reynolds et al., 2000b
Ciliated	Tubulin IV	NA	NA	NA	Borthwick et al., 2001
Mucous	Mucin 5AC, Mucin 5B	Yes	Basal, Mucous, Ciliated	NA	Engelhardt et al., 1995; Hook et al., 1987
AEC I	Aquaporin 5	No	NA	Bleomycin	Emura, 1997
AEC II	Lamellar body, SpA,SpB,SpC, Aquaporin 1	Yes	AEC I, AEC II	Bleomycin	Griffiths et al., 2005; Reddy et al., 2004
PNEC	CGRP	No?	Clara?	NA	Hong et al., 2001; Peake et al., 2000
SP	Sca-1, CD31 Hoechst 33342	Yes	NA	NA	Giangreco et al., 2004; Summer et al., 2003

NA: Unknown or under investigation.

TABLE III
SOME OF THE POLLUTANTS OR REGENTS USED TO GENERATE LUNG INJURY MODELS

Pollutants	Injury methods	Injury mechanism or targets	References
O$_2$, O$_3$, NO$_2$	Inhalation	Hyperoxic	Meulenbelt et al., 1992b; Smith, 1985
SO$_2$	Inhalation	Conducting airway	Asmundsson et al., 1973; Langley-Evans et al., 1996
Polidocanol	Intratracheal instillation	Surface of airway and alveoli	Borthwick et al., 2001; Suzuki et al., 2000
Naphthalene	Intraperitoneal injection	Clara cells	Reynolds et al., 2000b; West et al., 2001
Bleomycin	Intratracheal instillation	Alveolar Type I cells, alveolar Type II cells	Bigby et al., 1985
Elastase	Intratracheal instillation	Alveolar walls	Dubaybo et al., 1991
Radiation	Exposure	Exposed regions	Theise et al., 2002

through either apoptotic or nonapoptotic pathways, possibly via free oxygen radicals. The biochemical, cellular, and morphologic characterizations of hyperoxic lung injury have been extensively studied (Smith, 1985). Mice show mild damage after 3 days of exposure to high concentrations of O$_2$, damage that consists of alveolar septal thickening and increases in alveolar macrophages. Extensive damage can be induced by continuous prolonged exposure to O$_2$, resulting in the destruction of alveolar walls, proteinaceous exudates in alveoli, and large numbers of interstitial and alveolar polymorphonuclear leukocytes (PMNs) (Smith, 1985).

Inhalation exposure of mammals to NO$_2$ has also been used extensively as a model of controlled acute pulmonary injury and repair. Prolonged exposure of rodents to NO$_2$ at 20–30 ppm results in mild emphysema and a partially reversible decrease in lung elastin and collagen content. The mechanism by which NO$_2$ damages AEC I is thought to be oxidation of unsaturated fatty acids of the cell membranes (Evans et al., 1981). Subacute NO$_2$ exposure (20 ppm NO$_2$ for 28 days) in mice also results in swelling of the ciliary shaft, focal loss of cilia, and the formation of compound cilia in the airway epithelium. Such ciliary lesions appear to be reversible when NO$_2$ exposure is stopped (Ranga and Kleinerman, 1981). In addition to the mouse NO$_2$ injury lung models, rat (Meulenbelt et al., 1992a), rabbit (Meulenbelt et al., 1994) and sheep (Januszkiewicz and Mayorga, 1994) NO$_2$ injury lung models have also been described.

An alternative mode of lung injury includes radiation. Mice exposed to 1200 cGy of total body irradiation have lungs that appear hypocellular, with a breakdown of capillaries within alveolar septae and extravasation of erythrocytes into alveolar spaces within 3 days of exposure to radiation. Irradiation-induced alveolar epithelial damage is quickly repaired by AEC II proliferation followed by differentiation in AEC I (Theise et al., 2002). The notion that bone marrow-derived stem cells have the ability to contribute to repair of alveolar epithelium was also suggested by the transplantation of retrovirally tagged green fluorescent protein (GFP) bone marrow cells into irradiated mice (Grove et al., 2002).

Another mode of injury to alveolar regions of the lung includes bleomycin. Bleomycins are a family of compounds with antibiotic and antitumor activity that have the major side effect of pulmonary toxicity. Intratracheal instillation of bleomycin induces alveolar injury in the lung (Hay et al., 1991). The cellular toxicity of bleomycin is based on its bithiazole component that partially intercalates into the DNA helix and causes oxidative damage to DNA. However, lipid peroxidation caused by bleomycin may also account for alveolar damage and subsequent pulmonary inflammation. Bleomycin lung injury is characterized by pulmonary fibrosis due to increased production of collagen and other matrix components in the lung (Bigby et al., 1985; Hay et al., 1991; Izbicki et al., 2002). The AEC I are particularly sensitive to bleomycin and are the first to be injured, whereas AEC II have a more variable sensitivity. Bleomycin-resistant AEC II undergo metaplasia in the presence of the drug (Izbicki et al., 2002). More details regarding the generation of bleomycin injury for the characterization of stem cells have been described by Kim and colleagues studying the BASC niche at the BADJ (Kim et al., 2005).

Sulfur dioxide (SO_2), a common pollutant in the air responsible for many diseases of the respiratory system, has been used to induce injury in the proximal airways of animal models. Various species of animals have different sensitivities to SO_2—mice are most sensitive, whereas rats are the most tolerant. Inhalation of SO_2 has been used in pulmonary epithelial injury models of various species including mice, rats, sheep, guinea pigs, and swine. SO_2-induced lung injury in the rat and sheep has also been proposed as a model for bronchitis (Borthwick et al., 2001; Lamb and Reid, 1968). SO_2 is soluble in water, and therefore is easily absorbed into the wet mucous membranes of the airway, causing the greatest damage to the trachea and large bronchi, whereas the bronchioles are invariably spared (Asmundsson et al., 1973). This feature makes inhalation of SO_2 an important injury model for the proximal airway.

Another chemical used to induce injury in the proximal airway of mice includes polidocanol, a surface-active detergent clinically used to enhance absorption of small proteins/peptides. Other surface-active agents such as polyoxyethylene 9 lauryl ether (Laureth-9), sodium glycocholate, and Triton X-100 have also been shown to induce lung injury after intratracheal instillation (Borthwick *et al.*, 2001). Direct intratracheal instillation of 10 μl of 2% polidocanol in phosphate-buffered saline (PBS) to the mouse trachea causes widespread denudation of the airway epithelium by 24 h postinjury (Fig. 3D). Using this polidocanol injury model, Borthwick and colleagues have identified the existence of stem cell niches in SMGs of the proximal mouse airway (Borthwick *et al.*, 2001). A detailed method of this approach is described in a later section.

Naphthalene, a chemical agent that is toxic to Clara cells, can induce damage to both proximal and distal airways of mice. This injury method has been used to identify two distinct stem cell niches associated with NEBs in the proximal airway and BADJ (Giangreco *et al.*, 2002; Hong *et al.*, 2001; Reynolds *et al.*, 2000a,b). Naphthalene is metabolized by the 2F2 subtype of cytochrome *P*-450, resulting in the production of a cytotoxic

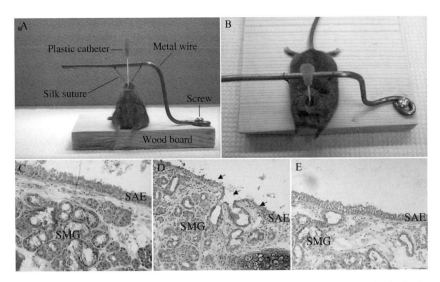

FIG. 3. Polidocanol-induced injury of the mouse tracheal epithelium. Anesthetized mice are hung on a self-made rack by placing the upper jaw incisor teeth within a suture before placing a plastic catheter into the trachea as shown in (A) (*front view*) and (B) (*top view*). The rack is made with a >3-mm metal wire fixed on a wooden board with a screw. After intratracheal instillation of 20 μl of 2% polidocanol, tracheas are harvested and sectioned. (C–E) H&E-stained sections from (C) control uninjured trachea, (D) injured trachea at 1 day, and (E) injured trachea at 14 days. Arrows mark injury to the surface airway epithelium at 1 day after polidocanol treatment in (D) as compared with injured controls (C) and regenerated epithelium (E). SAE, surface airway epithelium; SMG, submucosal gland. (See color insert.)

epoxide. Inhalation or intraperitoneal injection of naphthalene produces a dose-dependent cytotoxicity to Clara cells, whereas cells deficient in CYP2F2 are spared (Buckpitt *et al.*, 1992; West *et al.*, 2001). The bronchiolar epithelium of mice is a particularly sensitive site for naphthalene toxicity, given its high content of Clara cells (Giangreco *et al.*, 2002; Hong *et al.*, 2001). At higher doses of naphthalene (275 mg/kg), however, significant injury to the tracheobronchial epithelium can also be achieved (Fig. 2A). As discussed in more detail previously, a subpopulation of variant CCSP-expressing Clara cells deficient in CYP2F2 has been suggested as the principal stem/progenitor cell in the bronchiolar epithelium that facilitates repair after administration of naphthalene (Giangreco *et al.*, 2002; Hong *et al.*, 2001; Kim *et al.*, 2005; Peake *et al.*, 2000; Reynolds *et al.*, 2000a,b). Detailed methods for naphthalene-induced mouse lung injury and the identification of LRCs by BrdU labeling are presented in the following section.

Mouse Naphthalene Injury and BrdU-Labeling Protocol

1. Weigh the appropriate amount of naphthalene (cat. no. N143–500; Fisher Scientific, Barrington, IL) directly into a 15-ml polypropylene conical tube; make fresh on the day of use.

2. Calculate the volume of sterile corn oil (Mazola; density of corn oil, 0.9185 g/ml) needed to achieve the desired concentration of naphthalene (27.5 mg/ml). In general, the concentration of naphthalene used is 27.5 mg/ml; however, the dose can be adjusted to achieve different levels of injury to the proximal and/or distal airways. To generate a 27.5-mg/ml naphthalene stock sufficient for a 10-mouse experiment, 110 mg of naphthalene is weighed into a conical tube followed by the addition of 3.65 g of corn oil (4 ml × 0.9185 g/ml).

3. Naphthalene is difficult to dissolve. Vortex the solution at full speed for several minutes, and continue mixing on a tube rotator for at least 20 min at room temperature. Make sure that all the naphthalene has dissolved by verifying that the solution is clear of solid naphthalene.

4. Load the naphthalene solution into a 1-ml syringe, using an 18-gauge needle. After loading, remove the 18-gauge needle and use a 26-gauge needle for intraperitoneal injection.

5. Weigh the mice to the nearest 0.1 g to calculate the volume dose. It is particularly important that the dosing be exact for reproducible injury. In addition, it is important to inject the mice before 10:00 A.M. to take advantage of the minimal glutathione levels at this time of day. Sex differences exist in the extent of lung injury at a given dose of naphthalene (female mice are more sensitive than males), so make sure all animals are of the same sex.

6. Administering light anesthesia to the mice will help in obtaining an accurate weight and will also make the injection easier. Intraperitoneal

injections of the freshly made naphthalene solution described previously are performed at 10 μl/g body weight (i.e., 250 μl for a 25-g mouse given a total dose of 275 mg/kg). Naphthalene-injured animals will present a ruffled appearance with crusted eyes at approximately 20–36 h postinjury.

7. To identify LRCs, mice can be injected intraperitoneally with a 80- to 100-mg/kg body weight dose of BrdU (Roche, Indianapolis, IN)–PBS solution every 24 h, beginning 6 h after the naphthalene injection, for five consecutive days. Multiple injections (at least three) are required to label a significant percentage of LRCs.

8. Maximal injury of airway epithelium is typically observed 2 days after naphthalene injection, and newly regenerated airway epithelium can be seen by 1 week after injury (Fig. 2B). To localize LRC stem cell populations, mice are killed at various time points, such as 21, 42, and 90 days after injury and BrdU labeling.

9. Harvest the trachea and/or lung and remove blood by washing the organs in precooled PBS. The tissues can be either fixed in 10% formalin for paraffin embedding and sectioning or freshly frozen in optimal cutting temperature (O.C.T.) compound embedding medium (Tissue-Tek; Sakura Finetek USA, Torrance, CA).

10. BrdU incorporation is detected by immunofluorescence or immunohistochemical staining of 10-μm frozen sections or 6-μm paraffin sections, respectively.

11. Bring frozen sections to room temperature for 10–20 min to dry. Fix the sections in 4% paraformaldehyde in PBS for 20 min before processing for epitope retrieval.

12. Use an antigen-unmasking solution (cat. no. H-3300; Vector Laboratories, Burlingame, CA) to retrieve epitopes as specified by the manufacturer.

13. Rinse the sections in three changes of PBS (2 min for each), and then block nonspecific binding of immunoglobulin by incubating the sections in 5% rabbit serum (the same species to which the secondary antibody was produced) in PBS at room temperature for 30–60 min.

14. Directly apply a 1:100 dilution of mouse anti-BrdU antibody (cat. no. 1–299–946; Roche) in 5% rabbit serum–PBS to sections and incubate at room temperature for 1 h.

15. Rinse the sections in three changes of PBS (3 min for each) before incubating with a 1:200 dilution of fluorescein isothiocyanate (FITC)-conjugated rabbit anti-mouse IgG in 1.5% rabbit serum–PBS (cat. no. 315096045; Jackson ImmunoResearch, West Grove, PA) at room temperature for 45–60 min.

16. Rinse the sections in three changes of PBS (3 min for each) and mount them in VECTASHIELD mounting medium containing 4',

6-diamidino-2-phenylindole (DAPI, cat. no. H-1200; Vector Laboratories). Evaluate BrdU staining under a fluorescence microscope (Fig. 2C).

Mouse Polidocanol Injury and BrdU-Labeling Protocol

1. Tare the balance to zero with a 15-ml polypropylene conical tube and weigh 200 mg of polidocanol (cat. no. P-9641; Sigma, St. Louis, MO) in the tube. Add sterile PBS to bring the total volume up to 10 ml, making a 2% polidocanol (w/v) solution. Rotate the tube to ensure the polidocanol is completely dissolved.

2. Anesthetize the mice with ketamine (80 mg/kg) and xylazine (10 mg/kg) in sterile PBS. The dose of anesthesia should not be so deep as to hinder respiration, so it is advised to adjust this dose to the strain of mouse being used.

3. To facilitate intubation of mice for delivery of polidocanol into the trachea, hang the mouse on a self-made wire rack with a suture loop around the incisor teeth (Fig. 3A and B). A small flashlight can be placed under the neck of the mouse to aid in visualization of the tracheal orifice if needed.

4. Open the mouth of the mouse and pull its tongue out and down, using flat jaw forceps. Remove the needle from a 0.55-mm Angiocath (BD Angiocath; BD Medical, Sandy, UT) and insert the plastic catheter into the trachea.

5. Use a 25-gauge needle to apply 20 μl of 2% polidocanol into the trachea through the catheter. Leave the catheter in the trachea for 1 min after delivering the solution to prevent immediate expulsion into the mouth.

6. After the instillation, intraperitoneal injection of BrdU may be performed as described previously.

7. Massive injury to the tracheal epithelium is seen 24 h postinjury, and the epithelium is completely regenerated in 1–2 weeks (Fig. 3D and E). LRC location can also be performed as described for the naphthalene model.

Ex Vivo Epithelial Tracheal Xenograft Model to Study Stem Cell Expansion in Proximal Airway

Overview

The *ex vivo* epithelial xenograft model is an approach that has historically been used to study progenitor-progeny relationships in the adult airway (Duan *et al.*, 1998; Engelhardt *et al.*, 1991, 1992, 1995; Presente *et al.*, 1997; Sehgal *et al.*, 1996). This model seeds isolated airway epithelial cells onto graft tracheas that have been denuded of all endogenous airway epithelia by freeze–thawing. After the airway stem/progenitor cells are

seeded, the tracheal grafts are implanted subcutaneously into immunocompromised hosts, such as *nu/nu* or severely compromised immunodeficient
(SCID) mice. A fully differentiated epithelium regenerates approximately
3–4 weeks after transplantation (Filali *et al.*, 2002).

A major advantage of this approach is that it can be applied to airway epithelia from multiple species. This model has been used study airway
stem cell biology and/or lineage, using purified or enriched specific epithelial cell types to reconstitute the denuded tracheal grafts. These studies
have demonstrated the ability of basal and secretory cells to reconstitute a
fully differentiated surface airway epithelium in multiple species (Hook
et al., 1987; Inayama *et al.*, 1988; Randell, 1992). Rabbit basal cells purified
by centrifugal elutriation have the ability to repopulate epithelium containing basal, ciliated, and goblet cells after reconstitution of a tracheal xenograft (Inayama *et al.*, 1988). In addition, seeding of enriched rabbit Clara
cells into xenografts has demonstrated that this cell type has a limited
capacity for differentiation into Clara and ciliated cells (Hook *et al.*,
1987). Although enrichment of a given cell type is greater than 90–95%
in most instances, and can reach up to 98% purity after cell sorting with a
combination of cellular surface markers and light scattering, one major
limitation of such a cell enrichment approach is the potential for small
levels of contamination with unidentified airway stem cells.

To circumvent the limitations of cell enrichment approaches, retrovirus-
mediated gene transfer has been used to genetically tag epithelial stem/
progenitor cells and to follow lineage relationships, using histochemical
markers (Engelhardt *et al.*, 1992, 1995). Using this approach, primary human
airway cells can be infected with a variety of retroviral vectors before reconstituting the xenograft airway epithelium (Fig. 4A). Histochemical staining
for marker transgenes such as β-galactosidase and/or alkaline phosphatase
can then be used to visualize clonal expansion of stem/progenitor cells in the
reconstituted airway epithelium (Fig. 4B). Using two independent viral vectors with two independent marker transgenes (i.e., β-galactosidase and/or
alkaline phosphatase), studies have demonstrated that transgene-expressing
clones indeed arise from a single retrovirally infected progenitor/stem cell
(Engelhardt *et al.*, 1995). In reconstituted xenografts with retrovirally tagged
epithelial cells, the number of cells in each clone (i.e., size) can be used as an
index of proliferative capacity, with the notion that clones arising from adult
stem cell fractions would have a larger proliferative capacity. Phenotyping
of clones (i.e., the types of transgene-expressing cells in each clone: basal,
ciliated, goblet, or intermediate cells) can be used to assess the variety of
progenitor cell types in the airway with either limited or pluripotent capacity
for differentiation. These studies have suggested that a diverse repertoire of
progenitor cells likely exists in the adult human proximal airway, with

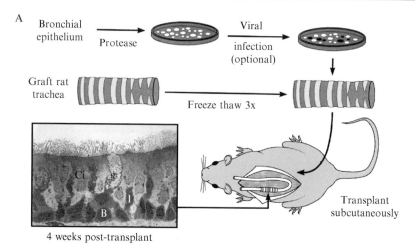

A

Bronchial epithelium → Protease → Viral infection (optional) →

Graft rat trachea → Freeze thaw 3x →

4 weeks post-transplant

Transplant subcutaneously

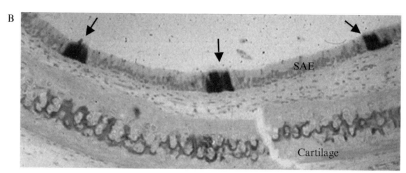

B

SAE

Cartilage

FIG. 4. Methods for generating tracheal xenografts to study clonal expansion. (A) Schematic methods for generating proximal airway epithelial xenograft models. Primary airway epithelial cells are cultured *in vitro* and may be infected with integrating recombinant viral vectors (lentivirus or retrovirus) before transplantation of epithelia into denuded rat tracheas. Fully differentiated epithelium is obtained by 4 weeks posttransplantation (B, basal cells; Ci, ciliated cells; I, intermediate cells; G, goblet cells). (B) Colony-forming efficiency (CFE) of stem/progenitor cells can be evaluated by detecting virally expressed transgenes in the reconstituted surface airway epithelium (SAE) (arrows). In this example a recombinant retroviral vector expressing the β-galactosidase transgene was used and detected by X-Gal staining. (C) Schematic view of various components of the xenograft cassette. The denuded rat trachea is connected to a sterile tubing cassette by a series of sutures as illustrated: *a*, 1-in. Silastic tubing (cat. no. 602-175; Dow Corning, Midland, MI); *b*, 3/4-in. Silastic tubing (cat. no. 602-175; Dow Corning); *c*, 1.75-in. Silastic tubing (cat. no. 602-175; Dow Corning); *d*, 1.25-in. Teflon tubing (cat. no. 9567-K10; Thomas Scientific, Swedesboro, NJ); *e* and *e'*, adapter (0.8-mm barb-to-barb connector; cat. no. 732–8300; Bio-Rad, Hercules, CA). A chrome wire plug (0.0035-in.-diameter Chromel A steel wire; Hoskins Manufacturing, Hamburg, MI) is

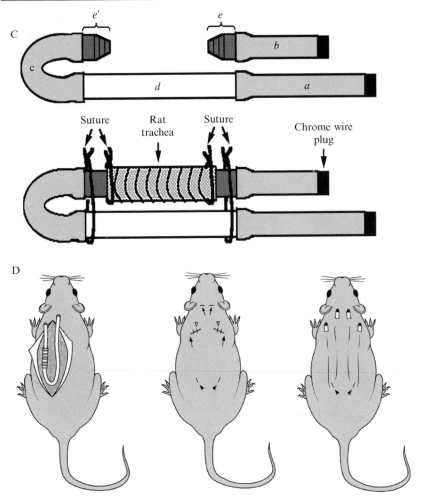

also used. (D) Subcutaneous transplantation of the xenograft cassettes in *nu/nu* athymic mice. *Left:* A subcutaneous view of the transplanted xenograft cassette. *Middle:* The four incisions (arrows) made on the back of the recipient *nu/nu* mouse before transplantation. The xenograft cassette is guided subcutaneously with forceps, so that one port exits through the back of the neck and the other port exits through the main incision. Surgical staples are then used to close the main incisions (marked by open arrowheads) and one additional staple is used to anchor the xenograft tubing near the tail end of the mouse (solid arrowheads). *Right:* The resultant xenograft cassette 1 week postsurgery, when the proximal staples are removed. The staples marked by solid arrowheads in the middle panel are used to maintain the position of the cassette and to prevent subcutaneous migration (it is necessary to leave these staples in for at least 2 weeks). (See color insert.)

differing capacities for proliferation and differentiation. These studies demonstrated that a stem/progenitor population with the ability to differentiate into SMGs and all surface airway epithelial cell types exists in the adult human airway (Engelhardt *et al.*, 1995). This cell type is thought to be a candidate stem cell in the proximal airway.

The protocol for the generation of human bronchial xenografts, described later, is also suitable for studies with other species such as rat, with modifications of cell culture conditions (Engelhardt *et al.*, 1991).

Human Bronchial Xenograft Protocol

1. Dissection of human airway tissue:
   a. The bronchial/lung cassette should be chilled immediately after removal from the donor in cold physiologic saline and should remain chilled during dissection. For ideal aseptic and safety conditions, the dissection is best done in a laminar flow hood. Handling human tissue and primary airway cultures may expose the user to blood-borne pathogens, and suitable health protection measures should be followed.
   b. The proximal airways are dissected from the lung on ice and placed immediately in chilled sterile Ca^{2+}- and Mg^{2+}-free PBS containing penicillin (50 U/ml), streptomycin (50 μg/ml), tobramycin (80 μg/ml), ceftazidime (100 μg/ml), imipenem/cilastatin (Primaxin, 100 μg/ml), and amphotericin B (5.0 μg/ml). Remove the blood, mucus, and extraneous connective tissues from the airways and cut the specimen into suitably sized half-rings and place them into 50-ml polypropylene conical tubes with the preceding antibiotic solution. Invert the tubes for several minutes to wash the tissue, and repeat four times by moving the tissue from one conical tube to a fresh tube with new antibiotic-containing cold PBS.
   c. It is important to avoid exposing specimens to drastic changes in temperature during processing and dissection to prevent epithelial degradation. In addition, segments of the proximal airway (trachea and bronchi) may be separately processed because their epithelial characteristics may vary.
   d. After washing in antibiotic solution, airway tissue is placed in a 50- or 15-ml conical tube filled with medium A (described at the end of this section) to dissociate airway epithelial cells. It is important that the conical tubes be filled to the top with medium A and tightly capped to avoid changes in pH caused by alteration in CO_2 concentration over time. Airway tissues are incubated in medium A at $4°$ for 24–96 h. Nasal turbinates require 24–48 h for dissociation whereas tracheobronchial tissues require 48–72 h (depending on the desired degree of separation into single cells versus cell clumps). However, it should be

noted that dissociation times longer than 72 h will decrease cell viability, and cell clumps appear to proliferate more rapidly in culture.

2. Harvesting airway epithelial cells:

   a. Add 10% fetal bovine serum (FBS) into the dissociation solution to inhibit the Pronase in medium A while transferring the solutions and tissue to a larger or greater number of conical tubes filled approximately one-third full. Invert the tube(s) several times vigorously to help dissociate the cells from the airway tissue. Mild shaking is helpful to increase yields, but avoid excessive force that may damage cells (i.e., do not shake hard enough to cause the FBS to foam).

   b. After shaking, allow 1 min for airway tissue to settle to the bottom of the tube (liberated airway cells will remain in suspension). Pipette the medium containing the airway cells into a fresh tube on ice, and place the remaining tissue into a 100-mm tissue culture plate.

   c. Using the blunt side of a scalpel, scrape the surface of the airways to remove the remaining epithelial cells. Rinse airway samples with 10% FBS–Ham's F12 medium, combine the washed cells into 50-ml conical tubes, and centrifuge the tubes at 120g for 5 min at 4°.

   d. Resuspend the cell pellets in medium B (described at the end of this section) and transfer the cell suspension to 100-mm Primaria tissue culture dishes (cat. no. 353803; BD Biosciences Discovery Labware, Bedford, MA).

   e. Incubate the suspension in a 5% CO_2 incubator at 37° for a minimum of 1–2 h to allow fibroblasts within the cell suspension to attach to the plastic surface. Airway epithelial cells will not attach rapidly to the plastic surface without collagen coating.

   f. Collect the nonattached cells by centrifugation, resuspend them in medium C (described at the end of this section), and count them with a hemacytometer.

3. Culture of airway epithelial cells:

   a. Plate 1–2 × 10^6 of the preceding airway cells per 100-mm collagen-coated tissue culture dish (BD Biosciences Discovery Labware) in 10 ml of airway culture medium C. Incubate the cells overnight in a 5% CO_2 incubator at 37°.

   b. On the day after plating, aspirate the culture medium and unattached cells. Wash the dish containing adherent cells with prewarmed Ham's F12 medium and refeed the cells with fresh medium C. Culture the cells for an additional 48 h.

   c. At 72 h postplating, feed the cells with medium D (containing lower levels of antibiotics and no amphotericin B). Typically, cells are ready for cryopreservation, passaging, or transplantation into xenograft models by 5 days postplating (~80% confluency). Care should be taken not to allow cells to become more than 80%

confluent or they will begin to differentiate and lose their capacity for subculturing. Unpassaged fresh cultures of airway epithelia also give consistently better reconstitution of epithelium in the xenograft model.

d. If primary airway cells are to be expanded (typically cells can be expanded once without the loss of the ability to differentiate in a xenograft model), they should be treated with 0.1% trypsin–EDTA for 1–3 min at 37° followed by neutralization with an equal volume of trypsin inhibitor buffer (recipe described at the end of this section). Cells should be closely monitored during trypsinization and harvested immediately once they are released by gentle tapping of the plate. Cell suspensions are centrifuged to remove trypsin and washed once in medium D, followed by plating at a 1:5 dilution.

e. For cryopreservation, epithelial cell pellets are resuspended in 4° medium D with 10% dimethyl sulfoxide (DMSO) and 10% FBS and aliquoted in 2-ml cryogenic vials for slow freezing at −80°, overnight. Slow freezing can be performed in isopropanol-containing cryopreservation containers (Nalgene; Nalge Nunc International, Rochester, NY). Cells are then moved to liquid nitrogen storage. Typically, one subconfluent 100-mm plate of cells is aliquoted per vial, and when subcultured placed into five 100-mm plates.

4. Retroviral infection of primary airway epithelial cells (optional): Retroviral infection of airway epithelial cells before seeding into tracheal xenograft has been useful in studying airway stem/progenitor cell biology (Engelhardt *et al.*, 1995) (Fig. 4A and B). Briefly, freshly isolated primary airway epithelial cells on the second day postseeding (approximately 10% confluency) can be incubated in the presence of serum containing conditioned retroviral or lentiviral producer supernatant (10 ml of retroviral supernatant per 100-mm plate of primary cells) for 2 h in the presence of Polybrene (2 μg/ml; Sigma). After each retroviral infection, the cells are washed twice with F12 medium before the addition of hormonally defined medium D. Cells can be infected up to three times on sequential days. Typically, viral titers of 1×10^6 colony-forming units/ml are capable of transducing primary airway cells at an efficiency of 10–30% after three serial infections. Serum-free concentrated stocks of vesicular stomatitis virus glycoprotein-pseudotyped retrovirus or lentivirus can also be used to achieve higher transduction efficiencies (our unpublished data).

5. Preparation and construction of xenograft cassettes: A xenograft cassette is assembled by connecting tubing and adapters fitted to the length of the donor rat trachea and secured with silk sutures as illustrated in Fig. 4C (materials and vendors are described in the caption to Fig. 4). Preassembled parts of the cassette are placed in a 100-mm tissue culture

plate, sealed in a sterilization pouch, and gas sterilized before anchoring tracheal xenografts.

6. Preparation of denuded rat tracheas:

   a. Rats are killed by CO_2 asphyxiation and pinned to a Styrofoam bed. Tracheas are excised from the pharynx to the carina under sterile conditions, using 70% ethanol to clean the site of incision. Place the tracheas in separate 2-ml screw cap tubes and place them on ice immediately after excision.

   b. After all tracheas have been harvested, denude the tracheas of all viable epithelium by freeze–thawing them three times: freezing at −80° and thawing at room temperature.

   c. After the third round of freeze–thawing, clean the tracheas of excessive fat and cut them to size (typically from the first to thirteenth cartilage ring). Rinse each tracheal lumen with 10 ml of precooled modified Eagle's medium (MEM).

   d. Pair tracheas of similar length in the same tube. The denuded tracheas can be stored at −80° for prolonged periods of time before transplantation.

7. Seeding tracheas with primary airway epithelial cells:

   a. Airway epithelial cells cultured in 100-mm dishes (at ∼80% confluence) are harvested with trypsin and trypsin inhibitor as described earlier, resuspended in medium D at a concentration of $1–2 \times 10^6$ cells per 20 μl, and kept on ice.

   b. Ligate the rat tracheas to the adaptor e attached to tubing b (as shown in Fig. 4C) with securely tied triple-knotted sutures, and loop the sutures around the tubing and trachea a total of three times.

   c. Inject 20–25 μl of medium containing $1–2 \times 10^6$ cells into the open end of the rat trachea, using a micropipettor under sterile conditions. Insert the pipette tip as deeply as possible into the trachea and slowly withdraw the pipette as cells are injected into the lumen of the trachea (taking care not to allow cells to leak out).

   d. Ligate the remaining open end of the rat trachea to adapter e' attached to tubing c as shown in Fig. 4C. Secure the length of the rat trachea by stretching it to physiologic length and clamping the tubing b and a with a hemostat. Tie the remaining two sutures as shown in Fig. 4C to secure the tracheas to adapter e' and adapter e' to tubing d.

   e. Place the cell-seeded xenograft cassettes into a 100-mm tissue culture dish with 1–2 ml of medium D overlaid on top of the trachea to keep it in moist.

   f. Incubate the cassette in a 5% CO_2 incubator at 37° for 1–2 h to equilibrate the pH before proceeding to transplantation.

   g. For transport, use a small, airtight, sterile container equilibrated in the incubator with the dishes of xenografts to maintain CO_2 before transplantation.

8. Transplanting xenografts into *nu/nu* mice:

   a. Anesthetize male *nu/nu* athymic mice (Harlan, Indianapolis, IN) by intraperitoneal injection of ketamine (80 mg/kg) and xylazine (10 mg/kg) in sterile PBS.

   b. Place the mice on a sterile drape after they are anesthetized. Clean the surgical incision sites with povidone-iodine followed by 70% ethanol. Make four incisions as shown in Fig. 4D (center panel), two small incisions on the neck of the mouse (~0.16 cm) with just enough width to pass the tubing, and two larger incisions on the flanks of the mouse (~1.0 cm).

   c. Separate the skin from the muscle by blunt dissection and place the xenograft cassette subcutaneously by tunneling the distal end of cassette (tubing *c*) under the skin toward the tail of the mouse, and the proximal end of the cassette (tubing *a*) out of the small incision behind the neck, using forceps to guide the xenograft tubing (Fig. 4D).

   d. After transplantation, use two or three staples to close each of the largest incisions and an additional staple to anchor each xenograft to the skin at the loop of tubing *c*. No staples are necessary for the exit port of tubing *a*. Then transfer the mouse to a sterile cage and monitor until it is awake.

   e. Irrigate the xenografts weekly with 1 ml of Ham's F12 medium, using a Surflo winged infusion set (Fisher Scientific) with an 0.75-in., 21-gauge needle to remove excess secretions in the first 3 weeks after transplantation. Beginning 3 weeks posttransplantation, irrigate the xenografts twice per week. Stratified xenograft epithelium usually is reconstituted by 4–6 weeks after transplantation (Engelhardt *et al.*, 1991, 1992, 1995).

Medium A

Modified Eagle's medium (MEM), Ca^{2+} and Mg^{2+} free
Penicillin (100 U/ml)
Streptomycin (100 μg/ml)
Tobramycin (80 μg/ml)
Ceftazidime (100 μg/ml)
Primaxin (100 μg/ml)
Amphotericin B (5.0 μg/ml)
Deoxyribonuclease I [DNase I (100 μg/ml), cat. no. DN25; Sigma]
Pronase (1.5 mg/ml) (cat. no. 1459643; Roche)
Make fresh and keep at 4°.

Medium B

5% FBS in Dulbecco's modified Eagle's medium (DMEM)–Ham's
    F12 (1:1 ratio)
1% MEM nonessential amino acids solution
Penicillin (100 U/ml)
Streptomycin (100 μg/ml)
Tobramycin (80 μg/ml)
Ceftazidime (100 μg/ml)
Primaxin (100 μg/ml)
Amphotericin B (5.0 μg/ml)

Medium C

DMEM–Ham's F12 (1:1 ratio)
15 mM HEPES
3.6 mM Na$_2$CO$_3$
Penicillin (100 U/ml)
Streptomycin (100 μg/ml)
Tobramycin (80 μg/ml)
Ceftazidime (100 μg/ml)
Primaxin (100 μg/ml)
Amphotericin B (0.25 μg/ml)
Clonetics BEGM SingleQuots kit (Cambrex Bio Science Walkersville,
    Walkersville, MD)

Combine all components of the SingleQuots kit with the preceding in-
gredients, resulting in the following final concentrations: insulin (10.0 μg/ml),
cholera toxin (1.0 μg/ml), bovine pituitary extract (40 μg/ml), human epidermal growth factor (hEGF, 1.0 μg/ml), epinephrine (1.0 μg/ml), transferrin
(20.0 μg/ml), and retinoic acid (0.0001 μg/ml).

Medium D

BEGM (bronchial epithelial growth medium)
BEGM SingleQuots kit
Penicillin (50 U/ml)
Streptomycin (50 μg/ml)
Tobramycin (40 μg/ml)
Ceftazidime (50 μg/ml)
Primaxin (50 μg/ml)

Trypsin Inhibitor Buffer

Trypsin inhibitor type I-S: from soybean (cat. no. T-6522; Sigma)
Ham's F12 medium

Combine trypsin inhibitor with Ham's F12 medium at 1 mg/ml. Filter through a 0.22-μm pore size filter, aliquot, and store at $-20°$.

In Vitro Colony-Forming Efficiency Assay to Characterize Stem Cell Populations in Conducting Airway Epithelium

Overview

Adult airway stem cells have the ability to generate highly proliferative TA cells by asymmetric cell division. This feature allows stem cells to undergo multiple rounds of clonal proliferation in the setting of injury/repair that can be measured using an *in vitro* colony-forming efficiency (CFE) assay in air–liquid interface (ALI) airway epithelial cell cultures (Randell, 1992) (Fig. 5). This CFE assay is similar to that used in the tracheal xenograft model with recombinant retroviral markers to assess clonal expansion of airway stem/progenitor cells (Engelhardt *et al.*, 1995; Zepeda *et al.*, 1995), but with the added advantage of being easier to carry out. Disadvantages of this *in vitro* CFE assay include less efficient differentiation of ciliated and goblet cell types in the ALI culture as compared with the tracheal xenograft method. Hence, this *in vitro* CFE assay has been used primarily to assess the proliferative potential of stem/progenitor cell populations, and less frequently to study progenitor–progeny relationships in the airway epithelium.

Using this ALI culture system, Schoch and colleagues evaluated the clonal growth potential of murine tracheal epithelial cells, using CFE as an index. They defined a subset of basal cells in mouse tracheal epithelium with the capacity to generate large colonies in ALI culture, suggesting they are derived from stem or TA cells (Schoch *et al.*, 2004). They diluted a single-cell suspension of mouse tracheal epithelial cells derived from β-galactosidase-expressing ROSA26 mice into non-ROSA26 tracheal epithelial cells and placed this mixture into the ALI culture model. After 3 weeks of ALI culture, 5-bromo-4-chloro-3-indolyl-β-D-galactopyranoside (X-Gal) staining revealed ROSA26 LacZ-positive colonies within the polarized airway epithelium. They observed that 1.7% of the tracheal epithelial cells formed colonies of varying size, with 0.1% of the clones forming large colonies (Fig. 5C–E). This subset of cells with larger proliferative potential was suggested as a possible stem or early TA cell population. Previous studies from the same group have demonstrated that high keratin 5 (K5) promoter activity exists in specific niches in the mouse trachea that correspond to the location of BrdU-labeled LRCs, thought to be stem cells (Borthwick *et al.*, 2001). Transgenic mice harboring a K5 promoter-driven EGFP transgene expressed EGFP in most basal cells of

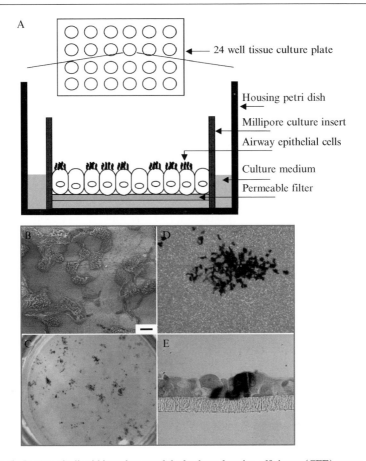

FIG. 5. *In vitro* air–liquid interface model of colony-forming efficiency (CFE) assays, using ROSA26 mouse tracheal epithelial cells. (A) Schematic view of the ALI culture model of primary airway epithelial cells growing on collagen-coated Millipore cell culture inserts housed in a 24-well tissue culture plate. (B) Scanning electron micrograph of fully differentiated mouse airway epithelial cultures (scale bar: 10 μm). (C) *En face* micrograph of X-Gal-stained ALI cultures seeded with a 1:50 mixture of ROSA26 to non-ROSA26 mouse tracheal epithelial cells. X-Gal-positive colonies of various sizes are seen. (D) Higher magnification of an individual X-Gal-positive colony from (C). (E) H&E-stained section of a small X-Gal-positive colony from (C).

the body, including a subset of tracheal basal cells apparently localized to positions similar to previously identified stem cell niches. Further studies using the same mouse model and FACS-facilitated purification of K5 promoter-driven EGFP-positive tracheal epithelial cells revealed an overall colony-forming efficiency 4.5-fold greater than that of EGFP-negative cells. In addition, these K5 promoter-driven EGFP-positive tracheal epithelial cells retained the ability to generate 12-fold larger colonies than EGFP-negative cells, consistent with the notion that K5-positive basal cells contain a proximal airway stem/progenitor cell subpopulation (Borthwick *et al.*, 2001; Schoch *et al.*, 2004). The following protocol presents the details of this CFE assay, using the mouse ALI culture model, and is based on the studies by Schoch and colleagues (2004).

In Vitro *Mouse Tracheal Epithelial Cell Colony-Forming Efficiency Assay Protocol*

1. Isolation of mouse tracheal epithelial cells: Four- to 8-week-old ROSA26 mice (B6.129S4-*Gt*(*ROSA*)*26*; Jackson Laboratory, Bar Harbor, ME) and non-ROSA26 mice (Harlan, Indianapolis, IN) are killed by CO_2 asphyxiation and pinned to a Styrofoam bed.

   a. Excise the tracheas from the pharynx to the bronchial main branches under sterile conditions and place the tracheas in ice-cold Ham's F12 medium containing penicillin (100 U/ml), streptomycin (100 μg/ml), and amphotericin B (Fungizone, 1.0 μg/ml) (Ham's F12 pen–strep–Fungizone). Isolate ROSA26 and non-ROSA26 tracheal epithelial cells separately.

   b. Open the tracheas longitudinally after the muscle and fat have been removed from the tracheas. Incubate the tracheas in the preceding Ham's F12 pen–strep–Fungizone medium containing Pronase (1.5 mg/ml; Roche Molecular Biochemicals, Indianapolis, IN) for 18–24 h at 4°, occasionally inverting the tube several times during dissociation (Davidson *et al.*, 2000; Liu *et al.*, 2006; You *et al.*, 2002).

   c. Add FBS to the digestion tube at a final concentration of 10%. Dissociate the tracheal epithelial cells by inverting the tube 10–20 times and transfer the cell suspension to a new tube. Wash the tracheas with Ham's F12 pen–strep–Fungizone medium twice and pool the cell suspensions. Collect the cells by centrifugation at 500*g* for 10 min at 4°.

   d. Resuspend the cell pellets in Ham's F12 medium containing penicillin (100 U/ml), streptomycin (100 μg/ml), Fungizone (1.0 μg/ml), crude pancreatic DNase I (0.5 mg/ml) (Sigma-Aldrich, St. Louis, MO), and bovine serum albumin (BSA, 10 mg/ml), and incubate them on ice for 5 min. Wash the cells with Ham's F12 pen–strep–Fungizone medium and centrifuge the cells at 500*g* for 5 min at 4°.

2. The cell pellet is then resuspended in Ham's F12 pen–strep–Fungizone medium with 10% FBS and cells are incubated in tissue culture plates

(Primaria; BD Biosciences Discovery Labware) for 2 h in 5% CO_2 at 37° to adhere fibroblasts. Nonadherent epithelial cells are collected by centrifugation and resuspended in 5% FBS–BEGM medium (Cambrex Bio Science Walkersville). Total cell yields are then counted with a hemacytometer. BEGM medium is made by adding reagents from one BEGM SingleQuots kit to 500 ml of 50% DMEM–50% Ham's F12 medium supplemented with 1% pen–strep, Fungizone (0.25 μg/ml), 15 mM HEPES, and 3.6 mM Na_2CO_3.

3. Preparation of ALI culture membranes: Primary airway epithelial cells will not attach to an ALI culture membrane support that is not coated with collagen. Use the following protocol to prepare ALI culture membranes.

    a. Mix 30 mg of human placental collagen type IV (cat. no. C-7521; Sigma) with 50 ml of deionized water and 100 μl of glacial acetic acid in a 100-ml sterile beaker.

    b. Cover the holding beaker with Parafilm and stir moderately at 37° until collagen strands are dissolved (~15–30 min).

    c. Dilute the collagen stock 1:10 with deionized sterile water and filter sterilize the solution through a 0.22-μm pore size membrane. The diluted sterile collagen stock (60 μg/ml) is the working solution for coating the plastic and membrane surfaces.

    d. Insert 12-mm-diameter Millicell-PCF membrane inserts (cat. no. PIHP 01250; Millipore, Bedford, MA) in a 24-well plate and apply 0.3–0.5 ml of collagen working solution on the apical surface of the inserts.

    e. Collagen coat the surface for a minimum of 18 h at room temperature. Remove the liquid collagen from the surface and air dry the membrane surface in a laminar flow hood. Once dried, rinse both sides of the membrane support three times with sterile PBS or DMEM to remove all traces of free collagen. Redry the membrane inserts in a laminar flow hood and store at 4° (they will be stable for several months).

4. Seeding cells on the membrane:

    a. Dilute an experimentally enriched subpopulation of ROSA26 tracheal epithelial cells into non-ROSA26 cells at an appropriate ratio. The ratio must be determined empirically to allow for outgrowth of isolated clones and will depend on the proliferative index of the cells being analyzed. Alternatively, one can also use retrovirally tagged cells expressing a detectable histochemical marker (i.e., EGFP, alkaline phosphatase, or β-galactosidase) for seeding onto membranes.

b. Seed a total of 1–2.5×10^5 cells/cm^2 in a 0.2-ml volume of 5% FBS–BEGM on the surface of a Millicell insert (about 1–1.5×10^5 cells per 12-mm insert). The volume should be sufficient to ensure a uniform distribution of cell settling on the membrane surface.

c. Apply 0.3–0.5 ml of 5% FBS–BEGM in the basal compartment of the insert to immerse the membrane without floating the insert; make sure that the membranes are level to ensure uniform cell attachment during the first 12 h.

d. Leave the plates containing the inserts undisturbed for a minimum of 18–24 h in an 8–9% CO_2 incubator at 37°. The higher CO_2 level has been shown to increases successful achievement of confluence (Karp et al., 2002).

5. Establishment of ALI culture:

a. The day after seeding, wash the membranes with prewarmed PBS to remove unattached cells and refeed them with fresh 5% FBS–BEGM. The medium is changed every 2 days until the transmembrane resistance (R_t) is greater than ~1000 Ω/cm^2 (this usually occurs 2–5 days after seeding). The electrical resistance across the membrane is measured with a Millicell-ERS system ohm meter (cat. no. MERS 000 01; Millipore).

b. Remove the medium inside the insert (apical) and refeed the outside (basal) chamber with BEGM without FBS to establish an ALI. Cells are refed with BEGM every 2–3 days. A polarized and differentiated airway epithelium will form ~10–15 days after moving the epithelium to an ALI.

6. X-Gal staining and evaluation: About 3 weeks (21 days) after seeding, the mouse ALI cultures should be ready for evaluation of CFE or other assays as appropriate. Depending on the purposes of the study, the membrane with polarized mouse tracheal epithelial cells can be either freshly embedded in O.C.T. compound for frozen sectioning, or directly fixed for en face immunostaining or X-Gal staining as previously described (Schoch et al., 2004). Figure 5C–E depicts X-Gal-stained clonal expansion of ROSA26 mouse tracheal epithelial progenitor cells in this model.

Models to Study Stem/Progenitor Cells of Airway Submucosal Glands

Overview

Airway submucosal glands (SMGs) are major secretory structures that reside in the cartilaginous airways of many mammalian species (Liu et al., 2004). SMGs play important roles in both innate immunity of the lung (Dajani et al., 2005) and cell biology of the proximal airways (Borthwick et al., 2001; Engelhardt, 2001; Engelhardt et al., 1995). As discussed in more detail in a previous section, SMGs serve as a protective niche for surface airway epithelial stem cells (Borthwick et al., 2001; Engelhardt, 2001). Furthermore, the pluripotent progenitor cells that exist in the human tracheobronchial surface airway epithelium (i.e., cells with the capacity to differentiate into ciliated, secretory, intermediate, and basal cells) also have a developmental capacity for SMGs (Engelhardt et al., 1995). Given that adult surface airway epithelial stem/progenitor cells have the capacity to develop SMGs, the biology that controls the morphogenesis of these structures in the airway is highly relevant to defining stem cell characteristics in the airway.

The development of three-dimensional culture substrates has enabled the creation of model systems to study certain features of SMG development, using a system that is far less complex than tracheal xenografts. Epithelial invasion of the extracellular matrix (ECM) is an important aspect of lung development and SMG morphogenesis. Infeld and colleagues initially developed an in vitro model for early airway gland development by culturing human tracheobronchial epithelial cells on a floating collagen gel substrate that contained fetal lung fibroblasts (Infeld et al., 1993). Similar studies by other groups have also demonstrated the capacity of human airway epithelial cells to invade a collagen gel matrix (Emura et al., 1996; Jacquot et al., 1994). Together with technical advances in the isolation and culture of mouse tracheal epithelial cells, we have adapted these protocols to evaluate bud formation, tubulogenesis, and branching morphogenesis of airway epithelia, using the collagen gel matrix model (Fig. 6). Although such models cannot reproduce the native cellular diversity founds in SMGs, they will be useful in defining fundamental epithelial–mesenchymal interactions required for airway gland morphogenesis.

In Vitro *Invasion and Tubulogenesis Assay Using Mouse Tracheal Epithelial Cells*

1. Collagen gel contraction:
   a. Primary mouse embryonic fibroblast cells (PMEFs): PMEFs are generated according to a previously described general protocol (Wassarman and DePamphilis, 1993). PMEFs generated from ICR or C57/BL6 mice can be used for this model with no differences in experimental outcome. Culture PMEFs in DMEM supplemented with 10% FBS. It is worth noting that early passages of PMEFs may give better results in gel contraction than later passage cells. However, for the most consistent results the same passages of PMEFs should always be used to generate contracted collagen cells.
   b. PMEFs are harvested by trypsinization, and the total cell number is quantified with a hemacytometer counting chamber.

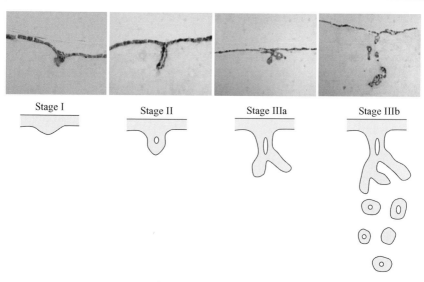

Fig. 6. Mouse tracheal epithelial cell tubulogenesis assay. Bud formation and tubulogenesis are two early events important in submucosal gland development and can be studied with a collagen gel matrix cultured at the air–liquid interface. Depicted here are various stages of gland-like structures formed from mouse tracheal epithelium (*top*) and schematic representations of various stages of tubulogenesis characteristic of early submucosal gland morphogenesis (*bottom*). (See color insert.)

c. To generate the collagen gel matrix support on which epithelial cells are seeded, add the following reagents, in order, for each 60-mm dish at room temperature: 3.5 ml of fibroblast culture medium, 1.8 ml of FBS, 0.3 ml of 0.1 N NaOH, 3–5 $\times$ 10^5 fibroblasts in 1.0 ml of medium, and 9.5 mg of rat tail collagen type I (BD Biosciences Discovery Labware) in 2.7 ml of H_2O to a final volume of 9.3 ml. Mix the solution gently by inverting the tube. The mixture should be light pink (if not, add 0.1 N NaOH to adjust the pH to turn the phenol red pink).

d. Apply ~9 ml of collagen gel to a 60-mm dish (or ~5 ml for a 35-mm dish) and incubate the gel at room temperature for 10 min before moving the dish into a 5% CO_2 incubator at 37° for 1 h. Add a small amount of fibroblast culture medium to the gel and run a sterile pipette around the outside rim of the gel to allow contraction. Continue incubation in 5% CO_2 at 37° and feed twice per week with fibroblast culture medium. The gel will be fully contracted in 2–4 weeks.

2. Isolation of mouse tracheal epithelial cells: Use the same protocol described in the previous section for *in vitro* CFE assays.

3. Mouse tracheal epithelial cell seeding:
   a. After gel contraction, fresh isolated or cultured mouse airway cells are resuspended in 5% FBS–BEGM at a density of 3–5 $\times$ 10^7 cells/ml. BEGM is made by adding reagents from one BEGM SingleQuot kit to 500 ml of 50% DMEM–50% Ham's F12 medium supplemented with 1% pen–strep, Fungizone (0.25 μg/ml), 15 mM HEPES, and 3.6 mM Na_2CO_3.
   b. Remove the medium from the fibroblast gel and apply 10 μl of the cell solution to the top of each gel while attempting to evenly spread the cells across the gel. Incubate the gels in 5% CO_2 at 37° for 30 min before adding a small amount of 5% FBS–BEGM around the edge of the gel. Continue incubation in 5% CO_2 at 37° overnight to allow adequate cell adhesion. Take care not to add so much medium that it covers the top of the gel and washes the cells off the surface.
   c. The following morning, add enough 5% FBS–BEGM to cover the top of the gel and let the gel sit for 3 days before placing the gels into 30-mm diameter Millicell-PCF membrane inserts (cat. no. PIHP 03012; Millipore) in a 6-well dish to establish an air–liquid interface (ALI). After the ALI is established, feed the cells from the bottom of the inserts with BEGM (without FBS). The apical side of the gel should be exposed to air. Bud formation and tubulogenesis of airway epithelial cells can be evaluated 1–2 weeks after establishment of the ALI (Fig. 6).

Acknowledgments

The authors gratefully acknowledge NIDDK research funding in the area of this review (DK47967) and the editorial assistance of Ms. Leah Williams.

References

Asmundsson, T., Kilburn, K. H., and McKenzie, W. N. (1973). Injury and metaplasia of airway cells due to SO_2. *Lab. Invest.* **29,** 41–53.

Berns, A. (2005). Stem cells for lung cancer? *Cell* **121,** 811–813.

Bigby, T. D., Allen, D., Leslie, C. G., Henson, P. M., and Cherniack, R. M. (1985). Bleomycin-induced lung injury in the rabbit: Analysis and correlation of bronchoalveolar lavage, morphometrics, and fibroblast stimulating activity. *Am. Rev. Respir. Dis.* **132,** 590–595.

Bishop, A. E. (2004). Pulmonary epithelial stem cells. *Cell Prolif.* **37,** 89–96.

Boers, J. E., Ambergen, A. W., and Thunnissen, F. B. (1998). Number and proliferation of basal and parabasal cells in normal human airway epithelium. *Am. J. Respir. Crit. Care Med.* **157,** 2000–2006.

Borges, M., Linnoila, R. I., van de Velde, H. J., Chen, H., Nelkin, B. D., Mabry, M., Baylin, S. B., and Ball, D. W. (1997). An achaete-scute homologue essential for neuroendocrine differentiation in the lung. *Nature* **386,** 852–855.

Borthwick, D. W., Shahbazian, M., Krantz, Q. T., Dorin, J. R., and Randell, S. H. (2001). Evidence for stem-cell niches in the tracheal epithelium. *Am. J. Respir. Cell. Mol. Biol.* **24,** 662–670.

Buckpitt, A., Buonarati, M., Avey, L. B., Chang, A. M., Morin, D., and Plopper, C. G. (1992). Relationship of cytochrome P450 activity to Clara cell cytotoxicity. II. Comparison of stereoselectivity of naphthalene epoxidation in lung and nasal mucosa of mouse, hamster, rat and rhesus monkey. *J. Pharmacol. Exp. Ther.* **261,** 364–372.

Castleman, W. L., Dungworth, D. L., and Tyler, W. S. (1975). Intrapulmonary airway morphology in three species of monkeys: A correlated scanning and transmission electron microscopic study. *Am. J. Anat.* **142,** 107–121.

Choi, H. K., Finkbeiner, W. E., and Widdicombe, J. H. (2000). A comparative study of mammalian tracheal mucous glands. *J. Anat.* **197,** 361–372.

Dajani, R., Zhang, Y., Taft, P. J., Travis, S. M., Starner, T. D., Olsen, A., Zabner, J., Welsh, M. J., and Engelhardt, J. F. (2005). Lysozyme secretion by submucosal glands protects the airway from bacterial infection. *Am. J. Respir. Cell. Mol. Biol.* **32,** 548–552.

Davidson, D. J., Kilanowski, F. M., Randell, S. H., Sheppard, D. N., and Dorin, J. R. (2000). A primary culture model of differentiated murine tracheal epithelium. *Am. J. Physiol. Lung Cell. Mol. Physiol.* **279,** L766–L778.

Duan, D., Sehgal, A., Yao, J., and Engelhardt, J. F. (1998). Lef1 transcription factor expression defines airway progenitor cell targets for *in utero* gene therapy of submucosal gland in cystic fibrosis. *Am. J. Respir. Cell. Mol. Biol.* **18,** 750–758.

Dubaybo, B. A., Crowell, L. A., and Thet, L. A. (1991). Changes in tissue fibronectin in elastase induced lung injury. *Cell Biol. Int. Rep.* **15,** 675–686.

Emura, M. (1997). Stem cells of the respiratory epithelium and their *in vitro* cultivation. *In Vitro Cell Dev. Biol. Anim.* **33,** 3–14.

Emura, M. (2002). Stem cells of the respiratory tract. *Paediatr. Respir. Rev.* **3,** 36–40.

Emura, M., Ochiai, A., and Hirohashi, S. (1996). *In vitro* reconstituted tissue as an alternative to human respiratory tract. *Toxicol. Lett.* **88,** 81–84.

Engelhardt, J. F. (2001). Stem cell niches in the mouse airway. *Am. J. Respir. Cell. Mol. Biol.* **24,** 649–652.

Engelhardt, J. F., Allen, E. D., and Wilson, J. M. (1991). Reconstitution of tracheal grafts with a genetically modified epithelium. *Proc. Natl. Acad. Sci. USA.* **88,** 11192–11196.

Engelhardt, J. F., Yankaskas, J. R., and Wilson, J. M. (1992). *In vivo* retroviral gene transfer into human bronchial epithelia of xenografts. *J. Clin. Invest.* **90,** 2598–2607.

Engelhardt, J. F., Schlossberg, H., Yankaskas, J. R., and Dudus, L. (1995). Progenitor cells of the adult human airway involved in submucosal gland development. *Development* **121,** 2031–2046.

Evans, M. J., Cabral-Anderson, L. J., Dekker, N. P., and Freeman, G. (1981). The effects of dietary antioxidants on NO_2-induced injury to type 1 alveolar cells. *Chest* **80,** 5–8.

Filali, M., Zhang, Y., Ritchie, T. C., and Engelhardt, J. F. (2002). Xenograft model of the CF airway. *Methods Mol. Med.* **70,** 537–550.

Giangreco, A., Reynolds, S. D., and Stripp, B. R. (2002). Terminal bronchioles harbor a unique airway stem cell population that localizes to the bronchoalveolar duct junction. *Am. J. Pathol.* **161,** 173–182.

Giangreco, A., Shen, H., Reynolds, S. D., and Stripp, B. R. (2004). Molecular phenotype of airway side population cells. *Am. J. Physiol. Lung Cell. Mol. Physiol.* **286,** L624–L630.

Griffiths, M. J., Bonnet, D., and Janes, S. M. (2005). Stem cells of the alveolar epithelium. *Lancet* **366,** 249–260.

Grove, J. E., Lutzko, C., Priller, J., Henegariu, O., Theise, N. D., Kohn, D. B., and Krause, D. S. (2002). Marrow-derived cells as vehicles for delivery of gene therapy to pulmonary epithelium. *Am. J. Respir. Cell. Mol. Biol.* **27,** 645–651.

Hansell, M. M., and Moretti, R. L. (1969). Ultrastructure of the mouse tracheal epithelium. *J. Morphol.* **128,** 159–169.

Hay, J., Shahzeidi, S., and Laurent, G. (1991). Mechanisms of bleomycin-induced lung damage. *Arch. Toxicol.* **65,** 81–94.

Hong, K. U., Reynolds, S. D., Giangreco, A., Hurley, C. M., and Stripp, B. R. (2001). Clara cell secretory protein-expressing cells of the airway neuroepithelial body microenvironment include a label-retaining subset and are critical for epithelial renewal after progenitor cell depletion. *Am. J. Respir. Cell. Mol. Biol.* **24,** 671–681.

Hong, K. U., Reynolds, S. D., Watkins, S., Fuchs, E., and Stripp, B. R. (2004a). Basal cells are a multipotent progenitor capable of renewing the bronchial epithelium. *Am. J. Pathol.* **164,** 577–588.

Hong, K. U., Reynolds, S. D., Watkins, S., Fuchs, E., and Stripp, B. R. (2004b). *In vivo* differentiation potential of tracheal basal cells: Evidence for multipotent and unipotent subpopulations. *Am. J. Physiol. Lung Cell. Mol. Physiol.* **286,** L643–L649.

Hook, G. E., Brody, A. R., Cameron, G. S., Jetten, A. M., Gilmore, L. B., and Nettesheim, P. (1987). Repopulation of denuded tracheas by Clara cells isolated from the lungs of rabbits. *Exp. Lung Res.* **12,** 311–329.

Inayama, Y., Hook, G. E., Brody, A. R., Cameron, G. S., Jetten, A. M., Gilmore, L. B., Gray, T., and Nettesheim, P. (1988). The differentiation potential of tracheal basal cells. *Lab. Invest.* **58,** 706–717.

Infeld, M. D., Brennan, J. A., and Davis, P. B. (1993). Human fetal lung fibroblasts promote invasion of extracellular matrix by normal human tracheobronchial epithelial cells *in vitro*: A model of early airway gland development. *Am. J. Respir. Cell. Mol. Biol.* **8,** 69–76.

Izbicki, G., Segel, M. J., Christensen, T. G., Conner, M. W., and Breuer, R. (2002). Time course of bleomycin-induced lung fibrosis. *Int. J. Exp. Pathol.* **83,** 111–119.

Jacquot, J., Spilmont, C., Burlet, H., Fuchey, C., Buisson, A. C., Tournier, J. M., Gaillard, D., and Puchelle, E. (1994). Glandular-like morphogenesis and secretory activity of human tracheal gland cells in a three-dimensional collagen gel matrix. *J. Cell. Physiol.* **161,** 407–418.

Januszkiewicz, A. J., and Mayorga, M. A. (1994). Nitrogen dioxide-induced acute lung injury in sheep. *Toxicology* **89,** 279–300.

Jeffery, P. K. (1983). Morphologic features of airway surface epithelial cells and glands. *Am. Rev. Respir. Dis.* **128,** S14–S20.

Karp, P. H., Moninger, T. O., Weber, S. P., Nesselhauf, T. S., Launspach, J. L., Zabner, J., and Welsh, M. J. (2002). An *in vitro* model of differentiated human airway epithelia: Methods for establishing primary cultures. *Methods Mol. Biol.* **188,** 115–137.

Kim, C. F., Jackson, E. L., Woolfenden, A. E., Lawrence, S., Babar, I., Vogel, S., Crowley, D., Bronson, R. T., and Jacks, T. (2005). Identification of bronchioalveolar stem cells in normal lung and lung cancer. *Cell* **121,** 823–835.

Lamb, D., and Reid, L. (1968). Mitotic rates, goblet cell increase and histochemical changes in mucus in rat bronchial epithelium during exposure to sulphur dioxide. *J. Pathol. Bacteriol.* **96,** 97–111.

Langley-Evans, S. C., Phillips, G. J., and Jackson, A. A. (1996). Sulphur dioxide: A potent glutathione depleting agent. *Comp. Biochem. Physiol. C Pharmacol. Toxicol. Endocrinol.* **114,** 89–98.

Liu, X., Driskell, R. R., and Engelhardt, J. F. (2004). Airway glandular development and stem cells. *Curr. Top. Dev. Biol.* **64,** 33–56.

Liu, X., Yan, Z., Luo, M., and Engelhardt, J. F. (2006). Species-specific differences in mouse and human airway epithelial biology of rAAV transduction. *Am. J. Respir. Cell. Mol. Biol.* **34,** 56–64.

Mason, R. J., Williams, M. C., Moses, H. L., Mohla, S., and Berberich, M. A. (1997). Stem cells in lung development, disease, and therapy. *Am. J. Respir. Cell. Mol. Biol.* **16,** 355–363.

Meulenbelt, J., Dormans, J. A., Marra, M., Rombout, P. J., and Sangster, B. (1992a). Rat model to investigate the treatment of acute nitrogen dioxide intoxication. *Hum. Exp. Toxicol.* **11,** 179–187.

Meulenbelt, J., van Bree, L., Dormans, J. A., Boink, A. B., and Sangster, B. (1992b). Biochemical and histological alterations in rats after acute nitrogen dioxide intoxication. *Hum. Exp. Toxicol.* **11,** 189–200.

Meulenbelt, J., van Bree, L., Dormans, J. A., Boink, A. B., and Sangster, B. (1994). Development of a rabbit model to investigate the effects of acute nitrogen dioxide intoxication. *Hum. Exp. Toxicol.* **13,** 749–758.

Neuringer, I. P., and Randell, S. H. (2004). Stem cells and repair of lung injuries. *Respir. Res.* **5,** 6.

Otto, W. R. (2002). Lung epithelial stem cells. *J. Pathol.* **197,** 527–535.

Pack, R. J., Al-Ugaily, L. H., Morris, G., and Widdicombe, J. G. (1980). The distribution and structure of cells in the tracheal epithelium of the mouse. *Cell Tissue Res.* **208,** 65–84.

Peake, J. L., Reynolds, S. D., Stripp, B. R., Stephens, K. E., and Pinkerton, K. E. (2000). Alteration of pulmonary neuroendocrine cells during epithelial repair of naphthalene-induced airway injury. *Am. J. Pathol.* **156,** 279–286.

Plopper, C. G., Hill, L. H., and Mariassy, A. T. (1980). Ultrastructure of the nonciliated bronchiolar epithelial (Clara) cell of mammalian lung. III. A study of man with comparison of 15 mammalian species. *Exp. Lung Res.* **1,** 171–180.

Plopper, C. G., Mariassy, A. T., Wilson, D. W., Alley, J. L., Nishio, S. J., and Nettesheim, P. (1983). Comparison of nonciliated tracheal epithelial cells in six mammalian species: Ultrastructure and population densities. *Exp. Lung Res.* **5,** 281–294.

Plopper, C. G., Weir, A. J., Nishio, S. J., Cranz, D. L., and St George, J. A. (1986). Tracheal submucosal gland development in the rhesus monkey, *Macaca mulatta*: Ultrastructure and histochemistry. *Anat. Embryol. (Berl.)* **174,** 167–178.

Presente, A., Sehgal, A., Dudus, L., and Engelhardt, J. F. (1997). Differentially regulated epithelial expression of an Eph family tyrosine kinase (fHek2) during tracheal surface airway and submucosal gland development. *Am. J. Respir. Cell. Mol. Biol.* **16,** 53–61.

Randell, S. H. (1992). Progenitor–progeny relationships in airway epithelium. *Chest* **101,** 11S–16S.

Ranga, V., and Kleinerman, J. (1981). A quantitative study of ciliary injury in the small airways of mice: The effects of nitrogen dioxide. *Exp. Lung Res.* **2,** 49–55.

Reddy, R., Buckley, S., Doerken, M., Barsky, L., Weinberg, K., Anderson, K. D., Warburton, D., and Driscoll, B. (2004). Isolation of a putative progenitor subpopulation of alveolar epithelial type 2 cells. *Am. J. Physiol. Lung Cell. Mol. Physiol.* **286,** L658–L667.

Reynolds, S. D., Giangreco, A., Power, J. H., and Stripp, B. R. (2000a). Neuroepithelial bodies of pulmonary airways serve as a reservoir of progenitor cells capable of epithelial regeneration. *Am. J. Pathol.* **156,** 269–278.

Reynolds, S. D., Hong, K. U., Giangreco, A., Mango, G. W., Guron, C., Morimoto, Y., and Stripp, B. R. (2000b). Conditional clara cell ablation reveals a self-renewing progenitor function of pulmonary neuroendocrine cells. *Am. J. Physiol. Lung Cell. Mol. Physiol.* **278,** L1256–L1263.

Reynolds, S. D., Giangreco, A., Hong, K. U., McGrath, K. E., Ortiz, L. A., and Stripp, B. R. (2004). Airway injury in lung disease pathophysiology: Selective depletion of airway stem and progenitor cell pools potentiates lung inflammation and alveolar dysfunction. *Am. J. Physiol. Lung Cell. Mol. Physiol.* **287,** L1256–L1265.

Schoch, K. G., Lori, A., Burns, K. A., Eldred, T., Olsen, J. C., and Randell, S. H. (2004). A subset of mouse tracheal epithelial basal cells generates large colonies *in vitro. Am. J. Physiol. Lung Cell. Mol. Physiol.* **286,** L631–L642.

Sehgal, A., Presente, A., and Engelhardt, J. F. (1996). Developmental expression patterns of CFTR in ferret tracheal surface airway and submucosal gland epithelia. *Am. J. Respir. Cell. Mol. Biol.* **15,** 122–131.

Smith, L. J. (1985). Hyperoxic lung injury: Biochemical, cellular, and morphologic characterization in the mouse. *J. Lab. Clin. Med.* **106,** 269–278.

Souma, T. (1987). The distribution and surface ultrastructure of airway epithelial cells in the rat lung: A scanning electron microscopic study. *Arch. Histol. Jpn.* **50,** 419–436.

Summer, R., Kotton, D. N., Sun, X., Ma, B., Fitzsimmons, K., and Fine, A. (2003). Side population cells and Bcrp1 expression in lung. *Am. J. Physiol. Lung Cell. Mol. Physiol.* **285,** L97–L104.

Suzuki, M., Machida, M., Adachi, K., Otabe, K., Sugimoto, T., Hayashi, M., and Awazu, S. (2000). Histopathological study of the effects of a single intratracheal instillation of surface active agents on lung in rats. *J. Toxicol. Sci.* **25,** 49–55.

Ten Have-Opbroek, A. A., and Plopper, C. G. (1992). Morphogenetic and functional activity of type II cells in early fetal rhesus monkey lungs: A comparison between primates and rodents. *Anat. Rec.* **234,** 93–104.

Theise, N. D., Henegariu, O., Grove, J., Jagirdar, J., Kao, P. N., Crawford, J. M., Badve, S., Saxena, R., and Krause, D. S. (2002). Radiation pneumonitis in mice: A severe injury model for pneumocyte engraftment from bone marrow. *Exp. Hematol.* **30,** 1333–1338.

Vacanti, M. P., Roy, A., Cortiella, J., Bonassar, L., and Vacanti, C. A. (2001). Identification and initial characterization of spore-like cells in adult mammals. *J. Cell. Biochem.* **80,** 455–460.

Van Winkle, L. S., Johnson, Z. A., Nishio, S. J., Brown, C. D., and Plopper, C. G. (1999). Early events in naphthalene-induced acute Clara cell toxicity: Comparison of membrane permeability and ultrastructure. *Am. J. Respir. Cell. Mol. Biol.* **21,** 44–53.

Wassarman, P. M., and DePamphilis, M. L. (1993). "Guide to Techniques in Mouse Development." Academic Press, San Diego, CA.

Watt, F. M., and Hogan, B. L. (2000). Out of Eden: Stem cells and their niches. *Science* **287,** 1427–1430.

West, J. A., Pakehham, G., Morin, D., Fleschner, C. A., Buckpitt, A. R., and Plopper, C. G. (2001). Inhaled naphthalene causes dose dependent Clara cell cytotoxicity in mice but not in rats. *Toxicol. Appl. Pharmacol.* **173,** 114–119.

Widdicombe, J. H., Chen, L. L., Sporer, H., Choi, H. K., Pecson, I. S., and Bastacky, S. J. (2001). Distribution of tracheal and laryngeal mucous glands in some rodents and the rabbit. *J. Anat.* **198,** 207–221.

You, Y., Richer, E. J., Huang, T., and Brody, S. L. (2002). Growth and differentiation of mouse tracheal epithelial cells: Selection of a proliferative population. *Am. J. Physiol. Lung Cell. Mol. Physiol.* **283,** L1315–L1321.

Zepeda, M. L., Chinoy, M. R., and Wilson, J. M. (1995). Characterization of stem cells in human airway capable of reconstituting a fully differentiated bronchial epithelium. *Somat. Cell Mol. Genet.* **21,** 61–73.

[13] Pancreatic Cells and Their Progenitors

By SETH J. SALPETER and YUVAL DOR

Abstract

Both type 1 and type 2 diabetes patients would greatly benefit from transplantation of insulin-producing pancreatic beta cells; however, a severe shortage of transplantable beta cells is a major current limitation in the use of such therapy. Understanding the mechanisms by which beta cells are naturally formed is therefore a central challenge for modern pancreas biology, in the hope that insights will be applicable for regenerative cell therapy strategies for diabetes. In particular, the cellular origins of pancreatic beta cells pose an important problem, with significant basic and therapeutic implications. This chapter discusses the current controversy regarding the identity of the cells that give rise to new beta cells in the adult mammal. Whereas numerous models suggest that beta cells can originate from adult stem cells, proposed to reside in the pancreas or in other locations, more recent work indicates that the major source for new beta cells during adult life is the proliferation of preexisting, differentiated beta cells. We present these different views, with emphasis on the methodologies employed. In particular, we focus on genetic lineage tracing using the Cre–*lox* system in transgenic mice, a technique considered the "gold standard" for addressing *in vivo* problems of cellular origins.

Introduction

The mystery of pancreatic beta cell origins poses a significant scientific question on several fronts. One pressing cause for investigation lies in the necessity to improve therapy for type I and type II diabetes, diseases that afflict 200 million people worldwide. In both diseases, insulin-producing beta cells, organized in clusters called the islets of Langerhans, fail to provide enough insulin to maintain glucose homeostasis. In type 1 diabetes, an autoimmune response leads to the destruction of beta cells. Type II diabetes, accounting for >90% of cases of diabetes, is typically associated with peripheral insulin resistance. However, evidence demonstrates that in addition to insulin resistance, this disease is also associated with defects in beta cell function and with a loss of as much as 50% of beta cell mass (Butler *et al.*, 2003). Thus, the possibility of regenerating or replacing beta cells offers immense therapeutic potential in both types of diabetes. Indeed, a clinical procedure for islet transplantation, developed at the

METHODS IN ENZYMOLOGY, VOL. 419   0076-6879/06 $35.00
Copyright 2006, Elsevier Inc. All rights reserved.   DOI: 10.1016/S0076-6879(06)19013-8

University of Alberta (Edmonton, AB, Canada; the Edmonton protocol), offers for the first time a satisfying cure for diabetes (Shapiro *et al.*, 2000). However, a major limitation of the Edmonton protocol is the severe shortage of donor islets. As a result of this problem, significant efforts are directed toward the development of strategies for enhancement of beta cell mass, *in vivo* (regenerative therapy) or *in vitro* for transplantation (cell therapy). Such efforts will greatly benefit from a better understanding of the normal process by which new beta cells are formed, and in particular the identification of the cell type(s) capable of giving rise to new beta cells. The search for the cellular origins of beta cells, with emphasis on the methodologies employed, is the subject of this chapter.

Beyond clinical importance, understanding the origins of beta cells represents a basic challenge for developmental biology. Although long considered static, islets of Langerhans (composed mainly of beta cells and constituting the endocrine pancreas) are now appreciated as an active and developing organ that maintains the ability to respond to external stimuli (Bonner-Weir *et al.*, 1989) and continues to grow throughout life (Montanya *et al.*, 2000; Skau *et al.*, 2001). What is the basic mechanism by which this tissue is maintained? Do beta cells rely on a continuous supply of differentiated cells derived from adult stem cells, and if so, what are the characteristics of these stem cells? Or, is beta cell expansion and maintenance based on the regulated proliferation of existing, terminally differentiated beta cells? We start by providing a brief overview of the dynamics of pancreas and beta cell formation during embryonic development. Later sections describe current views on the origins of new beta cells during normal adult life, as well as after injury. Throughout the discussion, we highlight the importance of one methodology, namely genetic lineage tracing (Fig. 1), which has emerged as the "gold standard" for addressing problems of cellular origins, in biology in general and in the field of pancreas research in particular.

Pancreas Development

To properly understand beta cell origins in the adult pancreas we must first review the development of the pancreas in its early stages. The adult pancreas is composed of two almost independent organs. The exocrine pancreas, occupying >95% of tissue mass, is responsible for secreting digestive enzymes into the duodenum. This is achieved by a network of converging tubes, called pancreatic ducts, which collect the enzymes from clusters of acinar cells organized in acini. The endocrine pancreas is organized in clusters of 100–1000 cells called the islets of Langerhans, and is mainly responsible for the regulation of systemic glucose homeostasis.

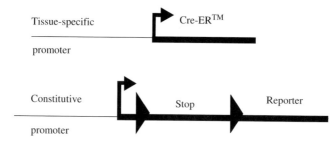

FIG. 1. Schematic of a lineage-tracing system using Cre–lox technology in transgenic mice. Tamoxifen-dependent Cre recombinase (Cre-ER) is expressed from a tissue-specific promoter (e.g., the insulin promoter driving expression specifically in beta cells). A second transgene expresses a reporter gene (e.g., EGFP or lacZ) from a constitutively active promoter. Expression of the reporter is blocked by a transcriptional stop sequence flanked by loxP sites. In the absence of tamoxifen, there is no expression of the reporter. Tamoxifen injection leads to transient activation of Cre-ER (lasting approximately 48 h), resulting in removal of the loxP-flanked stop sequence and activation of reporter expression. Recombined cells as well as their progeny will continue to express the reporter, allowing for retrospective identification of their origins. Specifically, one can conclude that cells expressing the reporter gene are the same cells or the progeny of cells that expressed Cre-ER at the time of tamoxifen injection.

Beta cells, composing 70% of islet cells, are responsible for accurate sensing of blood glucose levels and appropriate secretion of insulin to the bloodstream. Other, less abundant cell types in islets include alpha cells (producing the hormone glucagon, negating insulin action), delta cells (producing somatostatin), and PP cells (producing pancreatic polypeptide). There are about 1000 islets in the adult mouse and a total of ∼2,000,000 beta cells. An adult human pancreas contains about 2,000,000 islets.

We now know with certainty that all pancreatic cells derive from the embryonic primitive gut, a tissue derived from the endoderm. Careful embryological and molecular genetic studies have demonstrated that the pancreas develops from a small group of cells that bud from the gut tube, around embryonic day 9 in the mouse, that are marked by expression of the transcription factors pdx1 (pancreatic duodenal homeobox-1) and ptf1a/ p48 (p48 subunit of pancreas-specific transcription factor 1). The budding epithelium then undergoes a series of differentiation and morphogenesis steps, giving rise to the endocrine, hormone-producing cells, ducts, and acini, with the pancreas morphogenesis completing itself around birth (Edlund, 2001, 2002; Murtaugh and Melton, 2003; Slack, 1995). Particular attention was given to the elucidation of the molecular and cellular pathways by which beta cells are generated during development, in the hope that emerging principles will be applicable for the therapeutic generation of new beta cells, *in vitro* or *in vivo*. We describe some of these studies in relative detail because they provide an example of the methodological rigor essential in this field of research.

As mentioned previously, the earliest cells of the embryonic pancreas express the transcription factors pdx1 and ptf1a/p48. Moreover, mouse knockout studies demonstrated that these genes are essential for pancreas development, including beta cell development (Ahlgren *et al.*, 1996, 1998; Jonsson *et al.*, 1994; Kawaguchi *et al.*, 2002; Offield *et al.*, 1996). At first glance, this implies that adult beta cells at one time expressed pdx1 and ptf1a/p48. However, this is not necessarily the case: beta cells could have derived from mesenchymal or ectodermal cells, and still be dependent on the noncell autonomous expression of pdx1 and ptf1a/p48. To address this possibility, it is essential to genetically mark embryonic pdx1$^+$ and ptf1a/p48$^+$ cells, and to ask whether the direct progeny of these cells give rise to beta cells. Such lineage-tracing experiments are best performed with the Cre–*lox* system in transgenic mice, in which the expression of Cre recombinase in a particular cell type results in an indelible marking (usually the expression of a reporter gene) of this cell and its progeny (Branda and Dymecki, 2004; Danielian *et al.*, 1998; Hayashi and McMahon, 2002; Rossant and McMahon, 1999). Indeed, elegant studies have shown that beta cells are the progeny of cells that expressed pdx1 and ptf1a/p48 during early embryonic development (Kawaguchi *et al.*, 2002). These studies were based on the generation of transgenic mice expressing Cre recombinase under the control of either the pdx1 or ptf1a/p48 regulatory element. The role of pdx1 in beta cell development posed another challenge, because adult beta cells express pdx1. Therefore, simple expression of Cre from the *pdx1* promoter will result in the labeling of adult beta cells, regardless of their historical expression profile. To overcome this hurdle, Gu and colleagues used transgenic mice expressing a tamoxifen-inducible version of Cre recombinase under the control of the *pdx1* promoter (*pdx1-cre*-ER) (Gu *et al.*, 2002). Using such mice (together with a Cre reporter strain) they were able to demonstrate that pdx1-expressing cells in the early embryonic pancreas give rise to beta cells. Along similar lines, the role of the neurogenin-3 (ngn3) transcription factor was delineated. ngn3 is expressed transiently during embryonic development of the pancreas (in progenitor cells that do not express endocrine hormones such as insulin and glucagons), and in its absence the endocrine pancreas (islets of Langerhans) is not formed (Gradwohl *et al.*, 2000). Again, it required the indelible labeling of ngn3$^+$ cells and their progeny, using the Cre–*lox* system, to demonstrate that beta cells are derived from ngn3$^+$ cells (Gu *et al.*, 2002).

Another pioneering study on the cellular origins of beta cells has focused on a particularly interesting population of cells appearing early during pancreas development, expressing both insulin and glucagon (Herrera, 2000; Herrera *et al.*, 1994). From the initial observations of such cells it was concluded that insulin$^+$glucagon$^+$ "double hormone-positive" cells are the progenitors for both beta and alpha cells. This reasonable conclusion has important implications for the design of cell therapies in diabetes, because is suggests that efforts should be directed at increasing the numbers of insulin$^+$glucagon$^+$ cells, for example, during directed differentiation of embryonic stem cells. To critically test this notion, classic experiments by P. Herrera and colleagues used transgenic expression of diphtheria toxin to examine the fate of insulin$^+$ cells when glucagon$^+$ cells are eliminated and vice versa (Herrera *et al.*, 1994). Surprisingly, specific ablation of alpha cells (using glucagon–diphtheria toxin transgenic mice) generated embryos lacking alpha cells but contain nearly normal levels of beta cells. These experiments indicate that beta cells are not derived from insulin$^+$glucagon$^+$ cells. More recent lineage-tracing experiments, this time involving only "noninvasive" tagging of insulin$^+$ or glucagon$^+$ cells, support the view that adult beta cells have never expressed glucagon (Herrera, 2000).

As a result of these studies and others, we now know that pdx1$^+$ptf1a/ p48$^+$ cells of the embryonic pancreas give rise to all pancreatic lineages in a cell-autonomous manner (Kim and MacDonald, 2002). Some progeny of these cells express ngn3 transiently during late gestation, and give rise to islet cells including beta cells. Early double hormone-positive cells expressing both insulin and glucagon are largely considered developmental "dead ends," perhaps reflecting ancient stages in pancreas evolution but not participating in the main pathway for beta cell formation.

Origins of Beta Cells During Postnatal Life

After birth, the number of ngn3$^+$ progenitor cells declines almost to zero, indicating that new beta cells formed during postnatal life must be generated in a different pathway. What are the origins of postnatal beta cells? This seemingly simple question is the subject of current heated debate, not the least because of the potentially important practical implications for the design of regenerative therapies for diabetes. For example, if beta cells derive from stem cells residing in the spleen this almost immediately suggests exciting new approaches for the cure of diabetes. In the following sections we discuss some notable perceptions regarding the origins of adult beta cells, with emphasis on the methodologies employed.

Preexisting Beta Cells

Postnatal beta cells can proliferate, even though at a low rate, as demonstrated by the incorporation of BrdU or [^{3}H]thymidine (Kassem *et al.*, 2000;

Meier *et al.*, 2006; Messier and Leblond, 1960; Ritzel and Butler, 2003). Does this mean that beta cell replication is responsible for the generation of new beta cells? Not necessarily: beta cells could proliferate at a low rate, but still be derived mainly from stem cells. As described previously, lineage analysis has emerged as the gold standard for proving the origins of a mature cell. Therefore, a study was undertaken by Dor and colleagues to apply inducible lineage analysis to the adult beta cell, using a new method termed "genetic pulse–chase" (Dor *et al.*, 2004). Transgenic mice were generated in which tamoxifen-dependent Cre recombinase was placed downstream of the rat insulin promoter. This transgene was crossed with a *cre* reporter transgenic strain, expressing human placental alkaline phosphatase (HPAP) only after Cre-mediated removal of a *loxP*-flanked stop sequence. Adult mice were injected with tamoxifen to label existing, differentiated beta cells ("pulse"), and the fate of these labeled beta cells was monitored over time ("chase"). If beta cells derive from preexisting beta cells, the frequency of labeled beta cells should remain stable. If, however, dying beta cells are replaced by new beta cells derived from any type of stem cell (hence unlabeled), the frequency of labeled beta cells should decline with time. Strikingly, the frequency of labeled beta cells did not decline even 1 year after the original labeling. This led the authors to conclude that during adult life, new beta cells are largely the product of beta cell proliferation rather than stem cell differentiation. Similar experiments were performed with mice subject to partial pancreatectomy, a procedure known to result in some regeneration of beta cells. Here again, the results demonstrated that most new beta cells were the progeny of beta cells that existed before the operation. Importantly, this experimental system was designed to determine the major source of new beta cells during postnatal life. Stem cells could still exist and give rise to small numbers of beta cells. Such stem cells could even be recruited to action after specific injury conditions not tested so far. These are important considerations because even a small contribution of stem cells can in principle have significant biotechnological/therapeutic utility (Halban, 2004).

Other studies provide indirect support to the notion that beta cell proliferation is a dominant process in beta cell dynamics. For example, mice lacking the cell cycle components cyclin D2 or cyclin-dependent kinase 4 (CDK4) show a progressive failure of beta cell mass (Georgia and Bhushan, 2004; Kushner *et al.*, 2005; Rane *et al.*, 1999). Because these genes are expressed mainly in differentiated beta cells, this supports the primacy of beta cell proliferation. In addition, forced expression of the cyclin kinase inhibitor p27 in beta cells resulted in a progressive decline in beta cell mass, again suggesting that beta cell proliferation is essential (Uchida *et al.*, 2005).

Ducts

Pancreatic ducts are considered by many as a pool of adult stem cells constantly replenishing beta cells. This view is largely based on the fact that during embryonic development of the pancreas, primitive ducts give rise to ngn3[+] endocrine progenitors, which in turn differentiate into hormone-producing cells including beta cells (Kim and MacDonald, 2002).

According to one study, taking into account the rate of beta cell replication during the first month against the high amount of apoptosis occurring during this period, beta cell mass would drop dramatically if not for neogenesis (the formation of beta cells from stem cells) (Scaglia *et al.*, 1997). Therefore they concluded on the basis of their mathematical model that more than 30% of new beta cells at 31 days of age were not from replication of preexisting beta cells. The authors noted that the work of Dor *et al.* (2004) does not necessarily exclude the possibility of neogenesis in the 1-month-old pancreas, as the pulse–chase analysis was only undertaken starting at 6 weeks of age.

In addition to the "kinetic" argument, histological analyses of the adult pancreas, particularly in humans, show multiple islets adjacent to ducts, presumably "budding" from the ducts. Although it is tempting to extrapolate a dynamic process from such static observation, caution must be taken in the interpretation. For example, what appears to be an islet budding from a duct may in fact represent a mature islet, maintained by beta cell proliferation, which happened to reside adjacent to a duct (or even be derived from this duct earlier in its life) (Bouwens *et al.*, 1994). Similar to static histological observations, the expression in islets of genes typical of ducts cannot prove a lineal relationship between cell types. Finally, a number of *in vitro* studies have claimed that duct cultures can give rise to beta cells (Bonner-Weir *et al.*, 2000). However, the partial purity of these cultures and the possible "contamination" by endocrine cells prevent us from reaching a strong conclusion on the basis of such studies about the origins of beta cells (Gao *et al.*, 2005). Clearly, what is needed the most is a straightforward genetic lineage-tracing experiment, in which adult duct cells are labeled and their progeny are examined. Such experiments, although under way in many laboratories, have not been published so far.

Acini

Similar to the hypothesis of ductal contribution to beta cells, it has been suggested that acinar cells may undergo transdifferentiation and convert into beta cells in islets. However, evidence for this model is circumstantial; *in vivo* lineage-tracing experiments of the acinar pancreas have not so far provided a convincing demonstration of acinar to islet transdifferentiation. Interestingly, one study has used lineage tracing *in vitro* to label cultured acinar cells (by infecting cultured cells with an adenovirus expressing Cre recombinase from the elastase or amylase promoter) (Minami *et al.*, 2005). If verified and extended to the *in vivo* situation, this scenario will be of great interest.

Bone Marrow Cells

Studies have raised the possibility that bone marrow may contribute to islet mass both in the normal and diseased pancreas. In one notable study, bone marrow cells from insulin–green fluorescent protein (GFP) transgenic male mice were transplanted into irradiated female mice, and 2–3% of the recipient islets were reported to contain Y chromosome-positive, GFP-expressing beta cells (Ianus *et al.*, 2003). This elegantly designed experiment ruled out the possibility of fusion between marrow cells and beta cells as an explanation. However, more recent studies failed to repeat these experiments, casting doubts on the notion that bone marrow cells can give rise to beta cells (Lechner *et al.*, 2004; Mathews *et al.*, 2004). Interestingly, a similar study in a setting of regeneration from a diabetogenic injury has found that bone marrow cells do contribute to the formation of new beta cells (Hess *et al.*, 2003); however, the effect was strictly indirect: bone marrow cells were shown to differentiate into islet endothelial cells, which presumably enhanced the formation of new beta cells from other sources.

Another important study reported that in nonobese diabetic (NOD) mice, a model for autoimmune diabetes, neutralization of the autoimmune response resulted in dramatic regeneration of beta cells. The cellular origins of new beta cells in this case were claimed to be spleen cells that transdifferentiated to beta cells (Kodama *et al.*, 2003). However, more recent attempts by three independent groups to repeat this study failed to detect any contribution of spleen cells to beta cells (Chong *et al.*, 2006; Nishio *et al.*, 2006; Suri *et al.*, 2006). Interestingly, blockade of the autoimmune response did result in beta cell regeneration. Thus, the origins of new beta cells in this setting remain undefined, apart from the conclusion that they are not derived from marrow or spleen cells. Finally, one study of a human type 1 diabetes patient has found evidence for increased proliferation of beta cells, and not duct cells, in the early stages of the disease, strongly suggesting that beta cell regeneration in humans is initiated by beta cell proliferation rather than stem cell differentiation (Meier *et al.*, 2006).

Adult Stem Cells

Several studies suggested that beta cells can derive from stem cells residing in islets. An early report suggested that after administration of the beta cell-selective toxin streptozotocin, a subpopulation of cells expressing both somatostatin and pdx1 started to express insulin (Guz *et al.*, 2001). However, these studies did not demonstrate conclusively the stable formation of new beta cells. Other studies suggested that rare nestin[+] cells in islets could be adult islet stem or progenitor cells (Hunziker and Stein, 2000; Lechner *et al.*, 2002; Zulewski *et al.*, 2001). However, expression profiling and lineage-tracing experiments do not support this notion (Delacour *et al.*, 2004; Esni *et al.*, 2004; Means *et al.*, 2005; Selander and Edlund, 2002; Street *et al.*, 2004; Treutelaar *et al.*, 2003). More recently, it has been suggested that a small number of ngn3[+] embryonic endocrine progenitors remain during postnatal life, and that this population can be expanded and give rise to new beta cells under certain injury conditions (Gu *et al.*, 2002). Others have failed to detect such a contribution, leaving the issue open at this point (Lee *et al.*, 2006).

Dedifferentiation of Beta Cells

Perhaps most interestingly, it has been suggested that differentiated beta cells can undergo a process of epithelial to mesenchymal transition, giving rise to a proliferative population of insulin-negative cells capable in principle of redifferentiating into beta cells (Gershengorn *et al.*, 2004; Ouziel-Yahalom *et al.*, 2006). These studies are based on tissue culture experiments using human islets. Interestingly, *in vivo* evidence suggests that under certain conditions, beta cells can undergo significant phenotypic changes, including some that resemble epithelial to mesenchymal transition, as they proliferate (Kulkarni *et al.*, 2004). This proposal is exciting because it can potentially reconcile the view that beta cells originate from stem cells with the observation that beta cells derive from beta cells. Sophisticated lineage-tracing experiments will be needed to address this possibility. Because according to this model beta cells derive from cells that express insulin, simple irreversible labeling of beta cells cannot distinguish between beta cell proliferation and beta cell dedifferentiation, proliferation, and redifferentiation. Most importantly, it has not been shown so far that beta cells can lose insulin expression and proliferate.

Summary and Perspective

Insulin-producing beta cells are a key target for current efforts in regenerative medicine, as they could provide an effective cure for many diabetes patients. We believe that the cellular origins of beta cells must be identified for such efforts to be productive. It is clear that during embryonic development beta cells derive from progenitor cells expressing sequentially, among other genes, the transcription factors pdx1, ptf1a/p48, and ngn3. This understanding, based on robust lineage-tracing experiments, provides one conceptual framework in which to direct the differentiation of stem cells, be it embryonic stem cells (ESCs) or other types of embryonic cells, toward a beta cell fate. The cellular origins of new beta cells born postnatally are less clear. Multiple studies suggest that adult beta cells derive from stem cells residing in pancreatic ducts, acini, inside islets, or even in the bone marrow. However, these proposals are not supported so far by proper lineage-tracing evidence. On the contrary, lineage analysis and other approaches have suggested that most postnatal beta cells are formed by simple duplication of preexisting, differentiated beta cells. It remains to be demonstrated by lineage tracing if, under certain injury conditions, beta cells can derive from a pool of "facultative" stem cells. One particularly attractive suggestion is that beta cells may undergo, *in vitro* and *in vivo*, a process of transient dedifferentiation, perhaps via epithelial to mesenchymal transition.

Finally, all reliable evidence about the origins of beta cells has emerged from the use of transgenic mouse technology. It is acknowledged that human beta cells can in principle rely on a different mechanism for their renewal, or on a different balance between proliferation and stem cell differentiation. Performing lineage-tracing experiments on humans was long considered impossible. However, ingenious approaches suggest now that it will be possible one day to determine the life history of human beta cells and their origins (Frumkin *et al.*, 2005; Spalding *et al.*, 2005).

Methods: Design of a Lineage-Tracing Experiment in Mice

There are many approaches for lineage analysis, in mice and in other organisms, which use a wide variety of techniques. These include, among others:

• Direct labeling of particular cells with lipophilic dyes such as 1,1'-dioctadecyl-3,3,3',3'-tetramethylindocarbocyanine perchlorate (DiI), so that the fate of these cells and their immediate progenitors can be traced. An obvious disadvantage of this method is the 2-fold dilution of dye with every cell division.

• Labeling with retroviral infection, titrated such that each infection and integration event generates a unique identifiable mark. A major limitation of this method is the inability to prospectively control the identity of labeled cells.

• The injection of labeled cells into an embryo or adult animal. For example, the injection of labeled embryonic stem cells into a blastocyst generates a chimeric mouse. In such an animal, the contribution of injected cells to a particular tissue can be assessed. Cells can also be injected during later stages of development, as well as in adults. A commonly used example of the latter is the transplantation of labeled bone marrow or hematopoietic stem cells to an irradiated recipient. After engraftment of the cells, their contribution to specific cell lineages, in the blood system or in solid organs, can be assessed on the basis of marker expression. An alternative method for identification of donor cells, also applicable in humans, is the identification of Y chromosome by fluorescence *in situ* hybridization (FISH) when male cells are injected into female recipients.

These methods have provided numerous important insights into problems of cellular origins and tissue dynamics. However, the Cre–*lox* technology has become the gold standard for lineage-tracing experiments in mice, as it affords greater control over the temporal and spatial parameters of cell labeling, including a precise selection of the labeled cell population. Here we describe some considerations when using the Cre–*lox* system for addressing a question about a cell of origin, similar to problems discussed in the sections dealing with pancreatic beta cells.

Cre–lox System

The basic feature of the Cre–*lox* system, described in detail elsewhere (Branda and Dymecki, 2004; Nagy, 2000; Rossant and McMahon, 1999), is the ability of Cre recombinase to detect DNA sequences flanked by a specific 34-bp sequence called *loxP*. Cre activity removes the flanked sequence, and leaves a single *loxP* site in place. Importantly, Cre carries out this function without the need for any cofactors, allowing use of the system in mammalian cells. This technology is used mainly for modern tissue-specific knockouts of genes in mice, but the same tools can be used to delete a transcriptional stop sequence such that a reporter gene starts to be expressed after recombination. This "indelible labeling" forms the basis for Cre–*lox*-based lineage-tracing experiments. Such experiments typically involve the use of "double transgenic" mice, containing both a Cre-expressing transgene and a Cre reporter transgene.

Tissue Specificity of Cre Recombinase

A major strength of the Cre–*lox* system is the ability to select the cell type to be labeled based on its expression pattern, rather than its spatial location. Therefore, the system is best suited to address problems of the type "do cells expressing gene X contribute to cells of tissue Y?" This feature of the Cre–*lox* system relies on the use of tissue-specific promoters to drive expression of Cre. As a result, the usefulness of a particular Cre-expressing transgenic mouse is determined by the specificity of the promoter driving Cre expression: the more accurate the expression of Cre (in recapitulating expression of the endogenous gene) the better. In certain applications this is the key issue, for example, when attempting to label a rare cell type for which antibodies do not exist. In these cases, knocking-in Cre into the endogenous locus of the relevant gene is preferred. After such a manipulation, it is guaranteed that Cre expression will be subject to the same regulatory mechanisms as the endogenous gene. A good example for such an approach in the case of the pancreas was provided by Kawaguchi and colleagues, who knocked-in Cre into the ptf1a/p48 locus, and indeed were able to faithfully trace the fate of ptf1a/p48-expressing cells (Kawaguchi et al., 2002).

Temporal Control Over Cre Activity

A major advance in the Cre–lox system was introduced with the development of Cre variants whose activity depends on an injected ligand. The most popular version is Cre-ER, a fusion of Cre recombinase and a mutated, tamoxifen-responsive estrogen receptor (Danielian et al., 1998). This fusion protein resides in the cytoplasm, and can access DNA only in the presence of tamoxifen, which leads to a transient nuclear translocation. In transgenic mice that express Cre-ER, tamoxifen injection leads to a "pulse" of Cre activity, lasting about 48 h (Gu et al., 2003). The temporal control over recombination afforded by Cre-ER was used in the case of the pancreas to precisely define the fate of progenitor cells in the embryonic pancreas (Gu et al., 2003). In addition, pulse labeling of differentiated beta cells was used to demonstrate that new beta cells during adult life derive mainly from the proliferation of preexisting beta cells (Dor et al., 2004).

Cre Reporter

There are numerous transgenic strains that express a reporter gene (e.g., lacZ, EGFP, luciferase, and HPAP) depending on Cre activity. Reporters are designed such that only after Cre-mediated recombination will a "stop" cassette be removed, allowing for strong expression of the reporter. Some of the most popular reporters were inserted into the constitutively active ROSA26 locus, with a "lox-stop-lox" (LSL) cassette separating the reporter from the promoter (Novak et al., 2000). Other useful reporters include "Z/AP" and "Z/EG," where Cre-mediated recombination leads to removal of a lacZ gene, serving as a stop sequence, and the permanent activation of HPAP and EGFP, respectively (Lobe et al., 1999; Novak et al., 2000).

References

Ahlgren, U., Jonsson, J., and Edlund, H. (1996). The morphogenesis of the pancreatic mesenchyme is uncoupled from that of the pancreatic epithelium in IPF1/PDX1-deficient mice. *Development* **122,** 1409–1416.

Ahlgren, U., Jonsson, J., Jonsson, L., Simu, K., and Edlund, H. (1998). Beta-cell-specific inactivation of the mouse *Ipf1/Pdx1* gene results in loss of the beta-cell phenotype and maturity onset diabetes. *Genes Dev.* **12,** 1763–1768.

Bonner-Weir, S., Deery, D., Leahy, J. L., and Weir, G. C. (1989). Compensatory growth of pancreatic beta-cells in adult rats after short-term glucose infusion. *Diabetes* **38,** 49–53.

Bonner-Weir, S., Taneja, M., Weir, G. C., Tatarkiewicz, K., Song, K. H., Sharma, A., and O'Neil, J. J. (2000). *In vitro* cultivation of human islets from expanded ductal tissue. *Proc. Natl. Acad. Sci. USA* **97,** 7999–8004.

Bouwens, L., Wang, R. N., De Blay, E., Pipeleers, D. G., and Kloppel, G. (1994). Cytokeratins as markers of ductal cell differentiation and islet neogenesis in the neonatal rat pancreas. *Diabetes* **43,** 1279–1283.

Branda, C. S., and Dymecki, S. M. (2004). Talking about a revolution: The impact of site-specific recombinases on genetic analyses in mice. *Dev. Cell* **6,** 7–28.

Butler, A. E., Janson, J., Bonner-Weir, S., Ritzel, R., Rizza, R. A., and Butler, P. C. (2003). Beta-cell deficit and increased beta-cell apoptosis in humans with type 2 diabetes. *Diabetes* **52,** 102–110.

Chong, A. S., Shen, J., Tao, J., Yin, D., Kuznetsov, A., Hara, M., and Philipson, L. H. (2006). Reversal of diabetes in non-obese diabetic mice without spleen cell-derived beta cell regeneration. *Science* **311,** 1774–1775.

Danielian, P. S., Muccino, D., Rowitch, D. H., Michael, S. K., and McMahon, A. P. (1998). Modification of gene activity in mouse embryos *in utero* by a tamoxifen-inducible form of Cre recombinase. *Curr. Biol.* **8,** 1323–1326.

Delacour, A., Nepote, V., Trumpp, A., and Herrera, P. L. (2004). Nestin expression in pancreatic exocrine cell lineages. *Mech. Dev.* **121,** 3–14.

Dor, Y., Brown, J., Martinez, O. I., and Melton, D. A. (2004). Adult pancreatic beta-cells are formed by self-duplication rather than stem-cell differentiation. *Nature* **429,** 41–46.

Edlund, H. (2001). Developmental biology of the pancreas. *Diabetes* **50**(Suppl. 1), S5–S9.

Edlund, H. (2002). Pancreatic organogenesis: Developmental mechanisms and implications for therapy. *Nat. Rev. Genet.* **3,** 524–532.

Esni, F., Stoffers, D. A., Takeuchi, T., and Leach, S. D. (2004). Origin of exocrine pancreatic cells from nestin-positive precursors in developing mouse pancreas. *Mech. Dev.* **121,** 15–25.

Frumkin, D., Wasserstrom, A., Kaplan, S., Feige, U., and Shapiro, E. (2005). Genomic variability within an organism exposes its cell lineage tree. *PLoS Comput. Biol.* **1,** e50.

Gao, R., Ustinov, J., Korsgren, O., and Otonkoski, T. (2005). *In vitro* neogenesis of human islets reflects the plasticity of differentiated human pancreatic cells. *Diabetologia* **48,** 2296–2304.

Georgia, S., and Bhushan, A. (2004). Beta cell replication is the primary mechanism for maintaining postnatal beta cell mass. *J. Clin. Invest.* **114,** 963–968.

Gershengorn, M. C., Hardikar, A. A., Wei, C., Geras-Raaka, E., Marcus-Samuels, B., and Raaka, B. M. (2004). Epithelial-to-mesenchymal transition generates proliferative human islet precursor cells. *Science* **306,** 2261–2264.

Gradwohl, G., Dierich, A., LeMeur, M., and Guillemot, F. (2000). *Neurogenin3* is required for the development of the four endocrine cell lineages of the pancreas. *Proc. Natl. Acad. Sci. USA* **97,** 1607–1611.

Gu, G., Dubauskaite, J., and Melton, D. A. (2002). Direct evidence for the pancreatic lineage: NGN3[+] cells are islet progenitors and are distinct from duct progenitors. *Development* **129,** 2447–2457.

Gu, G., Brown, J. R., and Melton, D. A. (2003). Direct lineage tracing reveals the ontogeny of pancreatic cell fates during mouse embryogenesis. *Mech. Dev.* **120,** 35–43.

Guz, Y., Nasir, I., and Teitelman, G. (2001). Regeneration of pancreatic beta cells from intra-islet precursor cells in an experimental model of diabetes. *Endocrinology* **142,** 4956–4968.

Halban, P. A. (2004). Cellular sources of new pancreatic beta cells and therapeutic implications for regenerative medicine. *Nat. Cell Biol.* **6,** 1021–1025.

Hayashi, S., and McMahon, A. P. (2002). Efficient recombination in diverse tissues by a tamoxifen-inducible form of Cre: A tool for temporally regulated gene activation/inactivation in the mouse. *Dev. Biol.* **244,** 305–318.

Herrera, P. L. (2000). Adult insulin- and glucagon-producing cells differentiate from two independent cell lineages. *Development* **127,** 2317–2322.

Herrera, P. L., Huarte, J., Zufferey, R., Nichols, A., Mermillod, B., Philippe, J., Muniesa, P., Sanvito, F., Orci, L., and Vassalli, J. D. (1994). Ablation of islet endocrine cells by targeted expression of hormone-promoter-driven toxigenes. *Proc. Natl. Acad. Sci. USA* **91,** 12999–13003.

Hess, D., Li, L., Martin, M., Sakano, S., Hill, D., Strutt, B., Thyssen, S., Gray, D. A., and Bhatia, M. (2003). Bone marrow-derived stem cells initiate pancreatic regeneration. *Nat. Biotechnol.* **21,** 763–770.

Hunziker, E., and Stein, M. (2000). Nestin-expressing cells in the pancreatic islets of Langerhans. *Biochem. Biophys. Res. Commun.* **271,** 116–119.

Ianus, A., Holz, G. G., Theise, N. D., and Hussain, M. A. (2003). *In vivo* derivation of glucose-competent pancreatic endocrine cells from bone marrow without evidence of cell fusion. *J. Clin. Invest.* **111,** 843–850.

Jonsson, J., Carlsson, L., Edlund, T., and Edlund, H. (1994). Insulin-promoter-factor 1 is required for pancreas development in mice. *Nature* **371,** 606–609.

Kassem, S. A., Ariel, I., Thornton, P. S., Scheimberg, I., and Glaser, B. (2000). Beta-cell proliferation and apoptosis in the developing normal human pancreas and in hyperinsulinism of infancy. *Diabetes* **49,** 1325–1333.

Kawaguchi, Y., Cooper, B., Gannon, M., Ray, M., MacDonald, R. J., and Wright, C. V. (2002). The role of the transcriptional regulator Ptf1a in converting intestinal to pancreatic progenitors. *Nat. Genet.* **32,** 128–134.

Kim, S. K., and MacDonald, R. J. (2002). Signaling and transcriptional control of pancreatic organogenesis. *Curr. Opin. Genet. Dev.* **12,** 540–547.

Kodama, S., Kuhtreiber, W., Fujimura, S., Dale, E. A., and Faustman, D. L. (2003). Islet regeneration during the reversal of autoimmune diabetes in NOD mice. *Science* **302,** 1223–1227.

Kulkarni, R. N., Jhala, U. S., Winnay, J. N., Krajewski, S., Montminy, M., and Kahn, C. R. (2004). PDX-1 haploinsufficiency limits the compensatory islet hyperplasia that occurs in response to insulin resistance. *J. Clin. Invest.* **114,** 828–836.

Kushner, J. A., Ciemerych, M. A., Sicinska, E., Wartschow, L. M., Teta, M., Long, S. Y., Sicinski, P., and White, M. F. (2005). Cyclins D2 and D1 are essential for postnatal pancreatic beta-cell growth. *Mol. Cell. Biol.* **25,** 3752–3762.

Lechner, A., Leech, C. A., Abraham, E. J., Nolan, A. L., and Habener, J. F. (2002). Nestin-positive progenitor cells derived from adult human pancreatic islets of Langerhans contain side population (SP) cells defined by expression of the ABCG2 (BCRP1) ATP-binding cassette transporter. *Biochem. Biophys. Res. Commun.* **293,** 670–674.

Lechner, A., Yang, Y. G., Blacken, R. A., Wang, L., Nolan, A. L., and Habener, J. F. (2004). No evidence for significant transdifferentiation of bone marrow into pancreatic beta-cells *in vivo. Diabetes* **53,** 616–623.

Lee, C. S., De Leon, D. D., Kaestner, K. H., and Stoffers, D. A. (2006). Regeneration of pancreatic islets after partial pancreatectomy in mice does not involve the reactivation of neurogenin-3. *Diabetes* **55,** 269–272.

Lobe, C. G., Koop, K. E., Kreppner, W., Lomeli, H., Gertsenstein, M., and Nagy, A. (1999). Z/AP, a double reporter for cre-mediated recombination. *Dev. Biol.* **208,** 281–292.

Mathews, V., Hanson, P. T., Ford, E., Fujita, J., Polonsky, K. S., and Graubert, T. A. (2004). Recruitment of bone marrow-derived endothelial cells to sites of pancreatic beta-cell injury. *Diabetes* **53,** 91–98.

Means, A. L., Meszoely, I. M., Suzuki, K., Miyamoto, Y., Rustgi, A. K., Coffey, R. J., Jr., Wright, C. V., Stoffers, D. A., and Leach, S. D. (2005). Pancreatic epithelial plasticity mediated by acinar cell transdifferentiation and generation of nestin-positive intermediates. *Development* **132,** 3767–3776.

Meier, J. J., Lin, J. C., Butler, A. E., Galasso, R., Martinez, D. S., and Butler, P. C. (2006). Direct evidence of attempted beta cell regeneration in an 89-year-old patient with recent-onset type 1 diabetes. *Diabetologia* **49,** 1838–1844.

Messier, B., and Leblond, C. P. (1960). Cell proliferation and migration as revealed by radioautography after injection of thymidine-H3 into male rats and mice. *Am. J. Anat.* **106,** 247–285.

Minami, K., Okuno, M., Miyawaki, K., Okumachi, A., Ishizaki, K., Oyama, K., Kawaguchi, M., Ishizuka, N., Iwanaga, T., and Seino, S. (2005). Lineage tracing and characterization of insulin-secreting cells generated from adult pancreatic acinar cells. *Proc. Natl. Acad. Sci. USA* **102,** 15116–15121.

Montanya, E., Nacher, V., Biarnes, M., and Soler, J. (2000). Linear correlation between beta-cell mass and body weight throughout the lifespan in Lewis rats: Role of beta-cell hyperplasia and hypertrophy. *Diabetes* **49,** 1341–1346.

Murtaugh, L. C., and Melton, D. A. (2003). Genes, signals, and lineages in pancreas development. *Annu. Rev. Cell Dev. Biol.* **19,** 71–89.

Nagy, A. (2000). Cre recombinase: The universal reagent for genome tailoring. *Genesis* **26,** 99–109.

Nishio, J., Gaglia, J. L., Turvey, S. E., Campbell, C., Benoist, C., and Mathis, D. (2006). Islet recovery and reversal of murine type 1 diabetes in the absence of any infused spleen cell contribution. *Science* **311,** 1775–1778.

Novak, A., Guo, C., Yang, W., Nagy, A., and Lobe, C. G. (2000). Z/EG, a double reporter mouse line that expresses enhanced green fluorescent protein upon Cre–mediated excision. *Genesis* **28,** 147–155.

Offield, M. F., Jetton, T. L., Labosky, P. A., Ray, M., Stein, R. W., Magnuson, M. A., Hogan, B. L., and Wright, C. V. (1996). PDX-1 is required for pancreatic outgrowth and differentiation of the rostral duodenum. *Development* **122,** 983–995.

Ouziel-Yahalom, L., Zalzman, M., Anker-Kitai, L., Knoller, S., Bar, Y., Glandt, M., Herold, K., and Efrat, S. (2006). Expansion and redifferentiation of adult human pancreatic islet cells. *Biochem. Biophys. Res. Commun.* **341,** 291–298.

Rane, S. G., Dubus, P., Mettus, R. V., Galbreath, E. J., Boden, G., Reddy, E. P., and Barbacid, M. (1999). Loss of Cdk4 expression causes insulin-deficient diabetes and Cdk4 activation results in beta-islet cell hyperplasia. *Nat. Genet.* **22,** 44–52.

Ritzel, R. A., and Butler, P. C. (2003). Replication increases beta-cell vulnerability to human islet amyloid polypeptide-induced apoptosis. *Diabetes* **52,** 1701–1708.

Rossant, J., and McMahon, A. (1999). "Cre"-ating mouse mutants: A meeting review on conditional mouse genetics. *Genes Dev.* **13,** 142–145.

Scaglia, L., Cahill, C. J., Finegood, D. T., and Bonner-Weir, S. (1997). Apoptosis participates in the remodeling of the endocrine pancreas in the neonatal rat. *Endocrinology* **138,** 1736–1741.

Selander, L., and Edlund, H. (2002). Nestin is expressed in mesenchymal and not epithelial cells of the developing mouse pancreas. *Mech. Dev.* **113,** 189–192.

Shapiro, A. M., Lakey, J. R., Ryan, E. A., Korbutt, G. S., Toth, E., Warnock, G. L., Kneteman, N. M., and Rajotte, R. V. (2000). Islet transplantation in seven patients with type 1 diabetes mellitus using a glucocorticoid-free immunosuppressive regimen. *N. Engl. J. Med.* **343**, 230–238.

Skau, M., Pakkenberg, B., Buschard, K., and Bock, T. (2001). Linear correlation between the total islet mass and the volume-weighted mean islet volume. *Diabetes* **50**, 1763–1770.

Slack, J. M. (1995). Developmental biology of the pancreas. *Development* **121**, 1569–1580.

Spalding, K. L., Bhardwaj, R. D., Buchholz, B. A., Druid, H., and Frisen, J. (2005). Retrospective birth dating of cells in humans. *Cell* **122**, 133–143.

Street, C. N., Lakey, J. R., Seeberger, K., Helms, L., Rajotte, R. V., Shapiro, A. M., and Korbutt, G. S. (2004). Heterogenous expression of nestin in human pancreatic tissue precludes its use as an islet precursor marker. *J. Endocrinol.* **180**, 213–225.

Suri, A., Calderon, B., Esparza, T. J., Frederick, K., Bittner, P., and Unanue, E. R. (2006). Immunological reversal of autoimmune diabetes without hematopoietic replacement of beta cells. *Science* **311**, 1778–1780.

Treutelaar, M. K., Skidmore, J. M., Dias-Leme, C. L., Hara, M., Zhang, L., Simeone, D., Martin, D. M., and Burant, C. F. (2003). Nestin-lineage cells contribute to the microvasculature but not endocrine cells of the islet. *Diabetes* **52**, 2503–2512.

Uchida, T., Nakamura, T., Hashimoto, N., Matsuda, T., Kotani, K., Sakaue, H., Kido, Y., Hayashi, Y., Nakayama, K. I., White, M. F., and Kasuga, M. (2005). Deletion of Cdkn1b ameliorates hyperglycemia by maintaining compensatory hyperinsulinemia in diabetic mice. *Nat. Med.* **11**, 175–182.

Zulewski, H., Abraham, E. J., Gerlach, M. J., Daniel, P. B., Moritz, W., Muller, B., Vallejo, M., Thomas, M. K., and Habener, J. F. (2001). Multipotential nestin-positive stem cells isolated from adult pancreatic islets differentiate *ex vivo* into pancreatic endocrine, exocrine, and hepatic phenotypes. *Diabetes* **50**, 521–533.

[14] Intestinal Epithelial Stem Cells and Progenitors

By MATTHEW BJERKNES and HAZEL CHENG

Abstract

The adult intestinal epithelium contains a relatively simple, highly organized, and readily accessible stem cell system. Excellent methods exist for the isolation of intestinal epithelium from adults, and as a result collecting large quantities of intestinal stem and progenitor cells for study or culture and subsequent clinical applications should be routine. It is not, however, for two reasons: (1) adult intestinal epithelial cells rapidly initiate apoptosis on detachment from the basement membrane, and (2) *in vitro* conditions necessary for survival, proliferation, and differentiation are poorly understood. Thus to date the study of intestinal stem and progenitor cells has been largely dependent on *in vivo* approaches. We discuss existing

METHODS IN ENZYMOLOGY, VOL. 419
Copyright 2006, Elsevier Inc. All rights reserved.
0076-6879/06 $35.00
DOI: 10.1016/S0076-6879(06)19014-X

in vivo assays for stem and progenitor cell behavior as well as current methods for isolating and culturing the intestinal epithelium.

Introduction

The intestinal epithelium is the single layer of cells that lines the intestinal tract. It is the major intestinal barrier to the free flow of materials into and out of the body. Not only must the epithelium be selective, permitting or assisting the movement of some materials while excluding others, it must also orchestrate extensive digestive, immune, and homeo-static mechanisms allowing function while maintaining barrier integrity. Renewal of the epithelium occurs continuously at subcellular, cellular, and tissue levels in health and can be accelerated dramatically during injury-induced repair. Inhibition of these renewal processes results in death within a few days. Here we first briefly review current understanding of the stem and progenitor cell populations found within the epithelium and then focus on attempts to study these cell populations *in vitro*.

General Histology

The small intestinal epithelium is folded into a continuous series of villi, reminiscent of a mountain range rising up from the intestinal wall. In the valleys that encircle each villus are found the openings of deep cylindrical pits, the test tube shaped crypts of Lieberkühn (Fig. 1). The crypt and villus epithelia are contiguous but there are important differences in their detailed histology.

The villus epithelium is the major digestive agent of the epithelium. It is composed of three main differentiated cell types: columnar, mucous, and enteroendocrine cells. The most prevalent cells are the columnar cells (also known as absorptive cells or enterocytes) and they are involved in most intestinal functions. Next in frequency are the mucous cells (also known as goblet cells) that produce abundant mucus and its associated substances such as trefoil factors (Wong *et al.*, 1999). The enteroendocrine cell type is the least frequent, but contains multiple subtypes defined by the specific intestinal hormones produced (Schonhoff *et al.*, 2004), constituting in aggregate one of the largest populations of endocrine cells of the body. It seems likely that a similar functional diversity of the columnar and mucous cell types will be found as the expression patterns of new molecular markers are explored. Lesser known cell types such as the M cells (Gebert *et al.*, 1996; Neutra *et al.*, 1996), brush/tuft/caveolated cells (Nabeyama and Leblond, 1974), and cup cells (Madara, 1982) should also be mentioned, although little is known about their lineage relationships. Cells lost from the villus are replaced by a continuous flux of new cells supplied from the ring of adjoining crypts surrounding each villus (Fig. 1).

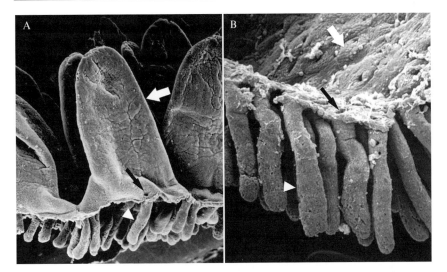

Fig. 1. Scanning electron micrographs of mouse intestinal epithelium stripped from underlying mesenchyme (taken from Magney et al., 1986). (A) Isolated small intestinal epithelium showing a test tube-shaped crypt (arrowhead) attached to the base of a villus (white arrow). The orifice of the crypt lumen is visible on the surface (black arrow). (B) Isolated colon epithelium. The lumenal surface of the colon is lined by the flat surface epithelium (white arrow); there are no villi. The black arrow indicates the surface epithelial orifice of the lumen of the crypt indicated by the white arrowhead.

All villus epithelial cells are normally made in the crypt. Accordingly, crypts contain columnar, mucous, and enteroendocrine cells in various stages of differentiation. In addition, the crypt epithelium contains a fourth major cell type, the Paneth cells, postmitotic cells that are usually found only in the crypt base. Paneth cells contribute to mucosal defense by producing a variety of antimicrobial substances including multiple defensins (Ouellette, 2005). The crypt top normally contains postmitotic maturing cells that are en route to the villus. They are derived from the actively proliferating cells found in the lower three-fourths of the crypt. The small intestinal stem cells, although multipotential, likely do not normally directly produce the principal mature cell types. Most output from the stem cell population feeds into the short-lived **Mix** progenitors (Bjerknes and Cheng, 1999). **Mix** then gives rise to a short-lived columnar progenitor, C_1, and a short-lived mucous progenitor, M_1 (probably via a Notch-dependent binary decision mechanism; Bjerknes and Cheng, 2005a), which then undergo multiple amplification divisions and feed into their respective columnar and mucous cell lineages. Similar committed progenitors also exist for the enteroendocrine and Paneth lineages (Bjerknes, 1985; Bjerknes and Cheng, 2006).

The crypts in colon are connected by a flat surface epithelium; there are no villi. The main cell types present in the distal colon are the columnar, vacuolated, mucous, and enteroendocrine cells (Colony, 1996). Vacuolated cells are largely restricted to the lower three-fourths of the crypt, columnar cells to the upper one-fourth and surface epithelium, and it has been proposed that the postmitotic columnar cells are derived from vacuolated cells, some of which are progenitors, migrating up from the crypt (Chang and Leblond, 1971a). The mucous and enteroendocrine cells are distributed throughout the crypt, but the enteroendocrine cells are more frequent in the crypt base and are postmitotic (Chang and Leblond, 1971b). Proliferative mucous cells are seen in the lower crypt and are thought to act as progenitors for the mucous lineage, although some may also be derived from immature vacuolated cells (Chang and Nadler, 1975). Caveolated and M cells are also present in the colon epithelium. Paneth cells are found in the crypt base of the proximal colon of mice and humans, but are not normally seen more distally. All cell lineages are ultimately derived from common multipotential stem cells, probably via a series of committed progenitors as in the small intestine, but the details are largely unknown. Species and regional subtleties exist that should be kept in mind when drawing comparisons.

Where Are the Stem Cells?

The most immature, undifferentiated cells in both small intestine (Cheng and Leblond, 1974) and colon (Chang and Leblond, 1971a; Lorenzsonn and Trier, 1968) are found in the crypt base. The stem cell populations of both regions are likely included among these undifferentiated crypt base cells (Bjerknes and Cheng, 2005a; Sancho et al., 2004). The precise location of small intestinal stem cells within the crypt base is contentious. In the classic model the crypt is thought of as a set of cell columns with Paneth cells in the bottom, proliferative cells in the middle, and differentiated cells in the top of the crypt. The stem cells are assumed to sit at the base of the proliferative column, just above the nonproliferative Paneth cells (e.g., Cairnie et al., 1965a,b). The most widely held version places the stem cells in a ring in cell positions 4–5, just above the Paneth cells; but in a more recent version of the model Marshman et al. (2002) propose that the stem cells sit immediately above the highest Paneth cell in their respective cell columns, ranging from cell positions 2–7 but on average in cell position 4 (Fig. 2). This postulated dependence on Paneth cell location seems unlikely; however, because small intestinal crypts occasionally do not have Paneth cells (Bjerknes, 1985), Paneth cell depletion has little untoward effect on cell renewal (Garabedian et al., 1997), and many mammalian species have no Paneth cells whatsoever. Furthermore, undifferentiated proliferative cells are

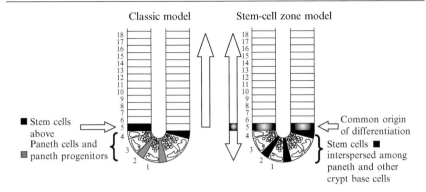

FIG. 2. Schematic diagram illustrating two models of crypt stem cell localization. The classic model proposes that stem cells (black) are located at the bottom of a column of proliferative cells, usually at cell position 4 or 5 of the crypt, just above the highest Paneth cell (cells containing secretory granules) in the column. Proliferative cells (gray) located in the crypt base among the postmitotic Paneth cells are Paneth cell progenitors. Stem cell progeny migrate upward in the crypt. In the stem cell zone model, stem cells (black) are located in the permissive microenvironment of about the first four or five cell positions of the crypt base. Stem cell progeny that leave the stem cell zone commence differentiation into the various epithelial cell lineages above the zone, in the common origin of differentiation in cell position 5 and above (gray with highlights). Most differentiating columnar, mucous, and enteroendocrine cells migrate upward, whereas all Paneth cells, and a few enteroendocrine, mucous, and columnar cells, migrate downward into the crypt base from the common origin of differentiation.

found among the Paneth cells (Cheng and Leblond, 1974), and these form the basis for our stem cell zone model (Bjerknes and Cheng, 1981a,b, 2005a). The stem cell zone model posits a stem cell-permissive microenvironment in about the first four or five cell positions; stem cells moving out of the zone are induced to differentiate, resulting in a common origin of differentiation for all lineages (Fig. 2). We found that the vast majority of cells arising in the common origin, including most of the columnar, mucous, and enteroendocrine cells, participate in an upward migration to populate the villus epithelium. However, Paneth cells and a few enteroendocrine, mucous, and columnar cells migrate downward into cell positions 1–4 from the common origin (Bjerknes and Cheng, 1981a,b). Furthermore, long-lived clones derived from multipotential progenitors (presumably stem cells) include crypt base columnar cells located in positions 1–5 (Bjerknes and Cheng, 1999; Fig. 3).

Assaying Intestinal Stem Cells, Progenitors, and Their Lineages *In Vivo*

It would be useful to measure the response of stem and progenitor cells to manipulation including changes in their numbers, molecular properties,

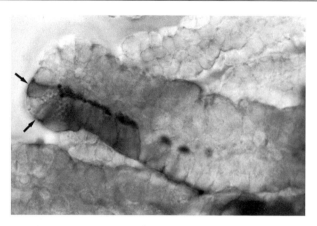

Fig. 3. A *Dlb-1*[+] clone (brown cells) in a crypt isolated from an SWR mouse and stained as outlined in the protocol on page 345. The clone continues onto the villus, where columnar and mucous cells were visible (data not shown). Crypt base columnar cells, potential stem cells, are visible (arrows; taken from Bjerknes and Cheng, 1999). (See color insert.)

proliferative activity, and the nature of their offspring. Unfortunately, at present there are few well-characterized molecular markers available with which to label and hence directly identify specific progenitors, and there are none for the stem cells. We discuss available *in vitro* tools later (p. 353), but as will be made clear, there are no useful *in vitro* stem or progenitor cell assays. Thus we are left with indirect *in vivo* assays that we now describe.

Regeneration Assays

Given a sufficient dose of radiation (or series of doses) all proliferating cells in the epithelium will be so injured that they undergo apoptosis or cell cycle arrest. As a result, the crypts empty out and within a few days most of the epithelium is gone (with the exception of some Paneth cells and a few other, ill-defined cells). At lower doses, progressively more epithelial cells survive and some of them are able to initiate regenerating colonies, recognized in sections after a few days as clusters of proliferating epithelial cells. With time these colonies grow into macroscopic nodules, many of which contain multiple cell types (Inoue *et al.*, 1988). If it is assumed that most of the colonies originated from a single surviving cell, then the observation of multiple cell types in the colonies demonstrates that crypts contain multipotential progenitors. However, it is far from clear that such colonies originate from a single surviving cell. The best evidence for clonality of the surviving colonies comes from an irradiation study using mice with X chromosome inactivation mosaicism. It was found that most colonies contained only one type of a marker gene, suggesting that the colonies were in fact derived from a single cell (Inoue *et al.*, 1988). Unfortunately, although this is suggestive, it does not prove a unicellular origin of the colonies because the sort of X chromosome inactivation mosaicism they relied on is known to give rise to patches of monoclonal crypts in adult mice (Griffiths *et al.*, 1988; Ponder *et al.*, 1985). Therefore, because most crypts, and in fact large patches of crypts, were already monoclonal before irradiation, the subsequent monoclonality of the regenerating nodules does not demonstrate origin from a single surviving clonogenic cell and hence a multicellular origin of the surviving colonies has not yet been excluded.

Similar nodules arise in the spleens of irradiated animals, and early attempts at the characterization of hematopoietic stem cells used these as the basis of the so-called spleen colony assay (Becker *et al.*, 1963; Siminovitch *et al.*, 1963; Till and McCulloch, 1961). Rough estimates of the frequency of clonogenic cells in bone marrow were obtained by fitting models of the survival probability to counts of the number of regenerating colonies after increasing doses of radiation. This approach was adapted to the intestinal epithelium by scoring the number of macroscopic nodules that had regenerated by 13 days after irradiation (Withers and Elkind, 1968, 1969). However, at higher doses many animals die before the nodules can be scored, motivating the introduction of the widely used microcolony assay, in which the number of microscopic colonies in tissue sections are scored 3.5 days after irradiation (Withers and Elkind, 1970). This change in timing has implications for the nature of the clonogenic cells being assayed. For comparison, it is now understood that in the spleen colony assay many of the colonies that arise early are actually derived from relatively short-lived progenitors and not the true hematopoietic stem cells (Weissman, 2002). This is a lesson to keep in mind when interpreting the results from clonogenic microcolony assays in the intestine. It should be noted that the estimates of the number of clonogenic cells obtained with these methods are heavily dependent on the specific survival model used to fit the data and have varied widely. The survival models are based on many strong assumptions, most of which are untested and some difficult to justify. Furthermore, the experimental measurements are not straightforward and contain many biases; some have been recognized and roughly compensated for whereas others are ignored. The potential difficulties underlying the approach are highlighted by work indicating the role of vascular elements in epithelial survival after irradiation (Paris *et al.*, 2001), further indicating that multiple, dosage-dependent mechanisms are operative (Ch'ang *et al.*, 2005). None of this is well understood, and hence conclusions about the number of clonogenic cells contained in a crypt derived with these methods should be considered with appropriate caution.

Detailed experimental advice (Potten and Hendry, 1985; Vidrich *et al.*, 2005) and examples of possible survival models (Hendry and Potten, 1974; Roberts *et al.*, 2003; Tucker and Thames, 1983) are available.

Similar regeneration assays have been used to study the response to various toxic chemicals, usually chemotherapeutic agents such as 5-fluorouracil. The interpretation of such studies involves difficulties similar to those outlined previously for radiation assays. We do not intend to suggest that such regeneration assays are not useful tools. On the contrary, radiation- and drug-induced damage of the intestine are important clinical issues. Comparison of survival curves generated with or without various doses of trophic or protective agents has frequently been used to study potential mitigating effects of these agents on radiation- or chemotherapy-induced damage (Houchen *et al.*, 1999; Farrell *et al.*, 2002; K. A. Kim *et al.*, 2005; Vidrich *et al.*, 2005). However, it should be noted that it is not possible solely on the basis of such studies to conclude that any observed effects indicate a change in stem cell numbers because there are too many alternate mechanisms that could impact survival in these assays.

Lineage-Tracing Assays

Stem and progenitor cell activity may be retrospectively assayed by investigating the cells they produce. For example, if it could be demonstrated that all the cells contained in a crypt, including all major cell types, were derived from a single precursor crypt cell then this would be evidence that the original cell was a stem cell. In this section we discuss various approaches that have been used to mark cells and follow their progeny.

Labeled Phagosomes

Cheng and Leblond (1974) found that crypt base columnar cells are radiosensitive and can be killed with relatively low doses of [^{3}H]thymidine. They are also actively phagocytic. Within a few hours of an injection of [^{3}H]thymidine some crypt base columnar cells die and are phagocytosed by neighboring viable crypt base columnar cells, producing a large radiolabeled phagosome in some crypt base columnar cells but not in any other cell type. Later, radiolabeled but more digested phagosomes appear in the other cell types. Because labeled phagosomes initially appeared in crypt base columnar cells and only later appeared in the four main cell types, they concluded that crypt base columnar cells had differentiated along each of the four cell lines. This was the first demonstration that the population of crypt base columnar cells includes members capable of generating each of the lineages. They further proposed that among the crypt base columnar cells are multipotential stem cells, each capable of generating the various epithelial lineages, their "unitarian theory of the origin of the four epithelial cell types," but it remained for later studies to prove their existence in the crypt.

Dlb-1 Mutation

The best evidence to date for the existence of multipotential stem cells capable of giving rise to all epithelial cell types comes from studies of the formation of genetically marked clones derived from single cells. Early studies used chimeric mice, chimeric either as a result of random X chromosome inactivation mosaicism or as a result of mixing blastomeres from different mouse strains. Crypts in the adult chimeric mice contained only a single marker. Hence it may be concluded that all cells in the crypt are derived from a single cell at some point during development, but no information is provided about the time of origin, type or location of the founder cell. More recent studies have used somatic mutation to generate genetically marked clones in the adult. These studies have shown definitively that adult crypts contain multipotential stem cells capable of giving rise to all cells in the crypt.

The $Dlb-1$ locus in mice generates an intestinal binding site for the lectin $Dolichos$ $biflorus$ agglutinin. SWR mice are $Dlb-1^{-/-}$, whereas C57BL/6 mice are $Dlb-1^{+/+}$. F_1 crosses of these strains yield heterozygous $Dlb-1^{+/-}$ mice. Mutagens such as N-nitroso-N-ethylurea (NEU) induce somatic mutations in the single $Dlb-1$ allele in random F_1 intestinal epithelial cells, yielding, if the affected cell is a progenitor, an unstained clone of cells. Using this approach, Winton and Ponder (1990) showed that a single mutated cell can give rise to all cells in a crypt in adult mice. Unfortunately, little can be said about the nature or location of the stem cell directly from their study. However, in combination with the results from the labeled phagosome study (p. 344), it supports the conclusion that there are multipotential stem cells among the crypt base columnar cells.

F_1 $Dlb-1^{+/-}$ mice yield unstained mutant cells on a background of stained normal cells, which make it difficult to identify small clones in such preparations. Positive clones on a negative background would be preferable for detailed lineage-tracing studies. This can be achieved with $Dlb-1^{-/-}$ SWR mice. NEU treatment results in positively staining clones (Bjerknes and Cheng, 1999). We have found that use of isolated crypt–villus units, rather than the usual intestinal whole mounts or sections, allows entire clones to be visualized at single-cell resolution in the crypt (Fig. 3).

Protocol

1. Induction of clones: NEU is mutagenic and appropriate precautions should be taken. Prepare a 20-mg/ml NEU solution by dissolving 100 mg of

NEU in 0.5 ml of dimethyl sulfoxide (DMSO) and then dilute with 4.5 ml of sterile phosphate-buffered saline (PBS). This should be prepared immediately before use because NEU has a short half-life in aqueous solution. SWR mice are given an intraperitoneal injection (250 mg/kg) of NEU to induce random somatic mutations, some of which involve the *Dlb-1* locus in occasional progenitors or stem cells. The animals will be killed days to weeks later, appropriate to the progenitor class under study.

2. Intestinal epithelial isolation: Various epithelial isolation protocols (we prefer the perfusion method) are outlined on page 369. Isolate the epithelium directly into a tube of cold 2.5% glutaraldehyde in PBS and fix on ice for 10 min. Wash the epithelium with three changes of cold PBS over 2 h and keep at 4° until use.

3. Processing for lectin *Dolichos biflorus* agglutinin (DBA) staining:

   a. The epithelium is first treated for 30 min at room temperature with sodium borohydride (0.5 mg/ml in PBS) to reduce unreacted aldehyde groups in the tissue. This helps reduce nonspecific background staining. The reaction generates gas that creates bubbles that cause the epithelium to float. This may be dealt with by placing the tube under vacuum until the tissue sinks to the bottom of the tube and can be further processed.

   b. Three PBS washes, 5 min each.

   c. Incubate in 20 mM D,L-dithiothreitol (DTT) and 20% ethanol in 150 mM Tris (pH 8.2) for 45 min at room temperature to strip surface mucus.

   d. Three PBS washes, 5 min each.

   e. To block endogenous peroxidase activity, incubate the tissue in 0.1% phenylhydrazine-HCl in PBS for 30 min at room temperature.

   f. Three PBS washes, 5 min each.

   g. Rinse the tissue for 10 min at room temperature in 0.5% bovine serum albumin (BSA) in PBS.

   h. Incubate the tissue with rotation overnight at room temperature with peroxidase-conjugated DBA (5 μg/ml in 0.5% BSA in PBS) in Sigmacote (Sigma, St. Louis, MO)-coated tubes (to minimize adherence of the tissue to the tube).

   i. The next day, process the tissue by a peroxidase-based histochemical staining procedure.

   j. Microdissect crypt–villus units under a stereomicroscope and mount them on slides. Screen for clones, using a ×100 oil immersion objective. It is necessary to work at high magnification for reliable identification of small clones and discrimination of neighboring cells during cell counts, and identification of specific cell types. In addition, high magnification helps differentiate between epithelial and lymphoid cells that also bind the lectin in SWR mice. An oil immersion condenser, if available, further improves resolution.

The assay can be adapted to identify clones derived from either stem cells or the various committed progenitors. Combining the clone marker with cell type-specific markers and differential interference contrast microscopy allows determination of detailed lineage progression (Bjerknes and Cheng, 2006). The assay can also be used to study the response of stem cells and progenitors to experimental manipulation (Bjerknes and Cheng, 2001). While the *Dlb-1* marker is useful for the study of small intestine, it displays variegated expression in colon and gastric epithelium and hence is of limited use in those regions. It also must be kept in mind that the NEU used to induce clones also has the potential to induce random mutations throughout the genome. Such additional mutations have the potential to alter the behavior of the resulting clones. Therefore it is important to be cautious when interpreting rare clones with unusual features. Furthermore, some cells are killed as a result of the NEU treatment and this may perturb normal cell behavior in the first day or two.

X-Linked Markers

The glucose-6-phosphate dehydrogenase (G6PD) gene, located on the X chromosome, can also be used as a lineage marker. Males possess only a single allele, and therefore loss-of-function mutations in that allele can be used to mark progenitor clones. The progeny of mutant cells appear unstained on a stained background, so visualization of small clones is difficult. The marker has been used primarily to study the long-term evolution of stem cell clones by observing the time course of the appearance of unstained crypts after mutagen treatment, again demonstrating that all epithelial cell types derive from a single stem cell in the adult colon and small intestine (Griffiths *et al.*, 1988; Park *et al.*, 1995). X-linked mosaicism of G6PD (and many other similar genes) has also been widely used as a lineage marker in females (human and mice), but the power of such assays for lineage tracing is limited because of the large patches of monoclonal crypts formed during the early developmental stage.

β-Galactosidase (lacZ) Mutation

The ROSA26 mouse strain has one copy of the bacterial gene *lacZ* inserted into chromosome 6. The associated mouse promoter drives β-galactosidase expression in all adult tissues. Mice hemizygous for *lacZ* contain only one transgene allele. Thus cells incurring inactivating mutations in the single *lacZ* allele lose β-galactosidase activity and will not stain for the enzyme. If the mutated cell is a progenitor, all its progeny inherit the nonfunctional allele and hence will also be unstained. NEU was used to mutate single stem cells, which then gave rise to all other cell types in the crypt, confirming the existence of multipotential stem cells in small intestinal, colon, and gastric epithelium (Bjerknes and Cheng, 2002).

O-Acetylated Sialoglycoproteins

Human colon epithelial cells usually contain O-acetylated sialomucins and hence do not stain by techniques such as mild periodic acid–Schiff (mPAS) that selectively stain non-O-acetylated forms. However, in about 9% of colons intense staining of the epithelium is observed (Sugihara and Jass, 1986). It is thought that this difference is due to expression of a single polymorphic autosomal gene responsible for the O-acetylation. Thus individuals homozygous or heterozygous for the active form have mPAS-negative epithelium whereas individuals homozygous for the inactive form have mPAS-positive epithelium. Relevant here is the appearance in some, presumably heterozygous and hence mPAS-negative, individuals of occasional mPAS-positive crypts, the frequency of which is increased after radiation exposure, demonstrating the existence of common stem cells in human colon, and also giving an estimate of the time required for clonal purification of a crypt (Campbell et al., 1996). The frequency of such mutant crypts was used to demonstrate an increased rate of stem cell mutation in ulcerative colitis (Okayasu et al., 2002).

Methylation

Gene methylation patterns have also been used as a lineage marker that can be used with human tissue (Yatabe et al., 2001). The methylation status of CpG sites of unexpressed genes can be inherited by the offspring of a mitosis. Therefore the clone of cells derived from a stem cell shares a common methylation pattern that will often differ from that of distantly related clones because of accumulated changes that occur over time at a relatively high rate in comparison with genetic mutation rates. Bisulfite treatment of genomic DNA extracted from single isolated crypts results in conversion of unmethylated cytosine bases to uracil. Methylated cytosines are protected from the conversion. The treated DNA is then amplified by polymerase chain reaction (PCR), cloned, and the DNA sequence of individual clones determined. It was found that multiple (more than can be explained by multiple alleles), but often seemingly related, sequences were derived from single crypts, confirming previous work suggesting that human colon crypts contain several stem cells. Yatabe et al. (2001) applied simple stochastic models of stem cell survival to the analysis of the resulting collections of methylation patterns derived from crypts and attempted to reach conclusions about the mutation process and stem cell dynamics that generated the observed patterns. There are many potential pitfalls that should be addressed before adapting the method. The method requires extensive PCR amplification of sample DNA. Accordingly the well-known issues of bias in the presence of multiple homologous templates (Suzuki and Giovannoni, 1996) and potential copy errors introduced by the PCR process itself need careful consideration. This is important because in this assay the sequences and their relative frequencies constitute the primary data. Furthermore, autosomal genes will have two alleles that will often differ in their methylation pattern. This must be accounted for. More troublesome is the fact that crypts contain proliferating cells. This is problematic because after DNA replication there will be a varying period during which many of the CpG sites of a gene will be hemimethylated and such sites, if PCR amplified, would be interpreted as related but distinct stem cell lineages. Another source of serious and potentially misleading sequence variability is the intraepithelial lymphocytes, each of which potentially contributes two distinct methylation alleles to the mix. The contribution of template from these cells might seem insignificant, but PCR homologous template amplification bias could result in such rare alleles having a disproportionate impact. Finally, the model used to analyze the data entails many strong assumptions that are difficult to test by independent means but that *a priori* constrain the conclusions that are reached.

Mitochondrial Genes

The mitochondrial genome is another inherited system that can be used as a lineage marker in the human intestinal epithelium (Greaves *et al.*, 2006; Taylor *et al.*, 2003). Mitochondria contain multiple copies of self-replicating DNA molecules encoding a subset of the proteins involved in the mitochondrial respiratory chain. Accumulated mutations impacting specific mitochondrial proteins may be assessed histochemically or immunocytochemically. In particular, monoclonal crypts displaying loss of cytochrome *c* oxidase on a background of normal succinate dehydrogenase were easily seen in sections and found to be a useful lineage marker in human colon crypts. The presence of mutations in such crypts was confirmed by sequencing the mitochondrial genome from adjacent microdissected crypt sections (Taylor *et al.*, 2003). One noteworthy aspect of this study is the high frequency of mutant crypts that was observed (1–34% of crypts contained mutant cells, that is, cells with reduced or absent cytochrome *c* oxidase staining). This was ascribed to a high mitochondrial mutation rate, but it would be advisable to be aware of potential contributions of staining artifacts in specimens if the method is to be used for stem cell assays. Interestingly, crypts containing specific mutations were often clustered (Greaves *et al.*, 2006), a finding consistent with the idea that crypt branching is an important aspect of the expansion of mutant stem cell clones (Bjerknes and Cheng, 1996).

Stem, Progenitor, and Lineage Markers

Dozens of potential lineage markers are available. In Table I we list widely used examples of markers useful for labeling the various epithelial lineages, whereas Table II lists examples of markers useful for labeling committed progenitors. Markers for stem cells, and other early progenitors such as **Mix**, are more contentious. As of the time of writing only a handful of potential intestinal epithelial stem cell markers have been proposed, but many groups are actively searching and therefore we expect a flood of candidate molecules to appear soon. All the candidates listed later are

TABLE I
LINEAGE MARKERS

Columnar	Mucous	Enteroendocrine	Paneth
Intestinal alkaline phosphatase, ALPI (human); Akp3 (mouse)	Mucin 2, MUC2	Chromogranin A, CHGA	Lysozyme, LYZ (human); Lzp-s (mouse)
Sucrase-isomaltase, SI	Mucopolysaccharide stains: periodic acid–Schiff (PAS), alcian blue, wheat germ agglutinin	Synaptophysin, SYP	Defensin α1, DEFA1 (cryptdin-1)
Solute carrier family 5 (sodium/glucose cotransporter), member 1, SLC5A1 (SGLT1)	Trefoil factor 3, TFF3	Paired box gene 6, PAX6; paired box gene 4, PAX4; neurogenic differentiation 1, Neurod1 (NeuroD, Beta2); various hormones	Matrix metallopeptidase 7, MMP7
Dipeptidyl-peptidase 4, DPP4 (CD26)	Atonal homolog 1 (*Drosophila*), ATOH1 (HATH1, Math1)	Atonal homolog 1 (*Drosophila*), ATOH1 (HATH1, Math1)	Atonal homolog 1 (*Drosophila*), ATOH1 (HATH1, Math1)

TABLE II
COMMITTED PROGENITOR MARKERS

C_1	M_1	E_1	$P_1?[a]$
	Mucin 2, MUC2	Neurogenin 3, Neurog3 (Ngn3)	
Hairy and enhancer of split 1, HES1	Atonal homolog 1 (*Drosophila*), ATOH1 (HATH1, Math1)	Atonal homolog 1 (*Drosophila*), ATOH1 (HATH1, Math1)	Atonal homolog 1 (*Drosophila*), ATOH1 (HATH1, Math1)?
Antigen identified by monoclonal antibody Ki-67, MKI67	Antigen identified by monoclonal antibody Ki-67, MKI67	Antigen identified by monoclonal antibody Ki-67, MKI67	Antigen identified by monoclonal antibody Ki-67, MKI67?

[a]A Paneth cell progenitor P_1 is indicated because Paneth cells are probably produced in pairs (Bjerknes, 1985).

positive markers, that is, cells expressing the marker are proposed to include stem cells. It also would be useful to identify a minimal set of strongly expressed lineage markers. Stem cells would then be enriched in the Lin⁻ population. Combining positive and negative markers, even if neither is specific, might prove sufficient to adequately identify the stem cell population *in situ* or by flow cytometry.

Label Retention

It has been proposed that small intestinal stem cells usually divide asymmetrically, selectively retaining those chromatids having the older or "immortal" parental DNA strand (Potten *et al.*, 2002), a clever but controversial idea originally suggested by Cairns (1975) as a means to protect the stem cell genome from replication errors. Accordingly, if a pulse of [³H]thymidine or bromodeoxyuridine (BrdU) is administered to an animal, initially labeled stem cells will become unlabeled after completion of the subsequent cell cycle. They further propose that if [³H]thymidine or BrdU is given when conditions are conducive to symmetric stem cell divisions (yielding two stem cells), such as during periods of rapid growth or injury-induced repair, then newly synthesized strands will also be immortalized. When these labeled stem cells revert to their usual asymmetric division the labeled immortal strands will be selectively retained and hence the stem cells will effectively be permanently labeled. Label-retaining cells have been documented in a number of systems, including intestinal crypts; however, the mechanism responsible for the label retention is still being debated. Regardless of the mechanism, label retention has been proposed as an intestinal stem cell marker (Kim *et al.*, 2004;

Potten *et al.*, 2002). In such experiments, label-retaining cells are distributed throughout the crypt, although there is a higher concentration in the lowest nine cell positions (Potten *et al.*, 2002). Kim *et al.* (2004), using flow cytometry, found that 4–7% of cells in the proliferative zone of colon crypts are label retaining. It should be noted, however, that to arrive at this range they considered "only the top 20–40% of the brighter end of the BrdU-positive peaks" (i.e., the top 20–40% most strongly stained of the BrdU-positive cells) to be label-retaining cells. Thus label-retaining cells are fairly numerous and spread throughout the crypt. The wide distribution of the label-retaining cells suggests that many nonstem cells may retain label in these experiments, potentially limiting the utility of label retention as a stem cell marker. Label-retaining cells might also include postmitotic cells such as Paneth cells, enteroendocrine cells, intraepithelial lymphocytes (Spyridonidis *et al.*, 2004), cycling cells arrested by the radiation treatment or BrdU incorporation, or cells labeled as a result of [^{3}H]thymidine or BrdU reutilization (Burns *et al.*, 2006). Kim *et al.* (2004) reported that there were essentially no enteroendocrine cells in the lower crypt of rat colon and hence felt that these cells were excluded as a confounding factor. However, their observation requires further investigation because it is well known that enteroendocrine cells are most frequent in the lower crypt of colon (Chang and Leblond, 1971b; Ho *et al.*, 1989; Sheinin *et al.*, 2000). Until intestinal label-retaining cells have been better characterized or, ideally, demonstrated to give rise to multilineage long-lived clones as has been shown in hair follicles (Blanpain *et al.*, 2004), their equivalence with intestinal stem cells remains uncertain.

Musashi-1

Musashi-1 is an RNA-binding protein thought to be important in stem cell maintenance and differentiation. In the adult intestine Musashi-1 is found in crypt base columnar cells as well as in a few cells above the Paneth cells (Asai *et al.*, 2005; Kayahara *et al.*, 2003; Potten *et al.*, 2003). Because *musashi-1* is expressed by cells located in the putative stem cell positions in the crypt, it was suggested that Musashi-1 may serve as an intestinal stem cell marker (Kayahara *et al.*, 2003; Potten *et al.*, 2003). However, others have observed Musashi-1 throughout the proliferative compartment of the crypt (Gregorieff *et al.*, 2005), and throughout the crypt–villus axis in mice with moderately increased expression of insulin-like growth factor II (Sakatani *et al.*, 2005).

β_1-Integrin

β_1-Integrin has been used as a marker to enrich for mammary and epidermal stem cells (Jones *et al.*, 1995; Shackleton *et al.*, 2006; Stingl *et al.*, 2006). In the intestinal epithelium, β_1-integrin is strongly expressed in the lower one-third of colon crypts and has been used to enrich for stem and progenitor cells by fluorescence-activated cell sorting (FACS) (Fujimoto *et al.*, 2002; see p. 368).

Phosphorylated PTEN, 14-3-3ζ, and Phosphorylated Akt

Phosphorylated PTEN is the inactive form of the tumor suppressor PTEN (phosphatase and tensin homolog) that He *et al.* (2004) claimed was present specifically in intestinal stem cells, and used as a stem cell marker to quantify intestinal epithelial stem cells. 14-3-3ζ and phosphory-lated Akt were also specifically expressed in these same cells, and hence were also proposed to be stem cell markers. However, we found that while most epithelial cells were faintly stained, the strongly phosphorylated PTEN-positive cells that they described were actually enteroendocrine cells (Bjerknes and Cheng, 2005b).

sFRP-5 and Dcamkl1

Two molecules expressed by solitary cells located at the base of the proliferative zone, just above the Paneth cells, have been proposed as potential stem cell markers: sFRP-5 (secreted frizzled-related protein 5), an antagonist of Wnt signaling (Gregorieff *et al.*, 2005), and the microtubule-associated kinase Dcamkl1 (double cortin and calcium/calmodulin-dependent protein kinase-like-1; Giannakis *et al.*, 2006). Positional information, although suggestive, is obviously insufficient evidence to define a molecule as a stem cell marker, and further studies are needed.

Identifying candidate stem cell markers by location in the crypt deserves further comment. Many investigators subscribe to the classic view that stem cells are located above the Paneth cells in cell positions 4 or 5 and expect as a result that stem cell markers should specifically identify such cells, and not cells further down in the crypt base among the Paneth cells. However, according to the stem cell zone model markers expressed in cells restricted to the common origin would more likely indicate committed progenitors, and candidate stem cell markers would be expressed in scat-tered crypt base columnar cells throughout the first four or five positions, including those among the Paneth cells. Both sets of markers will likely prove interesting and useful.

Intestinal Epithelium *In Vitro*

The abundant, simple, and well-structured nature of the intestinal epithelium should provide major advantages to the study of stem cell biology. Development of robust methods for the isolation and culture of adult intestinal stem cells and progenitors would provide an important tool set both for studies of the basic biology of the system and for potential clinical applications, including gene therapy or tissue engineering. Unfortunately, adult intestinal epithelium has proven to be relatively refractory to culture and it has been difficult to establish useful primary systems. There has been progress, but robust *in vitro* approaches to the culture of adult intestinal epithelium are still lacking.

Three fundamental issues continue to limit the utility of normal intestinal culture systems developed to date. First, untransformed intestinal epithelial cells are anchorage dependent. Extracting the epithelium from the intestine results, depending on the methods used (see later), in partial or complete loss of contact with the basal lamina, triggering a signal transduction cascade that can rapidly lead to apoptosis. Such apoptosis triggered by inadequate extracellular matrix contact is known as "anoikis," a transliteration of a Greek word meaning "homelessness" (Frisch and Francis, 1994). Sträter *et al.* (1996) demonstrated that anoikis is induced in isolated intestinal epithelium by the loss of interaction of β_1-integrin with matrix elements, thus explaining the repeated observation that some form of matrix (e.g., collagen-coated dishes or a feeder layer) is often beneficial to attachment and survival of intestinal epithelium in culture. Second, the survival and growth requirements of intestinal epithelial cells are poorly understood. For intermediate-term culture, isolated adult epithelium seems to require a feeder layer or other complex and ill-defined medium components to maintain growth and survival. Third, *in vitro* conditions supportive of normal lineage development have not been developed. There are no culture systems at present in which normal adult isolated epithelium has been shown to undergo lineage development. However, it has repeatedly been shown that organoids obtained from dissociated intestine, if implanted *in vivo*, can undergo organotypic reconstitution. Fetal, neonatal, and adult intestine have been dissociated by enzymatic or chelation methods, embedded into various collagen matrices, and then implanted into syngeneic or immunocompromised host animals. After several weeks, in conjunction with the host's mesenchymal tissue and establishment of vascular supply, the implanted tissues reorganize into remarkably normal-looking mucosa with crypt- and villus-like structures containing multiple epithelial cell lineages. Unfortunately, *in vitro* conditions supporting similar recapitulation of normal adult stem and progenitor cell lineage progression have not yet been demonstrated.

Later we discuss existing approaches to intestinal epithelial cell isolation and culture and include some general protocols that can be adapted to isolate epithelium for various purposes.

Established Intestinal Epithelial Cell Lines

Tumor Cell Lines

Hundreds of intestinal epithelial cell lines have been established from tumors, several of which have been extensively used and characterized. For example, a PubMed search for the term Caco-2, a human colon cell line derived from an adenocarcinoma (Fogh *et al.*, 1977), returned 5046 citations at the time of writing. Other widely used colon tumor-derived cell lines include HT-29, T84, LoVo, LS174T, DLD-1, SW480, and SW620.

Normal Cell Lines

Several intestinal epithelial cell lines have been produced from normal tissue from various species. These lines do not grow in soft agar and do not form tumors when implanted into animal hosts. They are relatively undifferentiated and the cells are usually described as "crypt-like." The most widely used of these, the IEC-6 and IEC-18 lines, were originally derived from explants of 18- to 24-day-old rat small intestine (Quaroni *et al.*, 1979). Also available are lines derived from piglet jejunum (IPEC-J2; Berschneider, 1989), dog small intestine (DIEC; Weng *et al.*, 2005), rat intestine (RIE-1; Blay and Brown, 1984), human fetal small intestine (HIEC; Perreault and Beaulieu, 1996), and human adult colon (NCM460, Moyer *et al.*, 1996; and NCOL-1, Deveney *et al.*, 1996). However, Melcher *et al.* (2005) conclude from a cytogenetic analysis that the NCOL-1 line may have been contaminated by LoVo carcinoma cells.

Immortalized Cell Lines

Another approach to intestinal cell line derivation uses a temperature-sensitive simian virus 40 (SV40) large T-antigen transgene. When cultured at the permissive temperature, proliferation is induced. Cell lines developed with this strategy from mouse (Whitehead *et al.*, 1993) and fetal human (Quaroni and Beaulieu, 1997) intestine are available.

Are They Stem Cells?

The various lines clearly contain cells with extensive proliferative capacity by virtue of the fact that they can be propagated, but do any of them contain intestinal stem cells or early lineage progenitors? The tumors from which many of the popular cell lines were derived might well have arisen from transformed stem cells, but do any of these lines retain the transformed stem cells? How would we know? In the near future we will likely have a suite of lineage-specific markers (as in the hematopoietic system) and it will then be possible to assign progenitor identity on the basis of gene expression patterns (assuming that gene expression patterns remain well defined and stable under culture conditions). In the absence of such knowledge, the ideal experiment would entail isolating single cells and transplanting them into an environment in which they can grow and reconstitute the various epithelial lineages.

Conditions have been found under which five tumor cell lines—Caco-2 (de Bruïne *et al.*, 1993), HRA-19 (Kirkland, 1988), NCI-H716 (de Bruïne *et al.*, 1992), LIM1863 (Whitehead *et al.*, 1987b), and HT-29 (Huet *et al.*, 1987)—produce heterogeneous mixtures of cells expressing different lineage markers. For example, embedding Caco-2 cells into a collagen matrix and implanting them subcutaneously results in cells that express either mucous, enteroendocrine, Paneth, or absorptive cell markers (de Bruïne *et al.*, 1993), suggesting that either some Caco-2 cells are multipotential or that the cell line contains various lineage progenitors.

IEC-17 cells, a normal rat cell line, were combined with fetal intestinal mesoderm and then implanted subcutaneously into syngeneic hosts. Remarkably, a fully differentiated epithelium developed containing four major cell lines, indicating that the IEC-17 line may contain multipotential cells (Kedinger *et al.*, 1986). It should be noted, however, that the IEC-17 line was not clonally derived, and an entire monolayer of cells was used to prepare the implants. Therefore the resulting epithelium could have been derived from multiple progenitor types resident within the IEC-17 line. Furthermore, the reconstituted epithelium was not convincingly demonstrated to be derived from the cell line and hence the possibility remains that the epithelium may have been derived from residual endoderm associated with the fetal mesoderm used to prepare the implants. More recent studies that implanted IEC-6 cells into a host did not observe cell differentiation (Kawaguchi *et al.*, 1998); however, when IEC-6 cells were transfected with vectors expressing the transcription factors Cdx-2 or Pdx-1 they underwent a degree of differentiation along columnar and mucous, or enteroendocrine lines, respectively (Suh and Traber, 1996; Yamada *et al.*, 2001). In light of these issues, similar experiments, if repeated, should be performed with clonal IEC sublines and the experiment should be designed to take advantage of the many currently available markers in order to definitively identify the source of the reconstituted epithelium, preferably using several independent tools. It will also be necessary to take account of the possibility that host nonepithelial cells, such as multipotent adult progenitor cells (Jiang *et al.*, 2002), may contribute to the regenerating epithelium.

Primary Cultures

Like most human endeavors, intestinal culture has its legends and myths. Some of these ring true; others with the benefit of hindsight more likely reflect circumstance and historical accident. For example, methods used to obtain epithelium have been a rich source of confusion and myth. Early studies found that isolated intestinal epithelial cells quickly died, but it was eventually found that matrix-coated dishes helped attachment and survival. Some found that enzymatically digested mucosa yielded viable cultures whereas epithelium isolated by EDTA chelation did not, and as a result it was concluded that EDTA was toxic. Others found that EDTA-isolated epithelium would in fact attach and survive if matrix or feeder layers were provided. We now understand that the apparent sensitivity of the epithelium to EDTA treatment is because loss of contact with the basement membrane induces onset of apoptosis, but that this can be reversed provided contact is rapidly reestablished. Thus, the myth that enzymatic isolation yields better viability may be explained by the fact that epithelium prepared by enzyme digestion is not pure. It usually contains residual attached basement membrane and mesenchymal elements, both of which are now understood to play an important role in cell survival. More recent studies have shown that if EDTA-isolated epithelium is quickly provided with these components then its viability is comparable to that of enzymatically prepared tissue (see later). Other similar and widely held myths, most probably containing elements of truth, include the benefits of age (youth is good), size (large is good), and region (distal is good). Thus human fetal colon is easier to culture than adult mouse small intestine.

Isolation Methods

Attempts to isolate and culture intestinal stem and progenitor cells might reasonably begin with isolation of the epithelium from its supporting tissues followed by dissociation and fractionation. Simple and effective methods to isolate viable intestinal epithelial cells free of nonepithelial mesenchymal elements have been developed and these are described later. Example protocols are provided on page 369. The application will determine the care with which the intestine needs to be handled during processing. Perusal of many protocols makes it clear that for many applications rough handling of the intestine does not appear to cause difficulty. However, it is best to keep in mind that even gentle handling of the gut can stimulate the enteric nervous system and activate epithelial responses. If tissue must be stored before preparing cultures, for example, if a surgical specimen must be transported, then effort must also be made to minimize degenerative changes. Preliminary exploration of optimal storage conditions to mitigate damage is described by Uchida *et al.* (2001).

EPITHELIAL ISOLATION USING MECHANICAL APPROACHES. The earliest method involved striping a sheet of mucosa by dragging a glass slide across the surface of the opened gut (Dickens and Šimer, 1930; Dickens and Weil-Malherbe, 1941). In specimens from species larger than mice this expediently removes the substantial muscle layers and submucosa, but the lamina propria with its attendant fibroblasts, vasculature, and neural and lymphoid elements remains attached. Many variations on such mechanical means of applying shear forces to strip off mucosal cells have been developed, perhaps reaching a pinnacle in the various apparatuses described by Sjöstrand (1968). Mechanical methods including stripping, spinning, scraping, shaking, vibrating, chopping, and trituration are widely used because of their simplicity and speed, and are often employed in various stages of more elaborate isolation protocols.

ORGAN CULTURE. Until recently the only *in vitro* approach maintaining epithelial lineage progression was the method originally used, the culture of intact fragments of intestine. Fischer (1922) cultured strips of intestine from 21-day-old chick embryos in embryo extract for 1 month and maintained a relatively normal epithelium. A similar approach was used subsequently to study the role of hydrocortisone in epithelial maturation (Dobbins *et al.*, 1967; Moog and Nehari, 1954). Others, hoping to develop systems to grow viruses, maintained explants of fetal human, chicken, and rabbit small intestine for 2–3 weeks (Dolin *et al.*, 1970; Enders *et al.*, 1949). The basic success of the approach, including differentiation along the various cell lineages, was confirmed with avian intestine (Mizuno and Yasugi, 1990; Yasugi and Mizuno, 1978), fetal rat intestine (DeRitis *et al.*, 1975; Fukamachi and Takayama, 1980; Kondo *et al.*, 1984; Quaroni, 1985), calf (Capdeville *et al.*, 1967), and fetal mouse intestine (Pyke and Gogerly, 1985).

Adult mucosa proved to be more problematic. The initial application was [^{3}H]thymidine incorporation into human colon to assay the distribution of proliferating cells in the crypt and for this purpose a few hours in culture was sufficient (Browning and Trier, 1969; Deschner *et al.*, 1963; Eastwood and Trier, 1973; Trier, 1976); after a few hours degenerative changes ensued. Ferland and Hugon (1979) established conditions for mouse small intestine that extended reasonable morphological preservation to 48 h, at which point some degenerative changes were noticeable. Later workers found, however, that after the initial degeneration, the epithelium will often regenerate to some extent and that mitosis and differentiation persist for weeks (Autrup, 1980; Autrup *et al.*, 1978; Defries and Franks, 1977; Moorghen *et al.*, 1996). Although physiologists and cell biologists have made extensive use of short-term adult intestinal organ culture (e.g., Ussing chamber and similar transport work), the instability and variability of long-term organ culture of adult intestine have limited its utility to date.

Microexplants. The intestinal fragments described previously are a few millimeters in size and are usually of full mucosal thickness. Microexplants obtained by forcing mucosa through the metal screen of a tissue press survive for months in suspension culture (Moyer, 1983; Moyer and

Aust, 1984). Others have confirmed that microexplants from colon, composed of a crypt surrounded by an intact lamina propria, maintained excellent morphology for at least 30 days, and survived for months in suspension culture (Wildrick *et al.*, 1997). The proliferative status of the epithelium in these cultures has not been well described, although the need for periodic subculturing suggests growth might be occurring in colon fragments (Moyer, 1983; Moyer and Aust, 1984). Mitotic figures were not seen in fragments cultured from small intestine (Boxberger *et al.*, 1997) albeit in a different culture medium from the complex mix of serum, crude extracts, conditioned medium, and growth factors used in the colon studies.

EPITHELIAL ISOLATION BY CHELATION. Embryologists have known since the 1800s that adhesion of cells to each other and to the extracellular matrix is weakened by exposure to medium free of divalent cations (reviewed in Chambers, 1940; Holtfreter, 1948). We now understand that this effect is due in large measure to the weakening of extracellular molecular interactions of integrins (Arnaout *et al.*, 2005; Leitinger *et al.*, 2000) and cadherins (Gooding *et al.*, 2004) on removal of these metals. It is likely that this is the basis for the ability of chelating agents such as EDTA to sufficiently weaken the epithelial attachment to the basement membrane that isolation is possible by applying mild shear forces.

The chelating agent EDTA was introduced as a water softener by Münz (1938). Clinical applications have also been found, including treating acute hypercalcemia and lead poisoning. The discovery of its utility in isolating the intestinal epithelium seems to have been serendipitous. Miller and Crane (1961) introduced EDTA to help inhibit enzyme activity in mucosal scrapings during preparation of brush borders. Subsequently, Harrison and Webster (1964), working to improve the brush border preparations, discovered that if they immersed the gut in 5 m*M* EDTA before scraping, then the epithelium would easily strip off. They applied shear forces by rapidly moving spiral rod-mounted intestinal segments through the fluid. At about the same time Stern (1966) introduced the chelating agent sodium citrate, but this is not widely used. Later, Webster and Harrison (1969) introduced minor variations of their method (mounting on a straight rod with continuous vibration) that Weiser (1973) adapted in his popular crypt–villus epithelial cell fractionation scheme. Incubation of intestine in EDTA is the basis for the most successful isolation techniques to date.

Dozens of protocols have been published, essentially varying the EDTA concentration, the incubation temperature, and the method of applying shear force. The duration of incubation necessary to loosen the epithelium is a function of both the temperature and EDTA concentration. We refer to these approaches collectively as "immersion protocols" because the gut (either segments of intestine or pieces of mucosa) is immersed in the EDTA solution during incubation. If shear forces are applied continuously during incubation then the epithelium is stripped off as single or small clumps of cells sequentially from upper villus to lower villus to crypts. This can be used to roughly fractionate the epithelium along the crypt–villus axis for comparative biochemical studies (Webster and Harrison, 1969; Weiser, 1973).

We find that the epithelium is least contaminated with mesenchymal elements and is most viable when it is stripped as intact structures rather than single cells (Fig. 4). This is achieved by first incubating the gut in EDTA before applying shear forces to dislodge the epithelium. This is

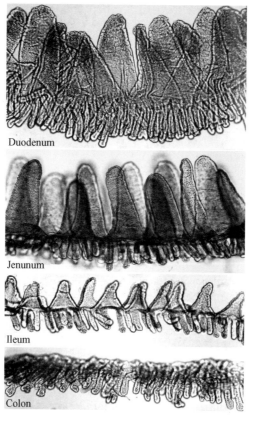

Duodenum

Jenunum

Ileum

Colon

FIG. 4. Sheets of epithelium isolated from various regions of mouse intestine by transcardiac perfusion with EDTA (taken from Bjerknes and Cheng, 1981c). All images are to the same scale, illustrating the regional variation in size and density of crypts and villi.

applicable to all regions of the mouse gastrointestinal tract (Bjerknes and Cheng, 1981c) and human colon (Cheng *et al.*, 1984). Bjorkman *et al.* (1986) showed by electron microscopy that intact epithelial sheet preparations retain better morphology and integrity than do cell preparations. Similarly, IEC cells have been found to be less sensitive to anoikis when grouped in clusters (Waterhouse *et al.*, 2001), a phenomenon referred to as synoikis (Shen and Kramer, 2004).

Culturing EDTA-Isolated Epithelium. Early attempts at culturing epithelium isolated by chelation were unsuccessful. Typically, most epithelial cells isolated with EDTA failed to attach and died within a few hours (Quaroni *et al.*, 1979). Similar difficulties with viability of EDTA-isolated epithelium were experienced by other groups (Deveney *et al.*, 1996; Evans *et al.*, 1992).

As mentioned previously, if steps are taken to override isolation-induced anoikis, chelation-isolated epithelium is viable and can be cultured. In particular, if the epithelium is allowed to quickly reestablish interaction with extracellular matrix, especially collagen, anoikis appears to be halted (Grossmann *et al.*, 2003; Sträter *et al.*, 1996). Epithelium isolated by chelation from human colon (Grossmann *et al.*, 2003; Pedersen *et al.*, 2000; Sträter *et al.*, 1996; Wildrick *et al.*, 1997; Whitehead *et al.*, 1987a), mouse (Branting *et al.*, 1991; Kalabis *et al.*, 2003), rat (Bartsch *et al.*, 2004; Horster *et al.*, 1994), rabbit (Vidrich *et al.*, 1988), and turkey (Ali and Reynolds, 1996) attached to and survived on collagen matrix, feeder layer, or both. Longer-term (>1–2 days) survival and proliferation seem to depend on additional factors supplied by feeder layers and/or complex media. Cellular differentiation *in vitro*, such as the appearance of mucous cells, seems to require mesenchymal elements. EDTA-isolated human fetal colon epithelium grown on a collagen gel containing embedded colon fibroblasts (Kalabis *et al.*, 2003) displayed some evidence of epithelial lineage progression (see also *Organ Culture*, p. 358). Thus in the absence of mesenchymal interaction no clear evidence of differentiation along the various lineages has been observed in normal cells *in vitro*.

EPITHELIAL ISOLATION WITH MATRISPERSE. MatriSperse (Becton Dickinson, Franklin Lakes, NJ) is a proprietary nonenzymatic product sold to solubilize collagen gels for recovery of cells grown on matrix. Perreault and Beaulieu (1998) found that incubating human fetal intestinal segments in MatriSperse at 4° followed by gentle shaking releases villus epithelium that is viable and can be readily cultured. They used this method to establish primary cultures of differentiated, nonproliferating cells from human fetal small intestine. The cells attached and spread on either plastic or collagen-coated dishes, but on plastic the cells began to die after 24 h. On collagen I the cells spread out for several days until confluent.

Similar approaches were successful with fetal gastric epithelial cultures, with all cell types adhering to plastic and surviving for at least 7 days without matrix, although proliferation subsided after 2 days (Basque *et al.*, 1999). MatriSperse has also been used to isolate villus epithelium from adult hamster (Haidari *et al.*, 2002) and mouse small intestine (Boudreau *et al.*, 2002; Fouquet *et al.*, 2004), and mouse colon epithelium (Boudreau *et al.*, 2002). In their hands MatriSperse-isolated adult mouse villus epithelium, contrary to fetal human villus epithelium, did not attach to matrix and rapidly underwent apoptosis. Both groups found that isolation of small intestinal crypts required subsequent treatment with a chelating agent, and used this fact to isolate separate crypt and villus fractions.

EPITHELIAL ISOLATION BY ENZYMATIC DIGESTION. Many enzymes have been employed in the isolation and culture of intestinal epithelium with varying degrees of success, including trypsin, pancreatin, various collagenases, hyaluronidase, neuraminidase, dispase, and thermolysin. Harrer *et al.* (1964) first adapted enzymatic digestion to the isolation of intestinal epithelial cells. They used a combination of trypsin and pancreatin in calcium- and magnesium-free (CMF) saline as introduced by Moscona and Moscona (1952) in their studies of the reconstitution of dissociated embryonic tissues. Collagenase treatment has long been used to help separate embryonic intestinal endoderm from mesenchyme (Haffen *et al.*, 1981; Kedinger *et al.*, 1981; Yasugi and Mizuno, 1978) and Quaroni *et al.* (1979) used it to isolate epithelium from young rats. Dispase, a bacterial neutral metalloprotease that acts on fibronectin and type IV collagen and seems to be gentler on the cell surface than other general proteases, was first used to prepare intestine for culture by Vidrich *et al.* (1988), but is most frequently used with collagenase, a combination introduced by Gibson *et al.* (1989). Another bacterial neutral metalloprotease, thermolysin, was used to dissociate human fetal intestine from which the HIEC cell line was established (Perreault and Beaulieu, 1996). DNA released from damaged nuclei during isolation procedures can cause cell aggregation and DNase I is often used to minimize cell clumping.

As mentioned previously, a significant amount of mesenchyme and basement membrane components often accompanies enzyme-isolated epithelium. This can be advantageous to epithelial survival because of their protective effects against immediate onset of anoikis. In addition, the mesenchymal elements may also promote longer-term survival, obviating the need for a feeder layer or conditioned medium. This is a mixed blessing, however, because if culture conditions are supportive of fibroblast growth, they soon overgrow. Thus most workers have found that minimizing or eliminating serum from the culture medium is advantageous for epithelial culture.

Culturing Epithelium Isolated Enzymatically

CULTURING DISSOCIATED CELLS. Relatively pure fetal rat endoderm obtained by digestion with collagenase has been repeatedly cultured. Collectively, the results indicate that matrix and mesenchymal interaction are also important for fetal cells in culture. Fetal endodermal cells grown on gut mesenchymal cells survive and differentiate better than those grown on plastic (Stallmach *et al.*, 1989). Later, it was found that the mesenchymal cells could be replaced by matrix-coated dishes (Fukamachi, 1992; Hahn *et al.*, 1990). Others have dissociated the entire gut from fetal or neonatal rats and found that the cells did well on plastic (Brubaker and Vranic, 1987; Chopra and Yeh, 1981; Haffen *et al.*, 1981). However, although some of the mature cells already present in the epithelium may survive (Brubaker and Vranic, 1987), there is little evidence of differentiation in cells propagated on plastic dishes (Chopra and Yeh, 1981; Haffen *et al.*, 1981). Monolayers prepared from 12-day fetal mouse mesenchyme supported the *in vitro* differentiation [as measured by reverse transcription-polymerase chain reaction (RT-PCR) of marker genes] of dissociated fetal endodermal cells cultured on top of the monolayer (B. M. Kim *et al.*, 2005). The mesenchymal interaction may also be important to maintaining continuation of normal epithelial capacities *in vitro*, because when monolayers of endodermal cells propagated on plastic were combined with embryonic mesenchyme and implanted into chick embryos the resulting epithelium appeared cuboidal rather than columnar, and mucous cells were not evident, suggesting that the culture period may have altered the cells (Haffen *et al.*, 1981). This is in contrast to the columnar epithelium with mucous cells that results when isolated fetal endoderm is combined with fetal mesenchyme and directly implanted without the intervening dissociation and propagation in culture (Kedinger *et al.*, 1981). As discussed later (p. 365), dissociated fetal gut cells, including both endoderm and mesenchyme, cultured in a Gelfoam (Pfizer, New York, NY) matrix for 7 days and then implanted into syngeneic rats developed into a well-organized and differentiated epithelium with multiple lineages (Montgomery *et al.*, 1983), indicating that it is not the culture period that is problematic, but rather that culture without mesenchymal interaction alters endodermal cell behavior.

Cells isolated after enzymatic digestion have also been cultured from adult human colon (Baten *et al.*, 1992; Deveney *et al.*, 1996), and some of these cultures were extensively propagated although it appears that the Deveney *et al.* (1996) cultures may have been contaminated with cells from a tumor cell line (Melcher *et al.*, 2005).

CULTURING ORGANOIDS. Limited digestion of mucosa results in a complex mixture of epithelial and mesenchymal fragments referred to as "organoids," frequently consisting of pieces of crypts with attached mesenchyme.

It is worth noting that in many respects these preparations resemble the mechanically prepared microexplant cultures described previously (p. 358), which survived in suspension culture for months. Organoids prepared from adult human colon or small intestine have been cultured. Initial studies found that a subset of organoids attaches to matrix-coated dishes, the cells spread, but proliferation ceases by 19 h (Friedman et al., 1984). Adjusting the matrix or culture medium achieved only minor improvements extending survival to a few days (Buset et al., 1987; Gibson et al., 1989; Pedersen et al., 2000). Coculturing with gut fibroblasts (Schörkhuber et al., 1998) or using complex culture medium on plastic (Brandsch et al., 1998) improved survival to 12–14 days. More recent studies achieved an additional week or two of survival with organoids prepared from suckling rats (Evans et al., 1992; Schörkhuber et al., 1998), adult mice (Booth et al., 1995, 1999), suckling mice (Macartney et al., 2000; Slorach et al., 1999), rabbits (Vidrich et al., 1988), and cows (Rusu et al., 2005). The cow organoids were subcultured up to seven times. However, in all cases no significant differentiation along the major cell lineages was observed although the cells remained epithelioid in nature.

Culture Systems for Studying Mesenchymal–Endodermal Interaction During Development

Organotypic regeneration of implants has often been used to study the role of mesenchymal influences in epithelial behavior (see Mizuno and Yasugi, 1990, for a review of early work on the independence of intestinal epithelial cell differentiation from intestinal mesenchyme; and Kedinger et al., 1998, for more recent work). Fukamachi and Takayama (1980) combined 16.5-day fetal rat duodenal endoderm with gastric or duodenal mesenchyme and then cultured the recombinants as explants in organ culture. The stratified endoderm differentiated into a simple columnar epithelium with mucous cells regardless of the origin of the mesenchyme, although the frequency of mucous cells depends on the origin of the mesenchyme. In addition, many explants contained crypt-like structures. Yasugi and Mizuno (1978) performed similar experiments with chick and quail gastrointestinal endoderm–mesenchyme recombinations implanted into 9-day chick embryos. They found that small intestinal endoderm retained its nature and developed as small intestine regardless of the origin of its partner mesenchyme. Endoderm from other regions was often more pliant. Similar combinations of fetal mesenchyme with fetal endoderm or with monolayers of cultured fetal endodermal cells implanted into chick embryo or adult rodent hosts have been extensively explored (Haffen et al., 1981; Kedinger et al., 1981, 1986). It is now understood that extensive molecular interactions occur between the epithelium and its underlying mesenchyme, through diffusible signaling molecules such as the Wnts, hedge-hogs, members of the transforming growth factor (TGF)/bone morpho-genetic protein (BMP) family, and so on (Bjerknes and Cheng, 2005a; Kedinger *et al.*, 1998; Sancho *et al.*, 2004). Importantly, gastric mesenchyme from 12-day fetal mice can be routinely dissociated, cultured as a mono-layer, transfected, and still support the *in vitro* differentiation (as measured by RT-PCR of marker genes) of dissociated 12-day fetal gastric or intest-inal endodermal cells cultured on top of the mesenchymal monolayer (B. M. Kim *et al.*, 2005). Similarly, organ culture of 12.5- to 13.5-day fetal intestine and stomach for 24 h allowed observation of changes in gene expression resulting from application of beads containing various signaling molecules (Ormestad *et al.*, 2006). If such observations can be reproduced these simple culture systems should provide useful tools with which to study the molecular mechanisms of early fetal development of intestine.

Implants: From Tissues to Tissue Engineering

Fragments of normal and cancerous intestine have long been xenografted (also called heterotransplants in the older literature) into various hosts, more recently usually into nude mice (Bhargava and Lipkin, 1981; Mach *et al.*, 1974; Povlsen and Rygaard, 1971; Valerio *et al.*, 1981). Implanted fragments of normal intestine initially degenerate, but once vascularized a relatively normal mucosa often regenerates. Syngeneic grafts have yielded the most dramatic results, as illustrated by two extreme examples. In the first example intact 3- to 8-cm-long segments of ileum from neonatal rats were successfully implanted into young rats by wrapping the implanted segments in the omen-tum (Uchida *et al.*, 1999). After 4 weeks most of the implants survived, were revascularized, and regenerated relatively normal mucosa. The implanted segments were functional as demonstrated by resecting most of the host's small intestine and then supplementing the residual small intestine by con-necting up the graft. Many of these animals survived while resected controls without the graft died. Also of note, these workers found that neonatal intestine can be frozen at $-180°$ for extended periods before implanting (Tahara *et al.*, 2001). At the other extreme, 18-day-old fetal rat intestine was trypsinized into single cells, injected into a Gelfoam matrix (pork skin gelatin sponge), and then cultured for 7 days (Montgomery *et al.*, 1983). After seeing no obvious organotypic reorganization in culture, they implanted the Gelfoam (after 7 days in culture) into syngeneic rats and harvested it 4 weeks later to find regenerated crypts and villi containing all major epithelial lineages. These studies illustrate the intrinsic hardiness of fetal and neonatal intestinal cells as well as their ability to regenerate organotypic structures given appropriate conditions.

Implantation of Organoids

Because conditions allowing isolated epithelial cells from adult animals to fully differentiate *in vitro* and to regenerate organotypic structure have not yet been demonstrated, implantation of dissociated intestinal material into syngeneic hosts or immunocompromised mice, either directly or with an intervening culture period, has been the only available tool with which to demonstrate the presence and potential of stem cells and progenitors in isolated or cultured intestinal cells. It was originally thought that only organoids (a crude mixture of epithelium and mesenchyme) prepared by enzymatic digestion of neonatal or adult mucosa could be successfully implanted (Booth *et al.*, 1999; Slorach *et al.*, 1999; Tait *et al.*, 1994). However, Del Buono *et al.* (2005) succeeded in implanting into nude mice single EDTA-isolated crypts, free of mesenchyme, from newborn rat colon or small intestine. After 5 weeks, in collaboration with the host's mesenchymal tissues, a relatively normal-appearing mucosa developed containing various lineages. Slorach *et al.* (1999) demonstrated that the regenerated epithelium was always derived from donor tissue whereas the mesenchymal elements were usually of mixed origin. The implantation and organotypic development of single characterized epithelial progenitor cells would be the ultimate extension of this approach and, in lieu of an adequate *in vitro* system, would be an alternative tool with which to probe the nature of epithelial stem and progenitor cells. Such single-cell *in vivo* implantation was achieved with mammary gland stem cells (Shackleton *et al.*, 2006; Stingl *et al.*, 2006).

Tissue Engineering

The implantation approach has been adapted to tissue engineering of intestine (Grikscheit *et al.*, 2003, 2004). A suspension of minced and enzymatically digested neonatal or adult colon or small intestine was placed in collagen I-coated polyglycolic acid tubes that were then wrapped in omentum of adult rats and retrieved 4–14 weeks later. They obtained remarkably organotypic "cysts" lined with an epithelium organized into crypts and villi and containing multiple cell types. There were also muscle layers with neurons, presumably components of the enteric nervous system. The cysts grew throughout the implantation period and were functional. When anastomosed to residual intestine the cysts were able to improve recovery from massive small bowel resection. New cysts could be created by digesting and reimplanting material from old cysts. The enzymatically dissociated cells could also be transfected before creating the cysts, indicating the feasibility of genetic engineering with its implications for clinical medicine and experimental biology.

Isolating Stem Cells and Progenitors by Cell Sorting

The development of methods for sorting specific progenitor cells and subsequently growing them under conditions in which they undertake their normal developmental program would be an important advance. Once successfully isolated and sorted, their growth in intestinal epithelial colony-forming assays would be valuable as this would allow ready quantification of the response of individual progenitor types to experimental treatments. Anchorage dependence is a fundamental problem that those wishing to isolate viable intestinal stem and progenitor cells must overcome because the cells must survive dissociation, staining, and sorting. As discussed previously, current understanding is that normal intestinal epithelial cells are anchorage dependent. If detached from the extracellular matrix (ECM) they undergo apoptosis and are dead within a few hours unless contact with matrix is reestablished. It might also be possible to halt the apoptotic pathway by providing sufficiently strong stimulation with growth factors or related means (Douma et al., 2004; Joseph et al., 2005; Waterhouse et al., 2001), but this is poorly understood in the intestine. Several groups have reported results from early attempts at colony-forming assays, with limited success.

Soft Agarose Colony-Forming Assay

The colony-forming assays that have been applied to the intestine are all based on embedding the cells in a matrix of soft agarose. Under these conditions normal human colon epithelial cells did not produce colonies unless the cells were intentionally transformed (Moyer and Aust, 1984). When transformed with SV40 they obtained about 14 colonies per 10^5 cells plated. Similarly, IEC-6 and -18 cells do not normally form colonies in soft agarose, but they did so when transformed with activated ras (Oldham et al., 1996; Rak et al., 1995). Normal fetal and adult mouse colon cells also did not grow in soft agarose unless an underlying feeder layer of fetal lung mesenchymal cells was included (Pyke and Gogerly, 1985). With the feeder layer they obtained about 10 colonies per 10^5 normal epithelial cells plated. Similar results were reported when human or mouse colon epithelial cells were isolated by EDTA, dissociated with pancreatin, and plated in soft agarose without a feeder layer (Whitehead et al., 1999). They obtained about 1 colony per 10^5 epithelial cells plated and 11 colonies per 10^5 if supplemented with LIM1863-conditioned medium. It is apparent that the rate of colony formation is depressingly low in these soft agarose assays, presumably a result of anoikis. One concern is that such a low colony formation rate might indicate that the few colonies seen were generated from transformed cells, while the rest of the culture underwent apoptosis. Checking for molecular signs of transformation seems an important next step to rule this out.

There have been two attempts, thus far, to isolate intestinal epithelial progenitors and stem cells by fluorescence-activated cell sorting (FACS).

β_1-Integrin

The first attempt used the fact that cells in the basal third of colonic crypts, the region of the crypt containing stem and progenitor cells, express more β_1-integrin than do cells higher up in the crypt. Thus epithelial cells expressing high levels of β_1-integrin collected by FACS should be enriched in crypt base cells. High β_1-integrin-expressing cells sorted from human colon generated about three times more colonies (7.6 versus 2.4 colonies per 10^5 cells plated) in soft agarose than did control unsorted epithelial cells (Fujimoto et al., 2002). This enrichment is consistent with the idea that the colony-forming cells originated from the lower crypt; however, addition of LIM1863-conditioned medium had little effect on the sorted cell colony count, increasing it to only 9.2 per 10^5 cells plated.

Side Population

Another study used the observation that early progenitors tend to express verapamil-sensitive ABC transporters to isolate so-called side population cells, using the vital DNA-binding dye Hoechst 33342. The side population cells are defined by the fact that they contain less Hoechst 33342 because it is pumped out of the cells by specific ABC transporters. The side population from many tissues is enriched for stem cells and progenitors (Challen and Little, 2006). However, this is not the case in all tissues (Shackleton et al., 2006; Stingl et al., 2006), perhaps a function of whether the cells are actively cycling (Uchida et al., 2004). Intestine contains side population cells, many of which are of hematopoietic lineage and hence express CD45 (Asakura and Rudnicki, 2003). Dekaney et al. (2005) stained intestinal cells with Hoechst 33342 and antibodies specific for CD45. The nonhematopoietic CD45$^-$ side population cells were collected by FACS and about 60% of the sorted cells were nonapoptotic as determined by annexin V staining. When cultured in plastic dishes the cells neither attached nor proliferated, although about 70% appeared "viable" by trypan blue exclusion assay after 14 days in culture. Most have found that isolated epithelial cells in suspension die within a few hours—are the side population cells different? Before reaching any conclusions, more critical assays of viability should be applied to confirm that the cells are indeed alive. If so, it would also be important to observe their behavior when plated on matrix or with feeder layers. If the side population cells are indeed anchorage independent this would be a major surprise and an important finding with many ramifications. For example, current theories of intestinal carcinogenesis include loss of anchorage dependence as a critical step leading to metastasis.

Conclusions

Anoikis has hampered successful intestinal epithelial cell isolation and culture, especially from adult animals, but only more recently have the mechanisms started to be understood (Dufour *et al.*, 2004; Dzierzewicz *et al.*, 2003; Grossmann *et al.*, 2003; Loza-Coll *et al.*, 2005; Sträter *et al.*, 1996). The duration of anchorage deprivation appears to be a critical factor determining cell survival *in vitro*. The isolation procedures using EDTA chelation are rapid and clean but the resulting epithelium is separated from its basement membrane, triggering onset of apoptosis within minutes. However, if contact with ECM elements is rapidly reestablished then EDTA-isolated cells seem to fare as well as cells obtained by the more widely used enzyme digestion procedures whether the tissue is grown under *in vitro* or implant conditions. Thus the method used to prepare the epithelium does not appear to be critical provided precautions necessary to prevent anoikis are recognized.

Although we now understand the basic conditions necessary for short-term survival, conditions fully supporting the growth and differentiation of intestinal epithelial stem cells and early progenitors *in vitro* have not been established. Full differentiation into the various epithelial cell types as well as organotypic development have only been demonstrated when cells have been implanted into syngeneic or immunocompromised hosts.

Epithelial Isolation Protocols

We have included a number of general protocols in the following sections that can be adapted for multiple end purposes. Measurements of mouse intestinal epithelial cell components, such as mRNA, often proceed directly from intact intestine rather than attempting first to purify the epithelium. This is not as unreasonable as it might seem because the epithelium comprises a substantial portion of the adult mouse gut, and thus for many purposes it has proven adequate (and in some circumstances preferable in order to minimize handling) to prepare extracts from entire gut segments without attempting to separate tissue elements. In larger animals including humans the mucosa is usually mechanically stripped from the underlying submucosa before further processing to eliminate the thick muscle layers. It must be kept in mind that a broad mixture of cell types is being assayed and this will often severely limit the conclusions that can be reached. Enzyme digestion of adult tissue can be used to enrich epithelium, but it does not yield pure preparations. Only chelation methods yield clean epithelium free of significant mesenchymal contamination; however, users should be aware that the isolated epithelium will contain intraepithelial lymphocytes. EDTA is the most common chelation agent used for this purpose. The optimal concentration, temperature, and time of exposure to EDTA are interdependent and can be adjusted to suit the nature of the experiments.

Perfusion Protocol to Obtain Large Amounts of Pure Epithelium for Biochemical or Physiological Studies

We have found that transcardiac perfusion of 30 mM EDTA at 37° is quick and gives good yield of clean and intact epithelium (Bjerknes and Cheng, 1981c; Fig. 4). Transcardiac perfusion, especially of a mouse, requires practice but yields more viable and intact epithelium in significantly greater quantities in a shorter time than immersion methods. However, immersion is almost universally used because of its simplicity and applicability to human material or specimens obtained from a slaughterhouse (see later).

Prepare 30 mM EDTA in calcium- and magnesium-free Hanks' balanced salt solution (CMF-HBSS); about 150–200 ml will be needed for each mouse. Prewarm the EDTA solution to 37° and then oxygenate it by gassing with O_2 for 10 min, adjust the pH to 7.4, and keep the solution at 37° until use. Perfusion can be accomplished in several ways; we usually use either an intravenous drip assembly or a pump.

1. Open the abdominal cavity of an anesthetized mouse and gently flush the lumen of the gut segment of interest with prewarmed saline. Note that although cleaning the lumen helps the epithelial isolation and reduces contamination, the requisite handling stimulates the gut and may lead to unwanted cellular reactions. Fasting the animals is an alternative, but this is stressful to the animals.

2. Open the thoracic cavity, cut the right atrium, and quickly insert the perfusion hypodermic needle into the left ventricle and perfuse the mouse with EDTA solution for 3–5 min depending on the specifics of the size of the mouse and the perfusion apparatus. These steps should be performed speedily to avoid intestinal vasoconstriction; drugs may be given to prevent the vasoconstriction, but we prefer to avoid their use.

3. Remove the segment of gut, gently evert it onto a rod, and attach the rod to a vibrating or spinning gadget (vibrating mixers or drills work well).

The idea is to rapidly move the tissue through a fluid in order to shear off the epithelium. It is also possible to strip the epithelium by placing the everted segment into a capped tube of medium and shaking, or by grasping the segment with forceps and rapidly moving it in a Petri dish or tube filled with medium.

Immersion Protocol Using Chelation to Obtain Crypts and Cell Clumps for Culture or Other Procedures

Most published protocols are variations of the same basic theme: cut mucosa into small pieces, incubate in EDTA, release crypts and epithelial fragments by shaking, and culture crypts on matrix-coated surfaces in supplemented medium. It should be recognized that the chopping of the mucosa into small pieces will result in the release of some mesenchymal elements. If this is an issue, the intestinal segment should be left as intact as possible to minimize exposed free edges. This described protocol requires the least equipment. Incubation at 37° will speed up the process and result in greater yield of more intact crypts.

1. Tissue preparation: The full thickness of the small intestine or colon can be used with tissue from mice or young rats; with larger specimens such as those from cow or human, the mucosa should be stripped from the underlying tissue before use. Cut the tissue into <5-mm pieces and wash thoroughly in a few changes of oxygenated CMF-HBSS; antibiotics may be added if the crypts will be used for culture or implantation.

2. Chelation: Transfer the tissue pieces to a 50-ml tube or small flask containing 1 mM EDTA and 0.5 mM dithiothreitol (to break down mucin and prevent clumping of crypts) in CMF-HBSS. Incubate for 90 min at room temperature with occasional gentle agitation (less time is needed if a higher EDTA concentration or incubation temperature is used). After the incubation, replace the EDTA with fresh CMF-HBSS and strip the epithelium by vigorous shaking by hand for a few minutes. Let the large tissue fragments settle for about 1 min on ice and remove the supernatant to a new tube. The crypts will settle in about 5–7 min on ice. Sedimentation by gravity is preferable to centrifugation, which may damage the isolated crypts. Small intestinal villus epithelium may be reduced by stripping and discarding the supernatant halfway through the incubation.

3. Crypt culture: The crypts should be transferred to a matrix (or feeder layer) as soon as possible, otherwise onset of anoikis will lead to apoptosis of the cells. Keep in mind that even successful plating of crypts onto a coated surface will result in rapid death of the portions of the crypts not in contact with the matrix.

Protocol Using Enzymatic Method to Obtain
Organoids for In Vitro *Culture*

Many proteases have been used but a mixture of dispase and collagenase is currently favored by most. This is a template protocol that users can adapt to suit their specific applications.

1. Tissue preparation: Prepare the intestine as described in step 1 of the immersion protocol. Finely mince the washed tissue pieces in preparation for digestion.

2. Digestion: The tissue mash is treated with a mixture of dispase I (100 μg/ml) and collagenase XI (300 U/ml) in HBSS or culture medium for 20 min at room temperature with shaking. The enzyme concentrations, incubation temperature, and duration of digestion are interdependent and can be adjusted accordingly. The goal is to obtain a preparation of organoids (preferably single crypts) without damaging the cells by overdigestion. Let the large tissue fragments settle for about 1 min on ice and remove the supernatant to a new tube. The crypts will settle in about 5–7 min on ice. Washing the crypts by gravity sedimentation or filtration is preferable to centrifugation, which may stress the cells. Resuspend the organoids with supplemented medium and repeat a few times to wash the preparation free of enzyme contamination, single cells, and other debris.

3. Culture: Organoids obtained by enzymatic digestion retain their basal lamina and the preparations contain significant amounts of mesenchyme. As a result the epithelial cells may be somewhat protected from anoikis. Such preparations can be grown directly on plastic although the cultures can be maintained for longer times when grown on matrix.

References

Ali, A., and Reynolds, D. L. (1996). Primary cell culture of turkey intestinal epithelial cells. *Avian Dis.* **40,** 103–108.

Arnaout, M. A., Mahalingam, B., and Xiong, J. P. (2005). Integrin structure, allostery, and bidirectional signaling. *Annu. Rev. Cell Dev. Biol.* **21,** 381–410.

Asai, R., Okano, H., and Yasugi, S. (2005). Correlation between Musashi-1 and c-hairy-1 expression and cell proliferation activity in the developing intestine and stomach of both chicken and mouse. *Dev. Growth Differ.* **47,** 501–510.

Asakura, A., and Rudnicki, M. A. (2003). Side population cells from diverse adult tissues are capable of *in vitro* hematopoietic differentiation. *Exp. Hematol.* **30,** 1339–1345.

Autrup, H. (1980). Explant culture of human colon. *Methods Cell Biol.* **21B,** 385–401.

Autrup, H., Stoner, G. D., Jackson, F., Harris, C. C., Shamsuddin, A. K., Barrett, L. A., and Trump, B. F. (1978). Explant culture of rat colon: A model system for studying metabolism of chemical carcinogens. *In Vitro* **14,** 868–877.

Bartsch, I., Zschaler, I., Haseloff, M., and Steinberg, P. (2004). Establishment of a long-term culture system for rat colon epithelial cells. *In Vitro Cell. Dev. Biol. Anim.* **40,** 278–284.

Basque, J. R., Chailler, P., Perreault, N., Beaulieu, J. F., and Menard, D. (1999). A new primary culture system representative of the human gastric epithelium. *Exp. Cell Res.* **253**, 493–502.

Baten, A., Sakamoto, K., and Shamsuddin, A. M. (1992). Long-term culture of normal human colonic epithelial cells *in vitro*. *FASEB J.* **6**, 2726–2734.

Becker, A., McCulloch, E., and Till, J. (1963). Cytological demonstration of the clonal nature of spleen colonies derived from transplanted mouse marrow cells. *Nature* **197**, 452–454.

Berschneider, H. M. (1989). Development of normal cultured small intestinal epithelial cell lines which transport Na and Cl [abstract]. *Gastroenterology* **96**, A41.

Bhargava, D. K., and Lipkin, M. (1981). Transplantation of adenomatous polyps, normal colonic mucosa and adenocarcinoma of colon into athymic mice. *Digestion* **21**, 225–231.

Bjerknes, M. (1985). Assessment of the symmetry of stem-cell mitoses. *Biophys. J.* **48**, 85–91.

Bjerknes, M., and Cheng, H. (1981a). The stem-cell zone of the small intestinal epithelium. I. Evidence from Paneth cells in the adult mouse. *Am. J. Anat.* **160**, 51–63.

Bjerknes, M., and Cheng, H. (1981b). The stem-cell zone of the small intestinal epithelium. III. Evidence from columnar, enteroendocrine, and mucous cells in the adult mouse. *Am. J. Anat.* **160**, 77–91.

Bjerknes, M., and Cheng, H. (1981c). Methods for the isolation of intact epithelium from the mouse intestine. *Anat. Rec.* **199**, 565–574.

Bjerknes, M., and Cheng, H. (1996). Mutant stem cells. *In* "The Gut as a Model in Cell and Molecular Biology" (F. Halter, D. Winton, and N. A. Wright, eds.) pp. 20–33. Proceedings of the Falk Symposium No. 94. Kluwer Academic Publishers, Boston.

Bjerknes, M., and Cheng, H. (1999). Clonal analysis of mouse intestinal epithelial progenitors. *Gastroenterology* **116**, 7–14.

Bjerknes, M., and Cheng, H. (2001). Modulation of specific intestinal epithelial progenitors by enteric neurons. *Proc. Natl. Acad. Sci. USA* **98**, 12497–12502.

Bjerknes, M., and Cheng, H. (2002). Multipotential stem cells in adult mouse gastric epithelium. *Am. J. Physiol.* **283**, G767–G777.

Bjerknes, M., and Cheng, H. (2005a). Gastrointestinal stem cells. II. Intestinal stem cells. *Am. J. Physiol. Gastrointest. Liver Physiol.* **289**, G381–G387.

Bjerknes, M., and Cheng, H. (2005b). Re-examination of P-PTEN staining patterns in the intestinal crypt. *Nat. Genet.* **37**, 1016–1017.

Bjerknes, M., and Cheng, H. (2006). Neurogenin 3 and the enteroendocrine cell lineage in the adult mouse small intestinal epithelium. *Dev. Biol.* (In press.)

Bjorkman, D. J., Allan, C. H., Hagen, S. J., and Trier, J. S. (1986). Structural features of absorptive cell and microvillus membrane preparations from rat small intestine. *Gastroenterology* **91**, 1401–1414.

Blanpain, C., Lowry, W. E., Geoghegan, A., Polak, L., and Fuchs, E. (2004). Self-renewal, multipotency, and the existence of two cell populations within an epithelial stem cell niche. *Cell* **118**, 635–648.

Blay, J., and Brown, K. D. (1984). Characterization of an epithelioid cell line derived from rat small intestine: Demonstration of cytokeratin filaments. *Cell Biol. Int. Rep.* **8**, 551–560.

Booth, C., Patel, S., Bennion, G. R., and Potten, C. S. (1995). The isolation and culture of adult mouse colonic epithelium. *Epithelial Cell Biol.* **4**, 76–86.

Booth, C., O'Shea, J. A., and Potten, C. S. (1999). Maintenance of functional stem cells in isolated and cultured adult intestinal epithelium. *Exp. Cell Res.* **249**, 359–366.

Boudreau, F., Rings, E. H., Swain, G. P., Sinclair, A. M., Suh, E. R., Silberg, D. G., Scheuermann, R. H., and Traber, P. G. (2002). A novel colonic repressor element regulates intestinal gene expression by interacting with Cux/CDP. *Mol. Cell. Biol.* **22**, 5467–5478.

Boxberger, H. J., Meyer, T. F., Grausam, M. C., Reich, K., Becker, H. D., and Sessler, M. J. (1997). Isolating and maintaining highly polarized primary epithelial cells from normal human duodenum for growth as spheroid-like vesicles. *In Vitro Cell. Dev. Biol. Anim.* **33,** 536–545.

Brandsch, C., Friedl, P., Lange, K., Richter, T., and Mothes, T. (1998). Primary culture and transfection of epithelial cells of human small intestine. *Scand. J. Gastroenterol.* **33,** 833–838.

Branting, C., Allinger, U. G., Toftgard, R., and Rafter, J. (1991). Proliferative potential and expression of cell type specific functions in primary mouse colonic epithelial cells. *In Vitro Cell. Dev. Biol.* **27A,** 927–932.

Browning, T. H., and Trier, J. S. (1969). Organ culture of mucosal biopsies of human small intestine. *J. Clin. Invest.* **48,** 1423–1432.

Brubaker, P. L., and Vranic, M. (1987). Fetal rat intestinal cells in monolayer culture: A new *in vitro* system to study the glucagon-like immunoreactive peptides. *Endocrinology* **120,** 1976–1985.

Burns, T. C., Ortiz-Gonzalez, X. R., Gutierrez-Perez, M., Keene, C. D., Sharda, R., Demorest, Z. L., Jiang, Y., Nelson-Holte, M., Soriano, M., Nakagawa, Y., Luquin, M. R., Garcia-Verdugo, J. M., Prosper, F., Low, W. C., and Verfaillie, C. M. (2006). Thymidine analogs are transferred from pre-labeled donor to host cells in the central nervous system after transplantation: A word of caution. *Stem Cells* **24,** 1121–1127.

Buset, M., Winawer, S., and Friedman, E. (1987). Defining conditions to promote the attachment of adult human colonic epithelial cells. *In Vitro Cell. Dev. Biol.* **23,** 403–412.

Cairnie, A. B., Lamerton, L. F., and Steel, G. G. (1965a). Cell proliferation studies in the small intestinal epithelium of the rat. I. Determination of kinetic parameters. *Exp. Cell Res.* **39,** 528–538.

Cairnie, A. B., Lamerton, L. F., and Steel, G. G. (1965b). Cell proliferation studies in the intestinal epithelium of the rat. II. Theoretical aspects. *Exp. Cell Res.* **39,** 539–553.

Cairns, J. (1975). Mutation selection and the natural history of cancer. *Nature* **255,** 197–200.

Campbell, F., Williams, G. T., Appleton, M. A., Dixon, M. F., Harris, M., and Williams, E. D. (1996). Post-irradiation somatic mutation and clonal stabilisation time in the human colon. *Gut* **39,** 569–573.

Capdeville, Y., Frézal, J., Jos, J., Rey, J., and Lamy, M. (1967). Culture de tissue intestinal de veau: Étude de la différenciation cellulair et des activities disaccarasiques. *C. R. Acad. Sci. Hebd. Seances Acad. Sci. D* **264,** 519–521.

Challen, G. A., and Little, M. H. (2006). A side order of stem cells: The SP phenotype. *Stem Cells* **24,** 3–12.

Chambers, R. (1940). The relation of extraneous coats to the organization and permeability of cellular membranes. *Cold Spring Harb. Symp. Quant. Biol.* **8,** 144–152.

Ch'ang, H. J., Maj, J. G., Paris, F., Xing, H. R., Zhang, J., Truman, J. P., Cardon-Cardo, C., Haimovitz-Friedman, A., Kolesnick, R., and Fuks, Z. (2005). ATM regulates target switching to escalating doses of radiation in the intestines. *Nat. Med.* **11,** 484–490.

Chang, W. W., and Nadler, N. J. (1975). Renewal of the epithelium in the descending colon of the mouse. IV. Cell population kinetics of vacuolated-columnar and mucous cells. *Am. J. Anat.* **144,** 39–56.

Chang, W. W. L., and Leblond, C. P. (1971a). Renewal of the epithelium in the descending colon of the mouse. I. Presence of three cell populations: Vacuolated-columnar, mucous, and argentaffin. *Am. J. Anat.* **131,** 73–100.

Chang, W. W. L., and Leblond, C. P. (1971b). Renewal of the epithelium in the descending colon of the mouse. II. Renewal of argentaffin cells. *Am. J. Anat.* **131,** 101–109.

Cheng, H., and Leblond, C. P. (1974). Origin, differentiation and renewal of the four main epithelial cell types in the mouse small intestine. V. Unitarian theory of the origin of the four epithelial cell types. *Am. J. Anat.* **141,** 537–562.

Cheng, H., Bjerknes, M., and Amar, J. (1984). Methods for the determination of epithelial cell kinetic parameters of human colonic epithelium isolated from surgical and biopsy specimens. *Gastroenterology* **86,** 78–85.

Chopra, D. P., and Yeh, K. Y. (1981). Long-term culture of epithelial cells from the normal rat colon. *In Vitro* **17,** 441–449.

Colony, P. C. (1996). Structural characterization of colonic cell types and correlations with specific functions. *Dig. Dis. Sci.* **41,** 88–104.

de Bruïne, A. P., Dinjens, W. N., Pijls, M. M., van der Linden, E. P., Rousch, M. J., Moerkerk, P. T., de Goeij, A. F., and Bosman, F. T. (1992). NCI-H716 cells as a model for endocrine differentiation in colorectal cancer. *Virchows Arch. B Cell Pathol. Mol. Pathol.* **62,** 311–320.

de Bruïne, A. P., de Vries, J. E., Dinjens, W. N., Moerkerk, P. T., van der Linden, E. P., Pijls, M. M., ten Kate, J., and Bosman, F. T. (1993). Human Caco-2 cells transfected with c-Ha-Ras as a model for endocrine differentiation in the large intestine. *Differentiation* **53,** 51–60.

Defries, E. A., and Franks, L. M. (1977). An organ culture method for adult colon from germfree and conventional mice: Effects of donor age and carcinogen treatment on epithelial mitotic activity. *J. Natl. Cancer Inst.* **58,** 1323–1328.

Dekaney, C. M., Rodriguez, J. M., Graul, M. C., and Henning, S. J. (2005). Isolation and characterization of a putative intestinal stem cell fraction from mouse jejunum. *Gastroenterology* **129,** 1567–1580.

Del Buono, R., Lee, C. Y., Hawkey, C. J., and Wright, N. A. (2005). Isolated crypts form spheres before full intestinal differentiation when grown as xenografts: An *in vivo* model for the study of intestinal differentiation and crypt neogenesis, and for the abnormal crypt architecture of juvenile polyposis coli. *J. Pathol.* **206,** 395–401.

DeRitis, G., Falchuk, Z. M., and Trier, J. S. (1975). Differentiation and maturation of cultured fetal rat jejunum. *Dev. Biol.* **45,** 304–317.

Deschner, E., Lewis, C. M., and Lipkin, M. (1963). *In vitro* study of human rectal epithelial cells. I. Atypical zone of ^{3}H thymidine incorporation in mucosa of multiple polyposis. *J. Clin. Invest.* **42,** 1922–1928.

Deveney, C. W., Rand-Luby, L., Rutten, M. J., Luttropp, C. A., Fowler, W. M., Land, J., Meichsner, C. L., Farahmand, M., Sheppard, B. C., Crass, R. A., and Deveney, K. E. (1996). Establishment of human colonic epithelial cells in long-term culture. *J. Surg. Res.* **64,** 161–169.

Dickens, F., and Šimer, F. (1930). The metabolism of normal and tumour tissue. II. The respiratory quotient and the relationship of respiration to glycolysis. *Biochem. J.* **24,** 1301–1326.

Dickens, F., and Weil-Malherbe, H. (1941). Metabolism of normal and tumour tissue. 19. The metabolism of intestinal mucous membrane. *Biochem. J.* **35,** 7–15.

Dobbins, W. O., III, Hijmans, J. C., and McCary, K. S. (1967). A light and electron microscopic study of the duodenal epithelium of chick embryos cultured in the presence and absence of hydrocortisone. *Gastroenterology* **53,** 557–574.

Dolin, R., Blacklow, N. R., Malmgren, R. A., and Chanock, R. M. (1970). Establishment of human fetal intestinal organ cultures for growth of viruses. *J. Infect. Dis.* **122,** 227–231.

Douma, S., Van Laar, T., Zevenhoven, J., Meuwissen, R., Van Garderen, E., and Peeper, D. S. (2004). Suppression of anoikis and induction of metastasis by the neurotrophic receptor TrkB. *Nature* **430,** 1034–1039.

Dufour, G., Demers, M. J., Gagne, D., Dydensborg, A. B., Teller, I. C., Bouchard, V., Degongre, I., Beaulieu, J. F., Cheng, J. Q., Fujita, N., Tsuruo, T., Vallee, K., and Vachon, P. H. (2004). Human intestinal epithelial cell survival and anoikis. Differentiation state-distinct regulation and roles of protein kinase B/Akt isoforms. *J. Biol. Chem.* **279,** 44113–44122.

Dzierzewicz, Z., Orchel, A., Parfiniewicz, B., Weglarz, L., Stojko, J., Swierczek-Zieba, G., and Wilczok, T. (2003). The delay of anoikis due to the inhibition of protein tyrosine dephosphorylation enables the maintenance of normal rat colonocyte primary culture. *Folia Histochem. Cytobiol.* **41,** 223–228.

Eastwood, G. L., and Trier, J. S. (1973). Organ culture of human rectal mucosa. *Gastroenterology* **64,** 375–382.

Enders, J. F., Weller, T. H., and Robbins, F. C. (1949). Cultivation of the Lansing strain of poliomyelitis virus in cultures of various human embryonic tissues. *Science* **109,** 85–89.

Evans, G. S., Flint, N., Somers, A. S., Eyden, B., and Potten, C. S. (1992). The development of a method for the preparation of rat intestinal epithelial cell primary cultures. *J. Cell Sci.* **101,** 219–231.

Farrell, C. L., Rex, K. L., Chen, J. N., Bready, J. V., DiPalma, C. R., Kaufman, S. A., Rattan, A., Scully, S., and Lacey, D. L. (2002). The effects of keratinocyte growth factor in preclinical models of mucositis. *Cell Prolif.* **35**(Suppl. 1), 78–85.

Ferland, S., and Hugon, J. S. (1979). Organ culture of adult mouse intestine. I. Morphological results after 24 and 48 hours of culture. *In Vitro* **15,** 278–287.

Fischer, A. (1922). Cultures of organized tissues. *J. Exp. Med.* **36,** 393–398.

Fogh, J., Fogh, J. M., and Orfeo, T. (1977). One hundred and twenty-seven cultured human tumor cell lines producing tumors in nude mice. *J. Natl. Cancer Inst.* **59,** 221–226.

Fouquet, S., Lugo-Martinez, V. H., Faussat, A. M., Renaud, F., Cardot, P., Chambaz, J., Pincon-Raymond, M., and Thenet, S. (2004). Early loss of E-cadherin from cell–cell contacts is involved in the onset of anoikis in enterocytes. *J. Biol. Chem.* **279,** 43061–43069.

Friedman, E., Gillin, S., and Lipkin, M. (1984). 12-*O*-Tetradecanoylphorbol-13-acetate stimulation of DNA synthesis in cultured preneoplastic familial polyposis colonic epithelial cells but not in normal colonic epithelial cells. *Cancer Res.* **44,** 4078–4086.

Frisch, S. M., and Francis, H. (1994). Disruption of epithelial cell–matrix interactions induces apoptosis. *J. Cell Biol.* **124,** 619–626.

Fujimoto, K., Beauchamp, R. D., and Whitehead, R. H. (2002). Identification and isolation of candidate human colonic clonogenic cells based on cell surface integrin expression. *Gastroenterology* **123,** 1941–1948.

Fukamachi, H. (1992). Proliferation and differentiation of fetal rat intestinal epithelial cells in primary serum-free culture. *J. Cell Sci.* **103,** 511–519.

Fukamachi, H., and Takayama, S. (1980). Epithelial–mesenchymal interaction in differentiation of duodenal epithelium of fetal rats in organ culture. *Experientia* **36,** 335–336.

Garabedian, E. M., Roberts, L. J., McNevin, M. S., and Gordon, J. I. (1997). Examining the role of Paneth cells in the small intestine by lineage ablation in transgenic mice. *J. Biol. Chem.* **272,** 23729–23740.

Gebert, A., Rothkotter, H. J., and Pabst, R. (1996). M cells in Peyer's patches of the intestine. *Int. Rev. Cytol.* **167,** 91–159.

Giannakis, M., Stappenbeck, T. S., Mills, J. C., Leip, D. G., Lovett, M., Clifton, S. W., Ippolito, J. E., Glasscock, J. I., Arumugam, M., Brent, M. R., and Gordon, J. I. (2006). Molecular properties of adult mouse gastric and intestinal epithelial progenitors in their niches. *J. Biol. Chem.* **281,** 11292–11300.

Gibson, P. R., van de Pol, E., Maxwell, L. E., Gabriel, A., and Doe, W. F. (1989). Isolation of colonic crypts that maintain structural and metabolic viability *in vitro. Gastroenterology* **96,** 283–291.

Gooding, J. M., Yap, K. L., and Ikura, M. I. (2004). The cadherin–catenin complex as a focal point of cell adhesion and signalling: New insights from three-dimensional structures. *Bioessays* **26,** 497–511.

Greaves, L. C., Preston, S. L., Tadrous, P. J., Taylor, R. W., Barron, M. J., Oukrif, D., Leedham, S. J., Deheragoda, M., Sasieni, P., Novelli, M. R., Jankowski, J. A., Turnbull, D. M., Wright, N. A., and McDonald, S. A. (2006). Mitochondrial DNA mutations are established in human colonic stem cells, and mutated clones expand by crypt fission. *Proc. Natl. Acad. Sci. USA* **103,** 714–719.

Gregorieff, A., Pinto, D., Begthel, H., Destree, O., Kielman, M., and Clevers, H. (2005). Expression pattern of Wnt signaling components in the adult intestine. *Gastroenterology* **129,** 626–638.

Griffiths, D. F. R., Davies, S. J., Williams, D., Williams, G. T., and Williams, E. D. (1988). Demonstration of somatic mutation and colonic crypt clonality by X-linked enzyme histochemistry. *Nature* **333,** 461–463.

Grikscheit, T. C., Ochoa, E. R., Ramsanahie, A., Alsberg, E., Mooney, D., Whang, E. E., and Vacanti, J. P. (2003). Tissue-engineered large intestine resembles native colon with appropriate *in vitro* physiology and architecture. *Ann. Surg.* **238,** 35–41.

Grikscheit, T. C., Siddique, A., Ochoa, E. R., Srinivasan, A., Alsberg, E., Hodin, R. A., and Vacanti, J. P. (2004). Tissue-engineered small intestine improves recovery after massive small bowel resection. *Ann. Surg.* **240,** 748–754.

Grossmann, J., Walther, K., Artinger, M., Kiessling, S., Steinkamp, M., Schmautz, W. K., Stadler, F., Bataille, F., Schultz, M., Scholmerich, J., and Rogler, G. (2003). Progress on isolation and short-term *ex-vivo* culture of highly purified nonapoptotic human intestinal epithelial cells (IEC). *Eur. J. Cell Biol.* **82,** 262–270.

Haffen, K., Kedinger, M., Simon, P. M., and Raul, F. (1981). Organogenetic potentialities of rat intestinal epithelioid cell cultures. *Differentiation* **18,** 97–103.

Hahn, U., Stallmach, A., Hahn, E. G., and Riecken, E. O. (1990). Basement membrane components are potent promoters of rat intestinal epithelial cell differentiation *in vitro*. *Gastroenterology* **98,** 322–335.

Haidari, M., Leung, N., Mahbub, F., Uffelman, K. D., Kohen-Avramoglu, R., Lewis, G. F., and Adeli, K. (2002). Fasting and postprandial overproduction of intestinally derived lipoproteins in an animal model of insulin resistance: Evidence that chronic fructose feeding in the hamster is accompanied by enhanced intestinal *de novo* lipogenesis and ApoB48-containing lipoprotein overproduction. *J. Biol. Chem.* **277,** 31646–31655.

Harrer, D. S., Stern, B. K., and Reilly, R. W. (1964). Removal and dissociation of epithelial cells from the rodent gastrointestinal tract. *Nature* **203,** 319–320.

Harrison, D. D., and Webster, H. L. (1964). An improved method for the isolation of brush borders from the rat intestine. *Biochim. Biophys. Acta* **93,** 662–664.

He, X. C., Zhang, J., Tong, W., Tawfik, O., Ross, J., Scoville, D. H., Tian, Q., Zeng, X., He, X., Wiedemann, L. M., Mishina, Y., and Li, L. (2004). BMP signaling inhibits intestinal stem cell self-renewal through suppression of Wnt–β-catenin signaling. *Nat. Genet.* **36,** 1117–1121.

Hendry, J. H., and Potten, C. S. (1974). Cryptogenic cells and proliferative cells in intestinal epithelium. *Int. J. Radiat. Biol.* **25,** 583–588.

Ho, S. B., Itzkowitz, S. H., Friera, A. M., Jiang, S. H., and Kim, Y. S. (1989). Cell lineage markers in premalignant and malignant colonic mucosa. *Gastroenterology* **97,** 392–404.

Holtfreter, J. (1948). Significance of the cell membrane in embryonic processes. *Ann. N. Y. Acad. Sci.* **49,** 709–760.

Horster, M., Fabritius, J., Buttner, M., Maul, R., and Weckwerth, P. (1994). Colonic-crypt-derived epithelia express induced ion transport differentiation in monolayer cultures on permeable matrix substrata. *Pflugers Arch.* **426,** 110–120.

Houchen, C. W., George, R. J., Sturmoski, M. A., and Cohn, S. M. (1999). FGF-2 enhances intestinal stem cell survival and its expression is induced after radiation injury. *Am. J. Physiol.* **276,** G249–G258.

Huet, C., Sahuquillo-Merino, C., Coudrier, E., and Louvard, D. (1987). Absorptive and mucus-secreting subclones isolated from a multipotent intestinal cell line (HT-29) provide new models for cell polarity and terminal differentiation. *J. Cell Biol.* **105,** 345–357.

Inoue, M., Imada, M., Fukushima, Y., Matsuura, N., Shiozaki, H., Mori, T., Kitamura, Y., and Fujita, H. (1988). Macroscopic intestinal colonies of mice as a tool for studying differentiation of multipotential intestinal stem cells. *Am. J. Pathol.* **132,** 49–58.

Jiang, Y., Jahagirdar, B. N., Reinhardt, R. L., Schwartz, R. E., Keene, C. D., Ortiz-Gonzalez, X. R., Reyes, M., Lenvik, T., Lund, T., Blackstad, M., Du, J., Aldrich, S., Lisberg, A., Low, W. C., Largaespada, D. A., and Verfaillie, C. M. (2002). Pluripotency of mesenchymal stem cells derived from adult marrow. *Nature* **418,** 41–49.

Jones, P. H., Harper, S., and Watt, F. M. (1995). Stem cell patterning and fate in human epidermis. *Cell* **80,** 83–93.

Joseph, R. R., Yazer, E., Hanakawa, Y., and Stadnyk, A. W. (2005). Prostaglandins and activation of AC/cAMP prevents anoikis in IEC-18. *Apoptosis* **10,** 1221–1233.

Kalabis, J., Patterson, M. J., Enders, G. H., Marian, B., Iozzo, R. V., Rogler, G., Gimotty, P. A., and Herlyn, M. (2003). Stimulation of human colonic epithelial cells by leukemia inhibitory factor is dependent on collagen-embedded fibroblasts in organotypic culture. *FASEB J.* **17,** 1115–1117.

Kawaguchi, A. L., Dunn, J. C., and Fonkalsrud, E. W. (1998). *In vivo* growth of transplanted genetically altered intestinal stem cells. *J. Pediatr. Surg.* **33,** 559–563.

Kayahara, T., Swawada, M., Takaishi, S., Fukui, H., Seno, H., Fukuzawa, H., Suzuki, K., Hiai, H., Kageyama, R., Okano, H., and Chiba, T. (2003). Candidate markers for stem and early progenitor cells, Musashi-1 and Hes1, are expressed in crypt base columnar cells of mouse small intestine. *FEBS Lett.* **535,** 131–135.

Kedinger, M., Simon, P. M., Grenier, J. F., and Haffen, K. (1981). Role of epithelial–mesenchymal interactions in the ontogenesis of intestinal brush-border enzymes. *Dev. Biol.* **86,** 339–347.

Kedinger, M., Simon-Assmann, P. M., Lacroix, B., Marxer, A., Hauri, H. P., and Haffen, K. (1986). Fetal gut mesenchyme induces differentiation of cultured intestinal endodermal and crypt cells. *Dev. Biol.* **113,** 474–483.

Kedinger, M., Duluc, I., Fritsch, C., Lorentz, O., Plateroti, M., and Freund, J. N. (1998). Intestinal epithelial–mesenchymal cell interactions. *Ann. N. Y. Acad. Sci.* **859,** 1–17.

Kim, B. M., Buchner, G., Miletich, I., Sharpe, P. T., and Shivdasani, R. A. (2005). The stomach mesenchymal transcription factor Barx1 specifies gastric epithelial identity through inhibition of transient Wnt signaling. *Dev. Cell* **8,** 611–622.

Kim, K. A., Kakitani, M., Zhao, J., Oshima, T., Tang, T., Binnerts, M., Liu, Y., Boyle, B., Park, E., Emtage, P., Funk, W. D., and Tomizuka, K. (2005). Mitogenic influence of human R-spondin1 on the intestinal epithelium. *Science* **309,** 1256–1259.

Kim, S. J., Cheung, S., and Hellerstein, M. K. (2004). Isolation of nuclei from label-retaining cells and measurement of their turnover rates in rat colon. *Am. J. Physiol. Cell Physiol.* **286,** C1464–C1473.

Kirkland, S. C. (1988). Clonal origin of columnar, mucous, and endocrine cell lineages in human colorectal epithelium. *Cancer* **61,** 1359–1363.

Kondo, Y., Rose, I., Young, G. P., and Whitehead, R. H. (1984). Growth and differentiation of fetal rat small intestinal epithelium in tissue culture: Relationship to fetal age. *Exp. Cell Res.* **153,** 121–134.

Leitinger, B., McDowall, A., and Hogg, N. (2000). The regulation of integrin function by Ca^{2+}. *Biochim. Biophys. Acta* **1498,** 91–98.

Lorenzsonn, V., and Trier, J. S. (1968). The fine structure of human rectal mucosa: The epithelial lining of the base of the crypt. *Gastroenterology* **55,** 88–101.

Loza-Coll, M. A., Perera, S., Shi, W., and Filmus, J. (2005). A transient increase in the activity of Src-famil kinases induced by cell detachment delays anoikis of intestinal epithelial cells. *Oncogene* **24,** 1727–1737.

Macartney, K. K., Baumgart, D. C., Carding, S. R., Brubaker, J. O., and Offit, P. A. (2000). Primary murine small intestinal epithelial cells, maintained in long-term culture, are susceptible to rotavirus infection. *J. Virol.* **74,** 5597–5603.

Mach, J. P., Carrel, S., Merenda, C., Sordat, B., and Cerottini, J. C. (1974). *In vivo* localisation of radiolabelled antibodies to carcinoembryonic antigen in human colon carcinoma grafted into nude mice. *Nature* **248,** 704–706.

Madara, J. L. (1982). Cup cells: Structure and distribution of a unique class of epithelial cells in guinea pig, rabbit, and monkey small intestine. *Gastroenterology* **83,** 981–994.

Magney, J. E., Erlandsen, S. L., Bjerknes, M. L., and Cheng, H. (1986). Scanning electron microscopy of isolated epithelium of the murine gastrointestinal tract: Morphology of the basal surface and evidence for paracrinelike cells. *Am. J. Anat.* **177,** 43–53.

Marshman, E., Booth, C., and Potten, C. S. (2002). The intestinal epithelial stem cell. *Bioessays* **24,** 91–98.

Melcher, R., Maisch, S., Koehler, S., Bauer, M., Steinlein, C., Schmid, M., Kudlich, T., Schauber, J., Luehrs, H., Menzel, T., and Scheppach, W. (2005). SKY and genetic fingerprinting reveal a cross-contamination of the putative normal colon epithelial cell line NCOL-1. *Cancer Genet. Cytogenet.* **158,** 84–87.

Miller, D., and Crane, R. K. (1961). The digestive function of the epithelium of the small intestine. II. Localization of disaccharide hydrolysis in the isolated brush border portion of intestinal epithelial cells. *Biochim. Biophys. Acta* **52,** 293–298.

Mizuno, T., and Yasugi, S. (1990). Susceptibility of epithelia to directive influences of mesenchymes during organogenesis: Uncoupling of morphogenesis and cytodifferentiation. *Cell Differ. Dev.* **31,** 151–159.

Montgomery, R. K., Zinman, H. M., and Smith, B. T. (1983). Organotypic differentiation of trypsin-dissociated fetal rat intestine. *Dev. Biol.* **100,** 181–189.

Moog, F., and Nehari, V. (1954). The influences of hydrocortisone on the epithelial phosphatase of embryonic intestine *in vitro*. *Science* **119,** 809–810.

Moorghen, M., Chapman, M., and Appleton, D. R. (1996). An organ-culture method for human colorectal mucosa using serum-free medium. *J. Pathol.* **180,** 102–105.

Moscona, A., and Moscona, H. (1952). The dissociation and aggregation of cells from organ rudiments of the early chick embryo. *J. Anat.* **86,** 287–301.

Moyer, M. P. (1983). Culture of human gastrointestinal epithelial cells. *Proc. Soc. Exp. Biol. Med.* **174,** 12–15.

Moyer, M. P., and Aust, J. B. (1984). Human colon cells: Culture and *in vitro* transformation. *Science* **224,** 1445–1447.

Moyer, M. P., Manzano, L. A., Merriman, R. L., Stauffer, J. S., and Tanzer, L. R. (1996). NCM 460, a normal human colon mucosal epithelial cell line. *In Vitro Cell. Dev. Biol. Anim.* **32,** 315–317.

Münz, F. (1938). Polyamino carboxylic acids and process of making same. U.S. Patent 2,130,505.

Nabeyama, A., and Leblond, C. P. (1974). "Caveolated cells" characterized by deep surface invaginations and abundant filaments in mouse gastro-intestinal epithelia. *Am. J. Anat.* **140,** 147–166.

Neutra, M. R., Frey, A., and Kraehenbuhl, J. P. (1996). Epithelial M cells: Gateways for mucosal infection and immunization. *Cell* **86,** 345–348.

Okayasu, I., Hana, K., Yoshida, T., Mikami, T., Kanno, J., and Fujiwara, M. (2002). Significant increase of colonic mutated crypts in ulcerative colitis correlatively with duration of illness. *Cancer Res.* **62,** 2236–2238.

Oldham, S. M., Clark, G. J., Gangarosa, L. M., Coffey, R. J., Jr., and Der, C. J. (1996). Activation of the Raf-1/MAP kinase cascade is not sufficient for Ras transformation of RIE-1 epithelial cells. *Proc. Natl. Acad. Sci. USA* **93,** 6924–6928.

Ormestad, M., Astorga, J., Landgren, H., Wang, T., Johansson, B. R., Miura, N., and Carlsson, P. (2006). Foxf1 and Foxf2 control murine gut development by limiting mesenchymal Wnt signaling and promoting extracellular matrix production. *Development* **133,** 833–843.

Ouellette, A. J. (2005). Paneth cell α-defensins: Peptide mediators of innate immunity in the small intestine. *Springer Semin. Immunopathol.* **27,** 133–146.

Paris, F., Fuks, Z., Kang, A., Capodieci, P., Juan, G., Ehleiter, D., Haimovitz-Friedman, A., Cordon-Cardo, C., and Kolesnick, R. (2001). Endothelial apoptosis as the primary lesion initiating intestinal radiation damage in mice. *Science* **293,** 293–297.

Park, H. S., Goodlad, R. A., and Wright, N. A. (1995). Crypt fission in the small intestine and colon: A mechanism for the emergence of G6PD locus-mutated crypts after treatment with mutagens. *Am. J. Pathol.* **147,** 1416–1427.

Pedersen, G., Saermark, T., Giese, B., Hansen, A., Drag, B., and Brynskov, J. (2000). A simple method to establish short-term cultures of normal human colonic epithelial cells from endoscopic biopsy specimens: Comparison of isolation methods, assessment of viability and metabolic activity. *Scand. J. Gastroenterol.* **35,** 772–780.

Perreault, N., and Beaulieu, J. F. (1996). Use of the dissociating enzyme thermolysin to generate viable human normal intestinal epithelial cell cultures. *Exp. Cell Res.* **224,** 354–364.

Perreault, N., and Beaulieu, J. F. (1998). Primary cultures of fully differentiated and pure human intestinal epithelial cells. *Exp. Cell Res.* **245,** 34–42.

Ponder, B. A. J., Schmidt, G. H., Wilkinson, M. M., Wood, M. J., Monk, M., and Reid, A. (1985). Derivation of mouse intestinal crypts from single progenitor cells. *Nature* **313,** 689–691.

Potten, C. S., and Hendry, J. H. (1985). The microcolony assay in mouse small intestine. *In* "Cell Clones: Manual of Mammalian Cell Techniques" (C. S. Potten and J. H. Hendry, eds.), pp. 155–159. Churchill-Livingstone, Edinburgh.

Potten, C. S., Owen, G., and Booth, D. (2002). Intestinal stem cells protect their genome by selective segregation of template DNA strands. *J. Cell Sci.* **115,** 2381–2388.

Potten, C. S., Booth, C., Tudor, G. L., Booth, D., Brady, G., Hurley, P., Ashton, G., Clarke, R., Sakakibara, S., and Okano, H. (2003). Identification of a putative intestinal stem cell and early lineage marker; musashi-1. *Differentiation* **71,** 28–41.

Povlsen, C. O., and Rygaard, J. (1971). Heterotransplantation of human adenocarcinomas of the colon and rectum to the mouse mutant Nude: A study of nine consecutive transplantations. *Acta Pathol. Microbiol. Scand. A* **79,** 159–169.

Pyke, K. W., and Gogerly, R. L. (1985). Murine fetal colon *in vitro*: Assays for growth factors. *Differentiation* **29,** 56–62.

Quaroni, A. (1985). Development of fetal rat intestine in organ and monolayer culture. *J. Cell Biol.* **100,** 1611–1622.

Quaroni, A., and Beaulieu, J. F. (1997). Cell dynamics and differentiation of conditionally immortalized human intestinal epithelial cells. *Gastroenterology* **113,** 1198–1213.

Quaroni, A., Wands, J., Trelstad, R. L., and Isselbacher, K. J. (1979). Epithelioid cell cultures from rat small intestine: Characterization by morphologic and immunologic criteria. *J. Cell Biol.* **80,** 248–265.

Rak, J., Mitsuhashi, Y., Erdos, V., Huang, S. N., Filmus, J., and Kerbel, R. S. (1995). Massive programmed cell death in intestinal epithelial cells induced by three-dimensional growth conditions: Suppression by mutant c-H-*ras* oncogene expression. *J. Cell Biol.* **131,** 1587–1598.

Roberts, S. A., Hendry, J. H., and Potten, C. S. (2003). Intestinal crypt clonogens: A new interpretation of radiation survival curve shape and clonogenic cell number. *Cell Prolif.* **36,** 215–231.

Rusu, D., Loret, S., Peulen, O., Mainil, J., and Dandrifosse, G. (2005). Immunochemical, biomolecular and biochemical characterization of bovine epithelial intestinal primocultures. *BMC Cell Biol.* **6,** 42.

Sakatani, T., Kaneda, A., Iacobuzio-Donahue, C. A., Carter, M. G., de Boom Witzel, S., Okano, H., Ko, M. S., Ohlsson, R., Longo, D. L., and Feinberg, A. P. (2005). Loss of imprinting of Igf2 alters intestinal maturation and tumorigenesis in mice. *Science* **307,** 1976–1978.

Sancho, E., Batlle, E., and Clevers, H. (2004). Signaling pathways in intestinal development and cancer. *Annu. Rev. Cell Dev. Biol.* **20,** 695–723.

Schonhoff, S. E., Giel-Moloney, M., and Leiter, A. B. (2004). Development and differentiation of gut endocrine cells. *Endocrinology* **145,** 2639–2644.

Schörkhuber, M., Karner-Hanusch, J., Sedivy, R., Ellinger, A., Armbruster, C., Schulte-Hermann, R., and Marian, B. (1998). Survival of normal colonic epithelial cells from both rats and humans is prolonged by coculture with rat embryo colonic fibroblasts. *Cell Biol. Toxicol.* **14,** 211–223.

Shackleton, M., Vaillant, F., Simpson, K. J., Stingl, J., Smyth, G. K., Asselin-Labat, M. L., Wu, L., Lindeman, G. J., and Visvader, J. E. (2006). Generation of a functional mammary gland from a single stem cell. *Nature* **439,** 84–88.

Sheinin, Y., Kallay, E., Wrba, F., Kriwanek, S., Peterlik, M., and Cross, H. S. (2000). Immunocytochemical localization of the extracellular calcium-sensing receptor in normal and malignant human large intestinal mucosa. *J. Histochem. Cytochem.* **48,** 595–602.

Shen, X., and Kramer, R. H. (2004). Adhesion-mediated squamous cell carcinoma survival through ligand-independent activation of epidermal growth factor receptor. *Am. J. Pathol.* **165,** 1315–1329.

Siminovitch, L., McCulloch, E. A., and Till, J. E. (1963). The distribution of colony-forming cells among spleen colonies. *J. Cell Comp. Physiol.* **62,** 327–336.

Sjöstrand, F. S. (1968). A simple and rapid method to prepare dispersions of columnar epithelial cells from the rat intestine. *J. Ultrastruct. Res.* **22,** 424–442.

Slorach, E. M., Campbell, F. C., and Dorin, J. R. (1999). A mouse model of intestinal stem cell function and regeneration. *J. Cell Sci.* **112,** 3029–3038.

Spyridonidis, A., Schmitt-Graff, A., Tomann, T., Dwenger, A., Follo, M., Behringer, D., and Finke, J. (2004). Epithelial tissue chimerism after human hematopoietic cell transplantation is a real phenomenon. *Am. J. Pathol.* **164,** 1147–1155.

Stallmach, A., Hahn, U., Merker, H. J., Hahn, E. G., and Riecken, E. O. (1989). Differentiation of rat intestinal epithelial cells is induced by organotypic mesenchymal cells *in vitro. Gut* **30,** 959–970.

Stern, B. K. (1966). Some biochemical propertines of suspensions of intestinal epithelial cells. *Gastroenterology* **51,** 855–864.

Stingl, J., Eirew, P., Ricketson, I., Shackleton, M., Vaillant, F., Choi, D., Li, H. I., and Eaves, C. J. (2006). Purification and unique properties of mammary epithelial stem cells. *Nature* **439,** 993–997.

Sträter, J., Wedding, U., Barth, T. F., Koretz, K., Elsing, C., and Möller, P. (1996). Rapid onset of apoptosis *in vitro* follows disruption of β_1-integrin/matrix interactions in human colonic crypt cells. *Gastroenterology* **110,** 1776–1784.

Sugihara, K., and Jass, J. R. (1986). Colorectal goblet cell sialomucin heterogeneity: Its relation to malignant disease. *J. Clin. Pathol.* **39,** 1088–1095.

Suh, E., and Traber, P. G. (1996). An intestine-specific homeobox gene regulates proliferation and differentiation. *Mol. Cell Biol.* **16,** 619–625.

Suzuki, M. T., and Giovannoni, S. J. (1996). Bias caused by template annealing in the amplification of mixtures of 16S rRNA genes by PCR. *Appl. Environ. Microbiol.* **62,** 625–630.

Tahara, K., Uchida, H., Kawarasaki, H., Hasizume, K., and Kobayashi, E. (2001). Experimental small bowel transplantation using newborn intestine in rats. III. Long-term cryopreservation of rat newborn intestine. *J. Pediatr. Surg.* **36,** 602–604.

Tait, I. S., Flint, N., Campbell, F. C., and Evans, G. S. (1994). Generation of neomucosa *in vivo* by transplantation of dissociated rat postnatal small intestinal epithelium. *Differentiation* **56,** 91–100.

Taylor, R. W., Barron, M. J., Borthwick, G. M., Gospel, A., Chinnery, P. F., Samuels, D. C., Taylor, G. A., Plusa, S. M., Needham, S. J., Greaves, L. C., Kirkwood, T. B., and Turnbull, D. M. (2003). Mitochondrial DNA mutations in human colonic crypt stem cells. *J. Clin. Invest.* **112,** 1351–1360.

Till, J. E., and McCulloch, E. A. (1961). A direct measurement of the radiation sensitivity of normal mouse bone marrow cells. *Radiat. Res.* **14,** 1419–1430.

Trier, J. S. (1976). Organ-culture methods in the study of gastrointestinal–mucosal function and development. *N. Engl. J. Med.* **295,** 150–155.

Tucker, S. L., and Thames, H. D., Jr. (1983). Optimal design of multifraction assays of colony survival *in vivo. Radiat. Res.* **94,** 280–294.

Uchida, H., Yoshida, T., Kobayashi, E., Mizuta, K., Fujimura, A., Miyata, M., Kawarasaki, H., and Hashizume, K. (1999). Experimental small bowel transplantation using newborn intestine in rats. I. Lipid absorption restored after transplantation of nonvascularized graft. *J. Pediatr. Surg.* **34,** 1007–1011.

Uchida, H., Tahara, K., Takizawa, T., Inose, K., Yashiro, T., Hashizume, K., Ikeda, H., Takahashi, M., and Kobayashi, E. (2001). Experimental small bowel transplantation using a newborn intestine in rats. IV. Effect of cold preservation on graft neovascularization. *J. Pediatr. Surg.* **36,** 1805–1810.

Uchida, N., Dykstra, B., Lyons, K., Leung, F., Kristiansen, M., and Eaves, C. (2004). ABC transporter activities of murine hematopoietic stem cells vary according to their developmental and activation status. *Blood* **103,** 4487–4495.

Valerio, M. G., Fineman, E. L., Bowman, R. L., Harris, C. C., Stoner, G. D., Autrup, H., Trump, B. F., McDowell, E. M., and Jones, R. T. (1981). Long-term survival of normal adult human tissues as xenografts in congenitally athymic nude mice. *J. Natl. Cancer Inst.* **66,** 849–858.

Vidrich, A., Ravindranath, R., Farsi, K., and Targan, S. (1988). A method for the rapid establishment of normal adult mammalian colonic epithelial cell cultures. *In Vitro Cell. Dev. Biol.* **24,** 188–194.

Vidrich, A., Buzan, J. M., Barnes, S., Reuter, B. K., Skaar, K., Ilo, C., Cominelli, F., Pizarro, T., and Cohn, S. M. (2005). Altered epithelial cell lineage allocation and global expansion of the crypt epithelial stem cell population are associated with ileitis in SAMP1/YitFc mice. *Am. J. Pathol.* **166,** 1055–1067.

Waterhouse, C. C., Joseph, R. R., and Stadnyk, A. W. (2001). Endogenous IL-1 and type II IL-1 receptor expression modulate anoikis in intestinal epithelial cells. *Exp. Cell Res.* **269,** 109–116.

Webster, H. L., and Harrison, D. D. (1969). Enzymatic activities during the transformation of crypt to columnar intestinal cells. *Exp. Cell Res.* **56,** 245–253.

Weiser, M. M. (1973). Intestinal epithelial cell surface membrane glycoprotein synthesis. I. An indicator of cellular differentiation. *J. Biol. Chem.* **248,** 2536–2541.

Weissman, I. L. (2002). The road ended up at stem cells. *Immunol. Rev.* **185,** 159–174.

Weng, X. H., Beyenbach, K. W., and Quaroni, A. (2005). Cultured monolayers of the dog jejunum with the structural and functional properties resembling the normal epithelium. *Am. J. Physiol.* **288,** G705–G717.

Whitehead, R. H., Brown, A., and Bhathal, P. S. (1987a). A method for the isolation and culture of human colonic crypts in collagen gels. *In Vitro Cell. Dev. Biol.* **23,** 436–442.

Whitehead, R. H., Jones, J. K., Gabriel, A., and Lukies, R. E. (1987b). A new colon carcinoma cell line (LIM1863) that grows as organoids with spontaneous differentiation into crypt-like structures *in vitro. Cancer Res.* **47,** 2683–2689.

Whitehead, R. H., VanEeden, P. E., Noble, M. D., Ataliotis, P., and Jat, P. S. (1993). Establishment of conditionally immortalized epithelial cell lines from both colon and small intestine of adult H-2Kb-tsA58 transgenic mice. *Proc. Natl. Acad. Sci. USA* **90,** 587–591.

Whitehead, R. H., Demmler, K., Rockman, S. P., and Watson, N. K. (1999). Clonogenic growth of epithelial cells from normal colonic mucosa from both mice and humans. *Gastroenterology* **117,** 858–865.

Wildrick, D. M., Lointier, P., Nichols, D. H., Roll, R., Quintanilla, B., and Boman, B. M. (1997). Isolation of normal human colonic mucosa: Comparison of methods. *In Vitro Cell. Dev. Biol. Anim.* **33,** 18–27.

Winton, D. J., and Ponder, B. A. (1990). Stem-cell organization in mouse small intestine. *Proc. R. Soc. London Ser. B Biol. Sci.* **241,** 13–18.

Withers, H. R., and Elkind, M. M. (1968). Dose–survival characteristics of epithelial cells of mouse intestinal mucosa. *Radiology* **91,** 998–1000.

Withers, H. R., and Elkind, M. M. (1969). Radiosensitivity and fractionation response of crypt cells of mouse jejunum. *Radiat. Res.* **38,** 598–613.

Withers, H. R., and Elkind, M. M. (1970). Microcolony survival assay for cells of mouse intestinal mucosa exposed to radiation. *Int. J. Radiat. Biol.* **17,** 261–267.

Wong, W. M., Poulsom, R., and Wright, N. A. (1999). Trefoil peptides. *Gut* **44,** 890–895.

Yamada, S., Kojima, H., Fujimiya, M., Nakamura, T., Kashiwagi, A., and Kikkawa, R. (2001). Differentiation of immature enterocytes into enteroendocrine cells by Pdx1 overexpression. *Am. J. Physiol. Gastrointest. Liver Physiol.* **281,** G229–G236.

Yasugi, S., and Mizuno, T. (1978). Differentiation of the digestive tract under the influence of heterologous mesenchyme of the digestive tract in the bird embryos. *Dev. Growth Differ.* **20,** 261–267.

Yatabe, Y., Tavare, S., and Shibata, D. (2001). Investigating stem cells in human colon by using methylation patterns. *Proc. Natl. Acad. Sci. USA* **98,** 10839–10844.

Section IV

Extraembryonic and Perinatal Stem Cells

[15]   Trophoblast Stem Cells

By MAYUMI ODA, KUNIO SHIOTA, and SATOSHI TANAKA

Abstract

At the first cell fate decision in mammalian development, the origins of trophoblast and embryonic cell lineages are established as the trophecto-derm and the inner cell mass (ICM) in the blastocyst. In the trophoblast cell lineage, a subset of the trophectoderm cells maintains the capacity to prolif-erate and contribute to the extraembryonic ectoderm, the ectoplacental cone, and the secondary giant cells of the early conceptus after implantation, and finally they produce the entire trophoblastic population in the mature placenta. The stem cell population of the trophectoderm lineage can be isolated and maintained *in vitro* in the presence of fibroblast growth factor 4, heparin, and a feeder layer of mouse embryonic fibroblast cells. These apparently immortal stem cells in culture are termed trophoblast stem (TS) cells, and exhibit the potential to differentiate into multiple trophoblastic cell types *in vitro*, as well as *in vivo*. Even after multiple passages, TS cells retain the ability to participate in the normal development of chimeras and contribute exclusively to the trophoblastic component of the placenta and of the parietal yolk sac. The fate of TS cells is strikingly in contrast to that of embryonic stem cells, which never contribute to these tissues. In this chapter, detailed protocols for the isolation and establishment of TS cell lines from blastocysts and their maintenance are described.

Introduction

In the mouse preimplantation development, the first irreversible segre-gation takes place between the inner cell mass (ICM) and trophectoderm cell lineages. The trophectoderm, which consists of a monolayer of polar-ized epithelial cells, forms a sphere surrounding the fluid-filled blastocoel and enclosing the ICM at one end of its inner surface. The polar trophec-toderm cells, which are adjacent to the ICM, maintain the capacity to proliferate, and are involved in the formation of a series of trophoblastic tissues including the extraembryonic ectoderm, the ectoplacental cone, and the trophoblastic component of placenta. These pluripotent cells have been identified as the trophoblast stem cells (Rossant and Cross, 2001).

It was found that the trophoblast stem cells can be isolated from mouse embryos and maintained in a proliferative, undifferentiated state in culture

METHODS IN ENZYMOLOGY, VOL. 419
Copyright 2006, Elsevier Inc. All rights reserved.
0076-6879/06 $35.00
DOI: 10.1016/S0076-6879(06)19015-1

(Tanaka *et al.*, 1998). It is speculated that the proliferative ability of tropho-blast stem cells is maintained by ICM-derived factors, and this environment seems to be recapitulated in culture by the supplement of fibroblast growth factor 4 (FGF4) with its cofactor, heparin, and a feeder layer of mouse embryonic fibroblast (MEF) cells. The apparently immortal trophoblast stem cells in culture, which are designated as TS cells, can differentiate after the withdrawal of FGF4–heparin or MEF cells, and mostly become the terminally differentiated secondary trophoblast giant cells. Some also dif-ferentiate into other trophoblastic subtypes including spongiotrophoblast and labyrinthine trophoblast cells (Hughes *et al.*, 2004; Tanaka *et al.*, 1998; Yan *et al.*, 2001). The comprehensive fates of TS cells can be evaluated in chimeras of whole early embryos, in which TS cells contribute to the tro-phoblastic component of the placenta and of the parietal yolk sac equally with the native recipient trophoblast cells, indicating that TS cells *in vitro* fully retain heritable multipotency as the trophoblast stem cells.

The trophoblast cell lineage plays a pivotal role in the fetal–maternal interaction: as the membrane-like vitelline placenta at the onset of implan-tation, and then as the mature chorioallantoic placenta. The placenta med-iates nutrient and gas transport between maternal and embryonic blood flow, and prevents rejection by the maternal immune system (see Rossant and Cross, 2001, for a review). In addition, the trophoblast cell lineage acts as the source of signaling factors in important embryonic developmental processes such as germ line specification (Lawson *et al.*, 1999; Ying *et al.*, 2001). Given that such cell properties were derived from the trophoblast stem cells, not from the embryonic cell lineage, the differences between trophoblast and embryonic stem cells are also of interest within the restricted differentiation abilities and characteristic gene expression traits of these stem-derived cells.

As shown in the chimera experiments, TS cells are strikingly different from embryonic stem (ES) cells, which never contribute to the trophoblast component in chimeras (Nagy and Rossant, 2001). A few genetic determi-nants between these two earliest cell lineages were identified by gene-targeted knockdown experiments (Deb *et al.*, 2006), although it has not been illustrated which are responsible for the different heritable multi-potencies in the two stem origins. To address this question, we have analyzed the genome-wide DNA methylation status (a heritable epigenetic mark) of TS, ES, and embryonic germ (EG) cells (Shiota *et al.*, 2002). The trophoblast cell lineage is also of interest in the evolution of mammals, especially in the similarities and differences in early development and in some epigenetic traits between the eutherians and noneutherians (Selwood and Johnson, 2006; Sharman, 1971). Methods for the establishment and manipulation of TS cells can provide increasingly important information about such interesting questions. This chapter, therefore, focuses on protocols for the establishment and maintenance of TS cell lines.

Preparations for Culture of TS Cells

Equipment

Tissue culture dishes (any brand)

Four-well multidishes (e.g., cat. no. 176740; Nalge Nunc International, Rochester, NY)

Non-tissue culture-treated, U-bottom, 96-well plates (e.g., cat. no. 351177; BD Biosciences Discovery Labware, Bedford, MA).

50-ml centrifuge tubes (any brand)

Cryovials (any brand)

Cell-freezing containers [e.g., Nalgene Cryo 1° freezing container (cat. no. 5100–0001); Nalge Nunc International]

Reagents

Calcium/magnesium-free phosphate-buffered saline [PBS(–)]: Dissolve one tablet (cat. no. P-4417; Sigma-Aldrich, St. Louis, MO) per 200 ml of deionized water. Autoclave to sterilize and store at 4°

0.05% trypsin–0.2 mM EDTA and 0.1% trypsin–0.4 mM EDTA: Dilute 0.25% trypsin–1 mM EDTA·4Na (GIBCO, cat. no. 27250–018; Invitrogen, Carlsbad, CA) with 4× and 1.5× volumes of PBS(–), respectively. Store at 4°.

FGF4 1000× stock solution (25 mg/ml): Add 1 ml of PBS(–) containing 0.1% (w/v) bovine serum albumin (BSA) to a vial of lyophilized human recombinant FGF4 (25 mg) (cat. no. 100–31, PeproTech, London, UK). Mix well by gentle pipetting and freeze in 100-μl aliquots at –80°. Sterilization of FGF4 solution is not necessary as long as sterile PBS(–) containing BSA (passed through 0.45-μm pore size filter) is used. Thaw each aliquot as needed and store at 4°; do not refreeze.

Heparin, 1000× stock solution (1 mg/ml): Resuspend heparin (cat. no. H-3149; Sigma-Aldrich) in PBS(–) to a concentration of 1.0 mg/ml and store at –80° in 100-μl aliquots. Thaw each aliquot as needed and store at 4°; do not refreeze.

Mitomycin C (MMC) (2 mg; cat. no. M-0503; Sigma-Aldrich)

10 mM 2-mercaptoethanol stock solution: Dilute pure liquid 14.3 M 2-mercaptoethanol (cat. no. M-7522; Sigma-Aldrich) with water. Pass through a 0.45-μm pore size filter to sterilize. Store at –20°.

Culture Media

Dulbecco's modified Eagle's medium (DMEM)–10% fetal bovine serum (FBS): DMEM (pH 7.2) supplemented with 10% FBS, 100 μM 2-mercaptoethanol, 2 mM L-glutamine, penicillin (50 U/ml), and streptomycin (50 mg/ml). For 1 liter, add 100 ml of FBS to 870 ml of DMEM (GIBCO, cat. no. 12800–017; Invitrogen). Add 10 ml each of 10 mM 2-mercaptoethanol stock solution, 200 mM L-glutamine (GIBCO, cat. no. 25030–081; Invitrogen), and 100× penicillin–streptomycin solution (GIBCO, cat. no. 15070–063; Invitrogen). Store at 4°.

TS medium: RPMI 1640 (GIBCO, pH 7.2; cat. no. 31800–022; Invitrogen) supplemented with 20% FBS, 100 μM 2-mercaptoethanol, 2 mM L-glutamine, 1 mM sodium pyruvate, penicillin (50 U/ml), and streptomycin (50 mg/ml). For 1 liter, add 200 ml of FBS to 760 ml of RPMI 1640 (pH 7.2). Add 10 ml each of 10 mM 2-mercaptoethanol stock solution, 200 mM L-glutamine, 100 mM sodium pyruvate (GIBCO, cat. no. 11360–070; Invitrogen), and 100× penicillin–streptomycin solution. Store at 4°.

MEF-CM: TS medium conditioned by mouse embryonic fibroblast (MEF) cells. Culture MMC-treated MEF (MMC-MEF) cells (see the next section, Preparation of Feeder Cell Stocks) in TS medium for 3 days (see Table I for adequate number of MEF cells to seed). Collect the medium and store at −20° while preparing additional batches. Prepare two more batches with the same dish of MMC-MEF cells. Combine three batches of conditioned medium and spin at ∼2300g at 4° for 20 min to remove debris. Filter (0.45-μm pore size) and store at −20° in 30- to 40-ml aliquots. Thaw each aliquot as needed and store at 4°; do not refreeze.

TABLE I

ADEQUATE NUMBERS OF MITOMYCIN C-TREATED MOUSE EMBRYONIC
FIBROBLAST CELLS TO PLATE

	Diameter of dish (mm)				
	15	35	60	100	150
For coculture (50% confluent)	4×10^4	2×10^5	4×10^5	1.2×10^6	3×10^6
For MEF-CM (∼fully confluent)				2.4×10^6	6×10^6

Abbreviation: MEF-CM, TS medium conditioned by mouse embryonic fibroblast (MEF) cells.

TABLE II
MEDIUM PREPARATIONS FOR TROPHOBLAST STEM CELLS

	TS medium (ml)	MEF-CM (ml)	FGF4 1000× stock (μl)	Heparin 1000× stock (μl)
TS+F4H	10	—	10	10
TS+1.5× F4H	10	—	15	15
70CM+F4H	3	7	10	10
70CM+1.5× F4H	3	7	15	15

Medium preparations for the establishment and maintenance of TS cell lines (see Table II).

2× freezing medium for MEF cells: 50% FBS, 20% dimethyl sulfoxide (DMSO) in DMEM.

2× freezing medium for TS cells: 50% FBS, 20% DMSO in TS medium.

Preparation of Feeder Cell Stocks

Overview

Because TS cells rapidly proliferate and constantly require the proper amount of feeder population for the next seeding, it is more convenient to prepare frozen stocks of mitotically inactivated MMC-MEF cells at several different densities for the various requirements, rather than treating cells with MMC just before use. We routinely freeze MMC-MEF cells at three different densities, that is, 6×10^6, 1.2×10^6, and 4×10^5 cells per vial, which are suitable for two 150-mm dishes, one 100-mm dish, and one 60-mm dish, respectively, as the feeder layer for TS cells (note that this density is 50% confluent; see Table I).

Preparation of MMC-Treated Feeder Cell Stocks

1. Prepare a vial containing primary MEF cells harvested from an 80–90% confluent 150-mm dish (see Nagy *et al.*, 2003). Quickly thaw a frozen vial of MEF cells at 37°.
2. Transfer the entire contents to 10 ml of DMEM–10% FBS in a 50-ml tube and centrifuge at 200g for 3 min.
3. Discard the supernatant. Loosen the cell pellet by gently tapping the bottom of the tube and then resuspend the cells in 25 ml of DMEM–10% FBS. Split the cells onto five 150-mm dishes, each containing 20 ml of DMEM–10% FBS.

4. Grow the cells at $37°$ in 5% CO_2–95% air to become ∼90% confluent (∼3 days after thawing). Change the medium the next day to remove cell debris.

5. When the cells are ∼90% confluent, remove the medium, rinse the cells with 10 ml of PBS(–) twice, and then add 3 ml of 0.05% trypsin–0.2 mM EDTA to each dish.

6. Incubate at $37°$ in 5% CO_2–95% air for 3–5 min.

7. Roughly dissociate the cell clumps by light tapping of the culture dish, and add 3 ml of DMEM–10% FBS to each dish.

8. Gather the cell suspensions in a single 50-ml tube, spin down (200g, 3 min), and discard the supernatant.

9. Loosen the cell pellet by gently tapping the bottom of the tube and then resuspend the cells in 40 ml of DMEM–10% FBS. Split the cells onto twenty 150-mm dishes, each containing 23 ml of DMEM–10% FBS.

10. Grow the cells at $37°$ in 5% CO_2–95% air to become almost confluent (2–3 days after step 9). Change the medium the second day (optional).

11. When the cells become almost confluent, remove the medium and add 10 ml of DMEM–10% FBS containing MMC (10 μg/ml) (a vial available from Sigma-Aldrich contains 2 mg of MMC, making it convenient to prepare an exact volume of MMC-containing medium for twenty 150-mm dishes).

12. Incubate the cells at $37°$ in 5% CO_2–95% air for 2 h.

13. Remove the medium and rinse the cells twice with 10 ml of PBS(–).

14. Trypsinize the cells as described in steps 5–8. Finally, resuspend the cell pellet in 20 ml of DMEM–10% FBS.

15. Count viable cells and dilute the cell suspension to two times the final desired density with DMEM–10% FBS. Another set of spinning down/resuspending steps may be required to achieve high density.

16. Freeze the cells as described in the section entitled Freezing, using 2× freezing medium for MEF cells (see Culture Media).

Establishment of TS Cell Lines

Overview

From the blastocyst to the early gastrula stages, the embryo contains trophoblast stem cells (Tanaka et al., 1998; Uy et al., 2002). In the postimplantation period, the multipotency of the trophoblast stem cells is also maintained for a while in the limited cell population of the extraembryonic and chorionic ectoderm (Uy et al., 2002). Such trophoblastic tissues, which can be mechanically and enzymatically isolated from postimplantation embryos, would greatly reduce the incidence of contamination with other cell populations in culture; however, higher technique is required to isolate the precise cell population. On the other hand, the culture of dissociated blastocyst outgrowths under TS culture conditions, while easier, is often accompanied by contamination with rapidly growing, unidentified cell types other than the TS cells. The following protocol, therefore, describes the isolation of TS cells from a blastocyst outgrowth, using relatively common culture techniques.

It should be noted that there will be well-to-well differences in growth rates even in a single experiment, likely due to the small initial number of cells in a blastocyst. There are also differences between mouse strains. The durations until zona pellucida "hatching" and embryo adhesion are longer and the proliferation rate of isolated cells themselves is lower in embryos from mouse strains containing a B6 background (C57BL/6, B6C3F1, and B6D2F1) than in those from CD-1 (ICR). Differences in the appearance of TS cell colonies are also observed between TS cell lines with CD-1 (ICR) and B6D2F1 backgrounds. It will take 4–6 weeks until enough TS cells are obtained to make frozen stocks (see Fig. 1).

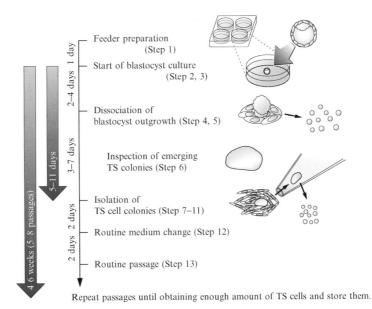

Fig. 1. Timetable for the establishment of a TS cell line.

TS Cell Line Establishment from Blastocyst-Stage Embryo

1. Thaw frozen stocks of MMC-MEF cells and seed on four-well plates (4×10^4 cells per well) in TS medium the day before collecting blastocysts (Fig. 1).

2. *Blastocyst culture*: Replace the medium of the four-well plates with TS+F4H (see Table II) before flushing blastocysts. Flush 3.5 dpc (days postcoitum) blastocysts from the uterine horns (Nagy *et al.*, 2003) and place one blastocyst per well in the four-well plates containing feeder cells; culture at $37°$ in 5% CO_2–95% air. For the next few days, check the adherence of blastocysts. Usually the blastocysts will have hatched and adhered to the well bottoms after 1 day, but embryos of certain strains may take more time and may still float for 2–3 days.

3. Check that a small outgrowth has formed from each embryo 2–4 days after the start of culture (Fig. 2A). Carefully replace the medium with 500 μl of fresh TS+F4H and culture for an additional 1 or 2 days to achieve enough outgrowth.

4. *Embryo dissociation*: When the outgrowth reaches a suitable size (Fig. 2B), remove the medium by aspiration and carefully wash the cells twice with 500 μl of PBS(–). Discard the PBS(–), add 100 μl of 0.1% trypsin–0.4 mM EDTA, and incubate for 5 min at $37°$ in 5% CO_2–95% air. Disaggregate the cell clump by thoroughly pipetting through a 200-μl tip (pipette up and down 20 times to ensure that the entire outgrowth is detached and thoroughly dissociated; view with a stereomicroscope in a tissue culture hood). Add 500 μl of 70CM+1.5× F4H (see Table II) and incubate at $37°$ in 5% CO_2–95% air.

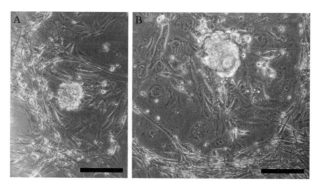

FIG. 2. (A) A blastocyst outgrowth on feeder cells (48 h). The embryo attached and the trophoblast cells migrated to the surface of the culture dish. (B) A blastocyst outgrowth on feeder cells (96 h) before trypsinization. The inner cell mass (ICM)-derived component is sitting on well-differentiated trophoblast giant cells. Scale bars: 200 μm.

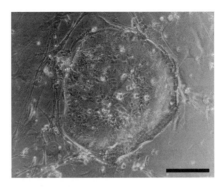

Fig. 3. A typical colony of trophoblast stem cells (5 days) after blastocyst dissociation. Note the epithelial cell-like appearance with a well-rounded, refractile, definitive colony boundary. Scale bar: 500 μm.

5. Change the medium 8 h after step 4 (500 μl of 70CM+1.5× F4H) and thereafter routinely change the medium at 2-day intervals.

6. TS cell colonies will begin to appear 3–7 days after the dissociation. They will look like flat, epithelial sheets with distinctive colony boundaries (Fig. 3). In many cases, colonies of morphologically distinct cells will also appear (Fig. 4). In such cases TS cell colonies should be selected as described in the following steps.

7. *TS colony isolation*: Prepare feeder cells in four-well plates 1 day before isolation. Replace the medium with 300 μl of TS+1.5× F4H (see Table II) just before picking TS cells.

8. Aliquot 50 μl of 0.1% trypsin–0.4 mM EDTA into the well of non-tissue culture-treated, U-bottom, 96-well plate immediately before picking TS cells.

9. Remove the medium from four-well plates containing TS cell colonies and wash the cells once with 500 μl of PBS(−); add 300 μl of PBS(−) and leave it.

10. Under the stereomicroscope in a tissue culture hood, pick a TS cell colony with a Pipetman tip (Gilson, Middleton, WI); pipette 5 μl of 0.1% trypsin–0.4 mM EDTA. Tearing off the surrounding feeder cells first makes it easier to pick up the colony. Place the lifted TS colony into the well with the trypsin (prepared in step 8), gently mix, and then incubate at 37° in 5% CO_2–95% air for 3–5 min.

11. Disaggregate the cell clumps by pipetting under the stereomicroscope. Add 150 μl of TS+1.5× F4H to stop trypsinization and then transfer the cells to four-well plates containing fresh feeder cells (step 7), and then incubate at 37° in 5% CO_2–95% for 8 h. After the incubation, change the medium (500 μl of TS+1.5× F4H) and continue the culture.

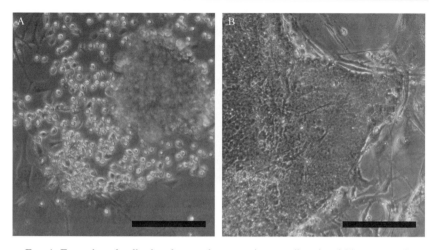

FIG. 4. Examples of cells that frequently contaminate a dissociated blastocyst culture. (A) "Endoderm-like" cells. Rounded, refractile, loosely attached cells appear similar to the endodermal cell type that forms when ES cells are induced to differentiate in culture. (B) "Epithelial-like" cells. These cells resemble TS colonies, but are slow growing. The constituent cells pack together to give a flatter, thinner edge than in TS cell colonies, although it is difficult to distinguish them on the basis of their appearance. They usually fail to proliferate after passaging. Scale bars: 200 μm.

12. Change the medium (500 μl of TS + 1.5× F4H) at 2-day intervals. TS cell colonies should reappear 3–4 days after colony isolation. If the TS cells reach half-confluency within 1 week, proceed to the next step. If only a few TS cell colonies reappear, the steps of colony pick-up should be repeated, this time gathering cells from multiple colonies into a single well.

13. Passage the half-confluent well of TS cells to a 35-mm dish of preplated feeders (2×10^5 cells per dish). After this step, the cells can be cultured in TS+F4H. Change the medium every other day. TS cell culture should gradually be expanded up to a 100-mm dish, keeping a 1:10 dilution ratio at each passage before the cells are frozen for storage (normally two or more additional passages to reach such an amount).

Passage and Maintenance of TS Cells

Overview

Because TS cells require coculture with feeder cells to maintain their proliferative ability and undifferentiated state, proper handling of feeder cell populations is important in the passage procedure. First, to leave enough space for the growth of TS cells, relatively smaller numbers of MEF cells will be used as the feeder layer than in ES cell culture. Feeder cell numbers that we found empirically to be optimal are described in Table I. To maximize TS cell growth, the prepared dishes with feeder cells should be checked for integrity and uniformity of the monolayer before use. In addition, our protocol is optimized for frozen stocks of MMC-treated MEF cells, in which the viable cells are counted before freezing. Cell numbers may be reduced if freshly prepared feeder cells are used. Second, cocultured MEF cells may be removed, taking advantage of the time gap between when MEF and TS cells adhere to the dish surface. Further purification is achieved by feeder-free culture of purified TS cells (see later, Feeder-Free Culture of TS Cells), because the gradual dilution of mitotically inactivated MEF cells takes place over several passages. Removal of feeder cells from culture is also recommended for routine passages, not just for the harvest of a pure TS cell population (e.g., for DNA/RNA extraction from TS cells), because the removal procedure also efficiently removes spontaneously differentiating cells.

The concentration of FGF4 in TS+F4H medium is also optimized. Along with the conditions for TS cell growth, we use a 50% higher amount of supplement ($1.5\times$ F4H) for the establishment of TS cell lines. Note that excess supplement may not work well for routine maintenance because FGF4 addition changes the adherence of feeder cells. It has been shown that other members of the FGF family, notably FGF1 and FGF2, also successfully replace FGF4 in maintaining TS cells in a proliferative, undifferentiated state (Kunath *et al.*, 2001; Uy *et al.*, 2002). To achieve more elaborate manipulation of TS cells, such as cloning with limiting dilution, Uy *et al.* have noted that Pronase works better in the dissociation of TS cell colonies (Uy *et al.*, 2002).

Passage of TS Cells

1. TS cells will be passaged when the colonies reach approximately 80% confluency (normally 4 days from previous passage). Seed MMC-MEF cells in new culture dishes in advance (see Table I for the correct number of cells to plate).

2. Discard the TS culture medium by aspiration, wash the cells twice with PBS(−), and add 0.05% trypsin–0.2 mM EDTA (2 ml per 100-mm dish).

3. Incubate for 3–5 min at 37° in 5% CO_2–95% air.

4. Roughly dissociate the cell clumps by light tapping of the culture dish, and check microscopically that the TS cell clumps are floating. Leave the remaining large cells, which are spontaneously differentiated trophoblast cells. Gently add TS medium (same volume as trypsin) to stop the trypsin reaction.

5. Mix by gently slanting the dish to allow the detached cells to float, and transfer the entire cell suspension into a 50-ml tube by pipetting up

with a glass pipette. Immediately, disaggregate the cell clumps thoroughly by pipetting up and down.

6. Centrifuge the cells (200g for 3 min) and discard the supernatant by aspiration, leaving a drop of medium.

7. *Feeder removal*: Loosen the cell pellet by gently tapping the bottom of the tube and then resuspend the cells in 70CM+F4H (see Table II) and seed in a new dish without MEF cells.

8. Incubate for 30–90 min at 37° in 5% CO_2–95% air. The longer time enhances the efficient removal of feeder cells and differentiating cells (while the number of cells may decrease).

9. *Optional*—to ensure the removal of feeder cells: Transfer the supernatant to a new dish without MEF cells and incubate for another 30 min at 37° in 5% CO_2–95% air.

10. After a gentle shake to make the dissociated cells float upward, collect the supernatant by pipetting up. This cell suspension should consist almost entirely of TS cells. Centrifuge the cells (200g for 3 min) and discard the supernatant by aspiration, leaving a drop of medium.

11. Seed the TS cells onto preplated MMC-MEF cells at a ratio of 1:10–1:20 (~5 $\times$ 10^4 cells per 35-mm dish). These cells may be cultured further under feeder-free conditions (see the next section).

12. Incubate at 37° in 5% CO_2–95% air. Change the medium every 2 days. Repeat steps 1–11 until the desired amount of cells has been produced.

Feeder-Free Culture of TS Cells

1. Prepare a gelatin-coated tissue culture dish. Briefly, fill the standard tissue culture dish with an adequate volume of 0.1% gelatin solution and incubate for 20 min. Discard the solution and dry. Gelatin coating of the dish is not always required, but may help growth and formation of "healthy" colonies of relatively unstable TS cell lines. Culture the TS cells in 70CM+F4H without feeder cells.

2. Passage the cells as described in the preceding section, skipping steps 7–10.

Freezing and Thawing of TS Cells

Overview

The general precaution for freezing and thawing (i.e., freeze slowly and thaw quickly) also applies to TS cells. TS cells may be frozen at a lower density. For example, nine vials may be made from an approximately 80% confluent 100-mm dish. The same procedures are applied for the freezing of MEF cells, with the modification of using 2× freezing medium for MEF cells. As with general primary cultured cells, it is advisable to record the passage number of TS cell lines.

Freezing

1. Prepare 2× freezing medium for TS cells (see Culture Media) and labeled cryovials.
2. Harvest TS cells from an ~80% confluent 100-mm dish by trypsinization and pellet them, as described in Passage of TS Cells.
3. Resuspend the cells in 1.5 ml of TS medium and add 1.5 ml of 2× freezing medium for TS cells; mix gently.
4. Aliquot 1 ml of cell suspension per cryovial (three vials for a 100-mm dish); close the cap firmly.
5. Place the cryovials into a cell-freezing container and immediately place the container in a −80° deep freezer overnight. The next day, transfer the cryovials to a liquid nitrogen tank.

Thawing

1. Quickly thaw a frozen vial of TS cells (containing one third of a 100-mm culture) in a water bath at 37°.
2. Transfer the entire contents to 10 ml of TS medium in a 50-ml tube and centrifuge at 200g for 3 min.
3. Discard the supernatant, loosen the cell pellet by gently tapping the bottom of the tube, and resuspend the cells in 30 ml of TS+F4H. Split the cells into three 100-mm dishes, each containing preplated MMC-MEF cells.
4. Incubate the cells at 37° in 5% CO_2–95% air. Change the medium the next day to remove cell debris.
5. Feed the cells with TS+F4H every other day. Passage the cells when they reach ~80% confluency (normally 3–4 days after thawing the cells) as described in Passage of TS Cells.

References

Deb, K., Sivaguru, M., Yong, H. Y., and Roberts, R. M. (2006). Cdx2 gene expression and trophectoderm lineage specification in mouse embryos. *Science* **311,** 992–996.

Hughes, M., Dobric, N., Scott, I. C., Su, L., Starovic, M., St-Pierre, B., Egan, S. E., Kingdom, J. C., and Cross, J. C. (2004). The Hand1, Stra13 and Gcm1 transcription factors override FGF signaling to promote terminal differentiation of trophoblast stem cells. *Dev. Biol.* **271,** 26–37.

Kunath, T., Strumpf, D., Rossant, J., and Tanaka, S. (2001). Trophoblast stem cells. *In* "Stem Cell Biology" (D. Gottlieb, ed.), pp. 267–286. Cold Spring Harbor Laboratory Press, New York.

Lawson, K. A., Dunn, N. R., Roelen, B. A., Zeinstra, L. M., Davis, A. M., Wright, C. V., Korving, J. P., and Hogan, B. L. (1999). Bmp4 is required for the generation of primordial germ cells in the mouse embryo. *Genes Dev.* **13,** 424–436.

Nagy, A., and Rossant, J. (2001). Chimaeras and mosaics for dissecting complex mutant phenotypes. *Int. J. Dev. Biol.* **45,** 577–582.

Nagy, A., Gertsenstein, M., Vintersten, K., and Behringer, R. (2003). Isolation, culture, and manipulation of embryonic stem cells. *In* "Manipulating the Mouse Embryo." Cold Spring Harbor Laboratory Press, New York.

Rossant, J., and Cross, J. C. (2001). Placental development: Lessons from mouse mutants. *Nat. Rev. Genet.* **2,** 538–548.

Selwood, L., and Johnson, M. H. (2006). Trophoblast and hypoblast in the monotreme, marsupial and eutherian mammal: Evolution and origins. *Bioessays* **28,** 128–145.

Sharman, G. B. (1971). Late DNA replication in the paternally derived X chromosome of female kangaroos. *Nature* **230,** 231–232.

Shiota, K., Kogo, Y., Ohgane, J., Imamura, T., Urano, A., Nishino, K., Tanaka, S., and Hattori, N. (2002). Epigenetic marks by DNA methylation specific to stem, germ and somatic cells in mice. *Genes Cells* **7,** 961–969.

Tanaka, S., Kunath, T., Hadjantonakis, A. K., Nagy, A., and Rossant, J. (1998). Promotion of trophoblast stem cell proliferation by FGF4. *Science* **282,** 2072–2075.

Uy, G. D., Downs, K. M., and Gardner, R. L. (2002). Inhibition of trophoblast stem cell potential in chorionic ectoderm coincides with occlusion of the ectoplacental cavity in the mouse. *Development* **129,** 3913–3924.

Yan, J., Tanaka, S., Oda, M., Makino, T., Ohgane, J., and Shiota, K. (2001). Retinoic acid promotes differentiation of trophoblast stem cells to a giant cell fate. *Dev. Biol.* **235,** 422–432.

Ying, Y., Qi, X., and Zhao, G. Q. (2001). Induction of primordial germ cells from murine epiblasts by synergistic action of BMP4 and BMP8B signaling pathways. *Proc. Natl. Acad. Sci. USA* **98,** 7858–7862.

[16] Pluripotent Stem Cells from Germ Cells

By CANDACE L. KERR, MICHAEL J. SHAMBLOTT, and JOHN D. GEARHART

Abstract

To date, stem cells have been derived from three sources of germ cells. These include embryonic germ cells (EGCs), embryonal carcinoma cells (ECCs), and multipotent germ line stem cells (GSCs). EGCs are derived from primordial germ cells that arise in the late embryonic and early fetal period of development. ECCs are derived from adult testicular tumors whereas GSCs have been derived by culturing spermatogonial stem cells from mouse neonates and adults. For each of these lines, their pluripotency has been demonstrated by their ability to differentiate into cell types derived from the three germ layers *in vitro* and *in vivo* and in chimeric

METHODS IN ENZYMOLOGY, VOL. 419   0076-6879/06 $35.00
Copyright 2006, Elsevier Inc. All rights reserved.   DOI: 10.1016/S0076-6879(06)19016-3

animals, including germ line transmission. These germ line-derived stem cells have been generated from many species including human, mice, porcine, and chicken albeit with only slight modifications. This chapter describes general considerations regarding critical aspects of their derivation compared with their counterpart, embryonic stem cells (ESCs). Detailed protocols for EGC derivation and maintenance from human and mouse primordial germ cells (PGCs) will be presented.

Introduction

At present, three types of germ line-derived stem cells have been identified including embryonic germ cells (EGCs), embryonal carcinoma cells (ECCs), and multipotent germ line stem cells (MGSCs) (for review see Donovan and Gearhart, 2001; Smith, 2001; Turnpenny et al., 2005b). Although not included in this chapter, more recent evidence has also suggested a germ cell lineage origin for embryonic stem cells (ESC) (Clark et al., 2004; Matsui and Okamura, 2005; Zwaka and Thomson, 2005). Importantly, all three cell types demonstrate the ability of unlimited self-renewal in vitro as well as the properties of pluripotency, as they can give rise to derivatives of all three embryonic germ layers in vitro in the form of embryoid bodies in culture and the formation of teratomas in vivo and through chimeric contributions after blastocyst injections including, in some cases, to the germ line. Embryonal carcinoma cells were the first to be identified in the 1960s from the mouse (Stevens, 1966) and then in human tissues (for review see Andrews, 1998). ECCs are pluripotent cells derived from adult testicular teratocarcinomas (or mixed germ cell tumors) from which genetic, immunological, and morphological evidence suggests a primordial germ cell (PGC) origin involving the ability of at least a subpopulation of PGCs to regain pluripotency given the appropriate cues (Stevens, 1967). However, unlike EGCs and GSCs, ECCs are karyotypically unstable, rendering their use in development studies problematic (Blelloch et al., 2004). Although ECCs injected into blastocysts have shown that the tumorigenic potential of these cells can be reversed by generating genetically normal chimeric mice (Mintz and Illmensee, 1975).

PGCs are the progenitors of the germ cell lineage and are lineage restricted in that they are unable to contribute to blastocyst-injected chimeras, and do not form embryoid bodies in culture (Donovan and de Miguel, 2003). Nevertheless, culturing conditions were defined that converted mouse PGCs from unipotent cells into pluripotent stem cells termed embryonic germ cells (EGCs) (Matsui et al., 1992; Resnick et al., 1992). These conditions were then successfully recapitulated with the derivation of EGCs in human tissue (Shamblott et al., 1998), pigs (Lee and Piedrahita, 2000; Mueller et al., 1999; Piedrahita et al., 1998; Shim et al., 1997; Tsung et al., 2003), and chickens

(Park and Han, 2000). The final source of germ line-derived stem cells has been produced by two groups reporting the conversion of spermatogonial stem cells isolated from mouse neonates and adults, multipotent germ line stem cells (GSCs) and multipotent adult germ cells (maGSCs), respectively, that under the appropriate culturing conditions display many properties of ESCs. In both cases, mouse GSCs demonstrated phenotypic characteristics of pluripotent stem cells, produced teratomas in mice, and formed germ line chimeras when injected into blastocysts (Kanatsu-Shinohara *et al.*, 2004, 2005). For the purposes of this chapter, the content will primarily on embryonic germ cell derivation from primordial germ cells and their maintenance in culture. Indeed, the significance of using human EGCs for cell-based therapies has been demonstrated in two landmark papers: in one, describing rats with diffuse motor neuron injury (Kerr *et al.*, 2003), injected human (h)EGC-derived cells were used to restore partial motor recovery; in the other, using a mouse excitotoxic brain damage model (Mueller *et al.*, 2005), hEGC-derived neural stem cells were able to partially restore the complement of striatum neurons in brain-damaged mice.

Germ Cell Development

Primordial germ cells are the progenitor cells of the germ cell lineage, which are the sole source of gametes in the adult. When strictly applied, the term PGCs refers to the diploid germ cell precursors that transiently exist in the embryo before they enter into associations with somatic cells of the gonad and become committed as germ cells. Much of the current knowledge of early mammalian germ cell development has been acquired by studies in the mouse. In the mouse, PGCs (mPGCs) are derived from a region of the epiblast that gives rise mainly to extraembryonic mesoderm. The fate of these cells as PGCs is determined by their proximity to the posterior extraembryonic ectoderm during gastrulation. Here external signals are secreted that regulate PGC differentiation as demonstrated by the observation that transplantation of cells from other parts of the epiblast to this region results in the acquisition of a PGC fate (Lawson and Hage, 1994). As a result, mPGCs are first detected at ~7 days postcoitum (dpc) as a small number of alkaline phosphatase-positive cells near the base of the allantois. By 8.5 dpc, actively proliferating mPGCs become associated with the endoderm as it invaginates to form the hindgut and by 10.5 dpc, mPGCs are associated with dorsal mesenteries and begin infiltrating the genital ridges such that the majority of the cells (~25,000) have localized to the gonad by 12.5 dpc (Mintz and Russell, 1957; Ozdzenski, 1967; Tam and Snow, 1981). Overall, this movement of mPGCs into these areas appears to be caused by both cellular migration and association with folding tissues (Donovan *et al.*, 1986). Male and female mPGCs are indistinguishable and continue proliferation via mitosis until they arrive at the genital ridge, at which time they are generally referred to as gonocytes. In female mice, more than 25,000 are detected by 13.5 dpc in the ovary, where the association with somatic cells of the ovary induces mPGCs to enter prophase of the first meiotic division. However, in males this entry into meiosis is inhibited during fetal life by signals from the developing testis (Francavilla and Zamboni, 1985; Upadhyay and Zamboni, 1982). Instead, male mPGCs, once they reach the gonad, continue to actively proliferate until around 15.5 dpc before undergoing mitotic arrest in G_1 phase until puberty.

A similar migration pattern is also seen for human PGCs (hPGCs). In humans, 50–100 PGCs are first distinguishable at $\sim$22 days in the endoderm of the dorsal wall of the yolk sac, near the allantois and in the mesenchyme of the stalk. From there, they proceed to migrate through the hindgut during the fourth week and dorsal mesentery in the fifth week to reach the genital ridge (Falin, 1969; McKay et al., 1953; Witschi, 1948, 1963). By the end of the fifth week or early in the sixth week, it has been estimated that approximately 1000 hPGCs begin to actively migrate from the dorsal mesentery into the gonadal anlage (Makabe et al., 1992; Motta et al., 1997; Pinkerton et al., 1961; Witschi, 1948). By the end of the seventh week the testis and ovary appear differentiated and germ cell proliferation is increased (Francavilla et al., 1990; Rabinovici and Jaffe, 1990). At this time, in the female, premeiotic hPGCs are called oogonia, which begin extensive mitotic expansion until their arrest in meiosis prophase 1 (Baker and Franchi, 1967; Makabe et al., 1992; Sun and Gondos, 1984). This proliferative expansion results in $\sim$10,000 germ cells at 5–6 weeks to $\sim$600,000 by week 8 of gestation. Between 11 and 12 weeks of gestation meiosis in the female begins and is concomitant with the loss of alkaline phosphatase activity in the primary oocytes (Francavilla et al., 1990; Gondos et al., 1986; Ohno et al., 1962; Pinkerton et al., 1961; Skrzypczak et al., 1981). In contrast, at 8 weeks male gonocytes are now called prospermatogonia, which exhibit much less mitotic expansion than in the female (at 9 weeks $\sim$30,000 have been reported) (Bendsen et al., 2003; Francavilla et al., 1990; Heyn et al., 1998) but continue to divide until they are arrested in mitosis between 16 and 18 weeks of gestation (Gondos, 1971).

Embryonic Germ Cell Derivation

Overview

Unlike their EGC descendents, PGCs are not pluripotent and do not survive past 1 week under standard tissue culture conditions. Early attempts to use various growth factors and feeder layers succeeded in prolonging mPGC survival, but proliferation was limited (Dolci *et al.*, 1991; Godin and Wylie, 1991; Matsui *et al.*, 1991). The culture conditions included using a combination of mitotically inactivated mouse fibroblast feeder cells and a synergistic action of stem cell factor (SCF) and leukemia inhibitory factor (LIF). Although these conditions promoted survival and proliferation of mPGCs, it was ultimately the addition of basic fibroblast growth factor (FGF-2) that resulted in the conversion of unipotent mPGCs into pluripotent mEGCs (Matsui *et al.*, 1992; Resnick *et al.*, 1992). FGF-2 functions as a potent mitogen in many cell types and induces telomerase activity in neural precursor cells (De Felici *et al.*, 1993). In this case, FGF-2 along with SCF and a feeder layer triggered differentiated mPGCs to proliferate from their single migratory state into multicellular colonies composed of pluripotent, self-renewing cells similar to mouse embryonic stem cells (mESCs), which had been established a decade earlier (Evans and Kaufman, 1981; Martin, 1981).

Six years after the first mouse EGC lines were described, Shamblott *et al.* (1998) reported the first case of EGC derivation from human tissue. To derive human EGCs (hEGCs), human primordial germ cells are isolated from the fetal gonad between 5 to 10 weeks of gestation. This time period coincides with peak PGC proliferation and encompasses the period in which the gonad undergoes sexual dimorphism into either an ovary or testis starting by week 7 (Bendsen *et al.*, 2003; Heyn *et al.*, 1998; Wartenberg, 1982; Witschi, 1963). To date, four laboratories have reported successful derivation of human EGC lines (Liu *et al.*, 2004; Park *et al.*, 2004; Shamblott *et al.*, 1998; Turnpenny *et al.*, 2003).

The derivation of EGCs from human tissue as well as from pigs and chickens has been performed by adapting methods based in part from the original EGC derivation in mice. For instance, all laboratories relied on culturing PGCs on a feeder layer and all species but chicken required one of two mouse embryonic fibroblast lines, MEF or STO cells, that were either irradiated or treated with mitomycin-C. For chicken EGCs, the gonadal tissue from the explants served as a feeder (Park and Han, 2000). The growth medium varies depending on the species but all derivations have reportedly relied on the addition of LIF and FGF-2. Soluble stem cell factor (sSCF), also known as c-kit ligand, Steel factor, or mast cell factor, was also added for mouse, pig, and chicken EGC derivation whereas forskolin was added for mouse and human EGC derivation. Forskolin is a pharmacological agent that raises intracellular cAMP levels and has been shown to stimulate mitosis in mPGC culture (Dolci *et al.*, 1993). LIF was originally known for its inhibitory role in liver cell differentiation (Smith *et al.*, 1988) and employed for the derivation of mouse embryonic stem cells (mESCs), where signaling via the LIF receptor (LIFR), gp130, and intracellular STAT3β (signal transducer and activator of transcription 3β) was necessary in maintaining mESC pluripotency (Niwa *et al.*, 1998). Interestingly, activation of this pathway does not maintain self-renewal in human ESCs but is required for human and mouse EGC culture. SCF mutations and mutations of the SCF receptor, c-kit, have also revealed direct roles in the proliferation and maintenance of mPGCs during their migration to the gonad as well as in culture (reviewed in Donovan and de Miguel, 2003). Although our laboratory has not seen an effect of soluble SCF in our cultures of hPGCs we have seen a positive correlation in hPGC proliferation and survival in culture with the expression of increasing concentrations of transmembrane SCF present on subcloned feeder cells (Shamblott *et al.*, 2004).

EGC derivation is most efficient from PGCs isolated before or during their migration, although it is possible to acquire lines from PGCs until 12.5 dpc in mice (McLaren and Durcova-Hills, 2001) and up to 10 weeks gestation in humans. This is significant given the ability of these cells to become EGCs in culture even after their gonadal niche has undergone sexual differentiation and development of the sex cords has occurred.

Protocol for Deriving Embryonic Germ Cells

Although many combinations of cytokines, feeder layers, and culture conditions have been evaluated by our laboratory, the following protocol for both human and mouse EGC derivation are described with some differences noted in text where they are applied.

Growth Medium Components

The growth medium used to derive and maintain both mouse and human EGCs is Dulbecco's modified Eagle's medium (Invitrogen, Carlsbad, CA) supplemented with 15% fetal bovine serum (FBS; HyClone, Logan, UT), 0.1 mM nonessential amino acids (Invitrogen), 0.1 mM 2-mercaptoethanol (Sigma, St. Louis, MO), 2 mM glutamine (Invitrogen), 1 mM sodium pyruvate (Invitrogen), penicillin (100 U/ml; Invitrogen), and streptomycin (100 μg/ml; Invitrogen). For human EGC derivation human recombinant LIF (hrLIF, 1000 U/ml; Chemicon International, Temecula, CA), human recombinant FGF-2 (hrFGF-2, 1–2 ng/ml; R&D Systems, Minneapolis, MN), and 10 μM forskolin (Sigma) prepared in dimethyl sulfoxide (DMSO) were added.

This medium was also applied in mouse EGC derivation except that mouse LIF (ESGRO, 1000 U/ml; Chemicon International) was used in place of hrLIF. Another notable difference between derivation in these two species is that whereas soluble stem cell factor (SCF) may also be added to enhance mouse EGC derivation there is no effect of soluble SCF on human EGC culture. However, our laboratory has noted the requirement for transmembrane SCF on the feeder layer for human EGC derivation (Shamblott et al., 2004).

Initial Disaggregation and Plating

1. For mouse EGC derivation, multiple mouse strains have been employed including 129/SV, ICR, and C57BL/6 from which mPGCs can be derived between 8 and 12.5 dpc (Durcova-Hills et al., 2001; Labosky et al., 1994; Matsui et al., 1992; Resnick et al., 1992; Stewart et al., 1994). At 8.5 dpc the lower one-third portion of mouse embryo, starting posteriorly from the last somites, is collected. At 9.5 dpc portions of the hindgut are isolated (Durcova-Hills et al., 2001), whereas at 10.5 and 12.5 dpc genital ridges are used (Matsui et al., 1992). The remainder of the protocol is identical to that for human EGC derivation, with some noted exceptions.

2. For human EGC derivation, gonadal ridges and mesenteries from 5–10 weeks of gestation are collected (as results of therapeutic termination of pregnancies, using a protocol approved by the Joint Committee on Clinical Investigation of the Johns Hopkins University School of Medicine, Baltimore, MD) in 1 ml of ice-cold growth medium.

3. The tissue is then placed in a three-well depression slide and soaked in calcium- and magnesium-free Dulbecco's phosphate-buffered saline (DPBS) for 5 min and then transferred to 100 μl of trypsin–EDTA solution. The concentration of trypsin and EDTA is varied such that at the earliest developmental stages, a gentler 0.05% trypsin–0.5 mM EDTA is used, and at later developmental stages a stronger 0.25% trypsin–0.5 mM EDTA solution is used.

4. The tissue is then mechanically disaggregated, using fine forceps and iris scissors, for 5–10 min at room temperature.

5. Next, sample is placed in a sterile 1.5-ml centrifuge tube and placed at 37° for 5–10 min (water bath or incubator). This disaggregation process often results in a single-cell suspension with large pieces of undigested tissue.

6. To stop the digestion, serum-containing growth medium is added to the tube. The digested tissue is then triturated again, ~30–50 times, with a 200-μl Pipetman (Gilson, Middleton, WI) and tip.

7. One hundred-microliter volumes of the sample are transferred to each well of a 96-well tissue culture plate previously prepared with the feeder layer (described later).

8. Usually the initial plating occupies ~4–10 wells of the 96-well plate per human gonad or 0.5–1 mouse embryo per well for mEGC derivation. Plating densities are critical. Density that is too low or too high reduces derivation efficiency.

9. The plate is incubated at 37° in 5% CO_2 with 95% humidity for 7 days. Approximately 90% of the growth medium is removed each day, and the plate is replenished with fresh growth medium containing LIF, forskolin, and FGF-2.

Subsequent Passage of EGC Cultures

In the first 7 days of derivation (passage 0), most mouse and human EGC cultures do not produce visible EGC colonies. Staining for tissue-nonspecific alkaline phosphatase (TnAP) activity demonstrates the presence of solitary PGCs with either stationary or migratory morphology (Fig. 1A). Often, colonies of cells that do not express TnAP activity are seen (Fig. 1B and C), as are small clumps of tissue remaining from the initial disaggregation (Fig. 1D and E). After 7 days, mouse and human PGC cultures are subcultured onto a

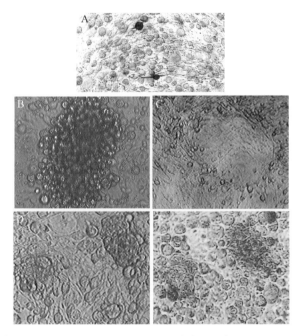

Fig. 1. (A) TnAP staining of a single stationary (on top) and migratory (on bottom) human PGC in primary culture on an STO feeder layer. (B) Multicellular clump of gonadal tissue. (C and D) Flat and rounded cell colonies that do not lead to human EGC colonies and are TnAP negative. (E) Negative staining for TnAP activity of colonies in (D).

new feeder. Timing is critical here because after 7 days their mitotically inactivated feeders begin to undergo significant apoptosis. Cell density of the specimens also increases at this time, leading to overcrowding.

1. First, medium is removed and the wells are gently rinsed with calcium- and magnesium-free DPBS.

2. Next, 40 μl of freshly thawed 0.05% trypsin–0.5 mM EDTA, 0.25% trypsin–0.5 mM EDTA, or a mixture of these two solutions is added to each well, and the plate is incubated on a heated platform or in a tissue culture incubator for 5 min at 37°. The important point at this stage is to facilitate the complete disaggregation of cells, which can be a significant challenge.

3. After incubation, a Pipetman and 200-μl tip is used to scrape the bottom of the wells, followed by gentle trituration $\sim$20–30 times.

4. After the samples have been loosened, fresh growth medium is added to each well, and the contents are triturated another 10–30 times. This phase is critical to successful disaggregation of STO feeder layer and EGCs.

5. Each well is divided in half and placed into twice the number of wells containing freshly prepared feeders.

All subsequent passages are repeated as described above. After 2 to 3 weeks (during passage 1 or 2), large and recognizable EGC colonies will arise in some of the wells from both mouse and human cultures (Fig. 2). At this point, wells that do not have EGC colonies are discarded. Interestingly, in our hands approximately 50% of the wells on average initially produce EGC colonies from both mouse and human cultures. To generate mouse EGCs it is critical that once EGC colony formation occurs, FGF-2 be taken out of the medium. Otherwise overgrowth of "non-EGCs" produces an adhesive monolayer of AP⁻ cells and in subsequent passages EGC colonies disappear. However, in humans, FGF-2 is required to sustain colonies, which do not seem to produce the same overgrowth of unwanted cells as seen in the mouse cultures.

Troubleshooting

Several common problems occur during the passage of human EGCs that do not occur with mEGCs. One observation is that hEGC colonies do not fully disaggregate. The consequences of poor disaggregation are that the large pieces differentiate or die and fewer hEGCs are available for continued culture expansion. Although much effort has been expended to find a solution to this problem, it remains the most difficult aspect of and challenging hurdle to human EGC biology. To gain some insight into this problem, a series of electron microscopic images were taken to compare the cell–cell interactions found in mouse ESC, mouse EGC, and human EGC colonies. It is evident from these images that cells within the human EGC colonies adhere more

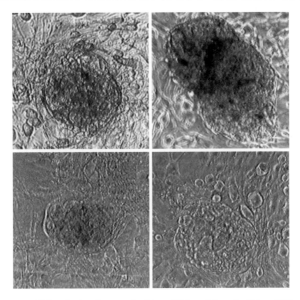

FIG. 2. Human EGC colonies from male (A and B) and female (C and D) gonads 6 to 10 weeks post conceptus on STO cell feeder layers. Colonies appear after 2–3 weeks of culture.

completely to each other than do cells within mouse ESC and EGC colonies (Fig. 3). It is possible that this tight association within the colony limits the access of disaggregation reagents. At this time, neither the nature of the cell–cell interactions nor an effective solution to this problem is evident.

Because of incomplete disaggregation and other intrinsic or extrinsic signals, many human EGC colonies (10–30% per passage) differentiate to form three-dimensional structures termed embryoid bodies (EBs) (Fig. 4A–C) or flatten into TnAP⁻ aggregates that no longer continue to proliferate (Fig. 4D and E). Human EGC colonies that are more fully disaggregated go on to produce new colonies that under the best circumstances can routinely exceed 20 passages. Inevitably, large hEGC colonies are removed from the culture as a result of EB formation and as the cultures become sparse are discontinued for practical considerations. Efforts employing standard DMSO cryopreservation techniques have so far been unsuccessful.

Feeder Layer

Overview

Embryonic germ cell derivation is highly dependent on a feeder layer. Mouse and human EGCs have been derived by using primary mouse embryonic fibroblast (PMEF) and Sandoz thioguanine- and ouabain (STO)-resistant

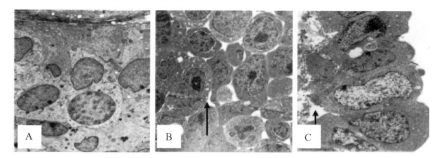

Fig. 3. Electron micrographs of EGC versus ECC colonies. (A) Human EGC colony; (B) mouse EGC colony; (C) mouse ESC colony. Arrows demonstrate areas of reduced adhesion that are not present in hEGC colonies, contributing to the complexity of culturing human EGCs. Scale bars = 10 μm (A and C) and 30 μm (B).

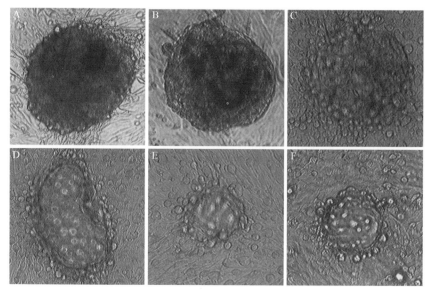

Fig. 4. Embryoid body formation and differentiation of human EGC colonies. (A and B) Human EGC colonies after incomplete disaggregation produce large three-dimensional structures called embryoid bodies (EBs) on top of STO feeder layers. (C) Subsequently, EBs begin to flatten and in (D–F) eventually form cystic embryoid bodies. These colonies lose TnAP activity and do not survive after disaggregation and replating.

mouse fibroblasts. The majority of reports for both mEGC and hEGC derivation and those on porcine EGC derivation have used STO cells as feeders. We have also been most successful with STO cells. The factor or factors provided exclusively by this cell line are not fully understood.

Although STO cells are a clonal cell line, individual isolates vary greatly in their ability to support human EGC derivation. This is further complicated by the known phenotypic variation of STO cells in continuous culture. Given the limited supply of human tissue, it is prudent to screen STO cells for suitability before use. The most reliable screening method is to produce a number of clonal STO lines (by limiting dilution or cloning cylinder) and to evaluate them for their ability to support the derivation of mouse ESCs. However, the growth of existing mouse EGC lines is not a sufficient method, as most lines become feeder layer independent after derivation. Therefore our laboratory has screened several STO cell lines for their ability to derive mouse EGCs and, as expected, the ability of a STO cell line to derive mouse EGCs directly reflected its ability to derive human EGCs. Furthermore, those lines that could not derive mouse EGCs could not derive human EGCs. Once a supportive STO fibroblast line is identified, it should be immediately cryopreserved in several low-passage aliquots. One of these aliquots can then be expanded into multiple replicates and frozen for later use. After thawing, each aliquot of feeder cells is then used with limited further expansion. Continuous passage of STO fibroblasts without frequent screening should be avoided. After five passages of continuous expansion we discard the feeder and start a new culture.

Plating an STO Feeder Layer

For mitotic inactivation, the STO feeder layer can be irradiated either before plating or after plating. Most human EGC cultures are derived by the later method, which requires a large γ radiation unit.

Plating Feeder Cells Before Inactivation

1. STO cells are passaged for short periods (not continuously) in the EGC growth medium without LIF, FGF-2, and forskolin and are disaggregated with 0.05% trypsin–EDTA solution.
2. One day before use, 96-well tissue culture dishes are coated with 0.1% gelatin for 30 min.
3. The gelatin is withdrawn, and 5×10^4 STO cells are plated per well of the 96-well plate in EGC growth medium without LIF, FGF-2, and forskolin. Similar cell densities ($\sim 1.5 \times 10^5$ cells/cm^2) can be achieved in other well configurations.
4. The cells are grown overnight and then exposed to 5000 rads (1 rad = 0.01 Gy) of γ radiation or X-rays. The cells are then returned to the tissue culture incubator until required.
5. Before use, the growth medium is removed, 100 μl of EGC growth medium with added factors is added to each well (i.e., half the

required well volume), and the dish is returned to the tissue culture incubator.

Plating Feeder Cells After Inactivation

This method of STO cell preparation is used when a large γ-radiation unit is not available, when large amounts of cells are required, or if better control of STO cell density is desired.

1. STO cells are grown as described previously, trypsinized, counted, and resuspended in growth medium without added factors.
2. The cells are placed into tubes and exposed to 5000 rads of γ radiation or X-rays.
3. After exposure, cells are adjusted to a convenient concentration in growth medium without added factors, counted, and plated into gelatinized tissue culture plates at $\sim$1.5 × 10^5 cells/cm^2.
4. Cells are allowed to adhere overnight.

Characterization of EG Cultures

Overview

Embryonic germ cells are assessed on the basis of morphological, biochemical, and/or immunocytological characteristics similar to embryonic stem cells. Morphologically, EGC cultures consist of tightly compacted multicellular colonies, which in humans double or triple in size over 7–10 days (Shamblott et al., 2004) and in mouse every 16 h (Lawson and Hage, 1994). In addition to morphological differences, EGCs as well as other germ line-derived stem cells also express a number of pluripotent markers as well as telomerase activity. These include the stage-specific embryonic antigens (SSEA-1, SSEA-3, and SSEA-4) and tumor rejection antigens 1-60 and 1-81 (TRA-1-60 and TRA-1-81). The pattern of expression is dependent in part on the species. For example, both mouse and human EGCs express SSEA-1 whereas hEGCs but not mEGCs also express SSEA-3 (albeit weakly), SSEA-4, TRA-1-60, and TRA-1-81 (Table I). High levels of tissue-nonspecific alkaline phosphatase (TnAP) activity are associated with all EGCs such that mouse and human EGC colonies are >70–90% AP$^+$ (Fig. 5A and C). As colonies differentiate, loss of TnAP activity appears to occur first in cells around the periphery of the colony, similar to the effect seen in ESC colony differentiation.

Nevertheless, TnAP activity and these cell surface markers are not exclusively expressed in pluripotent cells. As such the significance of the expression of these surface antigens remains unclear. Whereas the presence

TABLE I
EXPRESSION PROFILE OF EMBRYONIC GERM CELL MARKERS

Marker	Human EGCs, ECCs	Mouse EGCs, ECCs	Mouse MGSCs	Pig EGCs	Chicken EGCs
AP	+	+	+	+	+
SSEA-1	+ EGC	+	+	+	+
SSEA-3	+	−	ND	+	ND, + PGCs
SSEA-4	+	−	ND	+	ND, + PGCs
TRA-1-60	+	−	ND	ND	ND
TRA-1-81	+	−	ND	ND	ND
Oct4	+	+	+	+	ND
Nanog	+	+	+	ND	ND
hTERT	+	+	ND	ND	ND

Abbreviations: AP, alkaline phosphatase; ECCs, embryonal carcinoma cells; EGCs, embryonic germ cells; hTERT, human telomerase reverse transcriptase; MGSCs, multipotent germ line stem cells; ND, not determined; PGCs, primordial germ cells; SSEA, stage-specific embryonic antigen; TRA, tumor rejection antigen.

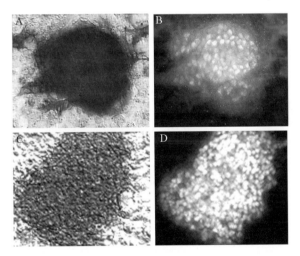

FIG. 5. Dual labeling of EGC colonies: TnAP and Oct4 staining of human and mouse colonies. (A) TnAP staining of a human EGC colony. (B) Indirect immunostaining of colony from (A), with Oct4 showing nuclear staining in most cells. (C) TnAP staining of a mouse EGC colony on a feeder layer. (D) Oct4 staining of a colony from (C). In both cases, the majority of cells in these colonies are AP+ and Oct4+.

of these markers is consistent in both mouse PGCs and EGCs, this does not appear to be the case for the human paradigm. Historically, the identification of human PGCs from somatic cells in the developing embryo has been largely restricted to TnAP activity and morphological characteristics including PGC rounded morphology, large eccentric nucleus, as well as the presence of glycogen particles and lipid droplets that stain with periodic acid (Bendsen *et al.*, 2003; McLaren, 2003; Motta *et al.*, 1997; Pinkerton *et al.*, 1961). Most importantly, TnAP activity for this lineage is not indicative of pluripotency but that of the early germ cell lineage before meiosis. Furthermore, preliminary evidence in the immunological characterization of human gonads with stem cell markers reveals that in fact human PGCs stain SSEA-1 and SSEA-4 but do not express TRA-1-60 and TRA-1-81. Instead, staining for the TRA antigens appears to be localized to the lining of the mesonephric ducts in both males and females (data not shown). This is exemplified in Fig. 6, which depicts a male gonad ~7.5 weeks post conceptus, when PGCs are localized in the gonad and the gonad undergoes sexual dimorphism. Here a large increase in the number of PGCs staining positive for SSEA-1 and SSEA-4 in the sex cords can be seen. It is not known, however, whether the TRA antigens are a marker of the pluripotent conversion of hPGCs into hEGCs or an arbitrary artifact of their cell culture. This is particularly interesting in that the expression of TRA antigens is also indicative of the

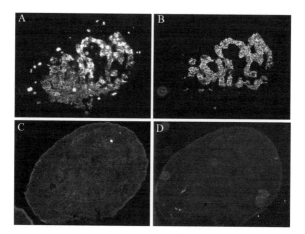

FIG. 6. Indirect immunostaining of a middle cross-section of a male gonad at 53 days post conceptus. Cryosections were stained with (A) Alexa 564 fluorescently labeled SSEA-4. Staining for SSEA4 is found in the developing rete and sex cords. (B) Section from (A) dually stained with Alexa 488 fluorescently labeled SSEA1 localized to the developing sex cords. Adjacent sections were stained with either (C) Alexa 564 fluorescently labeled Tra-1-80 or with (D) Alexa 564 fluorescently labeled Tra-1-61.

pluripotency shared by human ECCs and ESCs. Differences in marker expression between EGCs and PGCs have also been seen in pigs, where porcine EGCs appear to express SSEA-1, SSEA-3, and SSEA-4 whereas PGCs express only SSEA-1 (Takagi *et al.*, 1997; Tsung *et al.*, 2003).

In addition to cell surface markers, expression of certain transcription factors has also been associated with pluripotency. The most well-studied of these includes the POU domain transcription factor, Oct4, which has been associated with pluripotency in ESCs and its expression has also been detected in EGCs from human (Liu *et al.*, 2004; Park *et al.*, 2004; Turnpenny *et al.*, 2003) and mouse (Fig. 5B and D, respectively) (McLaren, 2003; Yeom *et al.*, 1996). To date, all the direct evidence for the role of Oct4 in pluripotency comes from studies done in the mouse (for review see Boiani and Scholer, 2005). Here, studies have shown that during development Oct4 is originally expressed in all blastomeres and then becomes restricted to the inner cell mass (ICM) with downregulation in the trophoectoderm and the primitive endoderm (Nichols *et al.*, 1998; Niwa *et al.*, 1998; Pesce and Scholer, 2000; Scholer *et al.*, 1990). Evidence that supports a role for Oct4 in pluripotency comes from gene-targeting studies in mice that resulted in embryos devoid of a pluripotent ICM and that failed to give rise to ES colonies *in vitro* (Nichols *et al.*, 1998). Furthermore, quantitative analysis of Oct4 expression revealed that the level of Oct4 expression caused differentiation of mESCs toward either extraembryonic mesoderm and endoderm lineages (when increased) or trophectodermal cells (when reduced) (Niwa *et al.*, 1998). During gastrulation, Oct4 expression is progressively repressed in the epiblast and by 7.5 dpc is confined exclusively to newly established primordial germ cells (Scholer *et al.*, 1990; Yeom *et al.*, 1996) and, at maturity, to the developing germ cells (Pesce and Scholer, 2001; Pesce *et al.*, 1998). More recently, a new role for Oct4 in PGC survival has been reported on the basis of a conditional genetargeting approach in which restricted loss of Oct4 function in mouse PGCs resulted in significant apoptosis (Kehler *et al.*, 2004). Oct4 expression has also been detected in hPGCs (Goto *et al.*, 1999). However, the significance of this expression in hPGC development and survival is not known.

Like Oct4, Nanog has also been associated with pluripotency in multiple species (Chambers *et al.*, 2003; Hatano *et al.*, 2005; Mitsui *et al.*, 2003; Yamaguchi *et al.*, 2005) including human EGCs (Turnpenny *et al.*, 2005a; C. L. Kerr and J. D. Gearhart, unpublished data). Nanog is another transcription factor that is expressed specifically in mouse preimplantation embryos, mESCs, and mEGCs as well as in monkey and human ESCs (Chambers *et al.*, 2003; Hart *et al.*, 2004; Hatano *et al.*, 2005; Mitsui *et al.*, 2003). The role of *nanog* in the maintenance of pluripotency is suggested by the loss of pluripotency in *nanog*-deficient mESCs and in *nanog*-null embryos shortly after implantation (Mitsui *et al.*, 2003). In addition, Nanog overexpression leads to the clonal expansion of mESCs by bypassing regulation of LIF–STAT3 signaling and maintenance of Oct4 levels (Chambers *et al.*, 2003). Evidence also shows that Sox2 and Oct4 bind to regulatory elements of *nanog* in mEGCs, suggesting the potential role of *nanog* in the action of Oct4 and Sox2 in these cells (Kuroda *et al.*, 2005). Although Nanog expression is also detected in human EGCs and in the developing human gonad (C. L. Kerr and J. D. Gearhart, unpublished data), a potential role in the action of Oct4 and Sox2 still remains to be determined.

Protocols for Immunostaining EGCs and Human Gonads

1. Gonads are cut into 5-μm sections, placed on slides (ProbeOn Plus; Fisher Scientific, Pittsburgh, PA), and immediately prepared for antibody staining.

2. Plated EGC colonies and gonad sections are first fixed in either 4% paraformaldehyde for 5 min for cell surface markers, or in 4% paraformaldehyde after a 10-min incubation with 0.2% Triton X in Dulbecco's PBS (DPBS; Invitrogen) to detect Oct-4 (clone YL8; BD Biosciences Pharmingen, San Diego, CA) and Nanog (R&D Systems) staining.

3. Cell surface glycolipid- and glycoprotein-specific monoclonal antibodies are used at 1:15 to 1:50 dilution. MC480 (SSEA-1), MC813–70 (SSEA-4) TRA-1-60, and TRA-1-81 antibodies are supplied by the Developmental Studies Hybridoma Bank (University of Iowa, Iowa City, IA).

4. Antibodies are diluted in 5% goat serum in DPBS and incubated on sections for 1 h at room temperature, except for Oct-4 and Nanog (1:50 dilution), which are incubated overnight at 4°.

5. Next, all antibodies are detected with fluorescently labeled goat anti-mouse secondary antibodies (1:200 dilution; Invitrogen Molecular Probes, Eugene, OR) in 5% goat serum in DPBS for 1 h at room temperature. Except for Nanog staining, rabbit anti-goat secondary antibodies (1:200 dilution; Invitrogen Molecular Probes) are used.

6. Sections are counterstained with 4′,6-diamidino-2-phenylindole (DAPI; Sigma) and mounted with antifade mounting medium (Invitrogen Molecular Probes). Negative controls for each procedure include incubations with secondary antibodies alone and with mouse ascites fluid.

Sex Determination of Human Gonads by Fluorescence
In Situ *Hybridization*

The sex of each gonad was determined with CEP X SpectrumOrange/ CEP Y SpectrumGreen DNA probe (Vysis, Downers Grove, IL). Five-micrometer sections were fixed in 1 part acetic acid:3 parts ethanol for 1 min, pretreated with 1 M sodium thiocyanate for 5 min at 75°, and postfixed in 100% methanol for 1 min. Sections were then denatured in 60% formamide, 2× saline–sodium citrate (SSC) buffer, pH 5.3 at 75° for 3 min, followed by 1 min in cold 70, 95, and 100% ethanol, and then incubated with CEP X,Y DNA probe overnight at 37°. The next day, sections underwent three posthybridization washes in 60% formamide (Sigma), 0.3% Nonidet P-40 [NP-40 (Igepal); Sigma], and 2× SSC (pH 5.3). Sections were then counterstained with DAPI and mounted as described previously.

Tissue-Nonspecific Alkaline Phosphatase and Telomerase Activity Assays

To detect TnAP activity, EGC colonies are fixed on plates in 66% acetone–3% formaldehyde for 5 min and then stained with naphthol/fast red violet (FRV)–alkaline AP substrate (Sigma) for 20 min according to the manufacturer's instructions. Telomerase assays are performed by a telomeric repeat amplification protocol followed by enzyme-linked immunosorbent assay (ELISA) detection of amplified products (TeloTAGGG Telomerase PCR ELISA[PLUS]; Roche, Indianapolis, IN).

Cytogenetic Analysis

EGC colonies are incubated in growth medium with KaryoMAX Colcemid (0.1 μg/ml; Invitrogen) for 3–4 h at 37° in 5% CO_2 with 95% humidity, isolated from the feeder with a 1-μl Pipetman and tip, and then disaggregated in 0.05% trypsin–EDTA at 37°. Cells are then resuspended in 2 ml of 0.75 M KCl hypotonic solution and incubated at 37° for 35 min. Next, cells are fixed by slowly adding 10 ml of cold Carnoy's fixative (methanol–acetic acid, 3:1) and rinsed twice with fixative before being plated. To prepare slides, ~20-μl cell suspensions are dropped onto microscope slides over a humidity chamber and then allowed to air dry. Metaphase chromosomes are stained in Giemsa staining solution (Invitrogen) and observed at ×600 magnification with oil immersion. Standard karyotyping includes approximately 20 metaphase spreads from each line to be examined for the presence of structural abnormalities of chromosomes and accurate chromosomal number.

Embryoid Body Formation and Analysis

Overview

EBs form spontaneously in human EGC cultures. Although this represents a loss of pluripotent EGCs from the culture, EBs have provided evidence for the pluripotent status of the EGCs and also provide cells for subsequent culture and experimentation (see the section Embryoid Body-Derived Cells).

Protocol for Embedding and Immunohistochemistry

The cells that comprise EBs can most reliably be identified by immunohistochemistry, using paraffin and staining sections with a variety of well-characterized antibodies. This process avoids the significant problem of antibody trapping that occurs on large three-dimensional structures when direct staining is attempted. Specifically, paraffin is preferred over cryosections of EBs as cryopreservation results in poorly defined cellular morphology.

First, human EBs are collected from cultures and placed into a small drop of molten 2% low melting point agarose (FMC BioPolymer, Philadelphia, PA), prepared in DPBS, and cooled to 42°. Solidified agarose-containing EBs are then fixed in 3% paraformaldehyde in DPBS and embedded in paraffin. Individual 6-μm sections are placed on microscope slides (ProbeOn Plus; Fisher Scientific) and antigen retrieval is performed. Antibodies used on paraffin sections include HHF35 (muscle-specific actin; Dako, Carpinteria, Ca), M 760 (desmin; Dako), CD34 (Immunotech Beckman Coulter, Marseille, France), Z311 (S-100; Dako), sm311 (panneurofilament; Sternberger Monoclonals, Lutherville, MD), A008 (α-1-fetoprotein), CKERAE1/AE3 (pancytokeratin; Roche), OV-TL 12/30 (cytokeratin 7; Dako), and Ks20.8 (cytokeratin 20; Dako). These antibodies are used in particular to demonstrate that when human EGC cells differentiate, they form EBs composed of endodermal, ectodermal, and mesodermal derivatives (Liu et al., 2004; Park et al., 2004; Shamblott et al., 1998; Turnpenny et al., 2003). Primary antibodies are detected with biotinylated anti-rabbit or anti-mouse secondary antibody, streptavidin-conjugated horseradish peroxidase, and diaminobenzidine (DAB) chromogen (Ventana Medical Systems, Tucson, AZ). Afterward, slides may also be counterstained with hematoxylin if desired.

Embryoid Body-Derived Cells

Overview

Human EGC-derived EBs can be used to produce multipotent populations capable of long-term and robust proliferation. These cells are referred to as EB-derived (EBD) cells. The method used to isolate cell populations from EBs is conceptually similar to microbiological selective medium experiments. EBs are disaggregated and plated into several different cellular growth environments. These environments consist of various growth media and matrix supports. Although many combinations have been evaluated by our laboratory (Shamblott et al., 2001), most of our EBD cell cultures have been derived from one of six environments generated by combinations of one each of two growth media and three plating surfaces. The growth media include an RPMI 1640 medium supplemented with either 15% FBS or a low 5% FBS in addition to FGF-2, epidermal growth factor (EGF), insulin-like growth factor I (IGF-I), and vascular endothelial growth factor (VEGF). The plating surfaces include bovine type I collagen, human extracellular matrix extract, and tissue culture-treated plastic alone. These are not intended to be highly selective environments; instead, they favor several basic themes: cells thriving in high serum and elevated glucose (10 mM) conditions versus cells proliferating in low glucose (5 mM) under the control of four mitogens. In general, the type I collagen and human extracellular matrices combined with the low-serum medium provide the most rapid and extensive cell proliferation. EBD cell lines and cultures are routinely maintained in the environment in which they were derived. To distinguish EBD cells from a culture that has developed into a lineage-restricted cell line, it is important to establish an extensive expression profile. This profile should include genetic screening using multiple markers for each of the five cell lineages (neuronal, glial, muscle, hematopoietic–vascular endothelia, and endoderm) combined with supportive immunocytochemical staining.

Using this approach, data have shown that rapidly proliferating EBD cell cultures simultaneously express a wide array of mRNA and protein markers normally associated with distinct developmental lineages (Shamblott et al., 2001). This is not a surprising property considering that EBD cells are, at least during the derivation stage, a mixed-cell population. More remarkable is the finding that human EBD cell lines isolated by dilution cloning also exhibit a broad multilineage gene expression profile and that this profile remains stable throughout the life span of the culture. This normally exceeds 70 population doublings but is not unlimited because human EBD cells are not immortal.

Interestingly, EBD cells can be genetically manipulated by lipofection and electroporation as well as with retroviral, adenoviral, and lentiviral vectors. Using these techniques, EBD lines have been generated that constitutively and tissue-specifically express enhanced green fluorescent protein (GFP) and contain many different genetic selection vectors. The proliferation and expression characteristics of EBD cells suggest they may be useful in the study of human cell differentiation and as a resource for cellular transplantation therapies. One important property in this regard is that no tumor has arisen in any animal receiving EBD cells from our laboratory, including hundreds of mice, rats, and African green monkeys that have received EBD transplants in a variety of anatomical locations, often consisting of more than 1 million cells injected (Frimberger et al., 2005; Kerr et al., 2003; Mueller et al., 2005). This is in contrast to the infrequent yet significant number of mixed germ cell tumors that have arisen after transplantation of cells produced through neural and hematopoietic differentiation of mouse ESCs.

Protocols for Human EBD Cells

EBD Cell Derivation

EBs are harvested in groups of 10 or more and are dissociated by digestion in collagenase/dispase (1 mg/ml; Roche) for 30 min to 1 h at 37°. Cells are then spun at 200g for 5 min and resuspended in either RPMI or EGM-2 MV growth medium (Cambrex Bio Science Walkersville, Walkersville, MD). RPMI growth medium includes RPMI 1640 (Invitrogen Life Technologies, Gaithersburg, MD), 15% FCS, 0.1 mM nonessential amino acids, 2 mM L-glutamine, penicillin (100 U/ml), and streptomycin (100 μg/ml). EGM-2 MV medium includes 5% FCS, hydrocortisone, FGF-2, hVEGF, Long R^3-IGF-I, ascorbic acid, hEGF, heparin, gentamicin, and amphotericin B. Tissue culture dishes are coated with either bovine collagen I (10 μg/cm^2; BD Biosciences Discovery Labware, Bedford, MA) or human extracellular matrix (5 μg/cm^2; BD Biosciences Discovery Labware), or uncoated (tissue culture plastic alone). EBD cells are cultured at 37° in 5% CO_2 and 95% humidity and are routinely passaged 1:10 to 1:40, using 0.025% trypsin–0.01% EDTA for 5 min at 37°. Low-serum cultures are treated with trypsin inhibitor (1 mg/ml) and then spun down and resuspended in growth medium.

Immunocytochemistry

Approximately 1 × 10^5 cells are plated in each well of an 8-well glass bottom chamber slide. Cells are fixed in either 4% paraformaldehyde in DPBS or in a 1:1 mixture of methanol–acetone for 10 min. Cells are permeabilized in 0.1% Triton X-100, followed by one DPBS rinse for 10 min, and then blocked in either Powerblock (BioGenex, San Ramon, CA), 5% goat serum, or 1–5% goat serum supplemented with 0.5% bovine serum albumin for 10–60 min. Primary antibodies and dilutions are as follows: neurofilament 68 kDa (diluted 1:4; Roche), neuron-specific enolase (diluted 1:100; BD Biosciences Pharmingen), tau (diluted 5 μg/ml; BD Biosciences Pharmingen), vimentin (diluted 1:10; Roche), nestin (diluted 1:250; National Institutes of Health, Bethesda, MD), galactocerebroside (diluted 1:500; Sigma), O4 (10 μg/ml; Roche), and SMI32 (diluted 1:5000; Sternberger Monoclonals). Detection was carried out with secondary antibodies conjugated to biotin, streptavidin-conjugated horseradish peroxidase, and 3-amino-9-ethylcarbazole chromogen (BioGenex).

mRNA Expression Profiles

RNA can be prepared with an RNeasy kit (Qiagen, Valencia, CA) according to the manufacturer's instructions from cells prepared on either 60- or 100-mm tissue culture plates. RNA preparations are digested with

RNase-free DNase (Roche) for 30 min at 37°, and then inactivated at 75° for 5 min. Synthesis of cDNA is performed on 5 μg of RNA by using oligo(dT) primers and a standard Moloney murine leukemia virus (MMLV; Invitrogen) reaction carried out at 42°. Thirty cycles of polymerase chain reaction (PCR) are carried out in the presence of 1.5 mM MgCl$_2$ at an annealing temperature of 55° and an incubation time of 30 s. PCRs are resolved on a 1.8% agarose gel. The efficacy of each PCR is established with human RNA tissue controls (Clontech, Palo Alto, CA). All amplimers are validated by Southern blot analysis with oligonucleotide probes end-labeled with [^{32}P] ATP, hybridized in 6× SSC, 5× Denhardt's solution, 0.1% sodium dodecyl sulfate (SDS), 0.05% sodium pyrophosphate, and sheared and denatured salmon sperm DNA (1000 μg/ml) at 45°. cDNA synthesis and genomic DNA contamination are monitored with primers specific to human phosphoglycerate kinase-1, which give products of ~250 base pairs (bp) and ~500 bp when amplifying cDNA and genomic DNA, respectively.

Cryopreservation of Mouse EGCs and Human EBD Cells

Although cryopreservation of mouse EGCs can be successful when using standard protocols, hEGCs are more fragile in this process. Instead, cryopreservation of human EGC-derived EBD cells has been performed. In mouse EGCs and human EBD cultures, choose cultures that are about three-fourths confluent in either 60- or 100-mm tissue culture dishes. To disaggregate the colonies, first wash with DPBS and add 1 ml of 0.05% trypsin–EDTA and then place the cells in an incubator for 5 min at 37°. After the cells detach from the dish, gently triturate up and down about three to five times with a 1-ml pipette and add a 2- to 10-fold excess of freezing medium [50% FCS–10% DMSO–40% Dulbecco's modified Eagle's medium (DMEM)] to stop the reaction. Spin the cells at 200g for 5 min and aspirate off the supernatant. Flick the pellet so that the cells will not clump, add 1 ml of freezing medium (one Petri/one cryotube vial), and freeze in cryotubes, using controlled-rate freezing vessels. Store the vials in liquid nitrogen and thaw according to standard procedures.

Acknowledgments

We like to thank Joyce Axelman for cell culture assistance and Christine M. Hill for immunological staining and FISH of human gonads.

References

Andrews, P. W. (1998). Teratocarcinomas and human embryology: Pluripotent human EC cell lines [review article]. *APMIS* **106,** 158–167; discussion 167–168.
Baker, T. G., and Franchi, L. L. (1967). The fine structure of oogonia and oocytes in human ovaries. *J. Cell Sci.* **2,** 213–224.

Bendsen, E., Byskov, A. G., Laursen, S. B., Larsen, H. P., Andersen, C. Y., and Westergaard, L. G. (2003). Number of germ cells and somatic cells in human fetal testes during the first weeks after sex differentiation. *Hum. Reprod.* **18,** 13–18.

Blelloch, R. H., Hochedlinger, K., Yamada, Y., Brennan, C., Kim, M., Mintz, B., Chin, L., and Jaenisch, R. (2004). Nuclear cloning of embryonal carcinoma cells. *Proc. Natl. Acad. Sci. USA* **101,** 13985–13990.

Boiani, M., and Scholer, H. R. (2005). Regulatory networks in embryo-derived pluripotent stem cells. *Nat. Rev. Mol. Cell. Biol.* **6,** 872–884.

Chambers, I., Colby, D., Robertson, M., Nichols, J., Lee, S., Tweedie, S., and Smith, A. (2003). Functional expression cloning of Nanog, a pluripotency sustaining factor in embryonic stem cells. *Cell* **113,** 643–655.

De Felici, M., Dolci, S., and Pesce, M. (1993). Proliferation of mouse primordial germ cells *in vitro*: A key role for cAMP. *Dev. Biol.* **157,** 277–280.

Dolci, S., Williams, D. E., Ernst, M. K., Resnick, J. L., Brannan, C. I., Lock, L. F., Lyman, S. D., Boswell, H. S., and Donovan, P. J. (1991). Requirement for mast cell growth factor for primordial germ cell survival in culture. *Nature* **352,** 809–811.

Dolci, S., Pesce, M., and De Felici, M. (1993). Combined action of stem cell factor, leukemia inhibitory factor, and cAMP on *in vitro* proliferation of mouse primordial germ cells. *Mol. Reprod. Dev.* **35,** 134–139.

Donovan, P. J., and de Miguel, M. P. (2003). Turning germ cells into stem cells. *Curr. Opin. Genet. Dev.* **13,** 463–471.

Donovan, P. J., and Gearhart, J. (2001). The end of the beginning for pluripotent stem cells. *Nature* **414,** 92–97.

Donovan, P. J., Stott, D., Cairns, L. A., Heasman, J., and Wylie, C. C. (1986). Migratory and postmigratory mouse primordial germ cells behave differently in culture. *Cell* **44,** 831–838.

Durcova-Hills, G., Ainscough, J., and McLaren, A. (2001). Pluripotential stem cells derived from migrating primordial germ cells. *Differentiation* **68,** 220–226.

Evans, M. J., and Kaufman, M. H. (1981). Establishment in culture of pluripotential cells from mouse embryos. *Nature* **292,** 154–156.

Falin, L. I. (1969). The development of genital glands and the origin of germ cells in human embryogenesis. *Acta Anat.(Basel)* **72,** 195–232.

Francavilla, S., and Zamboni, L. (1985). Differentiation of mouse ectopic germinal cells in intra- and perigonadal locations. *J. Exp. Zool.* **233,** 101–109.

Francavilla, S., Cordeschi, G., Properzi, G., Concordia, N., Cappa, F., and Pozzi, V. (1990). Ultrastructure of fetal human gonad before sexual differentiation and during early testicular and ovarian development. *J. Submicrosc. Cytol. Pathol.* **22,** 389–400.

Frimberger, D., Morales, N., Shamblott, M., Gearhart, J. D., Gearhart, J. P., and Lakshmanan, Y. (2005). Human embryoid body-derived stem cells in bladder regeneration using rodent model. *Urology* **65,** 827–832.

Godin, I., and Wylie, C. C. (1991). TGF β_1 inhibits proliferation and has a chemotropic effect on mouse primordial germ cells in culture. *Development* **113,** 1451–1457.

Gondos, B., and Hobel, C. J. (1971). Ultrastructure of germ cell development in the human fetal testis. *Z. Zellforsch Mikrosk Anat.* **119,** 1–20.

Gondos, B., Westergaard, L., and Byskov, A. G. (1986). Initiation of oogenesis in the human fetal ovary: Ultrastructural and squash preparation study. *Am. J. Obstet. Gynecol.* **155,** 189–195.

Goto, T., Adjaye, J., Rodeck, C. H., and Monk, M. (1999). Identification of genes expressed in human primordial germ cells at the time of entry of the female germ line into meiosis. *Mol. Hum. Reprod.* **5,** 851–860.

Hart, A. H., Hartley, L., Ibrahim, M., and Robb, L. (2004). Identification, cloning and expression analysis of the pluripotency promoting Nanog genes in mouse and human. *Dev. Dyn.* **230,** 187–198.

Hatano, S. Y., Tada, M., Kimura, H., Yamaguchi, S., Kono, T., Nakano, T., Suemori, H., Nakatsuji, N., and Tada, T. (2005). Pluripotential competence of cells associated with Nanog activity. *Mech. Dev.* **122**, 67–79.

Heyn, R., Makabe, S., and Motta, P. M. (1998). Ultrastructural dynamics of human testicular cords from 6 to 16 weeks of embryonic development: Study by transmission and high resolution scanning electron microscopy. *Ital. J. Anat. Embryol.* **103**, 17–29.

Kanatsu-Shinohara, M., Inoue, K., Lee, J., Yoshimoto, M., Ogonuki, N., Miki, H., Baba, S., Kato, T., Kazuki, Y., Toyokuni, S., Toyoshima, M., Niwa, O., Oshimura, M., Heike, T., Nakahata, T., Ishino, F., Ogura, A., and Shinohara, T. (2004). Generation of pluripotent stem cells from neonatal mouse testis. *Cell* **119**, 1001–1012.

Kanatsu-Shinohara, M., Miki, H., Inoue, K., Ogonuki, N., Toyokuni, S., Ogura, A., and Shinohara, T. (2005). Long-term culture of mouse male germline stem cells under serum- or feeder-free conditions. *Biol. Reprod.* **72**, 985–991.

Kehler, J., Tolkunova, E., Koschorz, B., Pesce, M., Gentile, L., Boiani, M., Lomeli, H., Nagy, A., McLaughlin, K. J., Scholer, H. R., and Tomilin, A. (2004). Oct4 is required for primordial germ cell survival. *EMBO Rep.* **5**, 1078–1083.

Kerr, D. A., Llado, J., Shamblott, M. J., Maragakis, N. J., Irani, D. N., Crawford, T. O., Krishnan, C., Dike, S., Gearhart, J. D., and Rothstein, J. D. (2003). Human embryonic germ cell derivatives facilitate motor recovery of rats with diffuse motor neuron injury. *J. Neurosci.* **23**, 5131–5140.

Kuroda, T., Tada, M., Kubota, H., Kimura, H., Hatano, S. Y., Suemori, H., Nakatsuji, N., and Tada, T. (2005). Octamer and Sox elements are required for transcriptional *cis* regulation of Nanog gene expression. *Mol. Cell. Biol.* **25**, 2475–2485.

Labosky, P. A., Barlow, D. P., and Hogan, B. L. (1994). Embryonic germ cell lines and their derivation from mouse primordial germ cells. *Ciba Found. Symp.* **182**, 157–168; discussion 168–178.

Lawson, K. A., and Hage, W. J. (1994). Clonal analysis of the origin of primordial germ cells in the mouse. *Ciba Found. Symp.* **182**, 68–84; discussion 84–91.

Lee, C. K., and Piedrahita, J. A. (2000). Effects of growth factors and feeder cells on porcine primordial germ cells *in vitro*. *Cloning* **2**, 197–205.

Liu, S., Liu, H., Pan, Y., Tang, S., Xiong, J., Hui, N., Wang, S., Qi, Z., and Li, L. (2004). Human embryonic germ cells isolation from early stages of post-implantation embryos. *Cell Tissue Res.* **318**, 525–531.

Makabe, S., Naguro, T., and Motta, P. M. (1992). A new approach to the study of ovarian follicles by scanning electron microscopy and ODO maceration. *Arch. Histol. Cytol.* **55**, (Suppl.), 183–190.

Martin, G. R. (1981). Isolation of a pluripotent cell line from early mouse embryos cultured in media conditioned by teratocarcinoma stem cells. *Proc. Natl. Acad. Sci. USA* **78**, 7634–7638.

Matsui, Y., Toksoz, D., Nishikawa, S., Nishikawa, S., Williams, D., Zsebo, K., and Hogan, B. L. (1991). Effect of Steel factor and leukaemia inhibitory factor on murine primordial germ cells in culture. *Nature* **353**, 750–752.

Matsui, Y., Zsebo, K., and Hogan, B. L. (1992). Derivation of pluripotential embryonic stem cells from murine primordial germ cells in culture. *Cell* **70**, 841–847.

McKay, D. G., Hertig, A. T., Adams, E. C., and Danziger, S. (1953). Histochemical observations on the germ cells of human embryos. *Anat. Rec.* **117**, 201–219.

McLaren, A. (2003). Primordial germ cells in the mouse. *Dev. Biol.* **262**, 1–15.

McLaren, A., and Durcova-Hills, G. (2001). Germ cells and pluripotent stem cells in the mouse. *Reprod. Fertil. Dev.* **13**, 661–664.

Mintz, B., and Russell, E. S. (1957). Gene-induced embryological modifications of primordial germ cells in the mouse. *J. Exp. Zool.* **134**, 207–237.

Mitsui, K., Tokuzawa, Y., Itoh, H., Segawa, K., Murakami, M., Takahashi, K., Maruyama, M., Maeda, M., and Yamanaka, S. (2003). The homeoprotein Nanog is required for maintenance of pluripotency in mouse epiblast and ES cells. *Cell* **113**, 631–642.

Motta, P. M., Makabe, S., and Nottola, S. A. (1997). The ultrastructure of human reproduction. I. The natural history of the female germ cell: Origin, migration and differentiation inside the developing ovary. *Hum. Reprod. Update* **3**, 281–295.

Mueller, D., Shamblott, M. J., Fox, H. E., Gearhart, J. D., and Martin, L. J. (2005). Transplanted human embryonic germ cell-derived neural stem cells replace neurons and oligodendrocytes in the forebrain of neonatal mice with excitotoxic brain damage. *J. Neurosci. Res.* **82**, 592–608.

Mueller, S., Prelle, K., Rieger, N., Petznek, H., Lassnig, C., Luksch, U., Aigner, B., Baetscher, M., Wolf, E., Mueller, M., and Brem, G. (1999). Chimeric pigs following blastocyst injection of transgenic porcine primordial germ cells. *Mol. Reprod. Dev.* **54**, 244–254.

Nichols, J., Zevnik, B., Anastassiadis, K., Niwa, H., Klewe-Nebenius, D., Chambers, I., Scholer, H., and Smith, A. (1998). Formation of pluripotent stem cells in the mammalian embryo depends on the POU transcription factor Oct4. *Cell* **95**, 379–391.

Niwa, H., Burdon, T., Chambers, I., and Smith, A. (1998). Self-renewal of pluripotent embryonic stem cells is mediated via activation of STAT3. *Genes Dev.* **12**, 2048–2060.

Ohno, S., Klinger, H. P., and Atkin, N. B. (1962). Human oogenesis. *Cytogenetics* **1**, 42–51.

Ozdzenski, W. (1967). Observations on the origin of primordial germ cells in the mouse. *Zool. Polon.* **17**, 367–379.

Park, J. H., Kim, S. J., Lee, J. B., Song, J. M., Kim, C. G., Roh, S., II, and Yoon, H. S. (2004). Establishment of a human embryonic germ cell line and comparison with mouse and human embryonic stem cells. *Mol. Cells* **17**, 309–315.

Park, T. S., and Han, J. Y. (2000). Derivation and characterization of pluripotent embryonic germ cells in chicken. *Mol. Reprod. Dev.* **56**, 475–482.

Pesce, M., and Scholer, H. R. (2000). Oct-4: Control of totipotency and germline determination. *Mol. Reprod. Dev.* **55**, 452–457.

Pesce, M., and Scholer, H. R. (2001). Oct-4: Gatekeeper in the beginnings of mammalian development. *Stem Cells* **19**, 271–278.

Pesce, M., Wang, X., Wolgemuth, D. J., and Scholer, H. (1998). Differential expression of the Oct-4 transcription factor during mouse germ cell differentiation. *Mech. Dev.* **71**, 89–98.

Piedrahita, J. A., Moore, K., Oetama, B., Lee, C. K., Scales, N., Ramsoondar, J., Bazer, F. W., and Ott, T. (1998). Generation of transgenic porcine chimeras using primordial germ cell-derived colonies. *Biol. Reprod.* **58**, 1321–1329.

Pinkerton, J. H., McKay, K. D., Adams, E. C., and Hertig, A. T. (1961). Development of the human ovary: A study using histochemical technics. *Obstet. Gynecol.* **18**, 152–181.

Rabinovici, J., and Jaffe, R. B. (1990). Development and regulation of growth and differentiated function in human and subhuman primate fetal gonads. *Endocr. Rev.* **11**, 532–557.

Resnick, J. L., Bixler, L. S., Cheng, L., and Donovan, P. J. (1992). Long-term proliferation of mouse primordial germ cells in culture. *Nature* **359**, 550–551.

Scholer, H. R., Ruppert, S., Suzuki, N., Chowdhury, K., and Gruss, P. (1990). New type of POU domain in germ line-specific protein Oct-4. *Nature* **344**, 435–439.

Shamblott, M. J., Axelman, J., Wang, S., Bugg, E. M., Littlefield, J. W., Donovan, P. J., Blumenthal, P. D., Huggins, G. R., and Gearhart, J. D. (1998). Derivation of pluripotent stem cells from cultured human primordial germ cells. *Proc. Natl. Acad. Sci. USA* **95**, 13726–13731.

Shamblott, M. J., Axelman, J., Littlefield, J. W., Blumenthal, P. D., Huggins, G. R., Cui, Y., Cheng, L., and Gearhart, J. D. (2001). Human embryonic germ cell derivatives express a broad range of developmentally distinct markers and proliferate extensively *in vitro*. *Proc. Natl. Acad. Sci. USA* **98**, 113–118.

Shamblott, M. J., Kerr, C. L., Axelman, J., Littlefield, Jl. W., Clark, G. O., Patterson, E. S., Addis, R. C., Kraszewski, J. L., Ket, K. C., and Gearhart, J. D. (2004). Derivation and differentiation of human embryonic germ cells. *In* "Handbook of Stem Cells"(R. Lanza, B. Hogan, D. Melton, R. Pedersen, J. Thomson, and M. West, eds.), Vol. 1, Elsevier Academic Press, New York, 459–469.

Shim, H., Gutierrez-Adan, A., Chen, L. R., BonDurant, R. H., Behboodi, E., and Anderson, G. B. (1997). Isolation of pluripotent stem cells from cultured porcine primordial germ cells. *Biol. Reprod.* **57**, 1089–1095.

Skrzypczak, J., Pisarski, T., Biczysko, W., and Kedzia, H. (1981). Evaluation of germ cell development in gonads of human fetuses and newborns. *Folia Histochem. Cytochem. (Krakow)* **19**, 17–24.

Smith, A. G. (2001). Embryo-derived stem cells: Of mice and men. *Annu. Rev. Cell. Dev. Biol.* **17**, 435–462.

Smith, A. G., Heath, J. K., Donaldson, D. D., Wong, G. G., Moreau, J., Stahl, M., and Rogers, D. (1988). Inhibition of pluripotential embryonic stem cell differentiation by purified polypeptides. *Nature* **336**, 688–690.

Stevens, L. C. (1966). Development of resistance to teratocarcinogenesis by primordial germ cells in mice. *J. Natl. Cancer Inst.* **37**, 859–867.

Stevens, L. C. (1967). Origin of testicular teratomas from primordial germ cells in mice. *J. Natl. Cancer Inst.* **38**, 549–552.

Stewart, C. L., Gadi, I., and Bhatt, H. (1994). Stem cells from primordial germ cells can reenter the germ line. *Dev. Biol.* **161**, 626–628.

Sun, E. L., and Gondos, B. (1984). Squash preparation studies of germ cells in human fetal testes. *J. Androl.* **5**, 334–338.

Takagi, Y., Talbot, N. C., Rexroad, C. E., Jr., and Pursel, V. G. (1997). Identification of pig primordial germ cells by immunocytochemistry and lectin binding. *Mol. Reprod. Dev.* **46**, 567–580.

Tam, P. P., and Snow, M. H. (1981). Proliferation and migration of primordial germ cells during compensatory growth in mouse embryos. *J. Embryol. Exp. Morphol.* **64**, 133–147.

Tsung, H. C., Du, Z. W., Rui, R., Li, X. L., Bao, L. P., Wu, J., Bao, S. M., and Yao, Z. (2003). The culture and establishment of embryonic germ (EG) cell lines from Chinese mini swine. *Cell Res.* **13**, 195–202.

Turnpenny, L., Brickwood, S., Spalluto, C. M., Piper, K., Cameron, I. T., Wilson, D. I., and Hanley, N. A. (2003). Derivation of human embryonic germ cells: An alternative source of pluripotent stem cells. *Stem Cells* **21**, 598–609.

Turnpenny, L., Cameron, I. T., Spalluto, C. M., Hanley, K. P., Wilson, D. I., and Hanley, N. A. (2005a). Human embryonic germ cells for future neuronal replacement therapy. *Brain Res. Bull.* **68**, 76–82.

Turnpenny, L., Spalluto, C. M., Perrett, R. M., O'Shea, M., Piper Hanley, K., Cameron, I. T., Wilson, D. I., and Hanley, N. A. (2005b). Evaluating human embryonic germ cells: Concord and conflict as pluripotent stem cells. *Stem Cells* **24**, 212–220.

Upadhyay, S., and Zamboni, L. (1982). Ectopic germ cells: Natural model for the study of germ cell sexual differentiation. *Proc. Natl. Acad. Sci. USA* **79**, 6584–6588.

Wartenberg, H. (1982). Development of the early human ovary and role of the mesonephros in the differentiation of the cortex. *Anat. Embryol. (Berl)* **165**, 253–280.

Witschi, E. (1948). Migration of the germ cells of human embryos from the yolk sac to the primitive gonadal folds. *In* "Contributions in Embryology," Vol. 32, pp. 67–80. Carnegie Institute, Washington, D.C.

Witschi, E. (1963). Embryology of the ovary. *In* "The Ovary" (H. G. Grady and D. E. Smith, eds.), Williams & Wilkins, Baltimore, MD.

Yamaguchi, S., Kimura, H., Tada, M., Nakatsuji, N., and Tada, T. (2005). Nanog expression in mouse germ cell development. *Gene Expr. Patterns* **5**, 639–646.

Yeom, Y. I., Fuhrmann, G., Ovitt, C. E., Brehm, A., Ohbo, K., Gross, M., Hubner, K., and Scholer, H. R. (1996). Germline regulatory element of Oct-4 specific for the totipotent cycle of embryonal cells. *Development* **122**, 881–894.

[17] Amniotic Fluid and Placental Stem Cells

By Dawn M. Delo, Paolo De Coppi,
Georg Bartsch, Jr., and Anthony Atala

Abstract

Human amniotic fluid has been used in prenatal diagnosis for more than 70 years. It has proven to be a safe, reliable, and simple screening tool for a wide variety of developmental and genetic diseases. However, there is now evidence that amniotic fluid may have more use than only as a diagnostic tool and may be the source of a powerful therapy for a multitude of congenital and adult disorders. A subset of cells found in amniotic fluid and placenta has been isolated and found to be capable of maintaining prolonged undifferentiated proliferation as well as able to differentiate into multiple tissue types encompassing the three germ layers. It is possible that in the near future, we will see the development of therapies using progenitor cells isolated from amniotic fluid and placenta for the treatment of newborns with congenital malformations as well as of adults, using cryopreserved amniotic fluid and placental stem cells. In this chapter, we describe a number of experiments that have isolated and characterized pluripotent progenitor cells from amniotic fluid and placenta. We also discuss various cell lines derived from amniotic fluid and placenta and future directions for this area of research.

Introduction

Amniotic fluid-derived progenitor cells can be obtained from a small amount of fluid during amniocentesis, a procedure that is already often performed in many of the pregnancies in which the fetus has a congenital

METHODS IN ENZYMOLOGY, VOL. 419
Copyright 2006, Elsevier Inc. All rights reserved.
0076-6879/06 $35.00
DOI: 10.1016/S0076-6879(06)19017-5

abnormality. Placenta-derived stem cells can be obtained from a small biopsy of the chorionic villi. Observations on cell cultures from these two sources provide evidence that they may represent new sources for the isolation of cells with the potency to differentiate into different cell types, suggesting a new source of cells for research and treatment.

Amniotic Fluid and Placenta in Developmental Biology

Gastrulation is a major milestone in early postimplantation development (Snow and Bennett, 1978). At about embryonic day 6.5 (E6.5), gastrulation begins in the posterior region of the embryo. Pluripotent epiblast cells are allocated to the three primary germ layers of the embryo (ectoderm, mesoderm, and endoderm) and germ cells, which are the progenitors of all fetal tissue lineages as well as the extraembryonic mesoderm of the yolk sac, amnion, and allantois (Downs and Harmann, 1997; Downs et al., 2004; Gardner and Beddington, 1988; Loebel et al., 2003). The latter forms the umbilical cord as well as the mesenchymal part of the labyrinthine layer in the mature chorioallantoic placenta (Downs and Harmann, 1997; Moser et al., 2004; Smith et al., 1994). The final positions of the fetal membranes result from the process of embryonic turning, which occurs around day 8.5 of gestation and "pulls" the amnion and yolk sac around the embryo (Kinder et al., 1999; Parameswaran and Tam, 1995). The specification of tissue lineages is accomplished by the restriction of developmental potency and the activation of lineage-specific gene expression (Parameswaran and Tam, 1995; Rathjen et al., 1999). This process is strongly influenced by cellular interactions and signaling (Dang et al., 2002; Li et al., 2004).

The amniotic sac is a tough but thin transparent pair of membranes that holds a developing embryo (and later fetus) until shortly before birth. The inner membrane, the amnion, contains the amniotic fluid and the fetus. The outer membrane, the chorion, contains the amnion and is part of the placenta (Kaviani et al., 2001; Kinder et al., 1999; Robinson et al., 2002). Amnion is derived from ectoderm and mesoderm, which grows and begins to fill, mainly with water (Robinson et al., 2002). Originally it is isotonic, containing proteins, carbohydrates, lipids and phospholipids, urea, and electrolytes. Later, urine excreted by the fetus increases its volume and changes its concentration (Bartha et al., 2000; Heidari et al., 1996; Sakuragawa et al., 1999; Srivastava et al., 1996). The fetus can breathe in the water, allowing normal growth and the development of lungs and the gastrointestinal tract. The fluid is swallowed by the fetus and passes via the fetal blood into the maternal blood. The functions of amniotic fluid are that it ensures symmetrical structure development and growth; cushions and protects the embryo; helps maintain consistent pressure and temperature; and permits freedom of fetal movement, important for musculoskeletal development and blood flow (Baschat and Hecher, 2004).

A wide variety of different origins has been suggested for the mixture of cells within amniotic fluid (Medina-Gomez and del Valle, 1988). Cells of different embryonic/fetal origins of all three germ layers have been reported to exist in amniotic fluid (In 't Anker et al., 2003; Prusa et al., 2004). These cells are thought to be sloughed from the fetal amnion, skin, and alimentary, respiratory, and urogenital tracts. In addition, it has been reported that cells cultured from amniotic fluid as well as placenta provide evidence that they may represent new stem cell sources with the potency to differentiate into different cell types (Prusa and Hengstschlager, 2002). Interestingly, it has been demonstrated that a subpopulation of cells in amniotic fluid produces Oct4 mRNA, which is used to maintain pluripotency (Prusa et al., 2003). Although research is at its early stages, these cells can be used to find treatments or even cures for many diseases in which irreplaceable cells are damaged.

Amniotic Fluid and Placenta for Cell Therapy

Ideal cells for regenerative medicine are the pluripotent stem cells, which have the capability to differentiate in stages into a huge number of different types of human cells. Amniotic fluid cells can be obtained from a small amount of fluid during amniocentesis at the second trimester, a procedure that is already often performed in many of the pregnancies in which the fetus has a congenital abnormality and to determine characteristics such as sex (Hoehn et al., 1975). Kaviani and co-workers reported that "just 2 milliliters of amniotic fluid" can provide up to 20,000 cells, 80% of which are viable (Kaviani et al., 2001, 2003). Because many pregnant women already undergo amniocentesis to screen for fetal abnormalities, cells can be simply isolated from the test fluid of infants with defects and saved for future use. With amniotic fluid cells, it takes 20 to 24 h to double the number of cells collected, which is faster than for umbilical cord stem cells (28 to 30 h) and bone marrow stem cells (more than 30 h) (Tsai et al., 2004). This phenomenon is an important feature for urgent medical conditions. In addition, while scientists have been able to isolate and differentiate on average only 30% of mesenchymal stem cells (MSCs) extracted from a child's umbilical cord shortly after birth, the success rate for amniotic fluid-derived MSCs is close to 100% (In 't Anker et al., 2003; Tsai et al., 2004). Furthermore, extracting the cells from amniotic fluid bypasses the problems associated with a technique called donor–recipient HLA matching, which involves transplanting cells (Tsai et al., 2004).

Isolation and Characterization of Progenitor Cells

Amniotic fluid progenitor cells are isolated by centrifugation of amniotic fluid from amniocentesis. Chorionic villi placental cells are isolated from single villi under light microscopy. Amniotic fluid cells and placental cells are allowed to proliferate *in vitro* and are maintained in culture for 4 weeks. The culture medium consists of modified α-modified Earl's medium (18% Chang medium B, 2% Chang medium C with 15% embryonic stem cell certified fetal bovine serum, antibiotics, and L-glutamine).

A pluripotential subpopulation of progenitor cells present in the amniotic fluid and placenta can be isolated through positive selection for cells expressing the membrane receptor c-kit (Fig. 1) (DeCoppi, 2001; Siddiqui and Atala, 2004). This receptor binds to the ligand stem cell factor. About 0.8 to 1.4% of cells present in amniotic fluid and placenta have been shown to be c-kit positive in analysis by fluorescence-activated cell sorting (FACS). Progenitor cells maintain a round shape for 1 week postisolation when cultured in nontreated culture dishes. In this state, they demonstrate low proliferative capability. After the first week the cells begin to adhere to the plate and change their morphology, becoming more elongated and proliferating more rapidly, reaching 80% confluence with a need for passage every 48–72 h. No feeder layers are required either for maintenance or expansion. The progenitor cells derived show a high self-renewal capacity with >300 population doublings, far exceeding Hayflick's limit. The doubling time of the undifferentiated cells is noted to be 36 h with little variation with passages.

These cells have been shown to maintain a normal karyotype at late passages and have normal G_1 and G_2 cell cycle checkpoints. They demonstrate

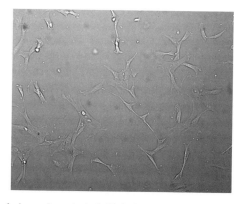

FIG. 1. Morphology of amniotic fluid-derived stem cells (AFSCs) in culture.

telomere length conservation while in the undifferentiated state as well as telomerase activity even in late passages (Bryan *et al.*, 1998). Analysis of surface markers shows that progenitor cells from amniotic fluid express human embryonic stage-specific marker SSEA4, and the stem cell marker Oct4, and did not express SSEA1, SSEA3, CD4, CD8, CD34, CD133, C-MET, ABCG2, NCAM, BMP4, TRA1-60, and TRA1-81, to name a few. This expression profile is of interest as it demonstrates expression by the amniotic fluid-derived progenitor cells of some key markers of the embryonic stem cell phenotype, but not the full complement of markers expressed by embryonic stem cells. This may indicate that the amniotic cells are not quite as primitive as embryonic cells, yet maintain greater potential than most adult stem cells. Another behavior showing similarities and differences between these amniotic fluid-derived cells and blastocyst-derived cells is that whereas the amniotic fluid progenitor cells do form embryoid bodies *in vitro*, which stain positive for markers of all three germ layers, these cells do not form teratomas *in vivo* when implanted in immunodeficient mice. Last, cells, when expanded from a single cell, maintain similar properties in growth and potential as the original mixed population of the progenitor cells.

Differentiation of Amniotic Fluid- and Placenta-Derived Progenitor Cells

The progenitor cells derived from amniotic fluid and placenta are pluripotent and have been shown to differentiate into osteogenic, adipogenic, myogenic, neurogenic, endothelial, and hepatic phenotypes *in vitro*. Each differentiation has been performed through proof of phenotypic and biochemical changes consistent with the differentiated tissue type of interest (Fig. 2). We discuss each set of differentiations separately.

Adipocytes

To promote adipogenic differentiation, progenitor cells can be induced in dexamethasone, 3-isobutyl-1-methylxanthine, insulin, and indomethacin. Progenitor cells cultured with adipogenic supplements change their morphology from elongated to round within 8 days. This coincides with the accumulation of intracellular droplets. After 16 days in culture, more than 95% of the cells have their cytoplasm filled with lipid-rich vacuoles. Adipogenic differentiation also demonstrates the expression of peroxisome proliferation-activated receptor $\gamma2$ (PPAR-$\gamma2$), a transcription factor that regulates adipogenesis, and of lipoprotein lipase through reverse

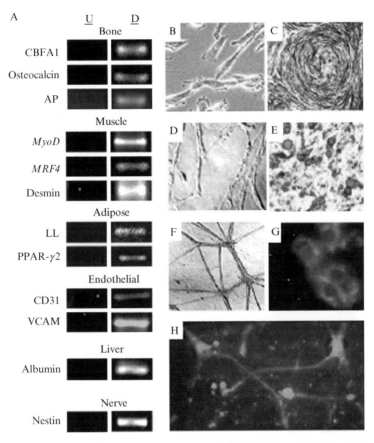

FIG. 2. Multilineage differentiation of hAFSCs. (A) RT-PCR analysis of mRNA. *Left*: Control undifferentiated cells. *Right*: Cells maintained under conditions for differentiation to bone (8 days), muscle (8 days), adipocyte (16 days), endothelial (8 days), hepatic (45 days), and neuronal (2 days) lineages. (B) Phase-contrast microscopy of control, undifferentiated cells. (b–h) Differentiated progenitor cells. (C) Bone: Histochemical staining for alkaline phosphatase. (D) Muscle: Phase-contrast microscopy showing fusion into multinucleated myotube-like cells. (E) Adipocyte: Staining with oil red O (day 8) shows intracellular oil aggregation. (F) Endothelial: Phase-contrast microscopy of capillary-like structures. (G) Hepatic: Fluorescent antibody staining (FITC, green) for albumin. (H) Neuronal: Fluorescent antibody staining of nestin (day 2). (See color insert.)

transcription-polymerase chain reaction (RT-PCR) analysis (Cremer *et al.*, 1981; Medina-Gomez and del Valle, 1988). Expression of these genes is noted in progenitor cells under adipogenic conditions but not in undifferentiated cells.

Osteocytes

Osteogenic differentiation was induced in progenitor cells with the use of dexamethasone, β-glycerophosphate, and ascorbic acid 2-phosphate (Jaiswal *et al.*, 1997). Progenitor cells maintained in this medium demonstrated phenotypic changes within 4 days, with a loss of spindle-shape phenotype and development of an osteoblast-like appearance with finger-like excavations into the cytoplasm. At 16 days, the cells aggregated, showing typical lamellar bone-like structures. In terms of functionality, these differentiated cells demonstrate a major feature of osteoblasts, which is to precipitate calcium. Differentiated osteoblasts from the progenitor cells are able to produce alkaline phosphatase (AP) and to deposit calcium, consistent with bone differentiation. Undifferentiated progenitor cells lacked this ability. Progenitor cells in osteogenic medium express specific genes implicated in mammalian bone development [AP, core-binding factor A1 (CBFA1), and osteocalcin] in a pattern consistent with the physiological analog. Progenitor cells grown in osteogenic medium show activation of the AP gene at each time point. Expression of CBFA1, a transcription factor specifically expressed in osteoblasts and hypertrophic chondrocytes and that regulates gene expression of structural proteins of the bone extracellular matrix, is highest in cells grown in osteogenic inducing medium on day 8 and decreases slightly on days 16, 24, and 32. Osteocalcin is expressed only in progenitor cells under osteogenic conditions at 8 days (Karsenty, 2000; Komori *et al.*, 1997).

Endothelial Cells

Amniotic fluid progenitor cells can be induced to form endothelial cells by culture in endothelial basal medium on gelatin-coated dishes. Full differentiation is achieved by 1 month in culture; however, phenotypic changes are noticed within 1 week of initiation of the protocol. Human-specific endothelial cell surface marker (P1H12), factor VIII (FVIII), and KDR (kinase insert domain receptor) are specific for differentiated endothelial cells. Differentiated cells stain positively for FVIII, KDR, and P1H12. Progenitor cells do not stain for endothelial-specific markers. Amniotic fluid progenitor-derived endothelial cells, once differentiated, are able to grow in culture and form capillary-like structures *in vitro*. These cells also express platelet endothelial cell adhesion molecule 1 (PECAM-1 or CD31) and vascular cell adhesion molecule (VCAM), which are not detected in the progenitor cells on RT-PCR analysis.

Hepatocytes

For hepatic differentiation, progenitor cells are seeded on Matrigel- or collagen-coated dishes at different stages and cultured in the presence of hepatocyte growth factor, insulin, oncostatin M, dexamethasone, fibroblast growth factor 4, and monothioglycerol for 45 days (Dunn *et al.*, 1989; Schwartz *et al.*, 2002). After 7 days of the differentiation process, cells exhibit morphological changes from an elongated to a cobblestone appearance. The cells show positive staining for albumin on day 45 postdifferentiation and also express the transcription factor HNF4α (hepatocyte nuclear factor 4α), the c-Met receptor, the multidrug resistance (MDR) membrane transporter, albumin, and α-fetoprotein. RT-PCR analysis further supports albumin production. The maximum rate of urea production for hepatic differentiation-induced cells is upregulated to 1.21×10^3 ng of urea per hour per cell from 5.0×10^1 ng of urea per hour per cell for the control progenitor cell populations (Hamazaki *et al.*, 2001).

Myocytes

Myogenic differentiation is induced in amniotic fluid-derived progenitor cells by culture in medium containing horse serum and chick embryo extract on a thin gel coat of Matrigel (Rosenblatt *et al.*, 1995). To initiate differentiation, the presence of 5-azacytidine in the medium for 24 h is necessary. Phenotypically, the cells can be seen to organize themselves into bundles that fuse to form multinucleated cells. These cells express sarcomeric tropomyosin and desmin, both of which are not expressed in the original progenitor population.

The development profile of cells differentiating into myogenic lineages interestingly mirrors a characteristic pattern of gene expression reflecting that seen with embryonic muscle development (Bailey *et al.*, 2001; Rohwedel *et al.*, 1994). With this protocol, *Myf6* is expressed on day 8 and suppressed on day 16. *MyoD* expression is detectable at 8 days and suppressed at 16 days in progenitor cells. Desmin expression is induced at 8 days and increases by 16 days in progenitor cells cultured in myogenic medium (Hinterberger *et al.*, 1991; Patapoutian *et al.*, 1995).

Neuronal Cells

For neurogenic induction, amniotic progenitor cells are induced in dimethyl sulfoxide (DMSO), butylated hydroxyanisole, and neuronal growth factor (Black and Woodbury, 2001; Woodbury *et al.*, 2000). Progenitor cells cultured under neurogenic conditions change their morphology within the first 24 h. Two different cell populations are apparent: morphologically large flat cells and small bipolar cells. The bipolar cell cytoplasm retracts toward the nucleus, forming contracted multipolar structures. Over subsequent hours, the cells display primary and secondary branches and cone-like terminal expansions. Induced progenitor cells show a characteristic sequence of expression of neural-specific proteins. At an early stage the intermediate filament protein nestin, which is specifically expressed in neuroepithelial stem cells, is highly expressed. Expression of βIII-tubulin and glial fibrillary acidic protein (GFAP), markers of neuron and glial differentiation, respectively, increases over time and seems to reach a plateau at about 6 days (Guan *et al.*, 2001). Progenitor cells cultured under neurogenic conditions show the presence of the neurotransmitter glutamic acid in the collected medium. Glutamic acid is usually secreted in culture by fully differentiated neurons (Carpenter *et al.*, 2001).

Amniotic Fluid and Placental Differentiation Protocols

Adipogenic Lineage Differentiation Protocol

1. Seed cells at a density of 3000 cells/cm^2 on tissue culture plates.
2. Culture in low-glucose Dulbecco's modified Eagle's medium (DMEM) with 10% fetal bovine serum (FBS), penicillin–streptomycin (Pen/Strep), and adipogenic supplements [1 μM dexamethasone, 1 mM 3-isobutyl-1-methylxanthine (Sigma-Aldrich, St. Louis, MO), insulin (10 μg/ml; Sigma-Aldrich), 60 μM indomethacin (Sigma-Aldrich)].

Osteogenic Lineage Differentiation Protocol

1. Seed cells at a density of 3000 cells/cm^2 on tissue culture plates.
2. Culture cells in low-glucose DMEM (GIBCO; Invitrogen, Carlsbad, CA) with 10% FBS (GIBCO; Invitrogen), penicillin (100 U/ml) and streptomycin (100 U/ml) (GIBCO; Invitrogen), and osteogenic supplements [100 nM dexamethasone (Sigma-Aldrich), 10 mM β-glycerophosphate (Sigma-Aldrich), and 0.05 mM ascorbic acid 2-phosphate (Wako Chemicals USA, Richmond, VA).

Endothelial Lineage Differentiation Protocol

1. Seed cells at a density of 3000 cells/cm^2 in 35-mm dishes precoated with gelatin.
2. Maintain the cells in culture for 1 month in endothelial cell medium-2 (Clonetics EGM-2; Cambrex Bio Science Walkersville, Walkersville, MD)
3. Add basic fibroblast growth factor (bFGF) at 2 ng/ml at intervals of 2 days.

Hepatic Lineage Differentiation Protocol

1. Seed cells at a density of 5000 cells/cm^2 on plates with Matrigel (BD Biosciences Discovery Labware, Bedford, MA), using the manufacturer's thin coat method.

2. Expand in AFS growth medium for 3 days to achieve a semiconfluent density.

3. Change the medium to DMEM low-glucose formulation containing 15% FBS, 300 μM monothioglycerol (Sigma-Aldrich), hepatocyte growth factor (20 ng/ml; Sigma-Aldrich), 10^{-7} M dexamethasone (Sigma-Aldrich), FGF4 (00 ng/ml; PeproTech, Rocky Hill, NJ), 1× ITS (insulin, transferrin, selenium; Roche, Indianapolis, IN), and Pen/Strep.

4. Maintain the cells in this medium for 2 weeks, with medium changes every third day.

Myogenic Lineage Differentiation Protocol

1. Seed cells at a density of 3000 cells/cm^2 on tissue culture plates precoated with Matrigel (BD Biosciences Discovery Labware; incubate for 1 h at 37° at 1 mg/ml in DMEM).

2. Culture cells in basic myogenic medium consisted of DMEM low-glucose formulation containing 10% horse serum (GIBCO; Invitrogen), 0.5% chick embryo extract (GIBCO; Invitrogen), and Pen/Strep.

3. Twelve hours after seeding, add 3 μM 5-aza-2-deoxycytidine (5-azaC; Sigma-Aldrich) to the culture medium for a period of 24 h.

4. After 24 h, change the medium completely. Incubation is then continued in complete medium without 5-azaC, with medium changes every 3 days.

Neurogenic Lineage Differentiation Protocol

1. Seed cells at a concentration of 3000 cells/cm^2 in 100-mm tissue culture plates.

2. Culture cells in DMEM low-glucose medium, Pen/Strep, supplemented with 2% DMSO, 200 μM butylated hydroxyanisole (BHA; Sigma-Aldrich), and nerve growth factor (NGF, 25 ng/ml).

3. After 2 days, return the cells to normal AFS growth medium lacking DMSO and BHA, but still containing NGF.

4. Add fresh NGF at intervals of 2 days.

Conclusion

Pluripotent progenitor cells isolated from amniotic fluid and placenta present an exciting possible contribution to the field of stem cell biology and regenerative medicine. These cells are an excellent source for research and therapeutic applications. The ability to isolate progenitor cells during gestation may also be advantageous for babies born with congenital malformations. Furthermore, progenitor cells can be cryopreserved for future self-use. Compared with embryonic stem cells, progenitor cells isolated from amniotic fluid have many similarities: they can differentiate into all three germ layers, they express common markers, and they preserve their telomere length. However, progenitor cells isolated from amniotic fluid and placenta have considerable advantages. They easily differentiate into specific cell lineages and they avoid the current controversies associated with the use of human embryonic stem cells. The discovery of these cells has been recent, and a considerable amount of work remains to be done on the characterization and use of these cells. In future, cells derived from amniotic fluid and placenta may represent an attractive and abundant, noncontroversial source of cells for regenerative medicine.

References

Bailey, P., Holowacz, T., and Lassar, A. B. (2001). The origin of skeletal muscle stem cells in the embryo and the adult. *Curr. Opin. Cell Biol.* **13,** 679–689.

Bartha, J. L., Romero-Carmona, R., Comino-Delgado, R., Arce, F., and Arrabal, J. (2000). α-Fetoprotein and hematopoietic growth factors in amniotic fluid. *Obstet. Gynecol.* **96,** 588–592.

Baschat, A. A., and Hecher, K. (2004). Fetal growth restriction due to placental disease. *Semin. Perinatol.* **28,** 67–80.

Black, I. B., and Woodbury, D. (2001). Adult rat and human bone marrow stromal stem cells differentiate into neurons. *Blood Cells Mol. Dis.* **27,** 632–636.

Bryan, T. M., Englezou, A., Dunham, M. A., and Reddel, R. R. (1998). Telomere length dynamics in telomerase-positive immortal human cell populations. *Exp. Cell Res.* **239,** 370–378.

Carpenter, M. K., Inokuma, M. S., Denham, J., Mujtaba, T., Chiu, C. P., and Rao, M. S. (2001). Enrichment of neurons and neural precursors from human embryonic stem cells. *Exp. Neurol.* **172,** 383–397.

Cremer, M., Schachner, M., Cremer, T., Schmidt, W., and Voigtlander, T. (1981). Demonstration of astrocytes in cultured amniotic fluid cells of three cases with neural-tube defect. *Hum. Genet.* **56,** 365–370.

Dang, S. M., Kyba, M., Perlingeiro, R., Daley, G. Q., and Zandstra, P. W. (2002). Efficiency of embryoid body formation and hematopoietic development from embryonic stem cells in different culture systems. *Biotechnol. Bioeng.* **78,** 442–453.

DeCoppi, P. (2001). Human fetal stem cell isolation from amniotic fluid [abstract]. Presented at the Proceedings of the American Academy of Pediatrics National Conference, San Francisco, CA, 2001, pp. 210–211.

Downs, K. M., and Harmann, C. (1997). Developmental potency of the murine allantois. *Development* **124,** 2769–2780.

Downs, K. M., Hellman, E. R., McHugh, J., Barrickman, K., and Inman, K. E. (2004). Investigation into a role for the primitive streak in development of the murine allantois. *Development* **131,** 37–55.

Dunn, J. C., Yarmush, M. L., Koebe, H. G., and Tompkins, R. G. (1989). Hepatocyte function and extracellular matrix geometry: Long-term culture in a sandwich configuration. *FASEB J.* **3,** 174–177.

Gardner, R. L., and Beddington, R. S. (1988). Multi-lineage "stem" cells in the mammalian embryo. *J. Cell Sci. Suppl.* **10,** 11–27.

Guan, K., Chang, H., Rolletschek, A., and Wobus, A. M. (2001). Embryonic stem cell-derived neurogenesis: Retinoic acid induction and lineage selection of neuronal cells. *Cell Tissue Res.* **305,** 171–176.

Hamazaki, T., Iiboshi, Y., Oka, M., Papst, P. J., Meacham, A. M., Zon, L. I., and Terada, N. (2001). Hepatic maturation in differentiating embryonic stem cells *in vitro. FEBS Lett.* **497,** 15–19.

Heidari, Z., Isobe, K., Goto, S., Nakashima, I., Kiuchi, K., and Tomoda, Y. (1996). Characterization of the growth factor activity of amniotic fluid on cells from hematopoietic and lymphoid organs of different life stages. *Microbiol. Immunol.* **40,** 583–589.

Hinterberger, T. J., Sassoon, D. A., Rhodes, S. J., and Konieczny, S. F. (1991). Expression of the muscle regulatory factor MRF4 during somite and skeletal myofiber development. *Dev. Biol.* **147,** 144–156.

Hoehn, H., Bryant, E. M., Fantel, A. G., and Martin, G. M. (1975). Cultivated cells from diagnostic amniocentesis in second trimester pregnancies. III. The fetal urine as a potential source of clonable cells. *Humangenetik* **29,** 285–290.

In 't Anker, P. S., Scherjon, S. A., Kleijburg-van der Keur, C., Noort, W. A., Claas, F. H., Willemze, R., Fibbe, W. E., and Kanhai, H. H. (2003). Amniotic fluid as a novel source of mesenchymal stem cells for therapeutic transplantation. *Blood* **102,** 1548–1549.

Jaiswal, N., Haynesworth, S. E., Caplan, A. I., and Bruder, S. P. (1997). Osteogenic differentiation of purified, culture-expanded human mesenchymal stem cells *in vitro. J. Cell Biochem.* **64,** 295–312.

Karsenty, G. (2000). Role of Cbfa1 in osteoblast differentiation and function. *Semin. Cell Dev. Biol.* **11,** 343–346.

Kaviani, A., Perry, T. E., Dzakovic, A., Jennings, R. W., Ziegler, M. M., and Fauza, D. O. (2001). The amniotic fluid as a source of cells for fetal tissue engineering. *J. Pediatr. Surg.* **36,** 1662–1665.

Kaviani, A., Guleserian, K., Perry, T. E., Jennings, R. W., Ziegler, M. M., and Fauza, D. O. (2003). Fetal tissue engineering from amniotic fluid. *J. Am. Coll. Surg.* **196,** 592–597.

Kinder, S. J., Tsang, T. E., Quinlan, G. A., Hadjantonakis, A. K., Nagy, A., and Tam, P. P. (1999). The orderly allocation of mesodermal cells to the extraembryonic structures and the anteroposterior axis during gastrulation of the mouse embryo. *Development* **126,** 4691–4701.

Komori, T., Yagi, H., Nomura, S., Yamaguchi, A., Sasaki, K., Deguchi, K., Shimizu, Y., Bronson, R. T., Gao, Y. H., Inada, M., Sato, M., Okamoto, R., Kitamura, Y., Yoshiki, S., and Kishimoto, T. (1997). Targeted disruption of Cbfa1 results in a complete lack of bone formation owing to maturational arrest of osteoblasts. *Cell* **89,** 755–764.

Li, L., Arman, E., Ekblom, P., Edgar, D., Murray, P., and Lonai, P. (2004). Distinct GATA6- and laminin-dependent mechanisms regulate endodermal and ectodermal embryonic stem cell fates. *Development* **131,** 5277–5286.

Loebel, D. A., Watson, C. M., De Young, R. A., and Tam, P. P. (2003). Lineage choice and differentiation in mouse embryos and embryonic stem cells. *Dev. Biol.* **264,** 1–14.

Medina-Gomez, P., and del Valle, M. (1988). The culture of amniotic fluid cells: An analysis of the colonies, metaphase and mitotic index for the purpose of ruling out maternal cell contamination. *Ginecol. Obstet. Mex.* **56,** 122–126.

Moser, M., Li, Y., Vaupel, K., Kretzschmar, D., Kluge, R., Glynn, P., and Buettner, R. (2004). Placental failure and impaired vasculogenesis result in embryonic lethality for neuropathy target esterase-deficient mice. *Mol. Cell. Biol.* **24,** 1667–1679.

Parameswaran, M., and Tam, P. P. (1995). Regionalisation of cell fate and morphogenetic movement of the mesoderm during mouse gastrulation. *Dev. Genet.* **17,** 16–28.

Patapoutian, A., Yoon, J. K., Miner, J. H., Wang, S., Stark, K., and Wold, B. (1995). Disruption of the mouse MRF4 gene identifies multiple waves of myogenesis in the myotome. *Development* **121,** 3347–3358.

Prusa, A. R., and Hengstschlager, M. (2002). Amniotic fluid cells and human stem cell research: A new connection. *Med. Sci. Monit.* **8,** RA253–RA257.

Prusa, A. R., Marton, E., Rosner, M., Bernaschek, G., and Hengstschlager, M. (2003). Oct-4-expressing cells in human amniotic fluid: A new source for stem cell research? *Hum. Reprod.* **18,** 1489–1493.

Prusa, A. R., Marton, E., Rosner, M., Bettelheim, D., Lubec, G., Pollack, A., Bernaschek, G., and Hengstschlager, M. (2004). Neurogenic cells in human amniotic fluid. *Am. J Obstet. Gynecol.* **191,** 309–314.

Rathjen, J., Lake, J. A., Bettess, M. D., Washington, J. M., Chapman, G., and Rathjen, P. D. (1999). Formation of a primitive ectoderm like cell population, EPL cells, from ES cells in response to biologically derived factors. *J. Cell Sci.* **112,** 601–612.

Robinson, W. P., McFadden, D. E., Barrett, I. J., Kuchinka, B., Penaherrera, M. S., Bruyere, H., Best, R. G., Pedreira, D. A., Langlois, S., and Kalousek, D. K. (2002). Origin of amnion and implications for evaluation of the fetal genotype in cases of mosaicism. *Prenat. Diagn.* **22,** 1076–1085.

Rohwedel, J., Maltsev, V., Bober, E., Arnold, H. H., Hescheler, J., and Wobus, A. M. (1994). Muscle cell differentiation of embryonic stem cells reflects myogenesis *in vivo*: Developmentally regulated expression of myogenic determination genes and functional expression of ionic currents. *Dev. Biol.* **164,** 87–101.

Rosenblatt, J. D., Lunt, A. I., Parry, D. J., and Partridge, T. A. (1995). Culturing satellite cells from living single muscle fiber explants. *In Vitro Cell Dev. Biol. Anim.* **31,** 773–779.

Sakuragawa, N., Elwan, M. A., Fujii, T., and Kawashima, K. (1999). Possible dynamic neurotransmitter metabolism surrounding the fetus. *J. Child. Neurol.* **14,** 265–266.

Schwartz, R. E., Reyes, M., Koodie, L., Jiang, Y., Blackstad, M., Lund, T., Lenvik, T., Johnson, S., Hu, W. S., and Verfaillie, C. M. (2002). Multipotent adult progenitor cells from bone marrow differentiate into functional hepatocyte-like cells. *J. Clin. Invest.* **109,** 1291–1302.

Siddiqui, M. J., and Atala, A. (2004). Amniotic fluid-derived pluripotential cells. *In* "Handbook of Stem Cells," Vol. 2, pp. 175–179. Elsevier Academic Press, San Diego, CA.

Smith, J. L., Gesteland, K. M., and Schoenwolf, G. C. (1994). Prospective fate map of the mouse primitive streak at 7.5 days of gestation. *Dev. Dyn.* **201,** 279–289.

Snow, M. H., and Bennett, D. (1978). Gastrulation in the mouse: Assessment of cell populations in the epiblast of t^{w18}/t^{w18} embryos. *J. Embryol. Exp. Morphol.* **47,** 39–52.

Srivastava, M. D., Lippes, J., and Srivastava, B. I. (1996). Cytokines of the human reproductive tract. *Am. J. Reprod. Immunol.* **36,** 157–166.

Tsai, M. S., Lee, J. L., Chang, Y. J., and Hwang, S. M. (2004). Isolation of human multipotent mesenchymal stem cells from second-trimester amniotic fluid using a novel two-stage culture protocol. *Hum. Reprod.* **19,** 1450–1456.

Woodbury, D., Schwarz, E. J., Prockop, D. J., and Black, I. B. (2000). Adult rat and human bone marrow stromal cells differentiate into neurons. *J. Neurosci. Res.* **61,** 364–370.

[18] Cord Blood Stem and Progenitor Cells

By HAL E. BROXMEYER, EDWARD SROUR, CHRISTIE ORSCHELL,
DAVID A. INGRAM, SCOTT COOPER, P. ARTUR PLETT,
LAURA E. MEAD, and MERVIN C. YODER

Abstract

Cord blood has served as a source of hematopoietic stem and progenitor cells for successful repopulation of the blood cell system in patients with malignant and nonmalignant disorders. It was information on these rare immature cells in cord blood that led to the first use of cord blood for transplantation. Further information on these cells and how they can be manipulated both *in vitro* and *in vivo* will likely enhance the utility and broadness of applicability of cord blood for treatment of human disease. This chapter reviews information on the clinical and biological properties of hematopoietic stem and progenitor cells, as well as the biology of endothelial progenitor cells, and serves as a source for the methods used to detect and quantitate these important functional cells. Specifically, methods are presented for enumerating human cord blood myeloid progenitor cells, including granulocyte-macrophage (CFU-GM), erythroid (BFU-E), and multipotential (CFU-GEMM or CFU-Mix) progenitors, and their replating potential; hematopoietic stem cells, as assessed *in vitro* for long-term culture-initiating cells (LTC-ICs), cobblestone area-forming cells (CAFCs), and myeloid–lymphoid-initiating cells (ML-ICs), and as assessed *in vivo* for nonobese diabetic (NOD)/severe combined immunodeficient (SCID) mouse repopulating cells (SRCs); and high and low proliferative potential endothelial progenitor cells (EPCs).

Introduction: Cord Blood Transplantation

Umbilical cord blood, collected at the birth of a baby, is a rich source of hematopoietic stem and progenitor cells (Broxmeyer, 2004, 2005). Cord blood stem and progenitor cells have been used for more than 6000 transplants to treat a wide range of malignant and nonmalignant disorders (Broxmeyer and Smith, 2004). The first cord blood transplant used sibling human leukocyte antigen (HLA)-matched cells to cure the hematological manifestations of Fanconi anemia (Gluckman *et al.*, 1989). The recipient of this transplant, which took place in October 1988, is still alive and well more than 17 years after the transplant. The studies and events that led up to this transplant have

METHODS IN ENZYMOLOGY, VOL. 419   0076-6879/06 $35.00
Copyright 2006, Elsevier Inc. All rights reserved.   DOI: 10.1016/S0076-6879(06)19018-7

been reported (Broxmeyer, 1998, 2000; Broxmeyer et al., 1989, 1990; Gluckman and Rocha, 2005).

The initial cord blood transplants were limited to HLA-matched sibling donors (Wagner et al., 1995), but the encouraging results soon led to the use of partially HLA-matched sibling and related transplants, and then to the use of unrelated allogeneic cord blood transplants that were first completely, and then subsequently partially, matched for HLA (Broxmeyer and Smith, 2004; Gluckman et al., 1997; Kurtzberg et al., 1996; Rubinstein et al., 1998; Wagner et al., 1996, 2002). Most of the original cord blood transplants were done in children or in low-weight recipients because of the fear that the numbers of hematopoietic stem and progenitor cells in single collections of cord blood were limiting in number, and thus might compromise the engraftment capability of adults and higher-weight recipients. However, as transplanters have gained more experience, increasing numbers of adults have been transplanted (Cornetta et al., 2005; Gluckman et al., 2004; Laughlin et al., 2001, 2004; Long et al., 2003; Ooi et al., 2002; Rocha et al., 2004; Rubinstein et al., 1998; Sanz et al., 2001a,b; Takahashi et al., 2004; Tse and Laughlin, 2005).

Limiting numbers of hematopoietic stem and progenitor cells still present a logistical problem for the ultimate broadness of applicability of cord blood transplantation in adults and in children. Current practice suggests the need for at least 2×10^7 nucleated cells/kg recipient body weight for successful transplantation, although there have been successful outcomes with fewer than this number of cells (Gluckman et al., 2004).

At present, there are three avenues of research that clinical and basic science investigators are pursuing in order to deal with the limited numbers of nucleated, as well as stem and progenitor, cells in single cord blood collections. A most recent clinical effort reflects the use of multiple cord blood units for transplantation in single recipients (Barker et al., 2005). The results here are encouraging but only one cord blood unit eventually "wins out" in the competition, and it is not clear yet what factors define the "winning" unit. There have not yet been controlled trials that definitively substantiate that multiple cord bloods are more efficacious than a single unit, and this remains to be established.

Another clinical effort has focused on the use of cells that have first been cultured ex vivo (out of the body) in attempts to expand the numbers of hematopoietic stem and progenitor cells beyond that collected from the single units before infusions of these cells in recipients (Jaroscak et al., 2003; Shpall et al., 2002). These results have not been encouraging thus far, most likely because either the hematopoietic stem cells that would allow for long-term stable and multilineage engraftment have not been expanded, or the engrafting capabilities of these cells have been compromised by the in vitro culture conditions.

A third effort has focused on enhancing the homing characteristics of the hematopoietic stem cells (Christopherson and Broxmeyer, 2004; Christopherson et al., 2004). A number of factors have been implicated in the homing of hematopoietic stem and progenitor cells to the environmental niches in the bone marrow that nurture these cells for survival, self-renewal, proliferation, and differentiation (Christopherson and Broxmeyer, 2004; Lapidot et al., 2005). It is believed that not all hematopoietic stem cells home with absolute efficiency to their microenvironmental niche(s) after infusion into transplant recipients. Thus, increasing the homing efficiency of limiting numbers of available hematopoietic stem and progenitor cells may enhance engraftment and repopulation of blood cells. One means to enhance the homing and engraftment of hematopoietic stem and progenitor cells is to inhibit the dipeptidylpeptidase IV activity of CD26. Such efforts have resulted in enhanced engraftment of limiting numbers of mouse bone marrow stem cells into lethally irradiated mice (Christopherson et al., 2004). These results have now been confirmed by a number of other groups for mouse bone marrow cell engraftment in adult mice (Tian et al., 2005) and also in utero (Peranteau et al., 2005). In addition, inhibition of CD26 on hematopoietic stem cells present in human cord blood resulted in enhanced engraftment of these stem cells in mice of the nonobese diabetic severe combined immunodeficiency (NOD/SCID) genotype (Campbell et al., 2005). It remains to be determined whether inhibition of CD26 peptidase activity found on human cord blood hematopoietic stem cells, or other means to enhance homing and engraftment of these stem cells, will result in enhanced engraftment in human recipients transplanted with limiting numbers of stem cells.

Hematopoietic Stem and Progenitor Cells

Hematopoietic stem cells are defined by their capacity to give rise to more of their own kind, a process termed self-renewal, and to differentiate down multiple blood lineages, including erythroid cells, neutrophils, monocytes, macrophages, platelets, T lymphocytes, B lymphocytes, other lymphocytes, natural killer T cells, dendritic cells, and so on. Hematopoietic progenitor cells have more limited capacity for self-renewal than stem cells, and they are more limited in their capacity to give rise to multiple blood cell types. The hierarchy of blood cell development begins with hematopoietic stem cells, which give rise to various hematopoietic progenitor cells, which then differentiate into precursor cells for the individual blood cell lineages. Regulation of the self-renewal, proliferation, and differentiation of these cells is modulated by cytokines, chemokines, and other growth factors (Shaheen and Broxmeyer, 2005) as well as by local cell–cell interactions involving stromal cells.

More is known about human hematopoietic progenitor cells than about human hematopoietic stem cells. This is in part because the stem cells are rarer in frequency than the progenitors, and assays for progenitor cells, performed *in vitro*, are easier to do than assays for engrafting stem cells, which are done *in vivo*. Hematopoietic progenitor cells in cord blood are increased in frequency compared with those in bone marrow (Broxmeyer *et al.*, 1989) and are enhanced in proliferative capacity, generation of progeny, and also in replating capacity *in vitro* (Broxmeyer *et al.*, 1989, 1992; Cardoso *et al.*, 1993; Carow *et al.*, 1991, 1993; Lansdorp *et al.*, 1993; Lu *et al.*, 1993); replating capacity offers an estimate of the limited self-renewal capacity of progenitor cells (Carow *et al.*, 1991, 1993). Hematopoietic progenitor cells are $CD34^+$ and are found enriched in the $CD34^+CD38^+$ subset of cells after cell sorting. Progenitors are ranked as those that are immature with enhanced proliferative capacity, and those that are more mature, with decreased proliferative and more limited and restricted differentiation capability. It is believed that the immature subsets of progenitors are responsive to stimulation by combinations of growth factors (Shaheen and Broxmeyer, 2005). Thus, multipotential progenitor cells (termed colony-forming unit-granulocyte, erythroid, macrophage, megakaryocyte; CFU-GEMM or CFU-Mix) are denoted by the colonies of mixed lineage blood cells they give rise to in semisolid culture medium when the cells are stimulated *in vitro* with combinations of cytokines such as erythropoietin (Epo), stem cell factor (SCF; also called Steel factor, SLF), granulocyte-macrophage colony-stimulating factor (GM-CSF), and interleukin-3 (IL-3), in the absence or presence of thrombopoietin (TPO). Immature subsets of granulocyte-macrophage progenitor cells (colony-forming unit-granulocyte, macrophage; CFU-GM) are detected by the colonies of granulocytes and macrophages they give rise to in the presence of either GM-CSF or IL-3 combined with potent costimulating cytokine SCF or Flt-3 ligand (FL). More mature subsets of CFU-GM, which give rise to smaller colonies *in vitro*, are detected after stimulation of the cells with either GM-CSF or IL-3. Granulocyte progenitors (colony-forming unit-granulocyte; CFU-G) and macrophage progenitors (colony-forming unit-macrophage; CFU-M) are detected after stimulation, respectively, with either granulocyte colony-stimulating factor (G-CSF) or macrophage colony-stimulating factor (M-CSF), and more immature subsets of CFU-G and CFU-M are detected by adding either SCF or FL with G-CSF or M-CSF. Erythroid progenitor cells (burst-forming unit-erythroid; BFU-E) or megakaryocyte progenitor cells (colony-forming unit-megakaryocyte; CFU-Meg) are detected by, respectively, stimulating these cells in semisolid culture medium with either Epo or TPO, and addition of costimulating factors induces larger colonies that derive from more immature subsets of these progenitor cells. Examples of CFU-GEMM-, BFU-E-, and CFU-GM-derived colonies are shown in Fig. 1. Cord blood cells have been stored frozen in cryopreserved form for at least 15 years without loss of functional hematopoietic stem and progenitor cell activity (Broxmeyer et al., 1992, 2003; Broxmeyer and Cooper, 1997).

Cord blood CD34^{+++} cells, those CD34$^+$ cells expressing the highest density of CD34 antigens, are highly enriched in more immature subsets of CFU-GEMM, BFU-E, and CFU-GM (Lu et al., 1993). In fact, the cloning efficiency of CD34^{+++} cord blood cells can be as high as 80% for these progenitors under optimal conditions when combinations of cytokines are added to the in vitro semisolid cultures. Less is known about lymphoid progenitor cells in human cord blood than of the myeloid progenitor (CFU-GEMM, BFU-E, and CFU-GM) cells.

Also, less is known phenotypically and functionally about hematopoietic stem cells from humans than from mice. In mice, a phenotype of Sca1$^+$ c-kit$^+$Lin$^-$ defines stem cells that under optimal conditions may engraft lethally irradiated mice at levels as low as one of these phenotyped cells. The human assay for hematopoietic stem cells is not as definitive. The assay for human hematopoietic stem cells takes advantage of the engraftment of sublethally irradiated SCID mice (usually NOD/SCID mice) with human cells. Human cord blood cells have greater capacity to repopulate these conditioned NOD/SCID mice than do human bone marrow or G-CSF-mobilized adult blood cells (Vormoor et al., 1994; Orazi et al., 1994; Bock et al., 1995; Bodine, 2004). These human SCID-repopulating cells (SRCs; considered to be engrafting hematopoietic stem cells) are enriched in the CD34$^+$CD38$^-$ population of human cord blood, but the relative inadequacy of phenotypically identifying human hematopoietic stem cells is highlighted by the finding that approximately only 1 in 700 CD34$^+$CD38$^-$ cord blood cells is an SRC. Nevertheless, this frequency of SRCs is much greater than for SRCs found in bone marrow or mobilized adult peripheral blood. Examples of flow cytometric readout of human cord blood cells infused and those that engrafted the bone marrow of sublethally irradiated NOD/SCID mice are shown in Fig. 2.

Endothelial Progenitor Cells

Endothelial progenitor cells have been detected, characterized, and isolated from human cord blood (Aoki et al., 2004; Bompais et al., 2004; Crisa et al., 1999; Eggermann et al., 2003; Fan et al., 2003; Hildbrand et al., 2004; Ingram et al., 2004; Kang et al., 2001; Murga et al., 2004; Peichev et al., 2000; Pesce et al., 2003). During embryogenesis, blood vessels are formed de novo by the patterned assembly of angioblasts in a process termed

CFU-GM colonies

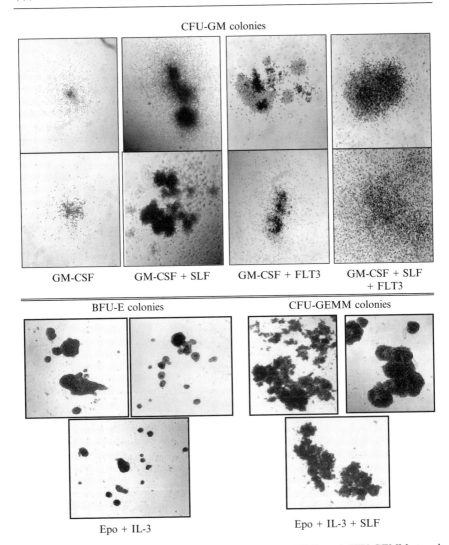

FIG. 1. Colonies formed from cord blood CFU-GM, BFU-E, and CFU-GEMM stored frozen for 15 years, defrosted, and plated in semisolid agar culture medium for 14 days in the presence of GM-CSF alone or in combination with SLF, FL, or SLF plus FL for CFU-GM, and in semisolid methyl cellulose culture medium for 14 days, respectively, in the presence of Epo plus IL-3 and of Epo and IL-3 plus SLF for BFU-E and CFU-GEMM. These are representative of the largest colonies formed. Reproduced from Fig. 1 of Broxmeyer *et al.* (2003) with permission from the *Proceedings of the National Academies of Science of the United States of America.*

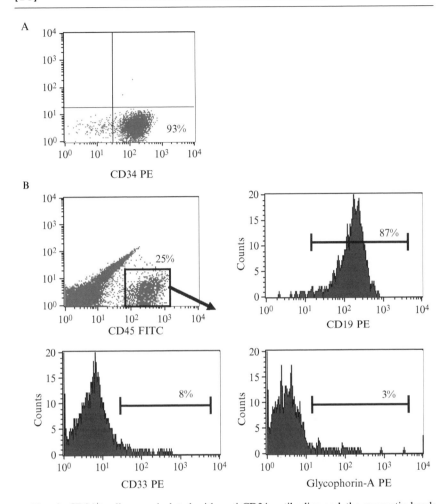

FIG. 2. CD34$^+$ cells were isolated with anti-CD34 antibodies and the magnetic beads system from Miltenyi Biotec (Bergisch Gladbach, Germany). Cells (1×10^5 to 3×10^5) were then infused into 300 cGy-conditioned NOD/SCID mice via tail vein injection. Mice were killed 8–9 weeks posttransplantation and BM cells were harvested and stained with anti-human CD45–fluorescein isothiocyanate (FITC) and the indicated lineage markers (CD19/lymphoid, CD33/myeloid, glycophorin-A/erythroid) conjugated to phycoerythrin (PE). Flow cytometry was used to assess percent human chimerism and lineage marker expression on engrafted cells. (A) Representative flow cytometric plot of purified CB CD34$^+$ cells used in *in vitro* experiments and as grafts in NOD/SCID mice. (B) Flow cytometric plots of human cells (CD45$^+$) in NOD/SCID mice and the gating used to obtain the expression of lineage markers on these cells. (See color insert.)

vasculogenesis (Flamme *et al.*, 1997). Once an intact vascular system has been established, the development of new blood vessels occurs primarily via the sprouting of endothelial cells from postcapillary venules or by the maturation and *de novo* growth of collateral conduits from larger diameter arteries (Simons, 2005; Skalak, 2005). These two mechanisms of new blood vessel formation are, respectively, termed angiogenesis and arteriogenesis. A population of human circulating $CD34^+$ cells was described that could differentiate *ex vivo* into cells with endothelial cell-like characteristics (Asahara *et al.*, 1997). These cells were termed "endothelial progenitor cells" (EPCs); this study challenged traditional understanding of angiogenesis by suggesting that circulating cells in adult peripheral blood may also contribute to new vessel formation. Subsequent studies showed that these cells were derived from bone marrow, circulate in peripheral blood, and home to sites of new blood vessel formation that include ischemic tissues and tumor microenvironments in a process termed "postnatal vasculogenesis" (reviewed in Hristov and Weber, 2004; Iwami *et al.*, 2004; Khakoo and Finkel, 2005; Murasawa and Asahara, 2005; Rafii and Lyden, 2003; Schatteman, 2004; Urbich and Dimmeler, 2004).

Using sophisticated cell-marking strategies, more recent studies indicate that marrow-derived EPCs may play a minimal, if any, role in neovascularization of tumors, vessel repair, or normal vessel growth and development (Gothert *et al.*, 2004; Stadtfeld and Graf, 2005). These conflicting reports have raised questions about the function of EPCs in vascular homeostasis and repair. The controversies surrounding these fundamental questions may in part originate from the heterogeneous phenotypic definitions of EPCs and a lack of functional clonogenic assays to isolate and accurately describe the proliferative potential of EPCs.

Some of the coauthors of this chapter described an approach that identifies a novel hierarchy of EPCs based on their clonogenic and proliferative potential, analogous to the hematopoietic stem/progenitor cell system (Ingram *et al.*, 2004). Using this approach, they identified a previously unrecognized population of EPCs in cord blood that achieved at least 100 population doublings, could be replated into at least secondary and tertiary colonies, and expressed levels of telomerase activity that correlated with the degree of cell autonomous proliferative potential. These studies described a clonogenic method that defined a hierarchy of EPCs based on their proliferative potential, and identified a unique population of high proliferative potential-endothelial colony-forming cells (HPP-ECFCs) in human umbilical cord blood (Ingram *et al.*, 2004). Examples of colonies that formed *in vitro* from human cord blood high- and low-proliferative EPCs are shown in Fig. 3.

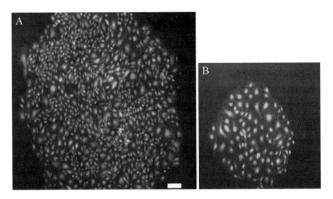

FIG. 3. Photomicrographs of colonies formed from human cord blood endothelial colony-forming cells (ECFCs). (A) High proliferative potential ECFC-derived colony and (B) low proliferative potential ECFC-derived colony. Scale bar in (A): 50 μm. (See color insert.)

Methodologies for Assessing Hematopoietic Progenitor and Stem Cells, as Well as Endothelial Progenitor Cells, Present in Human Cord Blood

The assays described below allow detection of functional progenitor and stem cells and have been used to great advantage to determine our current knowledge of the biology of these cells in umbilical cord blood as defined previously.

Methods for Enumerating Human Myeloid Progenitor Cells in Umbilical Cord Blood by In Vitro Colony Formation

Human myeloid progenitor cells are not clearly definable by phenotype with the precision of mouse myeloid progenitors. Phenotype does not always recapitulate function. Therefore, it is always wise when defining a progenitor or stem cell by phenotype to confirm the functional capacity of the pheno-typed cell. Functionally, myeloid progenitor cells are defined, in retrospect, at the end of proliferation and maturation by evaluating the types of cells formed within a colony at the end of a culture period and by the growth factors used for stimulation. By culturing a population of cells known to contain progenitor cells under defined conditions, one can obtain proliferation in the form of colonies (>40 cells per group) or clusters (3–40 cells). By using a semisolid material in these cultures such as agar, agarose, or methylcellulose, movement of the daughter cells produced by these progenitor cells can be restricted and thereby allow the formation of colonies.

These colonies can then be counted and the number and types of progenitors present in the starting population can be determined. This assay is used primarily to quantitate the numbers of progenitors in a given sample and to shed light on their proliferative potential (colony versus cluster, as well as the number of cells within a colony).

CFU-GM: Colony-Forming Unit-Granulocyte/Macrophage

Materials/Reagents

See Appendixes A and B at the end of this section for listings of suppliers and recipes. *Note*: Unless otherwise stated, all materials should be sterile.

Cells to be analyzed[a]
Agarose (0.8%) or agar (0.6%) solution (sterile)
2× McCoy's 5A complete medium
Fetal bovine serum (FBS): Heat inactivated at 37° for 30 min
Growth factors (recombinant)
Petri dishes, 35 mm
Assortment of disposable pipettes

[a]The cells used can be either whole, unseparated blood or enriched populations such as low-density (LD) mononuclear cells, FACS-sorted or bead-sorted CD34[+] cells, and so on.

Procedure

All steps are performed aseptically.

1. Melt the agarose/agar in a microwave oven and place at 45° until needed. This part of the culture mix is to be added last, just before plating.
2. For any given volume, combine the following reagents (except the agar/agarose) in the following proportions and mix well:

FBS	10%
Growth factors[a]	<1%
GM-CSF	(10–20 ng/ml)
GM-CSF	(10–20 ng/ml)
M-CSF	(10–20 ng/ml)
SCF	(50 ng/ml)
FL	(100 ng/ml)
Cells[b]	0–5%
2× McCoy's	35–40%, quantitation standard (QS) volume
Agar (0.6%)[c]	50%

[a]The growth factors used will determine the final colony output in terms of type; for example, GM-CSF yields granulocyte-macrophage colonies, M-CSF yields macrophage colonies, and so on. Single growth factors or combinations of growth factors may be used. Typically, to pick up earlier, more immature subsets of CFU-GM, we use GM-CSF (10 ng/ml) with either SCF (50 ng/ml) or FL (100 ng/ml).

[b]Depending on the type of cells used and final cell concentration of the cell preparation, the final cell concentration and volume of cells added to the culture mix (see later) will vary. Typically, we plate unseparated cord blood cells at $2.5–5.0 \times 10^4$ cells/ml, low-density cells at $1.25–2.5 \times 10^4$ cells/ml, and $CD34^+$ cells at 50–1000 cells/ml (depending on purity). It is important that the number of cells plated yield plates that have enough colonies to score accurately. This means colonies that do not overlap with each other, or do so in minimal fashion. This may require plating cells at several different concentrations. For example, when plating $CD34^+$ cells, at least three different concentrations are recommended. The volume added will be determined by the final concentration of the cell suspension after enrichment. When using unseparated blood, microliter volumes are added to the culture mix and therefore do not constitute a significant volume change with respect to the 2× McCoy's volume. However, when using low-density or $CD34^+$ cells, the final cell concentration may only be 10–20 times the desired culture cell concentration, thereby constituting 5–10% of the culture mix volume and thus requiring an adjustment of the 2× McCoy's volume.

[c]The agar is added last and just before plating. Make sure that it is not so hot that it will kill the cells.

2. Add agar (~40–42°) and mix well.

3. Pipette 1 ml into each dish (35-mm tissue culture dish) and swirl the plates to cover the bottom with culture. At least three plates per point should be plated.

4. Once all the plates are dispensed, leave at room temperature for 10 min or until all the plates have solidified.

5. Culture the plates in a humidified environment at 37° and 5% CO_2 for 7–14 days. We have found that culturing plates at lowered O_2 tension (e.g., 5% rather than 20%) enhances the number of colonies detected.

BFU-E/CFU-GEMM: Burst-Forming Unit-Erythroid/Colony-Forming Unit-Granulocyte/Erythroid/Macrophage/Megakaryocyte Colony-Forming Assay (Methylcellulose)

Materials/Reagents

See Appendixes A and B at the end of this section for listings of suppliers and recipes. *Note*: Unless otherwise stated, all materials should be sterile.

Cells to be analyzed[a]

Methylcellulose, 2.1%

Iscove's modified Dulbecco's medium (IMDM) with penicillin–streptomycin (Pen/Strep)

FBS (not heat inactivated)

Glutamine

2-Mercaptoethanol (2-ME), 2×10^{-5} M

Erythropoietin [Epogen (EPO); Amgen, Thousand Oaks, CA]

Source of growth factors (usually recombinant)

Petri dishes, 35 mm

Assortment of disposable pipettes

[a]The cells used can be either whole, unseparated blood or enriched populations such as low-density (LD) mononuclear cells, FACS-sorted or bead-sorted CD34[+] cells, and so on.

Procedure

All steps are performed aseptically.

1. Warm the methylcellulose in a 37° water bath.
2. The following quantities are for a 5-ml culture. However, by adjusting volumes proportionately, larger volumes can easily be obtained in order to test a variety of variables. Combine all the following reagents except for the methylcellulose and mix by swirling or gentle vortexing.

Methylcellulose (2.1%)	2.5 ml
FBS	1.5 ml
Glutamine	50 μl
2-ME (2×10^{-5} M)	16.7 μl
EPO (200 U/ml)	25 μl
Growth factors[a]	0.25 ml
Epo	(1–2 U/ml)
Epo	(1–2 U/ml)
IL-3	(10 ng/ml)
GM-CSF	(10 ng/ml)
SCF	(50 ng/ml)

TPO	(10 ng/ml)
Cells	0.25 ml
IMDM	QS to 5 ml

[a]The growth factors used will determine the final colony output in terms of type and colony content. If Epo is used alone, only erythroid colonies will be enumerated. But if a combination of growth factors is used, then multipotential progenitors will be enumerated. Typically, to pick up maximal numbers of multipotential colonies, we use IL-3, GM-CSF, SCF, Epo, with or without TPO. The addition of TPO increases the number of megakaryocytes in the mixed colonies.

3. Add methylcellulose and mix by vortexing.
4. Allow the tube to sit for several minutes to allow all bubbles to rise to the top.
5. Using a 3- to 5-ml syringe fitted with a 16-gauge needle, transfer 1 ml of culture mix into each dish (35-mm tissue culture dishes). Swirl the plates to cover the bottom with culture mix. At least three plates per point should be plated. Because methylcellulose does not solidify, the entire experiment can be plated before each point is mixed.
6. Culture the plates in a humidified environment at 37° and 5% CO_2–95% O_2 for 14 days.

Colony Replating: A Means to Estimate the Limited Self-Renewal Capacity of Progenitor Cells

Note: This technique can be adequately performed only when using primary cultures grown in methylcellulose. Agar cultures are far too solid and it is extremely difficult to remove the colonies from agar for replating into a secondary culture plate. However, to grow primary CFU-GM colonies for replating, all that is necessary is to replace the agar with methylcellulose and the 2× McCoy's with IMDM.

Materials/Reagents

Unless otherwise stated, all materials should be sterile.
  Clean bench with sterile air flow
  Inverted tissue culture microscope
  Micropipettor (20 μl)
  IMDM with Pen/Strep
  12 × 75 mm polypropylene tubes
  All reagents as previously described for desired culture conditions.

Procedure for Colony Harvest

All steps are performed aseptically.

1. Determine the number of colonies to be replated into secondary cultures. Prepare one tube for each colony by dispensing 100 μl of IMDM with 2% FBS into each tube. Recap.
2. Set up the microscope in a clean sterile air bench environment.
3. Place the plate containing the colonies to be harvested on the platform of the microscope and remove the dish lid. The plate should be immobilized. *Note*: If the microscope is not equipped with a small plate holder, immobilize the plate by means of two small loops of tape rolled sticky side out and placed on the stage. The plate can then be pushed up against the tape for the harvest.
4. Find the desired colony to be replated and, using a 20-μl pipettor, gently aspirate the colony into the pipettor. *Note*: The best colonies for harvest will be well isolated.
5. Transfer the colony to a tube containing 100 μl of IMDM (previously prepared) and rinse the tip with several up-and-down motions.
6. Once all desired colonies have been harvested, proceed to the plating procedure.

Plating Procedure for Harvested Colony Culture

All steps are performed aseptically.

1. Determine the number of colonies harvested. On the basis of 1.1 ml per colony harvested, combine a volume of culture medium and mix, including methylcellulose, and set aside to allow air bubble to rise.

Note: For example, if 100 colonies were harvested from plates containing IL-3, Epo, and SCF as growth factors, prepare 110 ml of culture mix (as described previously in the BFU-E/CFU-GEMM section) without the cells, containing all the growth factor as were in the primary plates and with the methylcellulose added before the cells.

2. Once the culture mixture has settled, use a 3- to 10-ml syringe fitted with a 16-gauge needle to dispense 1 ml into each tube containing the harvested colony; mix by vortexing.

3. Allow to settle.

4. Transfer as much of this culture mix as possible to a new 35-mm plate and swirl to cover the bottom of the plate. *Note*: Because methylcellulose is so viscous, it is virtually impossible to replate 100% of the harvested colony; some will remain behind. If a quantitative outcome is desired, simply plate a known percentage of the culture mix and do a

corrected calculation once the secondary colonies have been scored. For example, plate 0.9 ml of 1.1-ml cell mix (1 ml of mix plus 0.1 ml of harvested colony). Scored colonies would then represent ~82% of total colony-forming cells in the harvested colony.

5. Culture the plates in a humidified environment at $37°$ and 5% CO_2–95% O_2 for 14 days.

Appendix A: Suppliers

Agarose (FMC BioPolymer, Philadelphia, PA)

Agar (BD Bacto Agar; BD Biosciences, San Jose, CA)

McCoy's 5A (cat. no. M-4892; Sigma, St. Louis, MO)

Minimum essential medium (MEM) glutamine (cat. no. 17–605E; Cambrex Bio Science Walkersville, Walkersville, MD)

MEM Pen/Strep (cat. no. 17–603E; Cambrex Bio Science Walkersville)

MEM vitamin mixture (cat. no. 13–607C; Cambrex Bio Science Walkersville)

MEM essential amino acids (EAA) (GIBCO, cat. no. 11130–051; Invitrogen, Carlsbad, CA)

MEM nonessential amino acids (NEAA) (cat. no. 13–114E; Cambrex Bio Science Walkersville)

MEM sodium pyruvate (cat. no. 13–115E; Cambrex Bio Science Walkersville)

L-Asparagine (cat. no. A-0884; Sigma)

L-Serine (cat. no. S-4500; Sigma)

FBS, prescreened for colony growth (HyClone, Logan, UT)

Growth factors [R&D Systems (Minneapolis, MN), BioVision (Mountain View, CA), PeproTech (Rocky Hill, NJ); we have used products from all and all seem to have comparable activity]

IMDM (Iscove's modified Dulbecco's medium) (cat. no. 12–722F; Cambrex Bio Science Walkersville).

Appendix B: Recipes

SAG (Serine/Asparagine/Glutamine)

1. Dissolve 800 mg of L-asparagine and 420 mg of L-serine in 500 ml of double-distilled H_2O.
2. Add 200 ml of L-glutamine.
3. Filter sterilize with a 0.2-μm pore size vacuum filter and store frozen in 7.5-ml aliquots.

McCoy's 5A Complete

1. Dissolve 1 package (1 liter) of powdered medium into 400 ml of tissue-grade H_2O. Add Na_2HCO_2 as per the manufacturer's directions for 1 liter and stir until dissolved.
2. Add: Pen/Strep (10 ml), EAA (20 ml), NEAA (10 ml), vitamins (5 ml), pyruvate (10 ml), and SAG (15 ml).
3. Adjust to pH 7.0–7.2.
4. QS to 500 ml with tissue-grade H_2O.
5. Filter sterilize with a 0.2-μm pore size vacuum filter.

Methylcellulose (2.1%)

Methylcellulose can be rather tricky to make. The base medium, or medium complete with growth factors, may be purchased from StemCell Technologies (Vancouver, BC, Canada; https://www.stemcell.com).

Reagents/supplies

Sterile double-distilled H_2O
Methylcellulose, 4000 cP (cat. no. M-0512; Sigma)
IMDM powder (GIBCO, cat. no. 12200–036; Invitrogen)
Pen/Strep (cat. no. 17–603E; Cambrex Bio Science Walkersville)
2-liter flasks (2)
3-in. stir bar
Stir plate (strong)
50-ml conical tubes.

Procedure

1. Make 2× IMDM as per the vendor's instructions. Add Pen/Strep to 10 U/μg per milliliter. Filter sterilize and store at 40° until needed. *Note*: Must be warmed before adding to the slurry in step 6.
2. Place 21 g of dry methylcellulose in a sterilized, 2-liter flask containing a sterile 3-in. stir bar.
3. In a separate flask, bring 550 ml of sterile double-distilled H_2O to a boil and continue boiling for 5 min.
4. In a hood, add the boiled H_2O to the dry powder. Mix thoroughly, making sure that all the powder has been moistened.
5. Place the flask on a stir plate and mix well until the slurry has cooled to 37–40°. *Note*: Do not let it cool too much or it will solidify.
6. In the hood, add 500 ml of warmed 2× IMDM (as per the package instructions) to the slurry and mix vigorously by hand for 1 min.

7. Recap and place on the stir plate at room temperature for 1–2 h. Transfer to a cold room and stir overnight. *Note*: This matrix will begin to gel as it cools, so make sure the stir bar is big and spinning somewhat fast.

8. Aliquot into 50-ml tubes and store at –20° for no more than 6 months.

2-Mercaptoethanol

Make 2-mercaptoethanol (2-ME, 2×10^{-5} M) fresh every week.

1. Place 7 μl of 2-ME (undiluted) into 10 ml of IMDM.
2. Filter sterilize with a 0.2-μm pore size syringe filter.

Methods for Enumerating Human Hematopoietic Stem Cells in Umbilical Cord Blood

The most reliable method for assessing human hematopoietic stem cells is that involving engraftment of human cells into SCID (e.g., NOD/SCID) mice (Bodine, 2004). However, there are *in vitro* assays that have been used as surrogate assays for those stem cells detected by *in vivo* analysis. This section first covers *in vitro* assays that are believed to detect cells within the stem cell category, although these cells may not be equivalent to the long-term marrow *in vivo*-repopulating hematopoietic stem cell.

In Vitro *Hematopoietic Stem Cell Assays for Cord Blood*

Although bone marrow and, more recently, mobilized peripheral blood have been reliably used as sources of hematopoietic stem cells for transplantation, umbilical cord blood is, as noted previously, fast becoming recognized as a source of stem cells for transplantation of children and adults. The primary concern with cord blood is, however, the relatively lower numbers of total transplantable stem cells that can be isolated from single collections of cord blood compared with marrow or mobilized peripheral blood, and questions concerning whether sufficient numbers of hematopoietic stem cells are available in a typical cord blood collection for use in adults. For this reason, investigators have sought reliable *in vivo* and *in vitro* assays to estimate the stem cell content of different tissue sources.

Of the available *in vitro* assays, the long-term culture-initiating cell (LTC-IC) (Sutherland *et al.*, 1989), the extended (E) LTC-IC (Hao *et al.*, 1996), and cobblestone area-forming cell (CAFC) (Ploemacher *et al.*, 1989; Breems *et al.*, 1994) assays all measure primitive myeloid cells, whereas the myeloid–lymphoid initiating cell (ML-IC) assay measures cells with both myeloid and lymphoid [natural killer (NK) cell] activity (Punzel *et al.*, 1999).

In the LTC-IC assay (Sutherland *et al.*, 1989), putative stem cells are cultured in a modified Dexter-style culture system (Dexter, 1993) with or without stromal cell support for 5 weeks, at which time LTC-ICs will produce *de novo* colony-forming cells (CFCs) on secondary culture. The CAFC assay uses a similar approach but with a visual end point for the presence of cobblestone areas (CAs, tightly packed groups of small cells embedded in the stroma), eliminating the need for secondary culture. This assay can also be read sequentially at various time points, with the presence of CAFCs at later time points being indicative of more primitive hematopoietic progenitor cells (HPCs) (Ploemacher *et al.*, 1991).

Methods for CAFCs and LTC-ICs

1. A stromal feeder layer capable of supporting primitive hematopoiesis for up to 8 weeks is prepared. Primary stroma prepared from murine (Ploemacher *et al.*, 1989) or human bone marrow (Denning-Kendall *et al.*, 2003), as well as stromal cell lines such as FBMD-1 (Kusadasi *et al.*, 2000), MS5 and AFT024 (Theunissen and Verfaillie, 2005), and M210B4 (Traycoff *et al.*, 1995), have been used with success as feeder layers in these assays. Selection of stromal cell lines should be approached with caution because different lines may provide variable support of CAFCs and LTC-ICs depending on their particular complement of stromal cell phenotypes. Stromal layers are cultured to confluency in 96-well flat-bottomed plates, at which time cells are irradiated with 15 to 20 Gy of radiation to eliminate hematopoietic cells while preserving the ability of stroma to support hematopoiesis. An alternative method for LTC-ICs eliminates the feeder layer and instead cultures test cells in suspension in IMDM, 10% FBS, L-glutamine, and cytokines, which may effectively push more committed cells to differentiate, thereby enriching the week 5 culture with primitive HPCs scored as LTC-ICs (Podesta *et al.*, 2001).

2. One to 6 days later the irradiated stroma is overlaid with cells of interest in a limiting fashion, typically up to 12 successive 2-fold dilutions with 15–30 replicate wells per dilution. The first dilution should contain on the order of approximately 250 cord blood (CB) CD34$^+$ cells per well. Media consist of normal cell culture media, such as α-medium plus 10% FBS, 10% horse serum, human transferrin (0.5 mg/ml), hydrocortisone sodium succinate (10^{-5} mol/liter), and 2-mercaptoethanol (10^{-4} mol/liter) (Ploemacher *et al.*, 1989), or MyeloCult (StemCell Technologies) plus hydrocortisone (10^{-6} mol/liter) can also be used (Podesta *et al.*, 2001). IMDM plus 10% FBS and 1% L-glutamine has also been used with success in the LTC-IC assay (Traycoff *et al.*, 1995).

3. Cultures are maintained by weekly half or whole medium changes. For CAFC analysis, cultures are examined at predetermined time points (2, 4, 6, 8, or more weeks) for the presence of cobblestone areas (CAs, tightly packed groups of at least five small cells embedded beneath the stroma layer and visualized microscopically as "phase dark"). The percentage of wells with at least one CA clone is used to calculate CAFC frequency, using Poisson statistics. For LTC-IC analysis, the entire well is trypsinized, harvested, and then cultured in standard hematopoietic progenitor cell assays in methylcellulose (e.g., as noted previously in the section Methods for Enumerating Human Myeloid Progenitor Cells). For LTC-IC assays cultured in suspension without feeder layers, the supernatant is removed and a methylcellulose mixture is overlaid into the individual wells of the 96-well plate, and CFCs are assayed 2 weeks later (Traycoff et al., 1995). The frequency of LTC-ICs is also calculated, using Poisson statistics.

Notes

• It has been reported that CB CD34⁻lineage⁺ cells can produce CAFC-like colonies that are not analogous to LTC-ICs, suggesting the need for assaying purified CD34⁺ cells instead of unfractionated mononuclear cells, which would contain false CAFCs (Denning-Kendall et al., 2003).

• CAFCs and LTC-ICs are generally believed to assay similar, if not identical, primitive hematopoietic progenitor cells and thus could theoretically be used interchangeably. However, several reports show discrepancies between the frequencies of CAFCs and LTC-ICs (Gan et al., 1997; Pettengell et al., 1994; Weaver et al., 1997), and investigators should validate their own results for each type of sample and feeder layer used.

• LTC-IC frequency is reportedly lower in cord blood compared with bone marrow and mobilized peripheral blood, although each LTC-IC from cord blood produces significantly more CFCs compared with bone marrow and mobilized peripheral blood, illustrating the robust proliferative potential characteristic of cord blood primitive hematopoietic progenitor cells (Podesta et al., 2001; Theunissen and Verfaille, 2005; Traycoff et al., 1994). Of interest, whereas bone marrow LTC-ICs were insensitive to temperature of incubation, cord blood produced significantly more LTC-ICs at 37° than at 33° (Podesta et al., 2001). In addition, cord blood LTC-ICs also showed preference for type of stroma, producing 5-fold more LTC-ICs when cultured over NIH3T3 cells compared with M210B4. Bone marrow produced similar numbers of LTC-ICs over both types of stroma (Podesta et al., 2001).

ML-IC Methods

AFT024 feeder layers are overlaid with test cells in RPMI 1640 plus 20% FCS, 2-mercaptoethanol (100 μmol/liter), penicillin and streptomycin

(100 U/ml), and cytokines (see Theunissen and Verfaillie, 2005, for details). In contrast to CAFC and LTC-IC assays, test cells are seeded onto feeders in the ML-IC assay in a nonlimiting manner. After 2–4 weeks, the progeny of each well are divided into four to eight new AFT024-coated wells and half are assayed for LTC-IC content by overlaying with a methylcellulose mixture after 5 weeks in the presence of FL, SCF, and IL-7, and the other half are assayed for NK-IC content by staining with anti-CD56 antibody after 6–7 weeks of culture (Miller *et al.*, 1999). ML-ICs give rise to at least one LTC-IC and one NK-IC in secondary plates.

Notes

The frequency of ML-ICs in cord blood was not significantly different from that of bone marrow or mobilized blood, although the generative potential of cord blood ML-ICs was much higher compared with these other two sources of stem cells, again illustrating the high proliferative potential of primitive cord blood cells (Traycoff *et al.*, 1994).

In Vivo *Animal Models for Assay of Cord Blood-Derived Hematopoietic Stem Cells*

As in the case of murine hematopoietic stem cells, the *in vivo* repopulating potential of cord blood-derived hematopoietic stem cells represents the most stringent measure of their functional capacity and their ability to sustain long-term multilineage engraftment. However, unlike the murine system, marrow repopulating studies in the human system must rely on the use of different xenotransplantation models to adequately assess the *in vivo* function of putative HSCs (Bodine, 2004). Several of these transplantation models have been developed, each with its own advantages and disadvantages (Kamel-Reid and Dick, 1988; McCune *et al.*, 1988). Most, if not all, rely on the use of an immunocompromised host that offers a permissive microenvironment for the proliferation and differentiation of transplanted hematopoietic stem cells. Although some of these models use large animals as recipients (Bodine, 2004), most rely on the use of different strains of immunodeficient mice with (Bodine, 2004) or without a human graft of either bone or other tissue that harbors and supports the expansion of human hematopoietic stem cells in the animal. The NOD/SCID mouse has become the most accepted and widely used mouse model to assess human hematopoietic function *in vivo*. Here, we restrict the description of our experimental design to that used with NOD/SCID mice.

The general concept behind the use of NOD/SCID mice for the assessment of stem cell function is that the immunodeficient status of these mice allows for the engraftment of human cells in the bone marrow (BM) of recipient animals, thus mimicking normal human hematopoiesis. However, as is noted later in the notes section, several peculiar observations well documented in this system represent major deviations from normal human hematopoiesis. Although NOD/SCID mice are immunocompromised, further immunosuppression is still required before transplantation in order to facilitate or induce human stem cell engraftment. This immunosuppression is usually delivered in the form of sublethal dosages of total body irradiation.

Once all these conditions are met (in addition to some other important parameters discussed later in the notes section), it is almost always found that an adequate number of human stem cells engrafts and populates the bone marrow of recipient mice with human progeny. Engrafted animals, the level of chimerism in individual recipients, and the identity of human lineages detected in these mice can be used collectively or individually to evaluate several stem cell functions including, but not limited to, (1) the potential of candidate stem cell phenotypes to engraft, (2) the least required number of putative stem cells to support hematopoiesis, (3) the multilineage differentiation potential of a particular graft, (4) the frequency of engrafting stem cells in a given graft, a measurement that is more accurately defined as the SRC, (5) the success of long-term expression of genes transduced into human hematopoietic stem cells, (6) homing of cord blood cells (progenitors and stem cells) to the bone marrow of recipient mice both short and long term, and (7) the ability of human putative hematopoietic stem cells to engraft sites other than classical hematopoietic tissues and organs.

Because we are primarily concerned here with the assessment of hematopoietic potential and number of hematopoietic stem cells in cord blood, procedures outlined below are limited to the description of the first four assessments listed previously. As can be appreciated from the final goal of each of these assays, a substantial degree of overlap normally takes place in both the establishment of these assays and the collection and interpretation of the data.

Materials and Major Equipment for Transplantation

Cord blood cells fractionated or prepared as required for the assay
Flow cytometric or magnetic cell sorters if required for the fractionation of cord blood cells
Healthy NOD/SCID mice, preferably between 7 and 10 weeks of age
Cesium or X-ray irradiator
Holding chambers for irradiating mice
Holding chamber for tail vein injections
Tubes, syringes, and needles (27 or 30 gauge)
Surgical masks, bonnets, gloves, and clean robes.

Transplantation Procedure

1. All procedures involving handling and transplantation of NOD/SCID mice should be completed aseptically and all equipment used to restrain or hold these mice should be cleaned after each use with an antiseptic solution. Media in which cells are suspended, syringes, and needles should all be sterile.

2. The dose of radiation normally given to NOD/SCID mice ranges between 275 and 350 rad given in one dose at a rate of approximately 75 rad/min. Higher doses of radiation are lethal to NOD/SCID mice. It is preferable that each investigator assess the level of radiation, within the range given previously, that mice can tolerate without any morbidity and mortality. Transplantation can be completed any time after irradiation but preferably within 12 h.

3. Working inside a safety cabinet, mice should be removed from the cage and held in a clean chamber for tail vein injection.

4. Wipe the tail several times with an alcohol swab and locate the tail vein. If required, a topical anesthetic can be applied before this along the tail.

5. Inject test cells intravenously in a 200-μl volume of sterile medium or phosphate-buffered saline. Larger volumes can be used but these should not exceed 500 μl.

6. Return the transplanted mouse to the cage and observe for distress.

7. Wipe all the equipment and gloves with an antiseptic solution before transplantation of the next recipient.

Materials and Major Equipment for Assessment of Engraftment

Bone marrow cells from transplanted NOD/SCID mice, collected from at least two long bones from both legs (to avoid sampling errors)

Anti-human monoclonal antibodies recognizing at least CD45 and other lineage markers of interest as dictated by the overall goals of the study

Anti-murine CD45 monoclonal antibodies

Flow cytometer.

Assessment of engraftment is usually performed at least 6 weeks post-transplantation. It has been argued that at 6 weeks posttransplantation, progeny of short-term repopulating cells predominate the bone marrow of recipient mice and therefore assessments made at this time point reflect the function of short-term, but not long-term, repopulating cells. Because transplanted mice normally survive for several weeks after that, it is now more acceptable to analyze transplanted mice at 9 to 12 weeks posttransplantation in order to measure more accurately the function of long-term repopulating cells.

Measurement of Human Cell Engraftment

1. Collect the bone marrow from individual transplanted NOD/SCID mice.
2. Stain the cells with appropriate monoclonal antibodies. It is preferable (if the limitations of the flow cytometer used for analysis allow) to include anti-human CD45 and anti-murine CD45 in every stain combination to allow for accurate identification of chimeric human cells that can then be subjected to further linage analysis depending on the types of antibodies added to this combination.
3. Run samples on a flow cytometer and acquire at least 5000 events or until a minimum of 100 human CD45$^+$ cells are collected.
4. To clearly recognize engrafted chimeric human cells within the analyzed BM samples, the following controls are suggested.
   a. Prepare bone marrow cells from a control, untransplanted NOD/SCID mouse.
   b. Mix some cells from step 4.a with human peripheral blood cells to create a 90:10 ratio of murine to human cells.
   c. Stain samples from both steps 4.a and 4.b with anti-human CD45 and anti-murine CD45 individually and simultaneously.
   d. Use these samples to identify the exact position of human "chimeric" cells in a dotplot. Use this position to decide on the validity of chimeric human cells in test animals falling in or close to this position.

How to Use Collected Data for Human Cell Engraftment to Evaluate the Preceding Assay Assessments

There is no consensus as to what degree of human chimerism in a transplanted NOD/SCID mouse constitutes engraftment. It is therefore the responsibility of individual investigators to define engraftment in a consistent manner throughout the study. On the flow cytometric detection of human chimerism in the bone marrow of a transplanted mouse, the percentage of human CD45$^+$ cells (or human cells positive for other markers) among all cells analyzed can be quantified automatically. This percentage can be adjusted to reflect that contained within the lymphocyte gate (through the use of light scatter properties of analyzed cells) or within total bone marrow cells excluding erythrocytes and debris. The latter is more accurate because focusing on only the lymphocyte population usually decreases the denominator used in the calculation of engraftment and thus artificially increases the level of chimerism.

• Examining the ability and the minimum number of putative stem cells required for engraftment: Once chimerism is detected in transplanted mice, an average level of chimerism can be calculated from all mice receiving the same phenotype and number of putative stem cells. Assuming that chimerism is defined as a level of 0.1% human cells in the mouse bone marrow, comparisons can then be made between different phenotypes of putative stem cells to determine the levels of chimerism supported by each group of test cells. Similarly, with an escalating or dose-dependent assay, the minimum number of test cells required to achieve 0.1% chimerism can be established for different test cells.

• Multilineage engraftment can be assessed in mice harboring more than 0.5% human cells in most cases. Typically, analysis for the expression of lymphoid (B cells, using CD19 or CD20), myeloid (using CD14, CD15, or CD33), and erythroid (using glycophorin A) cells can be done. As indicated later in the notes section, percentages of these lineages may not reflect normal hematopoiesis; however, when two groups of engrafting cells are compared, a skewness toward one lineage or another can be detected and quantified.

• To calculate a frequency of SCID repopulating cells in a given graft, groups of NOD/SCID mice are transplanted with different numbers of test cells per group. This design generates a limiting dilution analysis and allows for the detection of engrafted and nonengrafted animals (again, determined by an arbitrarily set lower limit) within each group of mice. The number of cells in a graft across the range of doses chosen for the experiment should be selected in a way that generates, in more than two groups, a mix of engrafted and nonengrafted mice. Once the number of positive (engrafted) mice is determined within each group, the reciprocal number of negative (nonengrafted) mice in the group can be calculated. The percentages of negative mice in each group can then be used to calculate a frequency based on Poisson distribution (Taswell, 1981). Obviously, when two sets of groups of mice are used to examine two distinct stem cell candidates, different frequencies can be calculated and used to compare the relative abundance of repopulating cells in test grafts. It is recommended that at least four groups of mice be tested in a limiting dilution schema in which each group contains at least four mice. If the range of cell doses required to generate positive and negative results is not known, it is recommended to increase both the number of cell doses (to allow for a wider range of cell numbers) and the number of mice per group.

Notes

In this section, a few peculiarities of the NOD/SCID system are listed to enable investigators to avoid common problems with the use of this model in examining cord blood hematopoietic stem cells *in vivo*.

• NOD/SCID mice are "leaky." Therefore, some mice will develop mouse T cells that can be detected in the periphery as $CD3^+$ cells. These T cells can interfere with and possibly eliminate human engraftment. It is therefore critical to examine these mice for the presence of murine T cells in the periphery and to eliminate positive mice from the experimental design. Usually mice with >1% T cells are problematic.

• Even well-engrafted mice with high levels of chimerism in the bone marrow do not necessarily have detectable human cells in the periphery. Therefore monitoring engraftment in the blood of recipient mice over time may generate negative results.

• Human hematopoietic engraftment in NOD/SCID mice is preferentially skewed toward the B cell lineage, whereby multilineage assessment reveals a large percentage of B cells. In reverse, T cell engraftment in NOD/SCID mice is almost completely absent in most cases unless specific types of NOD/SCID mice are used.

• Human cytokines can be used to supplement transplanted mice. These cytokines both protected recipient mice from the effects of radiation and modulated the behavior of chimeric human hematopoietic cells.

• Use of antibiotics for a few days before and a few days after radiation was found to enhance survival and overall health of transplanted mice (E. F. Srour, unpublished observations).

• To reduce the impact of the reticuloendothelial system of NOD/SCID mice on transplanted human graft cells, injection of up to 10×10^7 nonadherent, irradiated CD34$^-$ cells a few hours before transplantation or mixed with graft cells has been practiced.

• A technique suitable for multiple samplings of bone marrow cells from the femur can be used to temporally examine the kinetics of human engraftment. Alternatively, this approach has been used to deliver graft cells directly into the bone marrow microenvironment and avoid the need for homing of cells into the bone marrow of recipient animals.

• When levels of engraftment are low, such that it cannot be detected by flow cytometric analysis, investigators have relied on polymerase chain reaction (PCR) analysis to detect human genes in the murine bone marrow. Although this method is suitable for the detection of small, and perhaps insignificant, levels of chimerism, PCR data cannot be reliably used for quantification of the level of chimerism.

Methods for Enumerating Endothelial Progenitor Cells in Umbilical Cord Blood

Information on the assay of high- and low-proliferative endothelial progenitor cells (Ingram *et al.*, 2004) follows.

Materials

Collection of Umbilical Cord Blood

Heparin sodium injection (1000 USP units/ml) (cat. no. NDC 0641–2440–41; Baxter, Deerfield, IL)

Sterile 60-cm^3 syringe (cat. no. 309653; Becton Dickinson, Franklin Lakes, NJ) fitted with a 16-gauge 1.5-in. needle (cat. no. 305198; Becton Dickinson).

Preparation and Culture of Mononuclear Cells

Phosphate-buffered saline, pH 7.2 (PBS) (cat. no. 20012–027; Invitrogen)
Ficoll-Paque PLUS (Ficoll) (cat. no. 17–1440–03; GE Healthcare, Piscataway, NJ)
Sterile 20-cm^3 syringe (cat. no. 309661; Becton Dickinson) fitted with a mixing cannula (cat. no. 500–11–012; Unomedical, Birkeroed, Denmark)
EBM-2 10:1: EBM-2 (cat. no. CC-3156; Cambrex Bio Science Walkersville) supplemented with 10% fetal bovine serum (FBS) (cat. no. SH30070.03; HyClone) and 1% penicillin (10,000 U/ml)–streptomycin (10,000 μg/ml)–amphotericin (25 μg/ml) (cat. no. 15240–062; Invitrogen)
0.4% trypan blue solution (cat. no. T-8154; Sigma) and a hemacytometer
cEGM-2: EGM-2 (cat. no. CC-3162; Cambrex Bio Science Walkersville) supplemented with the entire growth factor BulletKit, 10% FBS (HyClone), and 1% penicillin (10,000 U/ml)–streptomycin (10,000 μg/ml)–amphotericin (25 μg/ml) (cat. no. 15240–062; Invitrogen)
Collagen I-coated 6-well plates (cat. no. 356400; BD Biosciences).

Isolation and Culture of Clonal EPCs

Sterile cloning cylinders (cat. no. 07–907–10; Fisher Scientific, Pittsburgh, PA)
Vacuum grease (cat. no. 1658832; Dow Corning, Midland, MI), autoclave sterilized
Pasteur pipettes (cat. no. 13–678–20C; Fisher Scientific), autoclave sterilized
Trypsin–EDTA (cat. no. 25300–054; Invitrogen)
Sterile forceps.

Methods

Collection of Umbilical Cord Blood

1. A 60-cm^3 syringe is prepared by drawing up 2 ml of heparin, pulling the plunger to the 60-ml mark and swirling the heparin to coat the entire inner surface of the syringe. The plunger is depressed to expel all excess air, leaving only 2 ml of heparin in the syringe.

2. Umbilical cord blood (UCB) from the umbilical vein is collected into the prepared syringe, using a 16-gauge needle. Immediately mix the blood by inverting several times to prevent clotting (see Note 1).

Preparation and Culture of Mononuclear Cells

1. Fifteen milliliters of UCB is gently dispensed into each 50-ml conical tube and diluted with 20 ml of PBS.

2. Fifteen milliliters of Ficoll is drawn into a 20-cm^3 syringe and a mixing cannula is attached. The end of the mixing cannula is placed at the bottom of the tube containing the diluted UCB and 15 ml of Ficoll is carefully underlayed, maintaining a clean interphase.

3. Tubes are centrifuged at 740g at room temperature for 30 min with no brake (see Note 2).

4. After centrifugation, red blood cells will form a pellet, and mononuclear cells (MNCs) will form a hazy buffy coat at the interphase between the clear Ficoll layer below and the yellow serum layer above. The buffy coat MNCs are carefully removed with a transfer pipette by placing the tip of the pipette just above the buffy coat layer and drawing up. MNCs are dispensed into a 50-ml tube containing 10 ml of EBM-2 10:1. Care should be taken to collect all buffy coat MNCs while avoiding excess collection of the Ficoll layer or the serum layer.

5. MNCs are centrifuged at 515g for 10 min at room temperature with a high brake and the supernatant is discarded. Pelleted cells are gently tapped loose and resuspended in 10 ml of EBM-2 10:1.

6. Repeat step 5 one time.

7. The MNC suspension is mixed well by pipetting up and down several times. Thirty microliters of cells is removed and mixed with 30 μl of 0.4% trypan blue solution. Cells are loaded onto a hemacytometer and viable cells are counted.

8. The MNC suspension is again centrifuged at 515g for 10 min at room temperature with a high brake and the supernatant is discarded. Pelleted cells are gently tapped loose and resuspended in cEGM-2 at 1.25 $\times 10^7$ cells/ml.

9. MNCs (4 ml, 5 $\times 10^7$ cells) are seeded into each well of 6-well tissue culture plates precoated with rat tail collagen I and cultured in a 37°, 5% CO_2 humidified incubator (see Note 3).

10. After 24 h (day 1), medium is slowly removed from each well with a pipette and 2 ml of cEGM-2 is added to each well. The medium is again removed from each well and replaced with 4 ml of cEGM-2. Culture plates are returned to a 37°, 5% CO_2 humidified incubator.

11. After 24 h (day 2), medium is slowly removed from each well with a pipette and 4 ml of cEGM-2 is added to each well. Medium is changed in exactly this way each day from day 3 to day 7, and then every other day after day 7 (see Note 4).

12. Endothelial progenitor cell (EPC) colonies appear between day 4 and day 7 of culture as well-circumscribed areas of cobblestone-appearing cells. Individual colonies can be isolated and expanded on days 7–14. EPCs can also be allowed to grow to 80–90% confluency before subculturing to start a polyclonal EPC line.

Isolation and Culture of Clonal EPCs

1. EPC colonies are visualized by inverted microscopy (see, e.g., Fig. 3) and their location within the culture is well outlined with a fine-tipped marker.

2. Medium is aspirated and the culture well is washed two times with PBS.

3. After aspirating the final wash of PBS, cloning cylinders coated on the bottom surface with a thin bead of vacuum grease are placed around each colony and pressed firmly against the plate, using forceps.

4. Using a Pasteur pipette, one or two drops of warm trypsin–EDTA is added into each cloning cylinder. Plates are incubated at $37°$ for 1–5 min until the cells within the cylinder begin to ball up and detach (see Note 5).

5. When all the cells within the cylinder have balled up, place the tip of a Pasteur pipette containing 200–300 μl of cEGM-2 into the center of the cylinder and pipette up and down vigorously several times. Collect the entire volume to a tube. Serially wash the area within the cylinder one to three more times with cEGM-2 until all cells are collected.

6. Each EPC colony collected can be seeded into 1 well of a 24-well tissue culture plate precoated with rat tail collagen I (see Note 3) in a total volume of 1.5 ml of cEGM-2 and cultured in a $37°$, 5% CO_2 humidified incubator. Medium should be changed every other day.

Notes

Note 1: If UCB cannot be processed immediately, it can be kept at room temperature with gentle rocking for up to 16 h.

Note 2: Bringing the centrifuge up to $740g$ slowly over the course of 1–2 min helps result in a cleaner buffy coat.

Note 3: Six-well tissue culture-treated plates can also be coated with collagen I in the laboratory. To make the collagen coating solution, 0.575 ml of glacial acetic acid (17.4 N) (cat. no. A38–500; Fisher Scientific) is diluted in 495 ml of sterile distilled water (final concentration, 0.02 N). This solution is sterile filtered with a 0.22-μm pore size vacuum filtration system (cat. no. SCGPU05RE; Millipore, Bedford, MA), and then rat tail collagen I (cat. no. 354236; BD Biosciences) is added to a final concentration of 50 μg/ml. This solution can be kept at 4° for 1 month. One milliliter of the collagen-coating solution is placed in each well of a 6-well tissue culture-treated plate (500 μl/well for 24-well plates; or 4 ml/25-cm^2 flask, 9 ml/75-cm^2 flask) and incubated at 37° for at least 90 min. The collagen-coating solution is removed and wells are washed two times with PBS before seeding of cells.

Note 4: Removal and addition of medium on days 1–7 of culture should be done slowly (at a rate of about 1 ml per 4–5 s). After day 7 medium can be aspirated from the wells by vacuum. When changing the medium in the first few days, nonadherent MNCs will be removed along with the medium and can be discarded.

Note 5: Steps 3 and 4 must be done quickly to prevent the EPC colonies from drying out.

Uses of Preceding Assays

The assays for hematopoietic progenitor and stem cells, and for endothelial progenitor cells, can be used to quantitate these cells in cord blood, as well as other human tissue sources of these cells. Perhaps, even more importantly, they have been and can continue to be used for assessment of the regulation of these cells by cytokines, chemokines, and cell–cell interactions. The more we learn of the biology of these cells and how they can be manipulated for self-renewal, survival, proliferation, differentiation, and migration/homing/mobilization, the more likely we can use this information for clinical efficacy and benefit.

References

Aoki, M., Yasutake, M., and Murohara, T. (2004). Derivation of functional endothelial progenitor cells from human umbilical card blood mononuclear cells isolated by a novel cell filtration device. *Stem Cells* **22,** 994–1002.

Asahara, T., Murohara, T., Sullivan, A, Silver, M., van der Zee, R., Li, T., Witzenbichler, B., Schatteman, G., and Isner, J. M. (1997). Isolation of putative progenitor endothelial cells for angiogenesis. *Science* **275,** 964–967.

Barker, J. N., Weisdorf, D. J., DeFor, T. E., Blazar, B. R., McGlave, P. B., Miller, J. S., Verfaillie, C. M., and Wagner, J. E. (2005). Transplantation of 2 partially HLA-matched umbilical cord blood units to enhance engraftment in adults with hematologic malignancy. *Blood* **105,** 1343–1347.

Bock, T. A., Orlic, D., Dunbar, C. E., Broxmeyer, H. E., and Bodine, D. M. (1995). Improved engraftment of human hematopoietic cells in severe combined immunodeficient (SCID) mice carrying human cytokine transgenes. *J. Exp. Med.* **182,** 2037–2043.

Bodine, D. M. (2004). Animal models for the engraftment and differentiation of human hematopoietic stem and progenitor cells. *In* "Cord Blood: Biology, Immunology, and Clinical Transplantation" (H. E. Broxmeyer, ed.), pp. 47–64. American Association of Blood Banks Press, Bethesda, MD.

Bompais, H., Chagraoui, J., Canron, X., Crisan, M., Liu, X. H., Anjo, A., Tolla-Le Port, C., Leboeuf, M., Charbord, P., Bikfalvi, A., and Uzan, G. (2004). Human endothelial cells derived from circulating progenitors display specific functional properties compared with mature vessel wall endothelial cells. *Blood* **103,** 2577–2584.

Breems, D. A., Blokland, E. A., Neben, S., and Ploemacher, R. E. (1994). Frequency analysis of human primitive haematopoietic stem cell subsets using a cobblestone area-forming cell assay. *Leukemia* **8,** 1095–1104.

Broxmeyer, H. E. (1998). Introduction: The past, present, and future of cord blood transplantation. *In* "Cellular Characteristics of Cord Blood and Cord Blood Transplantation" (H. E. Broxmeyer, ed.), pp. 1–9. American Association of Blood Banks Press, Bethesda, MD.

Broxmeyer, H. E. (2000). Cord blood transplantation: Looking back and to the future [introduction]. *In* "Cord Blood Characteristics: Role in Stem Cell Transplantation" (S. B. A. Cohen, E. Gluckman, P. Rubinstein, and J. A. Madrigal, eds.), pp. 1–12. Martin Dunitz, London.

Broxmeyer, H. E. (2004). Proliferation, self-renewal, and survival characteristics of cord blood hematopoietic stem and progenitor cells. *In* "Cord Blood: Biology, Immunology, and Clinical Transplantation" (H. E. Broxmeyer, ed.), pp. 1–21. American Association of Blood Banks Press, Bethesda, MD.

Broxmeyer, H. E. (2005). Biology of cord blood cells and future prospects for enhanced clinical benefit. *Cytotherapy* **7,** 209–218.

Broxmeyer, H. E., and Cooper, S. (1997). High efficiency recovery of immature hematopoietic progenitor cells with extensive proliferative capacity from human cord blood cryopreserved for ten years. *Clin. Exp. Immunol.* **107,** 45–53.

Broxmeyer, H. E., and Smith, F. O. (2004). Cord blood hematopoietic cell transplantation. *In* "Thomas' Hematopoietic Cell Transplantation" (K. G. Blume, S. J. Forman, and F. R. Appelbaum, eds.), 3rd Ed., Chapter 43. Blackwell Scientific Publications, Cambridge, MA.

Broxmeyer, H. E., Douglas, G. W., Hangoc, G., Cooper, S., Bard, J., English, D., Arny, M., Thomas, L., and Boyse, E. A. (1989). Human umbilical cord blood as a potential source of transplantable hematopoietic stem/progenitor cells. *Proc. Natl. Acad. Sci. USA* **86,** 3828–3832.

Broxmeyer, H. E., Gluckman, E., Auerbach, A., Douglas, G. W., Friedman, H., Cooper, S., Hangoc, G., Kurtzberg, J., Bard, J., and Boyse, E. A. (1990). Human umbilical cord blood: A clinically useful source of transplantable hematopoietic stem/progenitor cells. *Int. J. Cell Cloning* **8,** 76–91.

Broxmeyer, H. E., Hangoc, G., Cooper, S., Ribeiro, R. C., Graves, V., Yoder, M., Wagner, J., Vadhan-Raj, S., Rubinstein, P., and Broun, E. R. (1992). Growth characteristics and expansion of human umbilical cord blood and estimation of its potential for transplantation of adults. *Proc. Natl. Acad. Sci. USA* **89,** 4109–4113.

Broxmeyer, H. E., Srour, E. F., Hangoc, G., Cooper, S., Anderson, J. A., and Bodine, D. (2003). High efficiency recovery of hematopoietic progenitor cells with extensive proliferative and *ex-vivo* expansion activity and of hematopoietic stem cells with

NOD/SCID mouse repopulation ability from human cord blood stored frozen for 15 years. *Proc. Natl. Acad. Sci. USA* **100,** 645–650.

Campbell, T. B., Hangoc, G., and Broxmeyer, H. E. (2005). A role of CD26 in human umbilical cord blood CD34$^+$ cell engraftment of NOD/SCID mice [abstract 1708]. *Blood* **106,** 487a.

Cardoso, A. A., Li, M. L., Batard, P., Hatzfeld, A., Brown, E. L., Levesque, J. P., Sookdeo, H., Panterne, B., Sansilvestri, P., Clark, S. C., and Hatzfeld, J. (1993). Release from quiescence of CD34$^+$CD38$^-$ human umbilical cord blood cells reveals their potentiality to engraft adults. *Proc. Natl. Acad. Sci. USA* **90,** 8707–8711.

Carow, C., Hangoc, G., Cooper, S., Williams, D. E., and Broxmeyer, H. E. (1991). Mast cell growth factor (c-kit ligand) supports the growth of human multipotential (CFU-GEMM) progenitor cells with a high replating potential. *Blood* **78,** 2216–2221.

Carow, C. E., Hangoc, G., and Broxmeyer, H. E. (1993). Human multipotential progenitor cells (CFU-GEMM) have extensive replating capacity for secondary CFU-GEMM: An effect enhanced by cord blood plasma. *Blood* **81,** 942–949.

Christopherson, K. W., II, and Broxmeyer, H. E. (2004). Hematopoietic stem and progenitor cell homing, engraftment, and mobilization in the context of the CXCL12/SDF-1–CXCR4 axis. *In* "Cord Blood: Biology, Immunology, and Clinical Transplantation" (H. E. Broxmeyer, ed.), pp. 65–86. American Association of Blood Banks Press, Bethesda, MD.

Christopherson, K. W., II, Hangoc, G., Mantel, C., and Broxmeyer, H. E. (2004). Modulation of hematopoietic stem cell homing and engraftment by CD26. *Science* **305,** 1000–1003.

Cornetta, K., Laughlin, M., Carter, S., Wall, D., Weinthal, J., Delaney, C., Wagner, J., Sweetman, R., McCarthy, P., and Chao, N. (2005). Umbilical cord blood transplantation in adults: Results of the prospective cord blood transplantation (COBLT). *Biol. Blood Marrow Transpl.* **11,** 149–160.

Crisa, L., Cirulli, V., Smith, K. A., Ellisman, M. H., Torbett, B. E., and Salomon, D. R. (1999). Human cord blood progenitors sustain thymic T-cell development and a novel form of angiogenesis. *Blood* **94,** 3928–3940.

Denning-Kendall, P., Singha, S., Bradley, B., and Hows, J. (2003). Cobblestone area-forming cells in human cord blood are heterogeneous and differ from long-term culture-initiating cells. *Stem Cells* **21,** 694–701.

Dexter, T. M. (1993). Synergistic interactions in haematopoiesis: Biological implications and clinical use. *Eur. J. Cancer* **29A,** (Suppl.) S6–S9.

Eggermann, J., Kliche, S., Jarmy, G., Hoffmann, K., Mayr-Beyrle, U., Debatin, K. M., Waltenberger, J., and Beltinger, C. (2003). Endothelial progenitor cell culture and differentiation *in vitro*: A methodological comparison using human umbilical cord blood. *Cardiovasc. Res.* **58,** 478–486.

Fan, C.-L., Li, Y., Gao, P.-J., Liu, J. J., Zhang, X. J., and Zhu, D. L. (2003). Differentiation of endothelial progenitor cells from human umbilical cord blood CD 34$^+$ cells *in vitro*. *Acta Pharmacol. Sin.* **24,** 212–218.

Flamme, I., Frolich, T., and Risau, W. (1997). Molecular mechanisms of vasculogenesis and embryonic angiogenesis. *J. Cell. Physiol.* **173,** 206–210.

Gan, O. I., Murdoch, B., Larochelle, A., and Dick, J. E. (1997). Differential maintenance of primitive human SCID-repopulating cells, clonogenic progenitors, and long-term culture-initiating cells after incubation on human bone marrow stromal cells. *Blood* **90,** 641–650.

Gluckman, E., and Rocha, V. (2005). History of the clinical use of umbilical cord blood hematopoietic cells. *Cytotherapy* **7,** 219–227.

Gluckman, E., Broxmeyer, H. E., Auerbach, A. D., Friedman, H. S., Douglas, G. W., Devergie, A., Esperou, H., Thierry, D., Socie, G., Lehn, P., Cooper, S., English, D., *et al.* (1989). Hematopoietic reconstitution in a patient with Fanconi anemia by means of umbilical-cord blood from an HLA-identical sibling. *N. Engl. J. Med.* **321,** 1174–1178.

Gluckman, E., Rocha, V., Boyer-Chammard, A., Locatelli, F., Arcese, W., Pasquini, R., Ortega, J., Souillet, G., Ferreira, E., Laporte, J. P., Fernandez, M., and Chastang, C. (1997). Outcome of cord-blood transplantation from related and unrelated donors. Eurocord Transplant Group and the European Blood and Marrow Transplantation Group. *N. Engl. J. Med.* **337,** 373–381.

Gluckman, E., Rocha, V., Arcese, W., Michel, G., Sanz, G., Chan, K. W., Takahashi, T. A., Ortega, J., Filipovich, A., Locatelli, F., Asano, S., Fagioli, F., *et al.* (2004). Factors associated with outcomes of unrelated cord blood transplant: Guidelines for donor choice. *Exp. Hematol.* **32,** 397–407.

Gothert, J. R., Gustin, S. E., van Eekelen, J. A., Schmidt, U., Hall, M. A., Jane, S. M., Green, A. R., Gottgens, B., Izon, D. J., and Begley, C. G. (2004). Genetically tagging endothelial cells *in vivo*: Bone marrow-derived cells do not contribute to tumor endothelium. *Blood* **104,** 1769–1777.

Hao, Q. L., Thiemann, F. T., Petersen, D., Smogorzewska, E. M., and Crooks, G. M. (1996). Extended long-term culture reveals a highly quiescent and primitive human hematopoietic progenitor population. *Blood* **88,** 3306–3313.

Hildbrand, P., Cirulli, V., Prinsen, R. C., Smith, K. A., Torbett, B. E., Salomon, D. R., and Crisa, L. (2004). The role of angiopoietins in the development of endothelial cells from cord blood CD34$^+$ progenitors. *Blood* **104,** 2010–2019.

Hristov, M., and Weber, C. (2004). Endothelial progenitor cells: Characterization, pathophysiology, and possible clinical relevance. *J. Cell. Mol. Med.* **8,** 498–508.

Ingram, D., Mead, L., Tanaka, H., Meade, V., Fenoglio, A., Mortell, K., Pollok, K., Ferkowicz, M. J., Gilley, D., and Yoder, M. C. (2004). Identification of a novel hierarchy of endothelial progenitor cells using human peripheral and umbilical cord blood. *Blood* **104,** 2752–2760.

Iwami, Y., Masuda, H., and Asahara, T. (2004). Endothelial progenitor cells: Past, state of the art, and future. *J. Cell. Mol. Med.* **8,** 488–497.

Jaroscak, J., Goltry, K., Smith, A., Waters-Pick, B., Martin, P. L., Driscoll, T. A., Howrey, R., Chao, N., Douville, J., Burhop, S., Fu, P., and Kurtzberg, J. (2003). Augmentation of umbilical cord blood (UCB) transplantation with *ex-vivo*-expanded UCB cells: Results of a phase I trial using the AstromReplicell system. *Blood* **101,** 5061–5067.

Kamel-Reid, S., and Dick, J. E. (1988). Engraftment of immune-deficient mice with human hematopoietic stem cells. *Science* **242,** 1706–1709.

Kang, H. J., Kim, S. C., Kim, Y. J., Kim, C. W., Kim, J. G., Ahn, H. S., Park, S. I., Jung, M. H., Choi, B. C., and Kimm, K. (2001). Short-term phytohaemagglutinin-activated mononuclear cells induce endothelial progenitor cells from cord blood CD34$^+$ cells. *Br. J. Haematol.* **113,** 962–969.

Khakoo, A. Y., and Finkel, T. (2005). Endothelial progenitor cells. *Annu. Rev. Med.* **56,** 79–101.

Kurtzberg, J., Laughlin, M., Graham, M. L., Smith, C., Olson, J. F., Halperin, E. C., Ciocci, G., Carrier, C., Stevens, C. E., and Rubinstein, P. (1996). Placental blood as a source of hematopoietic stem cells for transplantation into unrelated donors. *N. Engl. J. Med.* **335,** 157–166.

Kusadasi, N., van Soest, P. L., Mayen, A. E., Koevoet, J. L., and Ploemacher, R. E. (2000). Successful short-term *ex vivo* expansion of NOD/SCID repopulating ability and CAFC week 6 from umbilical cord blood. *Leukemia* **14,** 1944–1953.

Lansdorp, P. M., Dragowska, W., and Mayani, H. (1993). Ontogeny-related changes in proliferative potential of human hematopoietic cells. *J. Exp. Med.* **178,** 787–791.

Lapidot, T., Dar, A., and Kollet, O. (2005). How do stem cells find their way home? *Blood* **106**, 1901–1910.

Laughlin, M. J., Barker, J., Bambach, B., Koc, O. N., Rizzieri, D. A., Wagner, J. E., Gerson, S. L., Lazarus, H. M., Cairo, M., Stevens, C. E., Rubinstein, P., and Kurtzberg., J. (2001). Hematopoietic engraftment and survival in adult recipients of umbilical-cord blood from unrelated donors. *N. Engl. J. Med.* **344**, 1815–1822.

Laughlin, M. J., Eapen, M., Rubinstein, P., Wagner, J. E., Zhang, M. J., Champlin, R. E., Stevens, C., Barker, J. N., Gale, R. P., Lazarus, H. M., Marks, D. I., van Rood, J. J., *et al.* (2004). Outcomes after transplantation of cord blood or bone marrow from unrelated donors in adults with leukemia. *N. Engl. J. Med.* **351**, 2265–2275.

Long, G. D., Laughlin, M., Madan, B., Kurtzberg, J., Gasparetto, C., Morris, A., Rizzieri, D., Smith, C., Vredenburgh, J., Halperin, E. C., Broadwater, G., Niedzwiecki, D., *et al.* (2003). Unrelated umbilical cord blood transplantation in adult patients. *Biol. Blood Marrow Transplant.* **9**, 772–780.

Lu, L., Xiao, M., Shen, R. N., Grigsby, S., and Broxmeyer, H. E. (1993). Enrichment, characterization and responsiveness of single primitive CD34^{+++} human umbilical cord blood hematopoietic progenitor cells with high proliferative and replating potential. *Blood* **81**, 41–48.

McCune, J. M., Namikawa, R., Kaneshima, H., Shultz, L. D., Lieberman, M., and Weissman, I. L. (1988). The SCID-hu mouse: Murine model for the analysis of human hematolymphoid differentiation and function (severe combined immunodeficiency). *Science* **241**, 1632–1639.

Miller, J. S., McCullar, V., Punzel, M., Lemischka, I. R., and Moore, K. A. (1999). Single adult human CD34^{+}/Lin^{-}/CD38^{-} progenitors give rise to natural killer cells, B-lineage cells, dendritic cells, and myeloid cells. *Blood* **93**, 96–106.

Murasawa, S., and Asahara, T. (2005). Endothelial progenitor cells for vasculogenesis. *Physiology* (*Bethesda*). **20**, 36–42.

Murga, M., Yao, L., and Tosato, G. (2004). Derivation of endothelial cells from CD34^{-} umbilical cord blood. *Stem Cells* **22**, 385–395.

Ooi, J., Iseki, T., Takahashi, S., Tomonari, A., Nagayama, H., Ishii, K., Ito, K., Sato, H., Takahashi, T., Shindo, M., Sekine, R., Ohno, N., *et al.* (2002). A clinical comparison of unrelated cord blood transplantation and unrelated bone marrow transplantation for adult patients with acute leukaemia in complete remission. *Br. J. Haematol.* **118**, 140–143.

Orazi, A., Braun, S. E., and Broxmeyer, H. E. (1994). Immunohistochemistry represents a useful tool to study human cell engraftment in SCID mice transplantation models. *Blood Cells* **20**, 323–330.

Peichev, M., Naiyer, A., Pereira, D., Zhu, Z., Lane, W. J., Williams, M., Oz, M. C., Hicklin, D. J., Witte, L., Moore, M. A., and Rafii, S. (2000). Expression of VEGFR-2 and AC133 by circulating human CD34^{+} cells identifies a population of functional endothelial precursors. *Blood* **95**, 952–958.

Peranteau, W. H., Endo, M., Adibe, O. O., Merchant, A., Zoltick, P., and Flake, A. W. (2005). CD26 inhibition enhances allogeneic donor cell homing and engraftment after *in utero* bone marrow transplantation [abstract 1275]. *Blood* **106**, 371a.

Pesce, M., Orlandi, A., Iachininoto, M. G., Straino, S., Torella, A. R., Rizzuti, V., Pompilio, G., Bonanno, G., Scambia, G., and Capogrossi, M. C. (2003). Myoendothelial differentiation of human umbilical cord blood-derived stem cells in ischemic limb tissues. *Circ. Res.* **93**, e51–e62.

Pettengell, R., Luft, T., Henschler, R., Hows, J. M., Dexter, T. M., Ryder, D., and Testa, N.G (1994). Direct comparison by limiting dilution analysis of long-term culture-initiating cells in human bone marrow, umbilical cord blood, and blood stem cells. *Blood* **84,** 3653–3659.

Ploemacher, R. E., van der Sluijs, J. P., Voerman, J. S., and Brons, N. H. (1989). An *in vitro* limiting-dilution assay of long-term repopulating hematopoietic stem cells in the mouse. *Blood* **74,** 2755–2763.

Ploemacher, R. E., van der Sluijs, J. P., van Beurden, C. A., Baert, M. R., and Chan, P. L. (1991). Use of limiting-dilution type long-term marrow cultures in frequency analysis of marrow-repopulating and spleen colony-forming hematopoietic stem cells in the mouse. *Blood* **78,** 2527–2533.

Podesta, M., Piaggio, G., Pitto, A., Zocchi, E., Soracco, M., Frassoni, F., Luchetti, S., Painelli, E., and Bacigalupo, A. (2001). Modified *in vitro* conditions for cord blood-derived long-term culture-initiating cells. *Exp. Hematol.* **29,** 309–314.

Punzel, M., Wissink, S. D., Miller, J. S., Moore, K. A., Lemischka, I. R., and Verfaillie, C. M. (1999). The myeloid–lymphoid initiating cell (ML-IC) assay assesses the fate of multipotent human progenitors *in vitro*. *Blood* **93,** 3750–3756.

Rafii, S., and Lyden, D. (2003). Therapeutic stem and progenitor cell transplantation for organ vascularization and regeneration. *Nat. Med.* **9,** 702–712.

Rocha, V., Labopin, M., Sanz, G., Arcese, W., Schwerdtfeger, R., Bosi, A., Jacobsen, N., Ruutu, T., de Lima, M., Finke, J., Frassoni, F., and Gluckman, E. (2004). Transplants of umbilical-cord blood or bone marrow from unrelated donors in adults with acute leukemia. *N. Engl. J. Med.* **351,** 2276–2285.

Rubinstein, P., Carrier, C., Scaradavou, A., Kurtzberg, J., Adamson, J., Migliaccio, A. R., Berkowitz, R. L., Cabbad, M., Dobrila, N. L., Taylor, P. E., Rosenfield, R. E., and Stevens, C. E. (1998). Outcomes among 562 recipients of placental-blood transplants from unrelated donors. *N. Engl. J. Med.* **339,** 1565–1577.

Sanz, G. F., Saavedra, S., Jimenez, C., Senent, L., Cervera, J., Planelles, D., Bolufer, P., Larrea, L., Martin, G., Martinez, J., Jarque, I., Moscardo, F., *et al.* (2001a). Unrelated donor cord blood transplantation in adults with chronic myelogenous leukemia: results in nine patients from a single institution. *Bone Marrow Transplant.* **27,** 693–701.

Sanz, G. F., Saavedra, S., Planelles, D., Senent, L., Cervera, J., Barragan, E., Jimenez, C., Larrea, L., Martin, G., Martinez, J., Jarque, I., Moscardo, F., *et al.* (2001b). Standardized, unrelated donor cord blood transplantation in adults with hematologic malignancies. *Blood* **98,** 2332–2338.

Schatteman, G. C. (2004). Adult bone marrow-derived hemangioblasts, endothelial cell progenitors, and EPCs. *Curr. Top. Dev. Biol.* **64,** 141–180.

Shaheen, M., and Broxmeyer, H. E. (2005). The humoral regulation of hematopoiesis. *In* "Hematology: Basic Principles and Practice" (R. Hoffman, E. Benz, S. Shattil, B. Furie, H. Cohen, L. Silberstein, and P. McGlave, eds.) 4th Ed. pp. 233–265. Elsevier, Churchill Livingstone, New York.

Shpall, E. J., Quinones, R., Giller, R., Zeng, C., Baron, A. E., Jones, R. B., Bearman, S. I., Nieto, Y., Freed, B., Madinger, N., Hogan, C. J., Slat-Vasquez, V., *et al.* (2002). Transplantation of *ex-vivo* expanded cord blood. *Biol. Bone Marrow Transpl.* **8,** 368–376.

Simons, M. (2005). Angiogenesis: Where do we stand now? *Circulation* **111,** 1556–1566.

Skalak, T. (2005). Angiogenesis and microvascular remodeling: A brief history and future roadmap. *Microcirculation* **12,** 47–58.

Stadtfeld, M., and Graf, T. (2005). Assessing the role of hematopoietic plasticity for endothelial and hepatocyte development by non-invasive lineage tracing. *Development* **132,** 203.

Sutherland, H. J., Eaves, C. J., Eaves, A. C., Dragowska, W., and Lansdorp, P. M. (1989). Characterization and partial purification of human marrow cells capable of initiating long-term hematopoiesis *in vitro*. *Blood* **74,** 1563–1570.

Takahashi, S., Iseki, T., Ooi, J., Tomonari, A., Takasugi, K., Shimohakamada, Y., Yamada, T., Uchimaru, K., Tojo, A., Shirafuji, N., Kodo, H., Tani, K., *et al.* (2004). Single-institute comparative analysis of unrelated bone marrow transplantation and cord blood transplantation for adult patients with hematologic malignancies. *Blood* **104,** 3813–3820.

Taswell, C. (1981). Limiting dilution assays for the determination of immunocompetent cell frequencies. I. Data analysis. *J. Immunol.* **126,** 1614–1619.

Theunissen, K., and Verfaillie, C. M. (2005). A multifactorial analysis of umbilical cord blood, adult bone marrow and mobilized peripheral blood progenitors using the improved ML-IC assay. *Exp. Hematol.* **33,** 165–172.

Tian, C., Bagley, J., Forman, D., and Iacomini, J. (2005). Inhibition of CD26 peptidase activity significantly improves engraftment of retrovirally transduced hematopoietic progenitors. *Gene Ther.* **13,** 652–658.

Traycoff, C., Abboud, M., Laver, J., Clapp, D., and Srour, E. (1994). Rapid exit from G_0/G_1 phases of cell cycle in response to stem cell factor confers on umbilical cord blood CD34$^+$ cells an enhanced *ex vivo* expansion potential. *Exp. Hematol.* **22,** 1264–1272.

Traycoff, C. M., Kosak, S. T., Grigsby, S., and Srour, E. F. (1995). Evaluation of *ex vivo* expansion potential of cord blood and bone marrow hematopoietic progenitor cells using cell tracking and limiting dilution analysis. *Blood* **85,** 2059–2068.

Tse, W., and Laughlin, M. J. (2005). Cord blood transplantation in adult patients. *Cytotherapy* **7,** 228–242.

Urbich, C., and Dimmeler, S. (2004). Endothelial progenitor cells: Characterization and role in vascular biology. *Circ. Res.* **95,** 343–353.

Vormoor, J., Lapidot, T., Pflumio, F., Risdon, G., Patterson, B., Broxmeyer, H. E., and Dick, J. E. (1994). Immature human cord blood progenitors engraft and proliferate to high levels in immune-deficient SCID mice. *Blood* **83,** 2489–2497.

Wagner, J. E., Kernan, N. A., Steinbuch, M., Broxmeyer, H. E., and Gluckman, E. (1995). Allogeneic sibling umbilical cord blood transplantation in forty-four children with malignant and non-malignant disease. *Lancet* **346,** 214–219.

Wagner, J. E., Rosenthal, J., Sweetman, R., Shu, X. O., Davies, S. M., Ramsay, N. K., McGlave, P. B., Sender, L., and Cairo, M. S. (1996). Successful transplantation of HLA-matched and HLA-mismatched umbilical cord blood from unrelated donors: Analysis of engraftment and acute graft-versus-host disease. *Blood* **88,** 795–802.

Wagner, J. E., Barker, J. N., DeFor, T. E., Baker, K. S., Blazar, B. R., Eide, C., Goldman, A., Kersey, J., Krivit, W., MacMillan, M. L., Orchard, P. J., Peters, C., *et al.* (2002). Transplantation of unrelated donor umbilical cord blood in 102 patients with malignant and nonmalignant diseases: Influence of CD34 cell dose and HLA disparity on treatment-related mortality and survival. *Blood* **100,** 1611–1618.

Weaver, A., Ryder, W. D., and Testa, N. G. (1997). Measurement of long-term culture initiating cells (LTC-ICs) using limiting dilution: Comparison of endpoints and stromal support. *Exp. Hematol.* **25,** 1333–1338.

Author Index

A

Abbott, D. H., 215, 235
Abboud, M., 457, 458
Abdallah, B., 122
Aberg, T., 100, 102
About, I., 101, 102, 105, 107
Abraham, E. J., 330
Abramowicz, M. J., 33
Abramson, S., 149
Abu-Alfa, A., 203
Acil, Y., 131
Adachi, K., 295
Adamkiewicz, J., 181
Adams, E. C., 403, 414
Adams, G. B., 125, 237, 238, 240, 241
Adamson, J., 440
Addis, R. C., 405, 406, 412
Adeli, K., 362
Adibe, O. O., 441
Adjaye, J., 415
Adler, R., 65
Aglietta, M., 160, 165
Aguero, B., 152
Aguiar, D., 201
Aguzzi, A., 107
Ahlgren, U., 325
Ahmad, I., 54, 58
Ahmad, S., 153
Ahn, H. S., 443
Aigner, B., 401
Ainscough, J., 406
Ainsworth, C., 237, 243
Aitken, M. A., 214
Akagi, T., 55, 59, 60
Akai, J., 30
Akamine, A., 102
Akimov, S. S., 165, 166, 168
Akita, J., 59, 60
Akiyama, H., 124
Akiyama, M., 74, 82, 83
Akiyama, T., 35, 37
Al-Awqati, Q., 202

Albelda, S., 82, 83
Albertini, D., 237, 238, 241
Alberts, B., 209, 210
Albrecht, E. D., 215
Aldrich, S., 207, 356
Alexander, W. S., 159
Alexson, T., 30
Ali, A., 361
Allan, C. H., 361
Allen, D., 293, 295, 296
Allen, E., 237
Allen, E. D., 300, 304, 308
Allen, T. D., 119
Alley, J. L., 289
Allinger, U. G., 361
Allinquant, B., 27
Alliot-Licht, B., 102, 103
Almeida-Porada, G., 150, 151, 157, 179, 180, 201
Alsberg, E., 366
Alswaid, A., 33
Al-Ugaily, L. H., 289
Alvarez, J., 101
Alvarez-Buylla, A., 26, 27, 30, 31, 78
Alves, N. L., 157, 159, 160
Amalric, F., 34
Amar, J., 361
Ambergen, A. W., 291
Anastassiadis, K., 415
Anchan, R. M., 55, 60, 65
Andersen, C. Y., 237, 238, 241, 403, 404, 414
Anderson, D. J., 24, 37, 91
Anderson, G. B., 401
Anderson, J. A., 443, 444
Anderson, K. D., 292, 294
Anderson, L., 151
Anderson, M. J., 32
Anderson, R., 237, 238, 241
Ando, K., 159, 165
Andreu, G., 440
Andrews, P. W., 401
Angello, J., 55, 60, 65
Angelopoulou, M. K., 160

Subject Index

A

Adipocyte
 amniotic fluid progenitor cell
  differentiation, 430–431, 434
 bone marrow stromal cell differentiation,
  139–140
 formation from dental pulp stem cell,
  107–108
 placenta stem cell differentiation,
  430–431, 434
Akt, intestinal epithelial stem cell
 marker, 353
Aldehyde dehydrogenase, fluorescence assay
 for hematopoietic stem cell
 identification, 153–154
Alveolar type II cell, *see* Lung stem cells
γ-Aminobutyric acid, neural stem cell
 proliferation role, 31
Amniotic fluid progenitor cells
 cell therapy prospects, 428, 436
 collection, 426
 developmental biology, 427–428
 differentiation induction
  adipocytes, 430–431, 434
  endothelial cells, 432, 434
  hepatocytes, 432–435
  myocytes, 433, 435
  neural cells, 433–435
  osteocytes, 432, 434
 isolation and characterization, 429–430
Apoptosis, neural stem cell regulators, 31–32

B

Balbiani body, follicular renewal,
 220–221, 223
BDNF, *see* Brain-derived neurotrophic
 factor
Beta cell
 dedifferentiation, 330
 lineage tracing in mice
  Cre–*loxP* system

principles, 332
 reporters, 333
 temporal control over recombinase
  activity, 333
 tissue specificity of recombinase,
  332–333
 overview, 331–332
origins
 acini, 328–329
 adult stem cells, 329–330
 bone marrow cells, 329
 overview, 322–323
 pancreatic ducts, 328
 preexisting beta cells in postnatal life,
  326–328
pancreas development, 323–326
therapy prospects, 330–331
Bleomycin, lung injury model, 296
BMSC, *see* Bone marrow stromal cell
Bone marrow stromal cell
 colony-forming efficiency assay,
  136–137
 culture
  clonal culture
   feeder cell layer preparation, 135
   overview, 134–135
   plating, 135–136
   trypsinization, 136
  nonclonal culture
   bone marrow preparation, 133–134
   overview, 132–133
   passaging, 134
   plating, 134
 differentiation induction
  culture *in vitro*
   adipocyte differentiation, 139–140
   chondrocyte differentiation, 140
   osteogenic differentiation, 138–139
   overview, 138
  in vivo
   calvarial transplant, 145
   collagen sponge transplant, 144
   diffusion chamber, 143

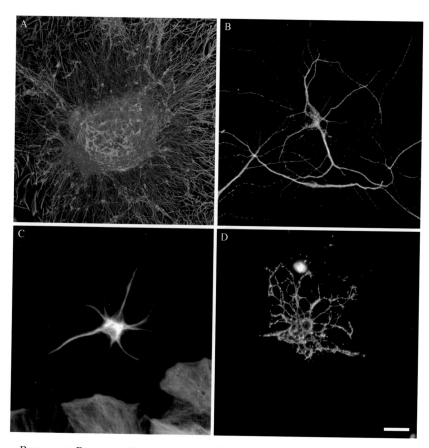

RIETZE AND REYNOLDS, CHAPTER 1, FIG. 3. When transferred to differentiating conditions for 7 DIV, neurospheres will lose their spherical shape and flatten to form essentially a monolayer. The greatest concentration of cells will remain in the center of the neurosphere [4′,6-diamidine-2-phenylindole (DAPI)-positive cells, blue], with astrocytes apparent throughout the sphere [glial fibrillary acidic protein (GFAP), green], and neurons (β-tubulin, red) surrounding the core of the sphere, lying on top of the astrocytes (A). Neurons are identified with a fluorescently labeled antibody raised against β-tubulin, a neuron-specific antigen found in cell bodies and processes (B). Both protoplasmic and stellate astrocytes are identified with a fluorescently tagged antibody against the astrocyte-specific protein GFAP (C). Oligodendrocytes are identified with an antibody against myelin basic protein (MBP) (D). Scale bar (B–D): 20 μm.

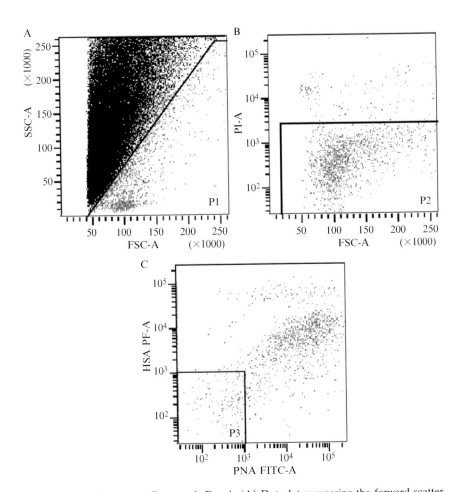

Rietze and Reynolds, Chapter 1, Fig. 4. (A) Dot plot comparing the forward scatter (FSC-A) and side scatter (SSC-A) attributes of periventricular cells harvested from the rostral periventricular region. Selecting cells in population 1 (P1) excludes the majority of cellular debris without affecting the number of neurospheres generated. (B) Viable cells are distinguished from those cells contained within P1 in (A), by comparing FSC-A and propidium iodide intensity, and then gating for those cells within the propidium iodide-negative population (P2). (C) Dot plot of viable periventricular cells comparing PNA and HSA staining intensities. Harvesting cells in the PNA^{lo}HSA^{lo} population (P3) will greatly enrich for stem cell activity.

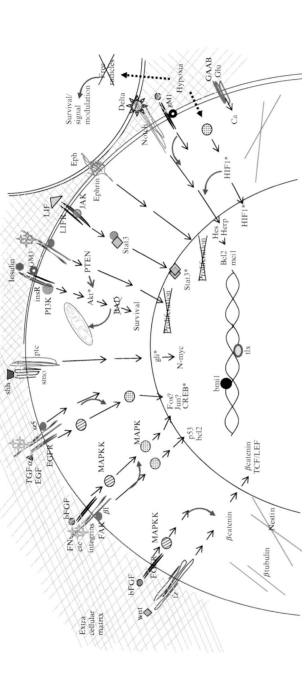

RAJAN AND SNYDER, CHAPTER 2, FIG. 2. Signals regulating proliferation of NSCs. Extracellular signals that regulate the proliferation of NSCs may be soluble ligands, extracellular matrix (ECM) molecules, and immobilized ligands present on neighboring cells. Interactions occur between intracellular signals generated by the growth factor receptor tyrosine kinases and the integrin receptors that respond to ECM molecules. Interactions also occur between the mitogen-activated protein (MAP) kinase pathway and wnt signals, and between hypoxia-induced signals and notch. Stat3 signals are activated by leukemia inhibitory factor (LIF), whereas sonic hedgehog possibly acts through the activation of myc in the nucleus. The epigenetic modifiers bmi1 and tlx are also involved in proliferation of NSCs. In addition to protein signaling intermediates, lipid signals are also involved, especially at the level of receptors at the cell surface [fibroblast growth factor (FGF) and insulin receptors]. It is interesting to note that almost all the transcription factors that are involved [T cell factor/lymphoid enhancing factor (TCF/LEF), Stat3, N-myc, p53, and bmi1], and proteins that are upregulated as a result of the signals (bcl2 and mci1), are associated with oncogenesis. *, activated transcription factor. See text for details and references.

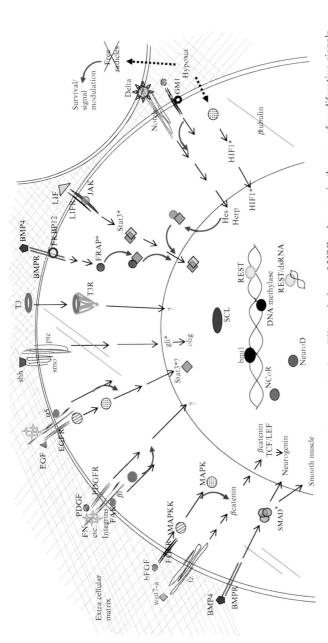

RAJAN AND SNYDER, CHAPTER 2, FIG. 5. Signals regulating differentiation of NSCs. As seen in the case of proliferative signals, extracellular signals that regulate the differentiation of NSCs may be soluble ligands, extracellular matrix molecules, and ligands present immobilized in neighboring cells. Almost all the ligands and receptor systems that were involved in proliferation are also involved in the regulation of differentiation, conspicuous in its absence is the Eph/ephrin system. Also, the transcription factors activated by ligands during the induction of differentiation is distinct from the nuclear effects that are elicited during proliferation. There are interactions of the receptor tyrosine kinases with the integrin receptors; however, in this case the signals lead to differentiation. Wnt and BMP activation leads to neuronal differentiation, possibly through the activation of neurogenin. Platelet-derived growth factor (PDGF) appears to lead to the proliferation of a neuronal precursor. Interaction of the notch and bone

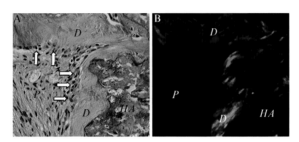

Liu *et al.*, Chapter 5, Fig. 2. Eight weeks posttransplantation, DPSCs are capable of differentiating into odontoblasts (arrows) that are responsible of dentin formation (*D*) on the surface of the hydroxyapatite/tricalcium (*HA*) carrier. Pulp-like tissue (*P*) is generated with newly formed dentin as shown by hematoxylin–eosin staining (A) and polarizing light microscopy (B).

---

morphogenetic protein (BMP) pathways leads to astrocytic differentiation, and so does leukemia inhibitory factor (LIF) receptor activation by LIF or ciliary neurotrophic factor (CNTF). Stat3 activation appears to be integrally involved in astrocytic differentiation. shh signals lead to oligodendrocyte differentiation, and so also do signals originating from the triiodothyronine (T_3) receptor. While hypoxia activates hypoxia inducible factor 1 (HIF1), which leads to specific differentiation effects, it also regulates free radical levels in the cell, thus modulating signals in general. The nuclear component of differentiation is more complex than proliferation, and there are several more epigenetic and cofactor molecules involved. In addition, helix–loop–helix proteins including neurogenin, stem cell leukemia protein (SCL), and NeuroD are involved in fate choice specification. The complexity of possible transcription events during differentiation may perhaps be explained by the number of fates into which an NSC can differentiate. *, activated transcription factor. See text for details and references.

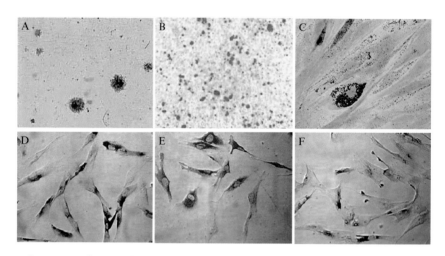

LIU *ET AL.*, CHAPTER 5, FIG. 3. Multipotent differentiation of DPSCs. DPSCs recovered from human dental pulp were capable of forming heterogeneous single-colony clusters after being plated at low density and cultured with regular culture medium for 10 days (A). DPSCs were cultured with L-ascorbate 2-phosphate, dexamethasone, and inorganic phosphate for 4 weeks. Alizarin red staining showed mineralized nodule formation (B). DPSCs were able to form oil red O-positive lipid clusters after 5 weeks of induction in the presence of 0.5 mM isobutylmethylxanthine, 0.5 μM hydrocortisone, and 60 μM indomethacin (C). Immunocyto-chemical staining depicts cultured DPSCs expressing nestin (D), GFAP (E), and neurofilament M (F), with culture medium containing Neurobasal A (Invitrogen GIBCO-BRL, Grand Island, NY), B27 supplement (GIBCO-BRL), 1% penicillin, epidermal growth factor (EGF, 20 ng/ml; BD Biosciences Discovery Labware, Bedford, MA), and FGF (40 ng/ml; BD Biosciences Discovery Labware).

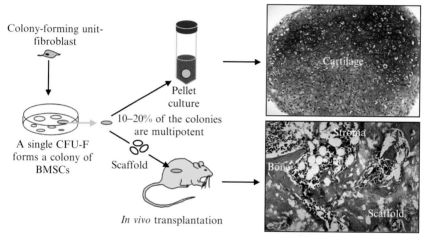

BIANCO *ET AL.*, CHAPTER 6, FIG. 1. Proof of the existence of a multipotential skeletal ("mesenchymal") stem cell in postnatal bone marrow. When plated at low density, the colony-forming unit-fibroblast (CFU-F) rapidly adheres, and proliferates to form a colony of bone marrow stromal cells (BMSCs). Approximately 10% of these colonies can be further expanded and, when placed in pellet cultures under relatively anaerobic conditions, they form cartilage. When attached to appropriate scaffolds and transplanted subcutaneously into immunocompromised mice, they form bone, myelosupportive stroma, and marrow adipocytes.

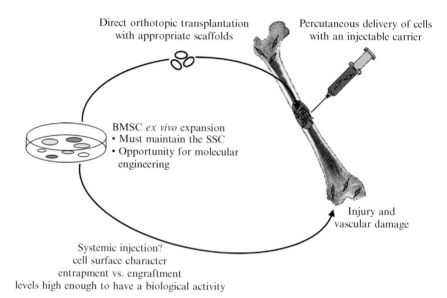

Direct orthotopic transplantation
with appropriate scaffolds

Percutaneous delivery of cells
with an injectable carrier

BMSC *ex vivo* expansion
• Must maintain the SSC
• Opportunity for molecular
  engineering

Injury and
vascular damage

Systemic injection?
cell surface character
entrapment vs. engraftment
levels high enough to have a biological activity

BIANCO *ET AL.*, CHAPTER 6, FIG. 2. Use of bone marrow stromal cells (BMSCs) in tissue engineering and regenerative medicine. Although *ex vivo* expansion cannot increase the numbers of skeletal stem cells contained within the BMSC population, the culture conditions must at least maintain them. Using lentiviral vectors, the *ex vivo* step also provides the opportunity to genetically modify them, to either make a deficient protein, or perhaps to silence a mutant protein by RNA interference. Cells attached to appropriate carriers can be used for direct orthotopic transplantation for bone regeneration of critical size defects through open surgery procedures, or via percutaneous delivery with an injectable carrier. Although systemic injection has been viewed as a delivery method for treatment of generalized skeletal disorders, it has yet to be demonstrated that BMSCs injected into intact animals are able to escape from the circulation; because of their cell surface characteristics they become entrapped within blood vessels. When there is vascular damage, BMSCs may be able to escape, but it is not clear that there are sufficient numbers of them to have a biological impact.

A Osteogenic
  differentiation

B Adipogenic
  differentiation

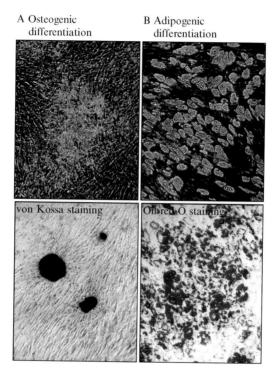

von Kossa staining

Oil red O staining

BIANCO ET AL., CHAPTER 6, FIG. 3. *In vitro* differentiation of bone marrow stromal cells (BMSCs). (A) When incubated under osteogenic conditions (see Osteogenic Differentiation Protocol) for extended periods of time, BMSCs focally multilayer and accumulate calcium, which is phase-bright by inverted light microscopy. These condensations are readily stainable with von Kossa as shown here, or with alizarin red S. (B) Incubation of BMSCs under adipogenic conditions (see Adipogenic Differentiation Protocol) induces accumulation of fat, again readily apparent by microscopic examination, and can be further visualized by staining with oil red O.

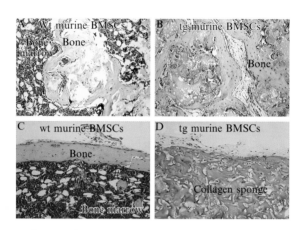

Bianco *et al.*, Chapter 6, Fig. 4. *In vivo* transplantation of bone marrow stromal cells (BMSCs) and regeneration of a diseased phenotype. When *ex vivo*-expanded murine cells are transplanted with hydroxyapatite/tricalcium phosphate (HA/TCP) (A) or collagen sponges (C) into subcutaneous pockets in immunocompromised mice, they completely regenerate a bone/marrow organ, with bone, hematopoietic stroma and adipocytes of donor origin, and hematopoietic cells of recipient origin. This model system can be used to determine the impact of abnormal gene expression on the development of the bone/marrow organ. As an example, BMSCs derived from transgenic mice in which a constitutively active PTH/PTHrP receptor is expressed under the control of a bone-specific promoter are capable of forming bone when transplanted with HA/TCP, but do not support formation of hematopoietic marrow (B), identical to what is observed in bone of young transgenic mice. However, when transplanted in collagen sponges, transgenic BMSCs not only do not support the formation of marrow, but also do not form bone (D), indicating depletion of skeletal stem cells. Thus, these two assays are able to probe the biological activity of both committed progenitors and skeletal stem cells within the BMSC population.

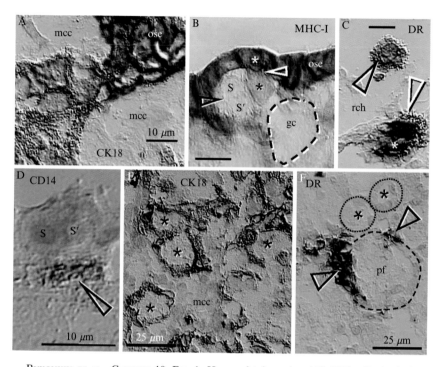

BUKOVSKY ET AL., CHAPTER 10, FIG. 1. Human fetal ovaries. (A) OSE cells (ose) show strong CK expression and descent among mesenchymal cell cords (mcc) to give rise to moderately CK⁺ primitive granulosa (pg) cells. (B) MHC class I⁺ OSE cell (white asterisk) undergoes asymmetric division (white arrowhead) to give rise to the MHC class I⁻ germ cell (black asterisk). After symmetric division (black arrowhead and s & s′ cells) the tadpole-like germ cells (gc, dashed line) enter the ovarian cortex. No hematoxylin counterstain. (C) Rete channels (rch) show HLA-DR monocytes (black arrowhead) interacting (white arrowhead) with resident cells (white asterisk). (D) Germ cells undergoing symmetric division (s and s′) in the fetal OSE and accompanying CD14 (primitive) tissue macrophage (arrowhead). (E) Primordial follicles (asterisks) adjacent to mesenchymal cell cord show CK⁺ granulosa cells but no Balbiani bodies (compare with Fig. 4C). (F) Development of primary follicle (pf) is accompanied by activated (HLA-DR⁺) tissue macrophages (arrowheads); asterisks indicate small primordial follicles. Reproduced from Bukovsky et al. (2005a), with permission.

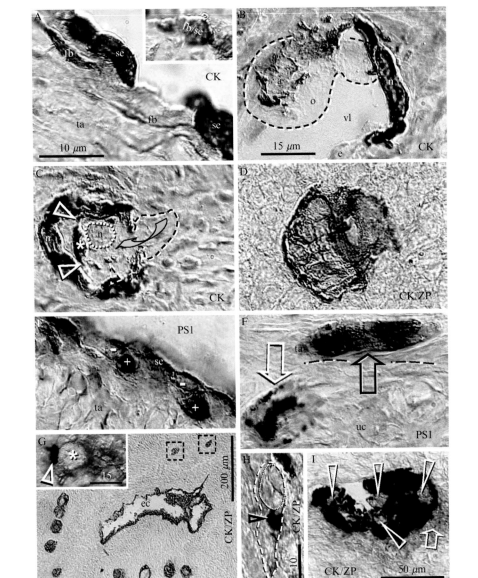

Bukovsky *et al.*, Chapter 10, Fig. 4. Follicular renewal in adult human ovaries. (A) Some fibroblasts (fb) in the tunica albuginea (ta) show CK18 expression and transition into OSE (se) cells. Inset: A cell in mesenchymal–epithelial transition (fb/se). (B) The CK^+ epithelial nest (n) wall inside the vascular lumen (vl) pocket, which is lined with endothelial cells (e), extends an arm (a) to catch the oocyte (o, dashed line). (C) The nest body (n) with closing "gate" and a portion of the oocyte (dashed line) still outside the nest–oocyte complex (arched arrow). The oocyte contains intraooplasmic CK^+ (brown color) extensions from the nest wall (arrowheads), which contribute to the formation of CK^+ paranuclear (Balbiani) body (asterisk); the nucleus is indicated by a dotted line. (D) Occupied "bird's nest" indicates a half-formed oocyte–nest assembly. CK (brown)/ZP (blue). (E) Segments of OSE show cytoplasmic PS1 (brown) expression (nuclei) and give rise to cells exhibiting nuclear PS1 (+ nuclei, asymmetric division), which descend into the tunica albuginea. (F) In the tunica albuginea, the putative germ cells show symmetric division (black arrow) and also exhibit development of cytoplasmic PS1 when entering (white arrow) the upper cortex (uc). (G) Primary follicles develop around the OSE cortical epithelial crypt (ec); dashed boxes indicate unassembled epithelial nests. Inset: The emergence of CK^- germ cell (asterisk) with ZP^+ intermediate segment (arrowhead) among crypt CK^+ epithelial cells. (H) Migrating tadpole-like germ cell shows no CK staining but ZP^+ intermediate segment (arrowhead). (I) Accumulation of multiple ZP^+ oocytes with unstained nuclei (arrowheads) in a medullary vessel. (A–C and E) Hematoxylin counterstain. Reprinted from Bukovsky *et al.* (2004), © Antonin Bukovsky.

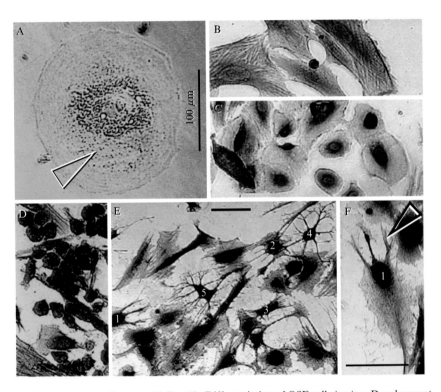

Bukovsky *et al.*, Chapter 10, Fig. 10. Differentiation of OSE cells *in vitro*. Development of oocyte (A), fibroblast (B), epithelial (C), granulosa (D), and neural-type cells (E and F) in primary OSE cultures. Arrowhead in (A) indicates initial breakdown of the germinal vesicle. Numbers in (E) show stages of neural cell differentiation. (F) Detail from (E) of transition of an epithelial cell type into a neural cell type (arrowhead). (A) live culture in phase contrast; (B–I) immunohistochemical staining for zona pellucida glycoproteins. Scale bars: 100 μm. Reproduced from Bukovsky *et al.* (2005b), © Antonin Bukovsky.

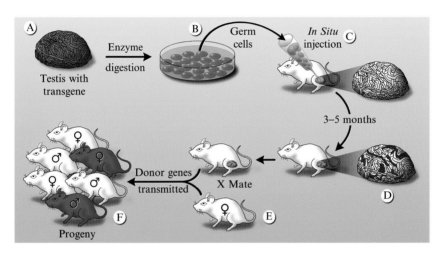

OATLEY AND BRINSTER, CHAPTER 11, FIG. 1. The spermatogonial stem cell transplantation technique in mice. (A) The testis from a male in which a reporter transgene is expressed in germ cells is digested to generate a single-cell suspension. (B and C) The isolated cells are subsequently cultured or microinjected as a fresh cell suspension into the seminiferous tubules of an infertile germ cell-depleted recipient. (D) Colonies of donor-derived spermatogenesis can be detected in the recipient testis on the basis of reporter transgene expression by donor germ cells. Only spermatogonial stem cells are capable of establishing colonies of spermatogenesis after transplantation. Each donor-derived colony of spermatogenesis in a recipient testis is generated from a single transplanted spermatogonial stem cell. (E and F) On mating of the recipient male to a wild-type female, offspring can be produced containing the donor haplotype and thus must have been produced by sperm generated from a donor spermatogonial stem cell-derived colony of spermatogenesis. Reproduced with permission from Brinster (2002) © American Association for the Advancement of Science.

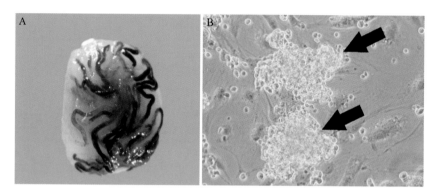

OATLEY AND BRINSTER, CHAPTER 11, FIG. 2. Transplantation and culture of mouse spermatogonial stem cells. (A) Busulfan-treated recipient testis transplanted with cultured Thy1$^+$ germ cells from ROSA donors. Donor-derived spermatogenesis is easily detectable by staining for β-galactosidase expression. Each blue colony of spermatogenesis represents the colonization of a single donor spermatogonial stem cell. Assessing the number and length of blue colonies provides unequivocal quantitative determination of spermatogonial stem cell presence, number, and biological function in the injected cell population. (B) An established spermatogonial stem cell-enriched germ cell culture from 6-day-old ROSA donor mice. The germ cells grow as clumps of stem cells with tightly adhering membranes (arrows) and are loosely attached to the underlying STO feeder cell monolayer. In healthy, robust cultures of self-renewing spermatogonial stem cells the germ cells form tight clumps where individual cells are difficult to distinguish.

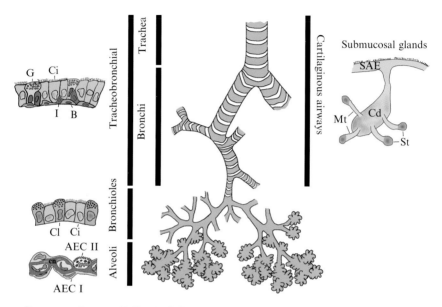

LIU *ET AL*., CHAPTER 12, FIG. 1. Cellular diversity of epithelial cell types in the human adult lung. When discussing progenitor/stem cells in adult airways, one must always consider the diversity of cell phenotypes that exist in spatially distinct epithelia of the lung. Three main levels of conducting airways exist in the lung, including the trachea, bronchi, and bronchioles. Predominant cell types in the human pseudostratified, columnar tracheal, and bronchial epithelia include basal (B), intermediate (I), goblet (G), and ciliated (Ci) cells; less abundant nonciliated and neuroendocrine cells are not shown. Submucosal glands are also present only in the cartilaginous airways of the trachea and bronchi. Predominant cell types in the human columnar bronchiolar epithelia include Clara cells (Cl) and ciliated cells; less abundant neuroendocrine cells are not shown. Predominant cell types in the gas-exchanging alveolar airspaces include alveolar epithelial type II cells (AEC II), alveolar epithelial type I cells (AEC I), and capillary endothelial cells of the capillary networks (cn). Mt, mucous tubule; St, serous tubule; SAE, surface airway epithelium; Cd, collecting duct.

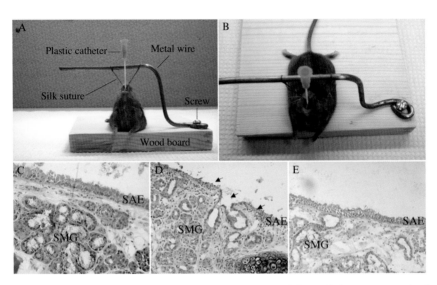

LIU *ET AL.*, CHAPTER 12, FIG. 3. Polidocanol-induced injury of the mouse tracheal epithelium. Anesthetized mice are hung on a self-made rack by placing the upper jaw incisor teeth within a suture before placing a plastic catheter into the trachea as shown in (A) (*front view*) and (B) (*top view*). The rack is made with a >3-mm metal wire fixed on a wooden board with a screw. After intratracheal instillation of 20 μl of 2% polidocanol, tracheas are harvested and sectioned. (C–E) H&E-stained sections from (C) control uninjured trachea, (D) injured trachea at 1 day, and (E) injured trachea at 14 days. Arrows mark injury to the surface airway epithelium at 1 day after polidocanol treatment in (D) as compared with injured controls (C) and regenerated epithelium (E). SAE, surface airway epithelium; SMG, submucosal gland.

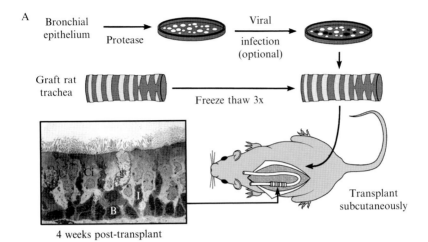

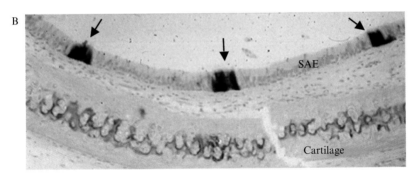

Liu ET AL., Chapter 12, Fig. 4. Methods for generating tracheal xenografts to study clonal expansion. (A) Schematic methods for generating proximal airway epithelial xenograft models. Primary airway epithelial cells are cultured *in vitro* and may be infected with integrating recombinant viral vectors (lentivirus or retrovirus) before transplantation of epithelia into denuded rat tracheas. Fully differentiated epithelium is obtained by 4 weeks posttransplantation (B, basal cells; Ci, ciliated cells; I, intermediate cells; G, goblet cells). (B) Colony-forming efficiency (CFE) of stem/progenitor cells can be evaluated by detecting virally expressed transgenes in the reconstituted surface airway epithelium (SAE) (arrows). In this example a recombinant retroviral vector expressing the β-galactosidase transgene was used and detected by X-Gal staining. (C) Schematic view of various components of the xenograft cassette. The denuded rat trachea is connected to a sterile tubing cassette by a series of sutures as illustrated: *a*, 1-in. Silastic tubing (cat. no. 602-175; Dow Corning, Midland, MI); *b*, 3/4-in. Silastic tubing (cat. no. 602-175; Dow Corning); *c*, 1.75-in. Silastic tubing (cat. no. 602-175; Dow Corning); *d*, 1.25-in. Teflon tubing (cat. no. 9567-K10; Thomas Scientific, Swedesboro, NJ); *e* and *e'*, adapter (0.8-mm barb-to-barb connector; cat. no. 732–8300; Bio-Rad, Hercules, CA). A chrome wire plug (0.0035-in.-diameter Chromel A steel wire; Hoskins

(*continued*)

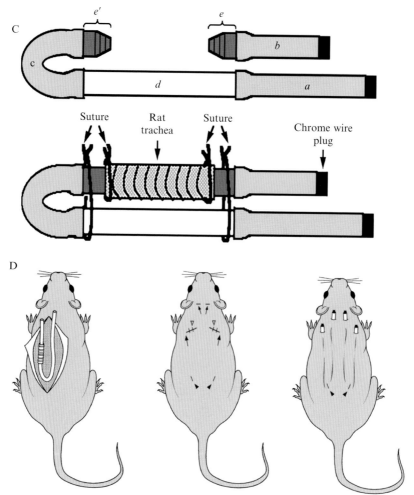

C

e′   e

c   d   b   a

Suture   Rat trachea   Suture   Chrome wire plug

D

Manufacturing, Hamburg, MI) is also used. (D) Subcutaneous transplantation of the xenograft cassettes in *nu/nu* athymic mice. *Left:* A subcutaneous view of the transplanted xenograft cassette. *Middle*: The four incisions (arrows) made on the back of the recipient *nu/nu* mouse before transplantation. The xenograft cassette is guided subcutaneously with forceps, so that one port exits through the back of the neck and the other port exits through the main incision. Surgical staples are then used to close the main incisions (marked by open arrowheads) and one additional staple is used to anchor the xenograft tubing near the tail end of the mouse (solid arrowheads). *Right:* The resultant xenograft cassette 1 week postsurgery, when the proximal staples are removed. The staples marked by solid arrowheads in the middle panel are used to maintain the position of the cassette and to prevent subcutaneous migration (it is necessary to leave these staples in for at least 2 weeks).

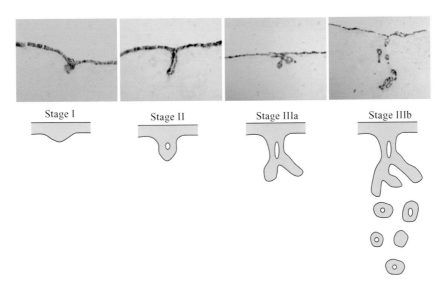

Stage I Stage II Stage IIIa Stage IIIb

LIU *ET AL.*, CHAPTER 12, FIG. 6. Mouse tracheal epithelial cell tubulogenesis assay. Bud formation and tubulogenesis are two early events important in submucosal gland development and can be studied with a collagen gel matrix cultured at the air–liquid interface. Depicted here are various stages of gland-like structures formed from mouse tracheal epithelium (*top*) and schematic representations of various stages of tubulogenesis characteristic of early submucosal gland morphogenesis (*bottom*).

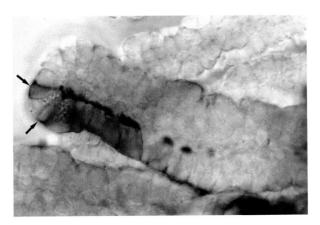

BJERKNES AND CHENG, CHAPTER 14, FIG. 3. A *Dlb-1*[+] clone (brown cells) in a crypt isolated from an SWR mouse and stained as outlined in the protocol on page 345. The clone continues onto the villus, where columnar and mucous cells were visible (data not shown). Crypt base columnar cells, potential stem cells, are visible (arrows; taken from Bjerknes and Cheng, 1999).

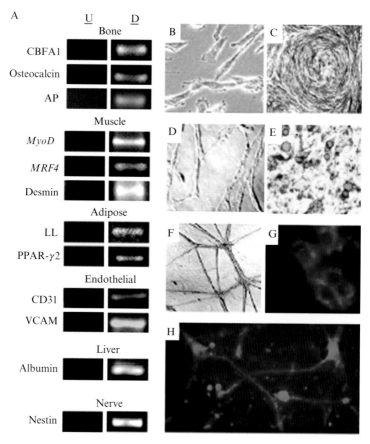

DELO *ET AL.*, CHAPTER 17, FIG. 2. Multilineage differentiation of hAFSCs. (A) RT-PCR analysis of mRNA. *Left*: Control undifferentiated cells. *Right*: Cells maintained under conditions for differentiation to bone (8 days), muscle (8 days), adipocyte (16 days), endothelial (8 days), hepatic (45 days), and neuronal (2 days) lineages. (B) Phase-contrast microscopy of control, undifferentiated cells. (B–H) Differentiated progenitor cells. (C) Bone: Histochemical staining for alkaline phosphatase. (D) Muscle: Phase-contrast microscopy showing fusion into multinucleated myotube-like cells. (E) Adipocyte: Staining with oil red O (day 8) shows intracellular oil aggregation. (F) Endothelial: Phase-contrast microscopy of capillary-like structures. (G) Hepatic: Fluorescent antibody staining (FITC, green) for albumin. (H) Neuronal: Fluorescent antibody staining of nestin (day 2).

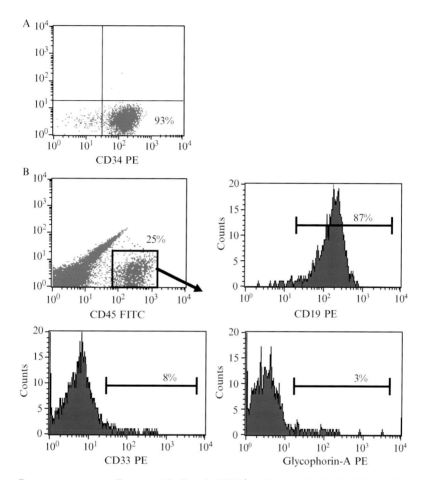

BROXMEYER *ET AL.*, CHAPTER 18, FIG. 2. CD34$^+$ cells were isolated with anti-CD34 antibodies and the magnetic beads system from Miltenyi Biotec (Bergisch Gladbach, Germany). Cells (1×10^5 to 3×10^5) were then infused into 300 cGy-conditioned NOD/SCID mice via tail vein injection. Mice were killed 8–9 weeks posttransplantation and BM cells were harvested and stained with anti-human CD45–fluorescein isothiocyanate (FITC) and the indicated lineage markers (CD19/lymphoid, CD33/myeloid, glycophorin-A/erythroid) con-jugated to phycoerythrin (PE). Flow cytometry was used to assess percent human chimerism and lineage marker expression on engrafted cells. (A) Representative flow cytometric plot of purified CB CD34$^+$ cells used in *in vitro* experiments and as grafts in NOD/SCID mice. (B) Flow cytometric plots of human cells (CD45$^+$) in NOD/SCID mice and the gating used to obtain the expression of lineage markers on these cells.

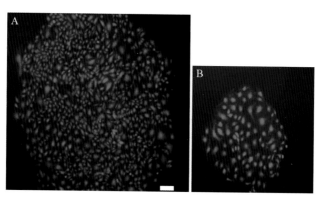

Broxmeyer *et al.*, Chapter 18, Fig. 3. Photomicrographs of colonies formed from human cord blood endothelial colony-forming cells (ECFCs). (A) High proliferative potential ECFC-derived colony and (B) low proliferative potential ECFC-derived colony. Scale bar in (A): 50 μm.